T0183084

Lecture Notes in Computer Science   9198

Commenced Publication in 1973
Founding and Former Series Editors:
Gerhard Goos, Juris Hartmanis, and Jan van Leeuwen

Editorial Board

David Hutchison
  Lancaster University, Lancaster, UK
Takeo Kanade
  Carnegie Mellon University, Pittsburgh, PA, USA
Josef Kittler
  University of Surrey, Guildford, UK
Jon M. Kleinberg
  Cornell University, Ithaca, NY, USA
Friedemann Mattern
  ETH Zurich, Zürich, Switzerland
John C. Mitchell
  Stanford University, Stanford, CA, USA
Moni Naor
  Weizmann Institute of Science, Rehovot, Israel
C. Pandu Rangan
  Indian Institute of Technology, Madras, India
Bernhard Steffen
  TU Dortmund University, Dortmund, Germany
Demetri Terzopoulos
  University of California, Los Angeles, CA, USA
Doug Tygar
  University of California, Berkeley, CA, USA
Gerhard Weikum
  Max Planck Institute for Informatics, Saarbrücken, Germany

More information about this series at http://www.springer.com/series/7407

Dachuan Xu · Donglei Du
Dingzhu Du (Eds.)

Computing and Combinatorics

21st International Conference, COCOON 2015
Beijing, China, August 4–6, 2015
Proceedings

 Springer

Editors
Dachuan Xu
Beijing University of Technology
Beijing
China

Dingzhu Du
University of Texas at Dallas
Richardson
USA

Donglei Du
University of New Brunswick
Fredericton
Canada

ISSN 0302-9743   ISSN 1611-3349   (electronic)
Lecture Notes in Computer Science
ISBN 978-3-319-21397-2   ISBN 978-3-319-21398-9   (eBook)
DOI 10.1007/978-3-319-21398-9

Library of Congress Control Number: 2015943451

LNCS Sublibrary: SL1 – Theoretical Computer Science and General Issues

Springer Cham Heidelberg New York Dordrecht London

© Springer International Publishing Switzerland 2015
This work is subject to copyright. All rights are reserved by the Publisher, whether the whole or part of the material is concerned, specifically the rights of translation, reprinting, reuse of illustrations, recitation, broadcasting, reproduction on microfilms or in any other physical way, and transmission or information storage and retrieval, electronic adaptation, computer software, or by similar or dissimilar methodology now known or hereafter developed.
The use of general descriptive names, registered names, trademarks, service marks, etc. in this publication does not imply, even in the absence of a specific statement, that such names are exempt from the relevant protective laws and regulations and therefore free for general use.
The publisher, the authors and the editors are safe to assume that the advice and information in this book are believed to be true and accurate at the date of publication. Neither the publisher nor the authors or the editors give a warranty, express or implied, with respect to the material contained herein or for any errors or omissions that may have been made.

Printed on acid-free paper

Springer International Publishing AG Switzerland is part of Springer Science+Business Media
(www.springer.com)

Preface

The 21st International Computing and Combinatorics Conference (COCOON 2015) was held during August 4-6, 2015, in Beijing, China. COCOON 2015 provided a forum for researchers working in the area of theoretical computer science and combinatorics.

The technical program of the conference included 49 contributed regular papers selected by the Program Committee from full submissions received in response to the call for papers. The accepting rate is 48%. In addition, to increase opportunities for exchange of research ideas, the conference also accepted 11 shorter papers. All the papers were peer reviewed by at least three Program Committee members or external reviewers. The papers cover various topics, including algorithms and data structures, algorithmic game theory, approximation algorithms and online algorithms, automata, languages, logic, and computability, complexity theory, computational learning theory, cryptography, reliability and security, database theory, computational biology and bioinformatics, computational algebra, geometry, number theory, graph drawing and information visualization, graph theory, communication networks, optimization, and parallel and distributed computing. Some of the papers will be selected for publication in special issues of *Algorithmica*, *Theoretical Computer Science* (TCS), and *Journal of Combinatorial Optimization* (JOCO). It is expected that the journal version papers will appear in a more complete form.

We would like to thank the Program Committee members and external reviewers for volunteering their time to review conference papers. We would like to extend special thanks to the publication, publicity, and local organization chairs for their hard work in making COCOON 2015 a successful event. Last but not least, we would like to thank all the authors for presenting their works at the conference.

August 2015

Dachuan Xu
Donglei Du
Dingzhu Du

Conference Organization

Program Chairs

Dingzhu Du   University of Texa at Dallas, USA
Donglei Du   University of New Brunswick, Canada
Dachuan Xu   Beijing University of Technology, China

Publication Chairs

Chenchen Wu   Tianjin University of Technology, China
Fengmin Wang   Beijing University of Technology, China

Publicity Chairs

Gaidi Li   Beijing University of Technology, China
Xuegang Chen   North China Electric Power University
Jianfeng Ren   Qufu Normal University, China

Local Organization Chairs

Dachuan Xu   Beijing University of Technology, China
Xianyuan Zhao   Beijing University of Technology, China

Program Committee

Hee-Kap Ahn   Pohang University of Science and Technology, Korea
Yossi Azar   Tel-Aviv University, Israel
Vladimir Braverman   Jonhs Hopkins University, USA
Zhipeng Cai   Georgia State University, USA
Yixin Cao   Hungarian Academy of Sciences, Hungary
Xi Chen   Columbia University, USA
Zhixiang Chen   University of Texas Pan American, USA
Janos Csirik   University of Szeged, Hungary
Dingzhu Du   University of Texas at Dallas, USA
Donglei Du   University of New Brunswick, Canada
Zachary Friggstad   University of Alberta, Canada
Xiaodong Hu   Chinese Academy of Sciences, China
Tsan-Sheng Hsu   Academia Sinica, Taiwan
Klaus Jansen   University of Kiel, Germany
Iyad Kanj   DePaul University, USA
Ming-Yang Kao   Northwestern University, USA

Donghyun Kim	North Carolina Central University, USA
Piotr Krysta	University of Liverpool, UK
Minming Li	City University of Hong Kong, SAR China
Guohui Lin	University of Alberta, Canada
Julian Mestre	University of Sydney, Australia
Rolf Moehring	Technische Universität Berlin, Germany
Benjamin Moseley	Toyota Technological Institute, Japan
Mitsunori Ogihara	University of Miami, USA
Desh Ranjan	Old Dominion University, USA
Yilin Shen	Samsung Research America, USA
Takeshi Tokuyama	Tohoku University, Japan
Marc Uetz	University of Twente, The Netherlands
Dachuan Xu	Beijing University of Technology, China
Jinhui Xu	State University of New York at Buffalo, USA

Additional Reviewers

Aaronson, Scott
Ahmad, Yanif
Anastasiadis, Eleftherios
Ateniese, Giuseppe
Aydinlioglu, Baris
Bei, Liu
Beisegel, Jesse
Bienkowski, Marcin
Biswas, Abhishek
Bonsma, Paul
Bouland, Adam
Broersma, Hajo
Buchbinder, Niv
Buffett, Scott
Burns, Randal
Byrka, Jarek
Chen, Danyang
Chen, Xujin
Chen, Zihe
Chestnut, Stephen
Cohen, Ilan
Cui, Lei
Dasari, Naga Shailaja
Eden, Alon
Epstein, Leah
Feldman, Michal
Fischer, Felix

Fotakis, Dimitris
Ghorbani, Ali
Gudmundsson, Joachim
Guo, Linke
Han, Xin
Huang, Ziyun
Hwang, Yoonho
Ivkin, Nikita
Jin, Yong
Kaluza, Maren
Kelly, Terence
Kim, Hyunbum
Kim, Min-Gyu
Klavzar, Sandi
Klein, Kim-Manuel
Klimm, Max
Köhler, Ekkehard
Kononov, Alexander
Korman, Matias
Kraft, Stefan
Kwok, Tsz Chiu
Land, Felix
Land, Kati
Levin, Keith
Li,Wenjun
Li, Yuchao
Liang, Dongyue

Liu, Xianliang
Liu, Yangwei
Liu, Zaoxing
Maack, Marten
Mahalanobis, Ayan
Mikkelsen, Jesper W.
Nadeem, Tamer
Nguyen, Nam P.
Nikhil Bansal
Obenshain, Daniel
Oh, Eunjin
Ono, Hirotaka
Pal, Marcin
Papamichail, Dimitris
Park, Dongwoo
Pluhár, András
Poon, Chung Keung
Pruhs, Kirk
Rahman, Md. Saidur
Rau, Malin
Rosenberg, Burton
Roytman, Alan
Sanders, Peter
Son, Junggab
Son, Wanbin
Song, Wei
Tong, Guanmo
Tu, Jianhua
Uma, RN
Vardi, Adi

Vasiliev, Saveliy
Vinar, Tomas
Vorsanger, Greg
Wang, Wei
Wang, Xiangyu
Wang, Yishui
Wang, Yu-Shuen
Wang, Zhenbo
Watrigant, Rémi
Wolff, Alexander
Wong, Prudence
Wu, Chenchen
Wu, Lidong
Wu, Weiwei
Xu, Yicheng
Yamanaka, Katsuhisa
Yang, Lin
Yoon, Sangduk
You, Jie
Yuan, Jing
Yuan, Jinjiang
Zhang, Jialin
Zhang, Jinshan
Zhang, Qiang
Zhao, Liang
Zhao, Yingchao
Zhong, Jiaofei
Zhou, Nanrun
Zhu, Yuqing

Contents

Graph Algorithms II

Knapsack and Allocation

Graph Algorithms III

Random

Geometric Cover

Complexity and Security

Encoding and Security

Network and Algorithms

Algorithm

Graph Algorithms I

Mining Preserving Structures
in a Graph Sequence

Takeaki Uno[1] and Yushi Uno[2]([✉])

[1] National Institute of Informatics, 2-1-2 Hitotsubashi, Chiyoda-ku,
Tokyo 101-8430, Japan
uno@nii.jp
[2] Graduate School of Science, Osaka Prefecture University, 1-1 Gakuen-cho,
Naka-ku, Sakai 599-8531, Japan
uno@mi.s.osakafu-u.ac.jp

Abstract. In the recent research of data mining, frequent structures in a sequence of graphs have been studied intensively, and one of the main concern is changing structures along a sequence of graphs that can capture dynamic properties of data. On the contrary, we newly focus on "preserving structures" in a graph sequence that satisfy a given property for a certain period, and mining such structures is studied. We bring up two structures of practical importance, a connected vertex subset and a clique that exist for a certain period. We consider the problem of enumerating these structures and present polynomial delay algorithms for the problems. Their running time may depend on the size of the representation, however, if each edge has at most one time interval in the representation, the running time is $O(|V||E|^3)$ for connected vertex subsets and $O(\min\{\Delta^5, |E|^2\Delta\})$ for cliques, where the input graph is $G = (V, E)$ with maximum degree Δ. To the best of our knowledge, this is the first systematic approach to the treatment of this notion, namely, preserving structures.

1 Introduction

Extracting useful information from *graph structured data* has become important in the era of explosive and complex data, and it is often achieved by specifying and/or finding frequent substructures in a graph, that is, *pattern mining* in graphs (or *graph mining*) [1,11,22]. In the case of hyperlink structure on the Web (i.e., the webgraph), for example, a clique is considered to be formed by a community, and finding it may be useful for tracing a social phenomenon on the Web [21]. These observations imply that one of the most promising approaches for graph mining is by *enumeration*, and efficient enumeration of crucial substructures has a rich history. For cliques, a theoretically efficient algorithm is presented in [15], and both [15] and [20] are state-of-the-art algorithms that performs well in practice. Enumerations of paths and matchings are studied in [17]

A part of this research is supported by JST CREST and Grant-in-Aid for Scientific Research (KAKENHI), No. 23500022 and 15H00853.

© Springer International Publishing Switzerland 2015
D. Xu et al. (Eds.): COCOON 2015, LNCS 9198, pp. 3–15, 2015.
DOI: 10.1007/978-3-319-21398-9_1

and [8], respectively, and enumeration of connected components is studied in [2]. Here, notice that all these algorithms work on a single (and thus "static") graph.

In a recent and practical situation, however, it is often the case that graph structures may change over time (i.e., "dynamic"), and such data is collected periodically along a time series. In this setting, graph patterns appearing sequentially could be more important. Along this direction, there are some topics of interest so far. Finding graph patterns that appear periodically in a graph sequence is studied in [9,13]. Graph patterns frequently appear during a certain period are also studied in [6]. On the other hand, some research address the change patterns that appear frequently in a graph sequence composed of graphs with edge insertions/deletions, such as changes between two time periods [3] and changes of subsequences [10]. Furthermore, there are several studies focusing on clustering of vertices by utilizing graph sequences [18,19]. However, these research only concern with the "changes" or their frequency and periodicity.

Objective. Taking these preceding research into account, we propose a new concept of graph mining; finding a part of a graph that satisfies a given property continuously for a long time in a series of dynamically changing graphs, that is, *capturing invariables in change*. More specifically, we consider the problem of enumerating all substructures that satisfy a given property during a prescribed period, i.e., those appearing in a consecutive subsequence of a graph sequence. We call such structures *preserving structures* in a graph sequence, and the problem for enumerating all such structures *preserving structure mining* in general. We consider connected vertex subsets and cliques for such properties. For example, a community on the Web that is active for a long time may correspond to a clique that exists in a consecutive sequence of webgraphs during a certain period. As another example, a group of a species in a wildlife environment may constitute a consecutive sequence of connected vertex subsets in a sequence of graphs that are constructed from its trajectory data [4,12,14]. To the best of our knowledge, this study is the first case in which a "long-lasting" or "unchanging" structure is regarded as the target structure to be captured.

Contributions. In this paper, we first propose a new concept, that is, a preserving structure in a graph sequence. By adopting this notion, we pose two problems of mining preserving structures of practical importance: cliques and connected vertex subsets.

We then propose efficient algorithms for solving the problems by enumerating all connected vertex subsets or cliques for a certain time period in a given graph sequence. For this purpose, we define a way of representing a graph sequence as the input format. In this model, instead of representing a graph at each time by the difference from the previous one which is used in the dynamic graph model [7], we represent a graph sequence by explicitly associating each edge with its time interval(s) during which it exists. Although there exist similar ways of representing such data (e.g., [5]), our model is new in the sense that it introduces a new parameter, namely the number of time intervals, which will be used to estimate the running time of these algorithms. That is, it would be used as a new measure in the complexity study.

Our enumeration algorithms for preserving connected vertex subsets is based on a recursive graph partition and for preserving cliques is based on the reverse search, which is a framework for efficient enumeration algorithm design. While a straightforward application of maximal clique enumeration may require a long delay per output, our algorithms achieve polynomial delay by exploiting properties of the time intervals of edges. Compared to a naive algorithm, this reduces the time complexity with a factor of the number of edges of an input graph. Although our problem setting is fundamendal, it gives a new perspectives for graphs that change over time, together with a way of data representations and analysis of algorithms, which will pioneer a new research field.

Organization of the Paper. We give definitions and representations of graph sequences and preserving structures together with basic terminology in Sect. 2. In Sect. 3, we deal with the enumeration problem of preserving connected vertex subsets. Then we discuss about the preserving clique enumeration problem in Sect. 4. We conclude this paper in Sect. 5.

2   Preliminaries

2.1   A Graph Sequence and Its Representation

A *graph* G is an ordered pair of a *vertex set* V and an *edge set* E, and is denoted by $G = (V, E)$. We assume that a vertex set V is $\{1, \ldots, n\}$ so that each vertex has an index and can be treated as an integer. The *neighborhood* of a vertex $v \in V$ is the set $N(v) = \{u \in V \mid \{u, v\} \in E\}$. The *degree* of a vertex v is $|N(v)|$, and is denoted by $\deg(v)$. We use Δ to denote the maximum degree of a graph. For a vertex subset U ($\subseteq V$), the *induced subgraph* $G[U]$ of G by U is the subgraph whose vertex set is U and edge set is composed of all edges in E that connect vertices in U. For an edge set F, let $V(F)$ denote the set of vertices that are endpoints of some edges in F. Then for an edge subset F ($\subseteq E$), we define the induced subgraph $G[F]$ of G by F by the subgraph $G[V(F)]$.

A *time stamp* is an integer representing a discrete time, and we denote by $\mathcal{T}$ the ground set of all possible time stamps during which our graph is supposed to exist. We assume $\mathcal{T} = \{1, \ldots, t_{\max}\}$ without loss of generality, and a subset T of $\mathcal{T}$ is called a *time stamp set*. We say that an edge of a graph is *active* at time stamp t if it exists at that moment. The edge set E of a supposed graph consists only of edges that are active at some time stamps. To represent a graph sequence, we associate a time stamp set with each edge on which it is active. We call it an *active time stamp set* of that edge, and is defined by a mapping $\tau : E \to 2^{\mathcal{T}}$. Then we define a *graph sequence* as a pair of a graph $G = (V, E)$ and a mapping τ, that is, (G, τ). Now the active time stamp set of an edge e is $\tau(e)$, and we define the active time stamp set of an edge set F to be $\tau(F) = \bigcap_{e \in F} \tau(e)$. Given a graph sequence (G, τ), we define a *closure graph* G_T of G for a time stamp set T ($\subseteq \mathcal{T}$) as the spanning subgraph in which its edge set consists of edges whose active time stamp sets includes T, that is, $G_T = (V, \{e \mid e \in E, T \subseteq \tau(e)\})$. Especially in case of $T = \{t\}$, a singleton, we sometimes denote the closure graph for T by G_t by convention. Intuitively, G_t represents a snapshot of G at time stamp t. By definition, G_T becomes G if $T = \emptyset$.

A time stamp set is an (*time*) *interval* if it constitutes a single interval $\{t, t + 1, \ldots, t + \ell\}$ ($\ell \geq 0$). In this paper, it is sometimes assumed that the active time stamp set of any edge is an interval, and we call this an *interval assumption*. Note that we can assume this without loss of generality, since if an active time stamp set of an edge is composed of multiple time intervals, we can replace it by a set of parallel edges each of which has one of those intervals, respectively. Unlike the existing ones, this way of representing a graph sequence has an advantage in its extendability.

2.2   Preserving Structures

Let (G, τ) be a graph sequence, where $G = (V, E)$ and $\tau : E \to 2^T$ with a ground time stamp set T. We consider preserving structures in a graph sequence, that is, a subgraph that consecutively satisfies certain properties, such as connected vertex subsets and cliques in this paper. Especially, we are interested in maximal ones in some sense, and we use the term "closed" which is usually employed in the pattern mining field [16]; a closed pattern is a maximal pattern that is not included in the other patterns with the same frequency.

A vertex subset U is *connected* if there exists a path between any two vertices of U. In this case we also say that $G[U]$ is connected. A vertex subset U is said to be *connected on a time stamp set* T if U is connected at any time stamp in T. Let $\gamma(U)$ be the set of time stamps at which U is connected. We say that a connected vertex subset U is *closed* if none of its superset U' satisfies $\gamma(U) = \gamma(U')$.

A *clique* is a complete subgraph of a graph. In this paper, we define a clique by its edge set, and thus we do not regard a single vertex as a clique. A clique is called *maximal* if none of its superset becomes a clique. An edge set F is called *active* if $\tau(F) \neq \emptyset$, and $\tau(F)$ equals T if $F = \emptyset$. An active clique K in a graph sequence is *closed* if no other clique K' such that $K \subset K'$ satisfies $\tau(K) = \tau(K')$.

3   Enumeration of Preserving Connected Vertex Subsets

In this section we study the closed connected vertex subsets in a graph sequence (G, τ), where $G = (V, E)$ and $\tau : E \to 2^T$ with a ground time stamp set T. We start by observing their properties, and then present how they are enumerated. We first have the following simple observations.

Property 1 (closed under union). For two vertex subsets U and U', if both U and U' are connected on a time stamp set T and $U \cap U' \neq \emptyset$, then $U \cup U'$ is also connected on T.

For two partitions $\mathcal{P}$ and $\mathcal{P}'$ of a universal set, let $\mathcal{P} \wedge \mathcal{P}'$ denote the partition composed of subsets given by the intersection of members of $\mathcal{P}$ and $\mathcal{P}'$, i.e., $\mathcal{P} \wedge \mathcal{P}' = \{I \mid I = H \cap H', H \in \mathcal{P}, H' \in \mathcal{P}'\}$. A *connected component* of G is a maximal vertex subset U such that $G[U]$ is connected. The set of connected components of G gives a partition of the vertex set, and we denoted it by $\mathcal{C}(G)$. For a time stamp set $T = \{t_{i_1}, \ldots, t_{i_k}\}$, let $\mathcal{P}(G, T)$ denote $\bigwedge_{j=1}^{k} \mathcal{C}(G_{t_{i_j}})$, which forms a partition of V.

Property 2 (partition). A connected vertex subset U on a time stamp set T is included in one of vertex subsets of $\mathcal{P}(G, T)$.

Property 3 (subdivision). A connected vertex subset U in W on a time stamp set T is contained in a vertex subset of $\mathcal{P}(G[W], T)$.

We denote the family of all maximal connected vertex subsets of G on a time stamp set T by $\mathcal{C}(G, T)$. Property 1 ensures that $\mathcal{C}(G, T)$ becomes a partition of V. In the subsequent discussions in this subsection, suppose the interval assumption holds for T, and let $T_{t,\ell}$ denote an interval time stamp set $T_{t,\ell} = \{t, t+1, \ldots, t+\ell\}$. In addition, we assume for simplicity that both ends of any interval time stamp set $T_{t,\ell}$ can be examined in $O(1)$ time by some appropriate pre-process and data structures. Then we have the following two lemmas.

Lemma 1. *For an interval time stamp set $T_{t,\ell}$ with a fixed time stamp t, $\mathcal{C}(G, T_{t,\ell})$ for all ℓ (≥ 0) can be computed in $O(|V||E|^2)$ time.*

Lemma 2. *Any member U in $\mathcal{C}(G, T)$ is a closed connected vertex subset of G on an interval time stamp set T.*

Lemma 2 motivates us to compute $\mathcal{C}(G, T)$ for all possible interval time stamp set T to enumerate all closed connected vertex subsets. For each time stamp t, we compute $\mathcal{C}(G, T_{t,\ell})$ for interval time stamp set $T = \{t, t+1, \ldots, t+\ell\}$ for all possible ℓ. From Lemma 1, this computation can be done in $O(|V||E|^2)$ time. Thus we obtain the following theorem, where we use $\ell = O(|E|)$ again.

Theorem 1. *In a graph sequence (G, τ), all closed connected vertex subsets can be enumerated in $O(|V||E|^3)$ time.* □

The correctness of this algorithm relies only on the above three properties, therefore the algorithm can be applied to similar connectivity conditions satisfying these properties, such as strong connectivity of a directed graph and two-edge connectivity of a graph.

Theorem 2. *In a graph sequence (G, τ) in which G is a directed graph, all closed strongly connected vertex subsets can be enumerated in $O(|V||E|^3)$ time.* □

Theorem 3. *In a graph sequence (G, τ), all closed two-edge connected vertex subsets in a graph can be enumerated in $O(|V||E|^3)$ time.* □

In the case of two-vertex connectivity, Property 1 holds only when the intersection size of two components is no less than two. Thus, $\mathcal{C}(G, T)$ could not be a partition of a vertex set. Instead of a vertex set, we represent a connected vertex subset by all vertex pairs included in the subset. Using this representation, when two subsets share at most one vertex, the intersection of their representations is the empty set. Obviously this representation satisfies the other two properties, thus we have the following theorem.

Theorem 4. *In a graph sequence (G, τ), all closed two-vertex connected vertex subsets can be enumerated in $O(|V|^2|E|^3)$ time.* $\square$

4   Enumeration of Closed Active Cliques

This section discusses about the enumeration of all closed active cliques in a graph sequence (G, τ). We first give some additional definitions for further arguments and observe some basic properties of closed active cliques. After that we state a simple output polynomial time algorithm as a warm-up, and then we present a more efficient algorithm based on the reverse search whose time complexity is much smaller than the simple algorithm.

For a time stamp set T, let $N_T(v) = \{w \mid w \in N(v), T \subseteq \tau(\{v, w\})\}$ and $N_T(F) = \bigcap_{v \in V(F)} N_T(v)$ for an edge set F, that is, $N_T(v)$ is the set of vertices adjacent to v at all time stamps in T and $N_T(F)$ is the set of vertices adjacent to *all* vertices in $V(F)$ at any time stamp in T. For an edge set F and a vertex set U, $F \setminus U$ denotes the edge set obtained from F by removing all edges incident to some vertices in U, and $F \cap U$ denotes $F \setminus (V \setminus U)$. For an edge set F and a vertex v, let $M(F, v)$ denote the set of edges connecting v and a vertex in $V(F)$. Let $\Gamma(F)$ be the set of vertices v such that $\tau(F) \subseteq \tau(M(F, v))$.

Now let $F_{\leq i}$ be the edge set obtained from F by removing edges incident to vertices whose index is greater than i. By definition, $F_{\leq i}$ is empty if $i < 1$, and is F if $i \geq n$. A *lexicographic order* on a family of sets is a total order defined in such a way that a set F is smaller than F' when the smallest element in their symmetric difference $F \triangle F'$ belongs to F. For an active clique K in a graph sequence, let $X(K)$ denote the lexicographically smallest closed clique including K among all closed cliques K' such that $\tau(K') = \tau(K)$.

4.1   A Simple Algorithm

Let (G, τ) be a graph sequence, where $G = (V, E)$ and $\tau : E \rightarrow 2^{\mathcal{T}}$ with a ground time stamp set $\mathcal{T}$. We observe a few basic properties of closed active cliques in a graph sequence. Remember that a clique is defined by an edge set in this paper.

Lemma 3. *For any active clique K, $X(K)$ can be computed in $O(\min\{|E|, \Delta^2\})$ time.*

Proof. We can obtain $X(K)$ by iteratively choosing the minimum vertex v in $N_{\tau(K)}(K)$ and adding edges of $M(K, v)$ to K, until $N_{\tau(K)}(K) = \emptyset$. $N_{\tau(K)}(K)$ can be computed in $O(\min\{|E|, \Delta^2\})$ time by scanning all edges adjacent to some edges in K. When we add $N_{\tau(K)}(K)$ to K, $N_{\tau(K)}(K \cup N_{\tau(K)}(K))$ can be computed in $O(\deg(v))$ time by checking whether $\tau(K) \subseteq \tau(\{u, v\})$ or not for each $u \in N_{\tau(K)}(K)$. Therefore the statement holds. $\square$

Lemma 4. *For any time stamp set T, any maximal clique K in G_T is closed.*

Proof. If K is not closed, $G_{\tau(K)}$ includes a clique K' such that $K \subset K'$. Since $T \subseteq \tau(K)$, $T \subseteq \tau(e)$ holds for any edge $e \in K'$. This implies that K' is a clique in G_T, which contradicts the assumption.   □

Conversely, we can easily see that any closed active clique K is maximal in the graph $G_{\tau(K)}$. This motivates us to compute all maximal cliques in all closure graphs of possible active time stamp sets for enumerating all closed active cliques.

Lemma 5. *All closed active cliques can be enumerated in $O(|V||E|^3)$ time for each, under the interval assumption.*

Proof. Under the interval assumption, the active time stamp set of any closed active clique is also an interval. These active time stamp sets satisfy that the both ends of the interval are given by the active time sets of some edges, thus their number is bounded by $|E|^2$. Let $\mathcal{K}$ be the family of cliques each of which is a maximal clique in a closure graph of some of those active time stamp sets. Then, from Lemma 4, we can see that $|\mathcal{K}|$ is bounded by the product of $|E|^2$ and the number of closed active cliques. By using the algorithm in [15], the maximal cliques can be enumerated in $O(|V| + |E|)$ time for each, and thus the maximal cliques in $\mathcal{K}$ can be enumerated in $O((|V| + |E|)|\mathcal{K}|)$ time. To check whether an enumerated clique K is closed or not, we compute $X(K)$ in $O(|V| + |E|)$ time. Since a closed active clique can be a maximal clique of G_T for at most $|E|^2$ time stamp sets T, the closed active cliques can be enumerated in $O(|V||E|^3)$ time for each.   □

4.2 An Efficient Algorithm Based on the Reverse Search

The reverse search is a scheme for constructing enumeration algorithms, and was originally proposed by Avis and Fukuda [2] for some problems such as enumeration of vertices of a polytope. The key idea of the reverse search is to define an acyclic relation among the objects including the ones to be enumerated. An acyclic relation induces a tree, which results in the so-called a parent-child relation, and we call the tree a *family tree*. Hence enumerating objects is realized by traversing the tree according to the parent-child relation to visit all the objects. In fact, the reverse search algorithm performs a depth-first search on the tree induced by the parent-child relation, and is implemented by a procedure for enumerating all children of a given object. It starts from the root object that has no parent and enumerates its children, and then it recursively enumerates children for each child.

It is easy to see the correctness of the algorithm; that is, the tree induced by the parent-child relation spans all the objects, and the algorithm visits all the vertices of the tree by a depth-first search. When a procedure for enumerating children takes at most $O(A)$ time for each child, the computation time of the reverse search algorithm is bounded by $O(AN)$, where N is the number of objects to be enumerated. Hence, if A is polynomial in terms of the input size, the entire reverse search algorithm takes output polynomial time. In the following, we carefully observe the properties of a graph sequence, and prove that enumeration of children can be done in polynomial time.

Now a more efficient algorithm for enumeration of closed active cliques can be designed based on the reverse search. We start with giving some definitions and fundamental observations. The scheme of the reverse search has already been applied to enumeration maximal cliques [15], and our algorithm for closed active cliques adopts their ideas. For an active clique K, let $i(K)$ be the minimum vertex i satisfying $X(K_{\leq i}) = K$. We define the parent $P(K)$ of closed active clique K by $X(K_{\leq i(K)-1})$, and $P(K)$ is not defined for $K = X(\emptyset)$, which is called the *root* of the family tree.

Lemma 6. *The parent-child relation defined by P is acyclic.*

Proof. Suppose that K is a closed active clique such that $P(K)$ is defined. $P(K)$ is generated by removing vertices one by one from K, and adding vertices so that the active time set does not change, thus $\tau(P(K))$ always includes $\tau(K)$. Since $X(K_{\leq i(K)-1}) \neq K$, $P(K)$ is lexicographically smaller than K when $\tau(P(K)) = \tau(K)$. Thus, either (a) $P(K)$ has a larger active time set than K, or (b) $P(K)$ has the same active time set as K and is lexicographically smaller than K. Therefore the statement holds.   □

Lemma 7. *Any vertex in $P(K)_{\leq i(K)} \setminus K$ does not belong to $N_{\tau(K)}(i(K))$, and therefore $K_{\leq i(K)-1} = P(K)_{\leq i(K)} \cap N_{\tau(K)}(i(K))$.*

Proof. Suppose that a vertex v in $P(K)_{\leq i(K)} \setminus K$ belongs to $N_{\tau(K)}(i(K))$. Then, $X(K_{\leq i(K)})$ has to include either v or another vertex $u < v$. It implies that $X(K_{\leq i(K)}) \cap \{1, \ldots, i(K)\} \neq K_{\leq i(K)}$, thereby $X(K_{\leq i(K)}) \neq K$. This contradicts the definition of $i(K)$.   □

A subset F of $M(K, v)$ is called *time maximal* if F is included in no other subset F' of $M(K, v)$ satisfying $\tau(F) \cap \tau(K) = \tau(F') \cap \tau(K)$. Let $I(K, v)$ be the set of all time maximal subsets of $M(K, v)$. For a time maximal subset $F \in I(K, v)$, we define $C(K, F) = X(K_{\leq v} \cap V(F) \cup F)$.

Lemma 8. *If K' is a child of non-root closed active clique K, then $K' = C(K, F)$ holds for some vertex v and $F \in I(K, i(K'))$.*

Proof. Let $F = M(K_{\leq i(K')}, i(K'))$. From Lemma 7, $K'_{\leq i(K')-1} = K_{\leq i(K')-1} \cap V(N_{\tau(K')}(i(K')))$ holds, and thus $K = X(K_{\leq i(K')-1} \cap V(F) \cup F)$. We next show that F is a member of $I(K, i(K'))$. Suppose that K' is a child of K, and F does not belong to $I(K, i(K'))$, i.e., F is properly included in an edge subset $F' \in I(K, i(K'))$ such that $\tau(F) = \tau(F')$. Then, the active time set of $K_{\leq i(K')} \cap V(F')$ is same as that of $K'_{\leq i(K')} = K_{\leq i(K')-1} \cap V(F) \cup F$. This implies that $X(K'_{\leq i(K')})$ includes several edges in F', which contradicts the definition of $i(K')$.   □

Since $X(K_{\leq v}) \neq K$ holds for any $v < i(K)$, we have the following corollary.

Corollary 1. *$C(K, F)$ is not a child of K for any $F \in I(K, v)$ satisfying $v < i(K)$.*

It is true that any child is $C(K, F)$ for some F. However, $C(K, F)$ cannot always be a child, that is, $C(K, F)$ is a child of K if and only if $P(K) = P(C(K, F))$. This implies that we can check whether $C(K, F)$ is a child or not by computing $P(K)$. Therefore, from Lemma 8, we obtain the following procedure to enumerate children of K. For avoiding the duplicated output of the same child K', we output K' only when K' is generated from $F \in I(K, i(K'))$.

Procedure. EnumChildren(K: non-root closed active clique)

1. **for each** $F \in I(K, v), v > i(K)$ **do**
2.   compute $C(K, F)$;
3.   compute $i(C(K, F))$ and $P(C(K, F))$;
4.   **if** $K = P(C(K, F))$ and $i(C(K, F)) = v$ **then output** $C(K, F)$;
5. **end for**

For analyzing the complexity of this procedure, which will later be used as a subroutine of the entire algorithm for enumerating closed active cliques, we show some technical lemmas.

Lemma 9. $P(K)$ *can be computed in* $O(|E|)$ *time.*

Proof. Suppose that K is not the root, i.e., $P(K)$ is defined. Let K' be initialized to the empty set, and we add vertices of K to K' one by one from the smallest vertices in the increasing order. In each addition, we maintain the change of $\tau(K')$ and $N_{\tau(K)}(K')$. Then, we can find the minimum vertex v satisfying $\tau(K_{\leq v}) = \tau(K)$, and the minimum vertex u satisfying $i = \min\{N_{\tau(K)}(K_{\leq i-1})\}$ for any $i \in K, i \geq u$. We have $i(K) = \max\{u, v\}$, since $X(K_{\leq j}) \neq K$ holds when either $\tau(K) \neq \tau(K_{\leq j})$ or $i \neq \min\{N_{\tau(K)}(K_{\leq i-1})\}$ holds for some $i \in K, i > j$. Under the assumption that both ends of any interval time stamp set can be examined in $O(1)$ time, $\tau(K \cup \{e\})$ can be computed in $O(1)$ time from $\tau(K)$ for any edge e. Thus, we can compute $i(K)$ in $O(\min\{|E|, \Delta^2\})$ time. Together with Lemma 3, the statement holds. □

Lemma 10. *If K is not the root, any child K' of K satisfies that $K_{\leq i(K')} \cap K'_{\leq i(K')} \neq \emptyset$.*

Proof. If $K_{\leq i(K')} \cap K'_{\leq i(K')} = \emptyset$, it holds that $K'_{\leq i(K')-1} \cap K = \emptyset$. Since $K'_{\leq i(K')-1}$ is always included in K, we have $K'_{\leq i(K')-1} = \emptyset$. Therefore, $P(K') = X(\emptyset)$, which implies that $P(K)$ is the root. □

Lemma 11. *If K is not the root, the children of K is enumerated by evaluating at most $\min\{\Delta|E|, \Delta^3\}$ edge sets under the interval assumption.*

Proof. By the interval assumption, the ends of the active time set of any subset F of $I(K, v)$ is given by the ends of some edges in F, and thus $|I(K, v)|$ is bounded from above by Δ^2. Lemma 10 ensures that if K is not the root, Step 2 of EnumChildren does not have to take care of vertices not adjacent to any vertex of $V(K)$. This means that we have to take care only of non-empty maximal subset in $I(K, v)$. Let I be the union of all non-empty subsets of $I(K, v)$. Since each edge in $F \in I(K, v)$ is incident to some vertices in K, we have $|I| \leq \min\{|E|, \Delta^2\}$. It implies that the number of possible choices of two edges from some non-empty $I(K, v)$ is bounded from above by $\Delta \cdot \min\{|E|, \Delta^2\}$. □

By the above lemmas, we can estimate the time complexity of the procedure of enumerating children.

Lemma 12. *Procedure EnumChildren enumerates all children of K in $O(\min \{\Delta^5, |E|^2\Delta\})$ time under the interval assumption.*

Proof. The correctness of the procedure comes from Lemma 8. We note that the procedure never output any child more than once, since each child is generated from its unique parent, a maximal subset included in $F \in I(K, i(K))$. We then observe that all non-empty subset $F \in I(K, v), v > i(K)$ can be computed in $O(\min\{|E|, \Delta^2\})$ time by scanning all edges adjacent to some edges in K, and $C(K, F)$ can be computed in $O(\min\{|E|, \Delta^2\})$ time in a straightforward manner. From Lemma 11, the procedure iterates the loop for $\min\{\Delta|E|, \Delta^3\}$ edge sets, and each edge set spends $O(\min\{|E|, \Delta^2\})$ time from Lemma 9. Thus, we conclude the lemma. □

Now we describe our algorithm for enumerating all closed active cliques in a graph sequence based on the reverse search as follows. It is presented in a slightly different form by introducing a threshold σ with respect to the length of active time stamp sets by observing that $\tau(K) \subseteq \tau(P(K))$ always holds. It enumerates all closed active cliques having active time sets larger than σ by giving $X(\emptyset)$ (thus enumerates all when σ is set to be 0).

Algorithm. EnumClosedActiveClique(K: closed active clique)

1. **output** K; $prv := nil$;
2. **if** $prv = nil$ **then** $K' :=$ the first clique found by EnumChildren(K);
   else $K' :=$ the clique found just after prv by EnumChildren(K);
3. **if** there is no such clique K' **go to** Step 8;
4. $K := K'$; free up the memory for K';
5. **if** $|P(K)| \geq \sigma$ **then** **call** EnumClosedActiveClique(K);
6. $K := P(K)$;
7. **go to** Step 2;
8. **if** K is not the root **then** return;
9. **for each** $e \in E$ **do**
10.   **if** e is lexicographically minimum in $X(e)$
                                    then EnumClosedActiveClique($X(e)$);
11. **end for**

Finally, we can establish the following theorem.

Theorem 5. *Under the interval assumption, Algorithm EnumClosedActiveClique enumerates all closed active cliques in a graph sequence in $O(N \min\{\Delta^5, |E|^2\Delta\})$ time and in $O(|V| + |E|)$ space, where N is the number of closed active cliques in a graph sequence.*

Proof. The correctness of the algorithm is easy to see from the framework of the reverse search and Lemma 6. The computation time of the reverse search is given by the product of the number of objects to be enumerated and the computation time on each object. From Lemma 12, an iteration requires $O(N \min\{\Delta^5, |E|^2 \Delta\})$ time for non-root closed active cliques. For the root $K = X(\emptyset)$, we can enumerate its children K' satisfying the condition of Lemma 10 in $O(N \min\{\Delta^5, |E|^2 \Delta\})$ time using procedure EnumChildren. When $K_{\leq i(K')} \cap K'_{\leq i(K')} = \emptyset$, we have $K'_{\leq i(K')-1} \cap K = \emptyset$. This implies that $K'_{\leq i(K')}$ is composed of an edge, thus by generating $X(\{e\})$ for all $e \in E$, we can enumerate the children that do not satisfy the condition of Lemma 10, in $O(\min\{|E|^2, |E|\Delta^2\})$ time. Note that the duplication can be avoided by outputting $X(\{e\})$ only when $e = \arg\min X(\{e\})$. Since $N \geq |E|/\Delta^2$, it holds that $\min\{|E|^2, |E|\Delta^2\} \leq N \min\{\Delta^5, |E|^2\Delta\}$. Therefore the time complexity of the algorithm is as stated.

In a straightforward implementation of the algorithm, each iteration may take $\omega(|V|+|E|)$ space for keeping the intermediate results of the computation in memory, especially for all $C(K, F)$. We can reduce this by restarting the iteration from the beginning. When we find a child K' of K, we immediately generate the recursive call with K', before the termination of the enumeration of the children. After the termination of the recursive call, we resume the enumeration of the children. To save the memory, we restart from the beginning of the iteration, and we pass through the children found before K', and reconstruct all the necessary variables. We note that the time complexity does not change by the restart, since the number of restarts is bounded by the number of recursive calls generated by the algorithm. A child is given by a maximal edge subset, and a maximal edge subset is given by two edges. Thus, we can memorize a child by a constant number of variables. The clique K is constructed by computing $P(K')$, thus it is also not necessary to have K in memory, and can be re-constructed without increasing the time complexity. The iteration with respect to the root takes $O(|V| + |E|)$ space, therefore we have the atatement of the theorem.   □

As we stated, since $\tau(K) \subseteq \tau(P(K))$ always holds, we have the following corollary.

Corollary 2. *Under the interval assumption, Algorithm EnumClosedActive-Clique enumerates all closed active cliques having active time sets no shorter than a given threshold σ in $O(\min\{\Delta^5, |E|^2\Delta\})$ time for each and in $O(|V|+|E|)$ space.*

Note again that the interval assumption can be set without loss of generality, since we can replace an edge with multiple time intervals by parallel edges having a single time interval for each, in their active time stamp sets. However, this transformation increases the degrees of the vertices, thus the time complexity may increase. If we set Δ to the maximum degree to the transformed graph, then the results hold.

5   Conclusion

In this paper, we focused on the structures preserved in a sequence of graphs continuously for a long time, which we call "preserving structures". We considered two structures, closed connected vertex subsets and closed active cliques, and proposed efficient algorithms for enumerating these structures preserved during a period no shorter than a prescribed length. An interesting future work is, of course, to develop efficient algorithms for preserving structure mining problems for other graph properties.

References

1. Arimura, H., Uno, T., Shimozono, S.: Time and space efficient discovery of maximal geometric graphs. In: Corruble, V., Takeda, M., Suzuki, E. (eds.) DS 2007. LNCS (LNAI), vol. 4755, pp. 42–55. Springer, Heidelberg (2007)
2. Avis, D., Fukuda, K.: Reverse search for enumeration. Discr. Appl. Math. **65**, 21–46 (1996)
3. Berlingerio, M., Bonchi, F., Bringmann, B., Gionis, A.: Mining graph evolution rules. In: Buntine, W., Grobelnik, M., Mladenić, D., Shawe-Taylor, J. (eds.) ECML PKDD 2009, Part I. LNCS, vol. 5781, pp. 115–130. Springer, Heidelberg (2009)
4. Buchin, K., Buchin, M., van Kreveld, M., Speckmann, B., Staals, F.: Trajectory grouping structure. In: Dehne, F., Solis-Oba, R., Sack, J.-R. (eds.) WADS 2013. LNCS, vol. 8037, pp. 219–230. Springer, Heidelberg (2013)
5. Xuan, B.B., Ferreira, A., Jarry, A.: Computing shortest, fastest, and foremost journeys in dynamic networks. Int. J. of Foundations of Computer Science **14**, 267–285 (2003)
6. Borgwardt, K.M., Kriegel, H.P., Wackersreuther, P.: Pattern mining in frequent dynamic subgraphs. In: Proc. 6th IEEE ICDM, pp. 818–822 (2006)
7. Eppstein, D., Galil, Z., Italiano, G.F., Nissenzweig, A.: Sparsification–A technique for speeding up dynamic graph algorithms. J. ACM **44**, 669–696 (1997)
8. Fukuda, K., Matsui, T.: Finding all the perfect matchings in bipartite graphs. Applied Mathematics Letters **7**, 15–18 (1994)
9. Han, J., Dong, G., Yin, Y.: Efficient mining of partial periodic patterns in time series database. In: Proc. 15th IEEE ICDE, pp. 106–115 (1999)
10. Inokuchi A., Washio, T.: A fast method to mine frequent subsequences from graph sequence data. In: Proc. 8th IEEE ICDM, pp. 303–312 (2008)
11. Mining graph data: A. Inokuchi T. Washio and H. Motoda. Complete mining of frequent patterns from graphs. Machine Learning **50**, 321–354 (2003)
12. Kalnis, P., Mamoulis, N., Bakiras, S.: On discovering moving clusters in spatio-temporal data. In: Medeiros, C.B., Egenhofer, M., Bertino, E. (eds.) SSTD 2005. LNCS, vol. 3633, pp. 364–381. Springer, Heidelberg (2005)
13. Lahiri, M. Berger-Wolf, T.Y.: Mining periodic behavior in dynamic social networks. In: Proc. 8th IEEE ICDM, pp. 373–382 (2008)
14. Li, Z., Ding, B., Han, J., Kays, R.: Swarm: Mining relaxed temporal moving object clusters. In: Proc. 36th Int'l Conf. on VLDB, pp. 723–734 (2010)
15. Makino, K., Uno, T.: New algorithms for enumerating all maximal cliques. In: Hagerup, T., Katajainen, J. (eds.) SWAT 2004. LNCS, vol. 3111, pp. 260–272. Springer, Heidelberg (2004)
16. Pasquier, N., Bastide, Y., Taouil, R., Lakhal, L.: Efficient mining of association rules using closed itemset lattices. J. Information Systems **24**, 25–46 (1999)
17. Read, R.C., Tarjan, R.E.: Bounds on backtrack algorithms for listing cycles, paths, and spanning trees. Networks **5**, 237–252 (1975)
18. Sun, J., Faloutsos, C., Papadimitriou, S., Yu, P.S.: GraphScope: Parameter-free mining of large time-evolving graphs. In: Proc. 13th ACM Int'l Conf. on KDD, pp. 687–696 (2007)

19. Tantipathananandh, C., Berger-Wolf, T.: Constant-factor approximation algorithms for identifying dynamic communities. In: Proc. 15th ACM Int'l Conf. on KDD, pp. 827–836 (2009)
20. Tomita, E., Tanaka, A., Takahashi, H.: The worst-case time complexity for generating all maximal cliques and computational experiments. Theor. Comp. Sci. **363**, 28–42 (2006)
21. Uno, Y., Ota, Y., Uemichi, A.: Web structure mining by isolated cliques. IEICE Transactions on Information and Systems **E90–D**, 1998–2006 (2007)
22. Yan, X., Han, J.: GSPAN: graph-based substructure pattern mining. In: Proc. 2nd IEEE ICDM, pp. 721–724 (2002)

On the Most Imbalanced Orientation of a Graph

Walid Ben-Ameur$^{(\boxtimes)}$, Antoine Glorieux, and José Neto

Institut Mines-Télécom, Télécom SudParis, CNRS Samovar UMR 5157,
9 Rue Charles Fourier, 91011 Evry Cedex, France
{walid.benameur,antoine.glorieux,jose.neto}@telecom-sudparis.eu

Abstract. We study the problem of orienting the edges of a graph such that the minimum over all the vertices of the absolute difference between the outdegree and the indegree of a vertex is maximized. We call this minimum *the imbalance* of the orientation, i.e. the higher it gets, the more imbalanced the orientation is. We study this problem denoted by MAXIM. We first present different characterizations of the graphs for which the optimal objective value of MAXIM is zero. Next we show that it is generally NP-complete and cannot be approximated within a ratio of $\frac{1}{2} + \varepsilon$ for any constant $\varepsilon > 0$ in polynomial time unless P = NP even if the minimum degree of the graph δ equals 2. Finally we describe a polynomial-time approximation algorithm whose ratio is equal to $\frac{1}{2}$ for graphs where $\delta \equiv 0[4]$ or $\delta \equiv 1[4]$ and $(\frac{1}{2} - \frac{1}{\delta})$ for general graphs.

Introduction and Notation

Let $G = (V, E)$ be an undirected simple graph, we denote by δ_G the minimum degree of the vertices of G. An orientation Λ of G is an assignment of a direction to each undirected edge $\{uv\}$ in E, i.e. any function on E of the form $\Lambda(\{uv\}) \in \{uv, vu\}, \forall \{uv\} \in E$. For each vertex v of G we denote by $d_G(v)$ or $d(v)$ the unoriented degree of v in G and by $d_\Lambda^+(v)$ or $d^+(v)$ (resp. $d_\Lambda^-(v)$ or $d^-(v)$) the outdegree (resp. indegree) of v in G w.r.t. Λ. Graph orientation is a well studied area in graph theory and combinatorial optimization and thus a large variety of constrained orientations as well as objective functions have been considered so far.

Among those arise the popular degree-constrained orientation problems: in 1976, Frank & Gyárfás [12] gave a simple characterization of the existence of an orientation such that the outdgree of every vertex is between a lower and an upper bound given for each vertex. Asahiro et al. in [1–3] proved the NP-hardness of the weighted version of the problem where the maximum outdegree is minimized, gave some inapproximability results, and studied similar problems for different classes of graphs. Chrobak & Eppstein proved that for every planar graph a 3-bounded outdegree orientation and a 5-bounded outdegree acyclic orientation can be constructed in linear time [6].

Other problems involving other criterion on the orientation have been studied such as acyclicity, diameter or connectivity. Robbins' theorem (1939) for example states that the graphs that have strong orientations are exactly the 2-edge-connected graphs [18] and later (1985), Chung et al. provided a linear time

© Springer International Publishing Switzerland 2015
D. Xu et al. (Eds.): COCOON 2015, LNCS 9198, pp. 16–29, 2015.
DOI: 10.1007/978-3-319-21398-9_2

algorithm for checking whether a graph has such an orientation and finding one if it does [7]. Then in 1960, Nash-Williams generalized Robbin's theorem showing that an undirected graph has a k-arc-connected orientation if and only if it is $2k$-edge-connected [17]. The problem called oriented diameter that consists in finding a strongly connected orientation with minimum diameter was introduced in 1978 by Chvátal & Thomassen: they proved that the problem is NP-hard for general graphs [8]. It was then proven to be NP-hard even if the graph is restricted to a subset of chordal graphs by Fomin et al. (2004) who gave also approximability and inapproximability results [10].

For an orientation Λ of $G = (V, E)$ and a vertex v we call $|d_\Lambda^+(v) - d_\Lambda^-(v)|$ the *imbalance* of v in G w.r.t Λ and thus we call $\min_{v \in V} |d_\Lambda^+(v) - d_\Lambda^-(v)|$ the *imbalance* of Λ. Biedl et al. studied the problem of finding an acyclic orientation of unweighted graphs minimizing the imbalance of each vertex: they proved that it is solvable in polynomial time for graphs with maximum degree at most three but NP-complete generally and for bipartite graphs with maximum degree six and gave a $\frac{13}{8}$-approximation algorithm [5]. Then Kára et al. closed the gap proving the NP-completeness for graphs with maximum degree four. Furthermore, they proved that the problem remains NP-complete for planar graphs with maximum degree four and for 5-regular graphs [14].

Landau's famous theorem [15] gives a condition for a sequence of non-negative integers to be the score sequence or outdegree sequence of some tournament (i.e. oriented complete graph) and later, Harary & Moser characterized score sequences of strongly connected tournaments [13]. Analogous results for the "imbalance sequences" of directed graphs are were given by Mubayi et al. [16]. In 1962, Ford & Fulkerson characterized the mixed graphs (i.e. partially oriented graphs) which orientation can be completed in a eulerian orientation, that is to say, an orientation for which the imbalance of each vertex equals zero [11]. Many other results related to orientation have been proposed. Some of them are reviewed in [4].

Let us denote by $\overrightarrow{O}(G)$ the set of all the orientations of G, we consider the problem of finding an orientation with maximized imbalance:

$$(\text{MaxIm}) \quad \text{MaxIm}(G) = \max_{\Lambda \in \overrightarrow{O}(G)} \min_{v \in V} |d_\Lambda^+(v) - d_\Lambda^-(v)|$$

and we call $\text{MaxIm}(G)$ the value of MaxIm for G. The minimum degree δ_G of a graph G is a trivial upper bound for $\text{MaxIm}(G)$.

The rest of this paper is organized as follows. In the first section, we give several characterizations of the the graphs verifying $\text{MaxIm}(G) = 0$. In section 2, we will show that MaxIm is generally NP-complete even for graphs with minimum degree 2 and inapproximable within a ratio $\frac{1}{2} + \varepsilon$ for any constant $\varepsilon > 0$ and then will give an approximation algorithm whose ratio is almost equal to $\frac{1}{2}$. Since the value of MaxIm for a graph is the minimum of the values of MaxIm on its connected component, from here on in, all the graphs we consider are assumed to be connected. For any graph G we will use the notations $V(G)$ and $E(G)$ to refer to the set of vertices of G and the set of edges of G respectively.

1   Characterizing the Graphs for which $\text{MaxIm}(G) = 0$

Now we ask ourselves which are the graphs verifying $\text{MaxIm}(G) = 0$. We will start by unveiling several necessary conditions and properties of such graphs. First we can show that concerning such a graph, we can find an orientation satisfying several additional properties.

Proposition 1. *Let G be a graph such that $\text{MaxIm}(G) = 0$ and $u \in V$. Then there exists an orientation $\Lambda \in \overrightarrow{O}(G)$ such that u is the only vertex of G with imbalance equal to zero w.r.t. Λ.*

Proof. Let $\Lambda \in \overrightarrow{O}(G)$ be an orientation minimizing $|\{v \in V/|d_\Lambda^+(v) - d_\Lambda^-(v)| = 0\}|$. We suppose that $|\{v \in V/|d_\Lambda^+(v) - d_\Lambda^-(v)| = 0\}| \geq 2$. We choose two distinct vertices v and w in $\{v \in V/|d_\Lambda^+(v) - d_\Lambda^-(v)| = 0\}$ and a path $p = (v = u_0, \cdots, u_n = w)$ between v and w. If we switch the orientation of the edge $\{u_0 u_1\}$, then the imbalance of u_0 becomes positive and necessarily the imbalance of u_1 becomes zero otherwise the resulting orientation would contradict the minimality of Λ. Using the same reasoning, if we switch the orientation of all the edges $\{u_0 u_1\}, \cdots, \{u_{n-2} u_{n-1}\}$, we obtain an orientation where both u_{n-1} and u_n have an imbalance equal to zero while the imbalance is positive on all the vertices $u_0, \cdots, u_{n-2}$ and unchanged on all other vertices. So now if we switch the orientation of the edge $\{u_{n-1} u_n\}$ as well, then the resulting orientation contradicts the minimality of Λ. Hence, $|\{v \in V/|d_\Lambda^+(v) - d_\Lambda^-(v)| = 0\}| = 1$.

Now let v be this unique vertex of G such that $|d_\Lambda^+(v) - d_\Lambda^-(v)| = 0$. Let $u \neq v$ be an arbitrary vertex and let $p = (v = u_0, \cdots, u_n = u)$ be a path between v and u. By switching the orientation of all the edges $\{u_0 u_1\}, \cdots, \{u_{n-2} u_{n-1}\}$, we obtain an orientation Λ' where u has an imbalance equal to zero while the imbalance is positive for u_0 and unchanged on all other vertices.   □

This yields the following necessary condition: if G is a graph such that $\text{MaxIm}(G) = 0$, then G is eulerian. For let $u \in V$, we know there exists $\Lambda \in \overrightarrow{O}(G)$ such that $\{v \in V/|d_\Lambda^+(v) - d_\Lambda^-(v)| = 0\} = \{u\}$. Then $d_\Lambda^+(u) = d_\Lambda^-(u)$, hence $d(u) = d_\Lambda^+(u) + d_\Lambda^-(u) = 2d_\Lambda^+(u)$ is even. The following lemma about eulerian graphs will prove useful for the proof of our characterization.

Lemma 2. *If G is an eulerian graph, then there exists an elementary cycle (hereafter just called cycle) C of G such that $G - E(C)$ has at most one connected component that is not an isolated vertex.*

Proof. Being G eulerian and connected, it can be decomposed into edge-disjoint cycles that we can order $C_1, \cdots, C_n$ according to the following condition: $\cup_{k=1}^{i} C_i$ is connected, $\forall i \in [\![1, n]\!]$. Then C_n is the cycle we are looking for.   □

Now let us define a certain family of graphs which will prove to be exactly the graphs for which the optimal objective value of MaxIm is zero. Intuitively they are the graphs for which every block is an odd cycle.

Theorem 3. *We define the class of graphs $\mathscr{C}^{odd}$ as follows: a simple graph G is in $\mathscr{C}^{odd}$ if there exists $C_1, \cdots, C_n$ odd cycles ($n \geq 1$) such that:*

- $\cup_{i=1}^{n} C_i = G$,
- $|V(\cup_{k=1}^{i-1} C_k) \cap V(C_i)| = 1, \; \forall i \in [\![2, n]\!]$.

$$(1)$$

Then for any simple graph G, $\mathrm{MAXIM}(G) = 0$ if and only if $G \in \mathscr{C}^{odd}$.

Proof.
- $\Leftarrow$ We will work by induction on the number of cycles n contained in the graph. Nothing is required for these cycles except that they must be elementary. If $n = 1$, then our graph is an odd cycle which implies $\mathrm{MAXIM}(G) = 0$. Let $n \geq 2$, we assume that all graphs of $\mathscr{C}^{odd}$ with $k \leq n-1$ cycles verify $\mathrm{MAXIM}(G) = 0$. Let $G \in \mathscr{C}^{odd}$ with n cycles $C_1, \cdots, C_n$ as in (1). Suppose there exists $\Lambda \in \overrightarrow{O}(G)$ with strictly positive imbalance. Let us call $G' = \cup_{i=1}^{n-1} C_i$ the graph obtained from G after removing C_n and let us take a look at $\Lambda_{|E(G')}$ the orientation of the edges of G' obtained from Λ as its restriction on $E(G')$. As G' is a graph of $n-1$ cycles in $\mathscr{C}^{odd}$, our inductive hypothesis implies that we have a vertex $u \in V(G')$ such that $|d^+_{\Lambda_{|E(G')}}(u) - d^-_{\Lambda_{|E(G')}}(u)| = 0$. Necessarily, $u = V(G') \cap V(C_n)$. Thus $|d^+_{\Lambda}(u) - d^-_{\Lambda}(u)| = |d^+_{\Lambda_{|E(C_n)}}(u) - d^-_{\Lambda_{|E(C_n)}}(u)| > 0$ implying that $\mathrm{MAXIM}(C_n) > 0$ which is absurd because C_n is an odd cycle.
- $\Rightarrow$ Since $\mathrm{MAXIM}(G) = 0$, we know that G is eulerian. We will again work by induction on the number of cycles n. If $n = 1$, then our graph is eulerian with a unique cycle, hence it is a cycle. Now as $\mathrm{MAXIM}(G) = 0$, necessarily it is an odd cycle and is therefore in $\mathscr{C}^{odd}$. Let $n \geq 2$, we assume that all graphs with $k \leq n-1$ cycles verifying $\mathrm{MAXIM}(G) = 0$ are in $\mathscr{C}^{odd}$. Let G be a graph with n cycles such that $\mathrm{MAXIM}(G) = 0$. Thanks to Lemma 2, there exists an cycle C of G such that $G - E(C)$ has at most one connected component G' that is not an isolated vertex.

  Suppose that $\mathrm{MAXIM}(G') > 0$, let $\Lambda \in \overrightarrow{O}(G')$ with strictly positive imbalance. Let $u_0 \in V(G') \cap V(C)$, we name the vertices of C as follows: $u_0, u_1, \cdots, u_k = u_0$. Without loss of generality, we can assume that $d^+_{\Lambda}(u_0) - d^-_{\Lambda}(u_0) > 0$; if it was not the case, replace Λ by its reverse. We complete Λ in an orientation of G by orienting the edges of C: we orient $u_0 u_1$ from u_0 to u_1 and go on as follows:

$$\forall i \in [\![1, k-1]\!], \begin{cases} \text{if } u_i \in V(G'), & \text{we orient } \{u_i u_{i+1}\} \text{ as } \{u_{i-1} u_i\}, \\ \text{otherwise,} & \text{we orient } \{u_i u_{i+1}\} \text{ as } \{u_i u_{i-1}\}. \end{cases}$$

Where orienting an edge $\{ab\}$ as another edge $\{cd\}$ means orienting it from a to b if $\{cd\}$ was oriented from c to d and from b to a otherwise. Let us have a look at the resulting orientation Λ' (cf Figure 1): when completing Λ in Λ', the imbalance of the vertices in $V(G') \backslash \{u_0\}$ was left unchanged, the imbalance of the vertices in $V(C) \backslash V(G')$ equals 2 and the imbalance of u_0

was either left unchanged or augmented by two. Hence Λ' has strictly positive imbalance which contraditcts $\text{MAXIM}(G) = 0$, therefore, $\text{MAXIM}(G') = 0$.

Suppose $|V(G') \cap V(C)| \geq 2$ and let u and v be 2 distinct vertices in $V(G') \cap V(C))$ such that $u \neq v$. Thanks to proposition 1, we know that there exists an orientation $\Lambda \in \overrightarrow{O}(G')$ such that $\{w \in V/|d_\Lambda^+(w) - d_\Lambda^-(w)| = 0\} = \{v\}$ and without loss of generality, $d_\Lambda^+(u) - d_\Lambda^-(u) > 0$. We name the vertices of C as follows: $u = u_0 u_1 \cdots u_k = u_0$, $v = u_l$ and we complete Λ in an orientation of G by orienting the edges of C: we orient $\{u_0 u_1\}$ from u_0 and u_1 and go on as follows:

$$\forall i \in [\![1, k-1]\!] \setminus \{l\}, \begin{cases} \text{if } u_i \in V(G'), & \text{we orient } \{u_i u_{i+1}\} \text{ as } \{u_{i-1} u_i\}, \\ \text{otherwise,} & \text{we orient } \{u_i u_{i+1}\} \text{ as } \{u_i u_{i-1}\}. \end{cases}$$

And we orient $\{u_l u_{l+1}\}$ as $\{u_l u_{l-1}\}$. In the resulting orientation Λ', the imbalance of the vertices in $V(G') \setminus \{u, v\}$ was left unchanged, the imbalance of the vertices in $V(C) \setminus V(G')$ equals 2, the imbalance of v was augmented by two and the imbalance of u was either left unchanged or augmented by two. Hence Λ' contradicts $\text{MAXIM}(G) = 0$, therefore, $|V(G') \cap V(C)| = 1$.

Suppose C is even. We call $u \in V(G')$ such that $V(G') \cap V(C) = \{u\}$, and $\Lambda \in \overrightarrow{O}(G')$ such that $\{v \in V/|d_\Lambda^+(v) - d_\Lambda^-(v)| = 0\} = \{u\}$. We name the vertices of C as follows: $u = u_0 u_1 \cdots u_k = u_0$ and we complete Λ in an orientation of G by orienting the edges of C: we orient $\{u_0 u_1\}$ from u_0 to u_1 and $\{u_i u_{i+1}\}$ as $\{u_i u_{i-1}\}$, $\forall i \in [\![1, k-1]\!]$. In the resulting orientation Λ', the imbalance of the vertices in $V(G') \setminus \{u\}$ was left unchanged, the imbalance of the vertices in $V(C) \setminus V(G')$ equals 2 and, C being even, the imbalance of u was augmented by two. Hence Λ' contradicts $\text{MAXIM}(G) = 0$, therefore, C is odd.

As G' is a graph with at most $n - 1$ cycles verifying $\text{MAXIM}(G) = 0$, by induction hypothesis, there exist $C_1, \cdots, C_{n-1}$ odd cycles such that:

○ $\cup_{i=1}^{n-1} C_i = G'$,
○ $|V(\cup_{k=1}^{i-1} C_k) \cap V(C_i)| = 1$, $\forall i \in [\![2, n-1]\!]$.

Adding the odd cycle $C_n = C$, we directly obtain that $G \in \mathscr{C}^{odd}$. $\qquad \square$

Now in order to widen our perception of those graphs, let us show another characterization.

Theorem 4. *For every simple graph G,*

$$G \in \mathscr{C}^{odd} \Leftrightarrow G \text{ is eulerian with no even cycle}$$

Proof.   • ⇒ By construction, every graph in $\mathscr{C}^{odd}$ is eulerian with no even cycle.

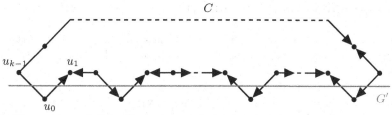

Fig. 1. The vertices of C in G' are left unchanged imbalance-wise, the other vertices of C are set to 2 and in the end $|d_{\Lambda'}^+(u_0) - d_{\Lambda'}^-(u_0)| \geq |d_{\Lambda}^+(u_0) - d_{\Lambda}^-(u_0)| > 0$

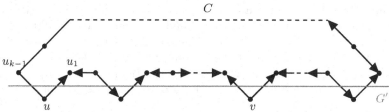

Fig. 2. The vertices of C in G' are left unchanged imbalance-wise except for v which is set to 2, like the other vertices of C and in the end $|d_{\Lambda'}^+(u_0) - d_{\Lambda'}^-(u_0)| \geq |d_{\Lambda}^+(u_0) - d_{\Lambda}^-(u_0)| > 0$

- $\Leftarrow$ We will once again work by induction on the number of cycles n.

  If $n = 1$, then our graph is eulerian with a unique odd cycle, hence it is an odd cycle and is therefore in $\mathscr{C}^{odd}$.

  Let $n \geq 2$, we assume that all eulerian graphs with no even cycle and $k \leq n-1$ odd cycles are in $\mathscr{C}^{odd}$. Let G be a graph with no even cycle and n odd cycles. Thanks to Lemma 2, there exists an odd cycle C of G such that $G - E(C)$ has only one connected component G' that is not an isolated vertex. As G' is eulerian and even-cycle-free with $n - 1$ odd cycles, by induction hypothesis, $G' \in \mathscr{C}^{odd}$, hence there exist $C_1, \cdots, C_{n-1}$ odd cycles such that:
  - $\cup_{i=1}^{n-1} C_i = G'$,
  - $|V(\cup_{k=1}^{i-1} C_k) \cap V(C_i)| = 1, \ \forall i \in [\![2, n-1]\!]$.

  Suppose there exist u and v ($u \neq v$) belonging to $V(\cup_{k=1}^{n-1} C_k) \cap V(C)$. Since G' is connected, let p be an elementary path in G' between u and v. We can assume that u and v are the only vertices of C contained in p, otherwise we could replace v by the first vertex of C encountered when travelling on p from u. C defines two other vertex-disjoint paths between u and v: one even that we will call p_{even} and one odd that we will call p_{odd}. p being vertex disjoint with either p_{even} or p_{odd}, by concatenating it with the one corresponding to its parity, we obtain an even cycle of G, contradicting our hypothesis on G. This yields that $|V(C) \cap V(G')| = 1$. From that we can conclude
  - $\cup_{i=1}^{n} C_i = G$,
  - $|V(\cup_{k=1}^{i-1} C_k) \cap V(C_i)| = 1, \ \forall i \in [\![2, n]\!]$.

  Hence $G \in \mathscr{C}^{odd}$.

  $\square$

2   Complexity, Inapproximability and Approximability

In this section we will prove the NP-completeness and inapproximability of our problem and give an approximation algorithm based on the special case of bipartite graphs.

Concerning the complexity of MAXIM, we will show that the problem is NP-complete. More precisely, that answering if MAXIM(G) equals 2 for a graph G such that $\delta_G = 2$ is NP-complete. For that purpose we will introduce a variant of the satisfiability problem that we will reduce to a MAXIM instance: the not-all-equal at most 3-SAT(3V).

Not-all-equal at most 3-SAT(3V) is a restriction of not-all-equal at most 3-SAT which is itself a restriction of 3-SAT known to be NP-complete [19] where each clause contains at most three literals and in each clause, not all the literals can be true. Since 2-SAT can be solved in polynomial time, we hereafter deal only with formulas having at least one three-literals clause. The added restriction of not-all-equal at most 3-SAT(3V) is that each variable (not literal) appears at most three times in a formula. The resulting problem is still NP-complete.

Lemma 5. *The not-all-equal at most 3-SAT(3V) problem is NP-complete.*

Proof. See Appendix A.

Now we will associate to a not-all-equal at most 3-SAT(3V) instance φ with n variables $\{x_1, \cdots, x_n\}$ and m clauses $\{c_1, \cdots, c_m\}$ a graph G_φ for which the value w.r.t. MAXIM will give the answer to whether φ is satisfiable or not. If a variable x_i occurs only in positive literals (resp. only in negative literals), it follows that a satisfying assignment of the variables of φ must necessarily give the value TRUE (resp. FALSE) to x_i, therefore x_i can be removed from φ with conservation of the satisfiability. Thus, without loss of generality, we can assume that in any not-all-equal at most 3-SAT(3V) formula, every variable occurs at least once as a positive literal and at least once as a negative literal. G_φ consists of gadgets that mimic the variables and the clauses of φ and additional edges that connect them together:

- the gadget corresponding to a variable x_i consists of two vertices labeled x_i and $\neg x_i$ and one edge connecting them;

- the gadget corresponding to a two-literals clause $c_j = (l^1 \vee l^2)$, where l^1 and l^2 are its literals, consists in two vertices labeled $a_{l^1}^j$ and $b_{l^2}^j$ corresponding to l^1 and l^2 respectively (the index "l^k" of the vertices labels stands for the literal they represent, i.e. x_i if l^k is the variable x_i and $\neg x_i$ if l^k is the negation of the variable x_i) and one edge connecting them;

- the gadget corresponding to a three-literals clause gadget consists in six vertices and six edges. For a clause $c_j = (l^1 \vee l^2 \vee l^3)$, where l^1, l^2 and l^3 are its literals (the order is arbitrary), three vertices labeled $a_{l^1}^j$, $b_{l^2}^j$ and $b_{l^3}'^j$ correspond to l^1, l^2 and l^3 respectively. Three additional vertices are labeled

u^j, v^j and w^j and the gadgets' edges are $\{a^j_{l_1} u_j\}$, $\{a^j_{l_1} v_j\}$, $\{u_j w_j\}$, $\{v_j w_j\}$, $\{w_j b^j_{l_2}\}$ and $\{w_j b'^j_{l_3}\}$;

- $\forall i \in [\![1, n]\!]$, the vertex labeled x_i (resp. $\neg x_i$) is connected to all the vertices labeled $a^j_{x_i}$, $b^j_{x_i}$ or $b'^j_{x_i}$ (resp. $a^j_{\neg x_i}$, $b^j_{\neg x_i}$ or $b'^j_{\neg x_i}$), $\forall j \in [\![1, m]\!]$.

As an example, for a formula

$$\varphi = (x_1 \vee \neg x_2 \vee x_3) \wedge (\neg x_1 \vee \neg x_3 \vee x_4) \wedge (x_1 \vee \neg x_2 \vee x_4) \wedge (x_2 \vee \neg x_4), \quad (2)$$

the corresponding graph G_φ is represented in Figure 3.

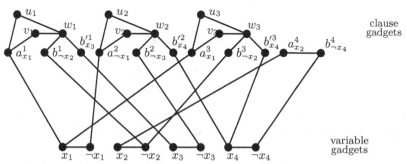

Fig. 3. G_φ for $\varphi = (x_1 \vee \neg x_2 \vee x_3) \wedge (\neg x_1 \vee \neg x_3 \vee x_4) \wedge (x_1 \vee \neg x_2 \vee x_4) \wedge (x_2 \vee \neg x_4)$

Theorem 6. *A not-all-equal at most 3-SAT(3V) formula φ is satisfiable if and only if* $\text{MaxIm}(G_\varphi) = 2$.

Proof. • $\Rightarrow$ Suppose φ is satisfiable and let $v : \{x_1, \cdots, x_n\} \rightarrow \{\text{TRUE}, \text{FALSE}\}$ be a satisfying assignment of $x_1, \cdots, x_n$. We know that $\delta_{G_\varphi} = 2$ which yields $\text{MaxIm}(G_\varphi) \leq 2$. So let us build an orientation $\Lambda \in \overrightarrow{O}(G_\varphi)$ which imbalance is greater than or equal to 2. First, we assign an orientation to the edges of the variable gadget:

$$\Lambda(\{x_i \neg x_i\}) = \begin{cases} x_i \neg x_i & \text{if } v(x_i) = \text{TRUE}; \\ \neg x_i x_i & \text{otherwise.} \end{cases}$$

For example, for the formula $\varphi = (x_1 \vee \neg x_2 \vee x_3) \wedge (\neg x_1 \vee \neg x_3 \vee x_4) \wedge (x_1 \vee \neg x_2 \vee x_4) \wedge (x_2 \vee \neg x_4)$ satisfied by the assignment $v(x_1, x_2, x_3, x_4) = (\text{FALSE}, \text{TRUE}, \text{TRUE}, \text{TRUE})$, the edges of the variable gadgets of graph G_φ are oriented as in figure 4(a). Since each variable x_i occurs at least once as a positive literal and at least once as a negative literal, $2 \leq d_{G_\varphi}(x_i) \leq 3$ and $2 \leq d_{G_\varphi}(\neg x_i) \leq 3$, $\forall i \in [\![1, n]\!]$. Then to ensure our objective on the imbalance of Λ, the orientation of the edges connecting vertex gadgets and clause gadgets must be such that $\forall i \in [\![1, n]\!]$, $|d^+_\Lambda(x_i) - d^-_\Lambda(x_i)| = d_{G_\varphi}(x_i)$

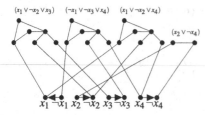

(a) orientation of the edges in the variable gadgets

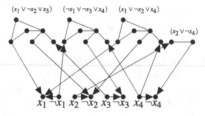

(b) orientation of the edges between the variable gadgets and the clause gadgets

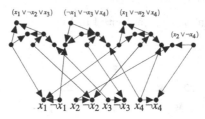

(c) orientation of the edges in the clause gadgets

Fig. 4. G_φ corresponding to $\varphi = (x_1 \vee \neg x_2 \vee x_3) \wedge (\neg x_1 \vee \neg x_3 \vee x_4) \wedge (x_1 \vee \neg x_2 \vee x_4) \wedge (x_2 \vee \neg x_4)$ satisfied by $v(x_1, x_2, x_3, x_4) = (\text{FALSE}, \text{TRUE}, \text{TRUE}, \text{TRUE})$

and $|d_\Lambda^+(\neg x_i) - d_\Lambda^-(\neg x_i)| = d_{G_\varphi}(\neg x_i)$. In other words, for $i \in [\![1, n]\!]$, if $v(x_i) = \text{TRUE}$ (resp. $v(x_i) = \text{FALSE}$), then the edges adjacent to the vertex x_i are oriented from x_i (resp. to x_i) and the edges adjacent to the vertex $\neg x_i$ are oriented to $\neg x_i$ (resp. from $\neg x_i$), e.g. Figure 4(b).

So far, all the edges in the variables gadgets and the edges connecting the vertex gadgets and the clause gadgets have been oriented and the vertices in the variables gadgets have imbalance greater than or equal to 2. In order to complete our orientation Λ we have to orient the edges in the clause gadgets.

Let $c_j = (l^1 \vee l^2)$ be a two-literals clause. Since v satisfies φ, we know that exactly one of the two literals is true w.r.t. v. Which, according to the way we oriented edges so far, means that exactly one of $a_{l^1}^j$ and $b_{l^2}^j$ has one incoming arc from a variable gadget and the other has one outgoing arc to a variable gadget. If $a_{l^1}^j$ is the one with the incoming arc from a variable gadget (meanings that $v(l^1) = \text{TRUE}$), then we assign $\Lambda(\{a_{l^1}^j b_{l^2}^j\}) = (b_{l^2}^j a_{l^1}^j)$, otherwise the opposite. We obtain $|d_\Lambda^+(a_{l^1}^j) - d_\Lambda^-(a_{l^1}^j)| = |d_\Lambda^+(b_{l^2}^j) - d_\Lambda^-(b_{l^2}^j)| = 2$.

Let $c_j = (l^1 \vee l^2 \vee l^3)$ (the order is identical to which was chosen to build the clause gadget, i.e. $d_{G_\varphi}(a_{l^1}^j) = 3$ and $d_{G_\varphi}(b_{l^2}^j) = d_{G_\varphi}(b_{l^3}^{ij}) = 2$) be at three-literals clause. If the edge connecting $a_{l^1}^j$ to a variable gadget is oriented to $a_{l^1}^j$ (meanings that $v(l^1) = \text{TRUE}$), then we assign $\Lambda(\{a_{l^1}^j u_j\}) = (u_j a_{l^1}^j)$, $\Lambda(\{a_{l^1}^j v_j\}) = (v_j a_{l^1}^j)$, $\Lambda(\{u_j w_j\}) = (u_j w_j)$ and $\Lambda(\{v_j w_j\}) = (v_j w_j)$. Since $v(l^1) = \text{TRUE}$, either both $v(l^2)$ and $v(l^3)$ are FALSE or exactly one of $v(l^2)$

and $v(l^3)$ is TRUE and one is FALSE. If both are FALSE then b_{l2}^j and $b_{l3}^{'j}$ have an outgoing arc to a variable gadget. In that case, we orient $w_j b_{l2}^j$ and $w_j b_{l3}^{'j}$ to w_j and we obtain $|d_\Lambda^+(a_{l1}^j) - d_\Lambda^-(a_{l1}^j)| = 3$, $|d_\Lambda^+(b_{l2}^j) - d_\Lambda^-(b_{l2}^j)| = |d_\Lambda^+((b_{l3}^{'j}) - d_\Lambda^-((b_{l3}^{'j})| = |d_\Lambda^+(u_j) - d_\Lambda^-(u_j)| = |d_\Lambda^+(v_j) - d_\Lambda^-(v_j)| = 2$ and $|d_\Lambda^+(w_j) - d_\Lambda^-(w_j)| = 4$. If exactly one of $v(l^2)$ and $v(l^3)$ is TRUE and one is FALSE, then exactly one of b_{l2}^j and $b_{l3}^{'j}$ has an incoming arc from a variable gadget and the other an outgoing arc to a variable gadget. If b_{l2}^j is the one with the incoming arc from a variable gadget (meanings that $v(l^2)$ = TRUE and $v(l^3)$ = FALSE), then we assign $\Lambda(\{w_j b_{l2}^j\}) = (w_j b_{l2}^j)$ and $\Lambda(\{w_j b_{l3}^{'j}\}) = (b_{l3}^{'j} w_j)$, otherwise the opposite. We obtain $|d_\Lambda^+(a_{l1}^j) - d_\Lambda^-(a_{l1}^j)| = 3$ and $|d_\Lambda^+(b_{l2}^j) - d_\Lambda^-(b_{l2}^j)| = |d_\Lambda^+(b_{l3}^{'j}) - d_\Lambda^-(b_{l3}^{'j})| = |d_\Lambda^+(u_j) - d_\Lambda^-(u_j)| = |d_\Lambda^+(v_j) - d_\Lambda^-(v_j)| = |d_\Lambda^+(w_j) - d_\Lambda^-(w_j)| = 2$.

If, on the other hand, the edge connecting a_{l1}^j to a variable gadget is oriented from a_{l1}^j (meanings that $v(l^1)$ = FALSE), then we assign $\Lambda(\{a_{l1}^j u_j\}) = (a_{l1}^j u_j)$, $\Lambda(\{a_{l1}^j v_j\}) = (a_{l1}^j v_j)$, $\Lambda(\{u_j w_j\}) = (w_j u_j)$ and $\Lambda(\{v_j w_j\}) = (w_j v_j)$. By symmetry, we conclude in the same way that $|d_\Lambda^+(a_{l1}^j) - d_\Lambda^-(a_{l1}^j)| = 3$ and $|d_\Lambda^+(b_{l2}^j) - d_\Lambda^-(b_{l2}^j)| = |d_\Lambda^+(b_{l3}^{'j}) - d_\Lambda^-(b_{l3}^{'j})| = |d_\Lambda^+(u_j) - d_\Lambda^-(u_j)| = |d_\Lambda^+(v_j) - d_\Lambda^-(v_j)| = |d_\Lambda^+(w_j) - d_\Lambda^-(w_j)| = 2$.

Consequently, the imbalance of the resulting orientation Λ is greater than or equal to 2, e.g. Figure 4(c).

- $\Leftarrow$ Now we assume that $\text{MaxIm}(G_\varphi) = 2$, let $\Lambda \in \overrightarrow{O}(G_\varphi)$ with optimal imbalance. Since all the vertices in the variable gadgets have degree at most 3, each vertex x_i (or $\neg x_i$) is necessarily adjacent to only incoming arcs or only outgoing arcs w.r.t. Λ. We will show that the assignment $v : \{x_1, \cdots, x_n\} \to \{\text{TRUE}, \text{FALSE}\}$ of $x_1, \cdots, x_n$ defined by

$$v(x_i) = \begin{cases} \text{TRUE} & \text{if } d_\Lambda^+(x_i) > d_\Lambda^-(x_i); \\ \text{FALSE} & \text{otherwise;} \end{cases}$$

satisfies φ. Suppose φ doesn't satisfy a clause c_j, $j \in [\![1, m]\!]$. If c_j is a two-literals clause $(l^1 \vee l^2)$ then either $v(l^1) = v(l^2) = \text{TRUE}$ or $v(l^1) = v(l^2) = \text{FALSE}$, i.e. either both a_{l1}^j and b_{l2}^j have an incoming arc from a variable gadget or both have an outgoing arc to a variable gadget and in both cases, whichever is the orientation assigned to $a_{l1}^j b_{l2}^j$ by Λ, either a_{l1}^j or b_{l2}^j have a zero imbalance which contradicts our assumption. So c_j is a three-literals clause $(l^1 \vee l^2 \vee l^3)$ (the order is identical to which was chosen to build the clause gadget, i.e. $d_{G_\varphi}(a_{l1}^j) = 3$ and $d_{G_\varphi}(b_{l2}^j) = d_{G_\varphi}(b_{l3}^{'j}) = 2$). Then either $v(l^1) = v(l^2) = v(l^3) = \text{TRUE}$ or $v(l^1) = v(l^2) = v(l^3) = \text{FALSE}$, i.e. either all a_{l1}^j, b_{l2}^j and $b_{l3}^{'j}$ have an incoming arc from a variable gadget or they all have an outgoing arc to a variable gadget. In the first case, it implies $\Lambda(\{a_{l1}^j u_j\}) = (u_j a_{l1}^j)$, $\Lambda(\{a_{l1}^j v_j\}) = (v_j a_{l1}^j)$, $\Lambda(\{u_j w_j\}) = (u_j w_j)$, $\Lambda(\{v_j w_j\}) = (v_j w_j)$, $\Lambda(\{w_j b_{l2}^j\}) = (w_j b_{l2}^j)$ and $\Lambda(\{w_j b_{l3}^{'j}\}) = (w_j b_{l3}^{'j})$, and we obtain $|d_\Lambda^+(w_j) - d_\Lambda^-(w_j)| = 0$ which contradicts the optimality of Λ. Similarly, in the second case it implies that the orientations assigned to the edges of the clause gadgets are the opposite from the previous ones and we obtain the same contradiction.

So we can conclude that v does satisfy φ.

□

Corollary 7. MAXIM *is* NP-*complete and inapproximable within* $\frac{1}{2} + \varepsilon$ *where* $\varepsilon \in \mathbb{R}_+^*$, *unless* P = NP.

Proof. Let $\varepsilon \in \mathbb{R}_+^*$, suppose that there existed a polynomial approximation algorithm giving $val \geq (\frac{1}{2} + \varepsilon)$ MAXIM(G) for an input graph G. Let φ be a not-all-equal at most 3-SAT(3V) formula and G_φ its associated graph. Since G_φ contains at least one three-literals clause gadget, we know that G_φ contains an even cycle and $\delta_{G_\varphi} = 2$. This leads to MAXIM$(G_\varphi) \in \{1, 2\}$ and since $(\frac{1}{2} + \varepsilon)$ MAXIM$(G_\varphi) \leq val \leq$ MAXIM(G_φ), if the polynomial approximation algorithm returns a value less than or equal to 1 then

$$(\frac{1}{2} + \varepsilon) \text{ MAXIM}(G_\varphi) \leq 1 \Rightarrow \text{MAXIM}(G_\varphi) < 2 \Rightarrow \text{MAXIM}(G_\varphi) = 1;$$

and if it returns a value greater than 1, then MAXIM(G_φ) is greater than 1 hence equal to 2. In other words the polynomial approximation algorithm output answers whether φ is satisfiable or not which is absurd unless P = NP.   □

Now we consider the case of bipartite graphs: if $G = (V1 \bigsqcup V2, E)$ is a bipartite graph, the orientation that consists in assigning to each edge in E the orientation from its extremity in V_1 to its extremity in V_2 has an imbalance equal to δ_G, i.e. optimal. This simple case permits us to obtain the following lower bound:

Theorem 8. *For every graph* G,

$$\text{MAXIM}(G) \geq \lceil \frac{\delta_G}{2} \rceil - 1.$$

Proof. Let (V_1, V_2) be a partition of V corresponding to a cut $C \subset E$ such that we have $|\delta(\{v\}) \cap C| \geq \lceil \frac{d(v)}{2} \rceil$, $\forall v \in V$. Such a cut exists: for example a maximum cardinality cut verifies this property, otherwise we could find a higher cardinality cut by switching a vertex $v \in V$ s.t. $|\delta(\{v\}) \cap C| < \lceil \frac{d(v)}{2} \rceil$ from V_1 to V_2 (or the contrary). Moreover, if we iterated this process starting from a random cut, we would converge in polynomial time time to a such a cut. Now we define $\Lambda \in \overrightarrow{O}(G)$ as follows. We begin by orienting all edges in C from V_1 to V_2. Then for any $i \in \{1, 2\}$, we orient the edges of the induced subgraph $G[V_i]$. We add a new vertex v_0 and an edge between v_0 and each vertex with an odd degree in $G[V_i]$ if it isn't eulerian and we consider a decomposition of its edges into edge-disjoint cycles. we orient each of these cycles as a directed cycle. Removing v_0 if necessary, the imbalance of each vertex in $G[V_i]$ is now in $\{-1, 0, 1\}$ which implies that $\forall v \in V$ we have $|d_\Lambda^+(v) - d_\Lambda^-(v)| \geq \lceil \frac{d(v)}{2} \rceil - 1$, hence, MAXIM$(G) \geq \lceil \frac{\delta_G}{2} \rceil - 1$.   □

Since the imbalance of the orientation defined in that proof is at least $\frac{\lceil \frac{\delta_G}{2} \rceil - 1}{\delta_G}$. MaxIm($G$), we can derive a polynomial $(\frac{1}{2} - \frac{1}{\delta_G})$-approximation algorithm. It is also easy to see that when $\delta_G \equiv 0[4]$ then MaxIm(G) $\geq \lceil \frac{\delta_G}{2} \rceil$ while MaxIm(G) $\geq \lceil \frac{\delta_G+1}{2} \rceil$ when $\delta_G \equiv 1[4]$. This leads to a $(\frac{1}{2})$-approximation algorithm when either $\delta_G \equiv 0[4]$ or $\delta_G \equiv 1[4]$.

3   Further Research

While computing the most imbalanced orientation of a graph is generally diffi-cult, the problem turns out to be easy for cactus graphs. It may be the same for other graph classes, characterizing such graph classes would be interesting.

We are currently looking for efficient mathematical programming formula-tions to solve the problem for large size graphs. Details will follow in the extended version of the paper. One can also study the weighted version of the problem.

Appendix A   Proof of Lemma 5

Let φ be a not-all-equal at most 3-SAT formula with n variables $\{x_1, \cdots, x_n\}$ and m clauses $\{c_1, \cdots, c_m\}$ and for all $i \in [\![1, n]\!]$, let $k_i \in \mathbb{N}$ be the number of occurences of x_i in φ. We assume that there is at least one variable x_i that has at least 4 occurences in φ (otherwise φ is already a not-all-equal at most 3-SAT(3V) formula) and we will build from φ a not-all-equal at most 3-SAT(3V) φ' such that φ and φ' are equisatisfiable as follows.

- For all $i \in [\![1, n]\!]$, if $k_i \geq 4$ then we introduce k_i new variables $\{x_i^1, \cdots, x_i^{k_i}\}$ and for $l \in [\![1, k_i]\!]$ we replace the l-th occurence of x_i in φ with x_i^l.

- For all $i \in [\![1, n]\!]$, if $k_i \geq 4$ then we add k_i new clauses $\{c_{x_i}^1, \cdots, c_{x_i}^{k_i}\}$ where for $l \in [\![1, k_i - 1]\!]$, $c_{x_i}^l = (x_i^l \vee \neg x_i^{l+1})$ and $c_{x_i}^{k_i} = (x_i^l \vee \neg x_i^1)$.

Suppose there exists an assignment $v : \{x_1, \cdots, x_n\} \rightarrow \{\text{TRUE}, \text{FALSE}\}$ of $x_1, \cdots, x_n$ satisfying φ. Then

$$v' : \begin{array}{ll} x_i \mapsto v(x_i) & \forall i \in [\![1, n]\!] \text{ s.t. } k_i \leq 3; \\ x_i^l \mapsto v(x_i) & \forall i \in [\![1, n]\!] \text{ s.t. } k_i \geq 4 \text{ and } \forall l \in [\![1, k_i]\!]; \end{array}$$

is an assignment of the variables x_i and x_i^l satisfying φ' for

- $\forall j \in [\![1, m]\!]$, the values of the literals of c_j w.r.t. v and v' are piecewise equal so $v'(c_j) = v(c_j) = \text{TRUE}$ and v' is not-all-equal for c_j as well as v is;

- $\forall i \in [\![1, n]\!]$ s.t. $k_i \geq 4$, $\forall l \in [\![1, k_i - 1]\!]$, $v'(x_i^l) = v'(x_i^{l+1}) = v(x_i)$ and $v'(x_i^{k_i}) = v'(x_i^1) = v(x_i)$ so we directly have $\forall l \in [\![1, k_i - 1]\!]$, $v'(c_{x_i}^l) = \text{TRUE}$ and $v'(c_{x_i}^{k_i}) = \text{TRUE}$ and v' is not-all-equal for each of these clauses since they all consist of two literals having opposite values w.r.t. v'.

As an example, for a formula

$$\varphi = (x_1 \vee \neg x_2 \vee x_3) \wedge (\neg x_1 \vee \neg x_3 \vee x_4) \wedge (x_1 \vee \neg x_2) \wedge (\neg x_1 \vee \neg x_3 \vee \neg x_4) \wedge (x_1 \vee x_3),$$

where x_1 occurs five times and x_3 four so we add nine new variables x_1^1, x_1^2, x_1^3, x_1^4, x_1^5, x_3^1, x_3^2, x_3^3, x_3^4 and x_3^4 and nine new clauses:

$$\varphi' = (x_1^1 \vee \neg x_2 \vee x_3^1) \wedge (\neg x_1^2 \vee \neg x_3^3 \vee x_4) \wedge (x_1^3 \vee \neg x_2) \wedge (\neg x_1^4 \vee \neg x_3^3 \vee \neg x_4) \wedge (x_1^5 \vee x_3^4)$$
$$\wedge (x_1^1 \vee \neg x_1^2) \wedge (x_1^2 \vee \neg x_1^3) \wedge (x_1^3 \vee \neg x_1^4) \wedge (x_1^4 \vee \neg x_1^5) \wedge (x_1^5 \vee \neg x_1^1)$$
$$\wedge (x_3^1 \vee \neg x_3^2) \wedge (x_3^2 \vee \neg x_3^3) \wedge (x_3^3 \vee \neg x_3^4) \wedge (x_3^4 \vee \neg x_3^1).$$

Now suppose there exists an assignment v' of the x_i and x_i^l satisfying φ' and let $i \in [\![1, n]\!]$ such that $k_i \geq 4$. If we take a look at the clauses $c_{x_i}^1, \cdots, c_{x_i}^{k_i}$, we notice that if $v'(x_i^1) = $ FALSE then for $c_{x_i}^1$ to be satisfied, $v'(\neg x_i^2) = $ TRUE, i.e. $v'(x_i^2) = $ FALSE, then for $c_{x_i}^2$ to be satisfied, $v'(\neg x_i^3) = $ TRUE ...etc. Repeating this argument, we obtain that if $v'(x_i^1) = $ FALSE then $v'(x_i^1) = v'(x_i^2) = \cdots = v'(x_i^{k_i}) = $ FALSE. Similarly, if $v'(x_i^{k_i}) = $ TRUE then for $c_{x_i}^{k_i}$ to be satisfied, $v'(\neg x_i^{k_i-1}) = $ FALSE, i.e. $v'(x_i^{k_i-1}) = $ TRUE, then for $c_{x_i}^{k_i-1}$ to be satisfied, $v'(\neg x_i^{k_i-2}) = $ FALSE ...etc. Hence if $v'(x_i^{k_i}) = $ TRUE then $v'(x_i^{k_i}) = v'(x_i^{k_i-1}) = \cdots = v'(x_i^1) = $ TRUE. This yields that

$$\forall i \in [\![1, n]\!] \text{ s.t. } k_i \geq 4, \ v'(x_i^1) = v'(x_i^2) = \cdots = v'(x_i^{k_i}).$$

Hence for all $i \in [\![1, n]\!]$ such that $k_i \geq 4$, we can replace $x_i^1, \cdots, x_i^{k_i}$ by a unique variable x_i and doing so the clauses $c_{x_i}^1, \cdots, c_{x_i}^{k_i}$ become trivial and can be removed and only φ remains. So the following assignment of $x_1, \cdots, x_n$:

$$v : \begin{array}{l} x_i \mapsto v'(x_i) \quad \forall i \in [\![1, n]\!] \text{ s.t. } k_i \leq 3; \\ x_i \mapsto v'(x_i^1) \quad \forall i \in [\![1, n]\!] \text{ s.t. } k_i \geq 4; \end{array}$$

satisfies φ. We have just shown that φ and φ' are equisatisfiable. $\qquad \square$

References

1. Asahiro, Y., Jansson, J., Miyano, E., Ono, H., Zenmyo, K.: Approximation algorithms for the graph orientation minimizing the maximum weighted outdegree. In: Kao, M.-Y., Li, X.-Y. (eds.) AAIM 2007. LNCS, vol. 4508, pp. 167–177. Springer, Heidelberg (2007)
2. Asahiro, Y., Miyano, E., Ono, H.: Graph classes and the complexity of the graph orientation minimizing the maximum weighted outdegree. In: Proceedings of the Fourteenth Computing: the Australasian Theory Symposium(CATS2008), Wollongong, NSW, Australia (2008)
3. Asahiro, Y., Jansson, J., Miyano, E., Ono, H.: Degree-constrained graph orientation: maximum satisfaction and minimum violation. In: Kaklamanis, C., Pruhs, K. (eds.) WAOA 2013. LNCS, vol. 8447, pp. 24–36. Springer, Heidelberg (2014)
4. Bang-Jensen, J., Gutin, G.: Orientations of graphs and digraphs in Digraphs: Theory, Algorithms and applications, 2nd edition, pp. 417–472. Springer (2009)

5. Biedl, T., Chan, T., Ganjali, Y., Hajiaghayi, M., Wood, D.R.: Balanced vertex-orderings of graphs. Discrete Applied Mathematics **48**(1), 27–48 (2005)
6. Chrobak, M., Eppstein, D.: Planar orientations with low out-degree and compaction of adjacency matrices. Theoretical Computer Sciences **86**, 243–266 (1991)
7. Chung, F., Garey, M., Tarjan, R.: Strongly connected orientations of mixed multigraphs. Networks **15**, 477–484 (1985)
8. Chvátal, V., Thomassen, C.: Distances in orientation of graphs. Journal of Combinatorial Theory, Series B **24**, 61–75 (1978)
9. Diestel, R.: Graph Theory, 4th edn. Springer (2010)
10. Fomin, F., Matamala, M., Rapaport, I.: Complexity of approximating the oriented diameter of chordal graphs. Journal of Graph Theory **45**(4), 255–269 (2004)
11. Ford, L.R., Fulkerson, D.R.: Flows in networks. Princeton University Press, Princeton (1962)
12. Frank, A., Gyárfás, A.: How to orient the edges of a graph? Colloquia Mathematica Societatis János Bolyai **18**, 353–364 (1976)
13. Harary, F., Krarup, J., Schwenk, A.: Graphs suppressible to an edge. Canadian Mathematical Bulletin **15**, 201–204 (1971)
14. Kára, J., Kratochvíl, J., Wood, D.R.: On the complexity of the balanced vertex ordering problem. In: Wang, L. (ed.) COCOON 2005. LNCS, vol. 3595, pp. 849–858. Springer, Heidelberg (2005)
15. Landau, H.G.: On dominance relations and the structure of animal societies III. the condition for a score structure. The Bulletin of Mathematical Biophysics **15**, 143–148 (1953)
16. Mubayi, D., Will, T.G., West, D.B.: Realizing Degree Imbalances in Directed Graphs. Discrete Mathematics **239**(173), 147–153 (2001)
17. Nash-Williams, C.: On orientations, connectivity and odd vertex pairings in finite graphs. Canadian Journal of Mathematics **12**, 555–567 (1960)
18. Robbins, H.: A theorem on graphs with an application to a problem of traffic control. American Mathematical Monthly **46**, 281–283 (1939)
19. Schaefer, T.J.: The complexity of satisfiability problems. In: Proceedings of the 10th Annual ACM Symposium on Theory of Computing, pp. 216–226 (1978)

Cheeger Inequalities for General Edge-Weighted Directed Graphs

T.-H. Hubert Chan, Zhihao Gavin Tang, and Chenzi Zhang[(✉)]

The University of Hong Kong, Hong Kong, China
{hubert,zhtang,czzhang}@cs.hku.hk

Abstract. We consider Cheeger Inequalities for general edge-weighted directed graphs. Previously the directed case was considered by Chung for a probability transition matrix corresponding to a strongly connected graph with weights induced by a stationary distribution. An Eulerian property of these special weights reduces these instances to the undirected case, for which recent results on multi-way spectral partitioning and higher-order Cheeger Inequalities can be applied.

We extend Chung's approach to general directed graphs. In particular, we obtain higher-order Cheeger Inequalities for the following scenarios:
(1) The underlying graph needs not be strongly connected.
(2) The weights can deviate (slightly) from a stationary distribution.

1 Introduction

There have been numerous works relating the expansion properties of an undirected graph with the eigenvalues of its Laplacian [1,3,9]. Given an undirected graph with non-negative edge weights, the weight of a vertex is the sum of the weights of its incident edges. Then, the *expansion* $\rho(S)$ of a subset S of vertices is the ratio of the sum of the weights of edges having only one end-point in S to the sum of the weights of vertices in S. The celebrated Cheeger's Inequality [1,6] relates the smallest expansion of a subset of vertices having at most half the sum of vertex weights with the second smallest eigenvalue of the corresponding normalized Laplacian. Recently, there have been extensions to the case where the expansions of k disjoint subsets are related to the k-th smallest eigenvalue [13].

The notion of expansion can be extended to directed graphs, where the weight of a vertex is the sum of the weights of its out-going edges. Then, the expansion of a subset S is defined with respect to the sum of the weights of edges going out of S. Chung [8] considered the special case for a probability transition matrix whose non-zero entries correspond to the edges of a **strongly connected graph**. The weights of the vertices are chosen according to the (unique) stationary distribution, and the weight of an edge is the probability mass going along the edge under this stationary distribution. Under this specific choice of weights, Chung has proved an analogous Cheeger's Inequality [8] for directed graphs.

This research is partially funded by a grant from Hong Kong RGC under the contract HKU17200214E.

© Springer International Publishing Switzerland 2015
D. Xu et al. (Eds.): COCOON 2015, LNCS 9198, pp. 30–41, 2015.
DOI: 10.1007/978-3-319-21398-9_3

In this paper, we explore how this relationship between expansion and spectral properties can be extended to more general cases for directed graphs. In particular, we consider the following cases.

1. The directed graph is not strongly connected.
2. The weights of vertices deviate (slightly) from the stationary distribution.

As we shall explain, each of these cases violates the technical assumptions that are used by Chung to derive the Cheeger Inequality for directed graphs. We explore what expansion notions are relevant in these scenarios, and how to define Laplacians whose eigenvalues can capture these notions.

1.1   Overview of Chung's Approach [8]

All spectral arguments rely on some symmetric matrix, which has the desirable properties of having real eigenvalues and an orthonormal basis of eigenvectors. For an undirected graph (with non-negative edge weights), its normalized Laplacian is a symmetric matrix. To apply spectral analysis on directed graphs, one should consider what the natural candidates for symmetric matrices should be and whether they have any significance. We explain the importance of the technical assumptions made by Chung in the analysis of the transition matrix $\mathbf{P}$ associated with the random walk on some directed graph $G(V, E)$.

(1) Choice of Weights. Suppose $\phi : V \to \mathbb{R}_+$ is a stationary distribution of the transition matrix $\mathbf{P}$. Then, the weights are chosen such that each vertex u has weight $\phi(u)$, and each (directed) edge (u, v) has weight $\phi(u) \cdot P(u, v)$, which is the probability mass going from u to v in one step of the random walk starting from distribution ϕ.

Suppose the starting vertex u of a random walk is chosen according to distribution ϕ. The expansion of a subset S has the following meaning: conditioning on the event that u is in S, it is the probability that the next step of the random walk goes out of S.

This notion of expansion can be defined with respect to any distribution on the vertex set V, but the edge weights induced by a stationary distribution has the following *Eulerian* property: for any subset S of vertices, the sum of weights of edges going out of S is the same as that of edges going into S.

Hence, one can consider the underlying undirected graph such that each undirected edge has weight that is the average of those for the corresponding directed edges in each direction. Then, because of the Eulerian property, for any subset S, its expansion in the directed graph defined with respect to the out-going edges is exactly the same as its expansion defined with respect to the undirected graph (with edge weights defined above). Therefore, it suffices to consider the normalized Laplacian of the undirected graph to analyze the expansion properties of the directed graph.

(2) Irreducibility of Transition Matrix. This means that the underlying directed graph with edges corresponding to transitions with non-zero probabilities is strongly connected. Under this assumption, the stationary distribution is unique, and every vertex has a positive mass.

If the directed graph is not strongly connected, a strongly connected component is known as a *sink* if there is no edge going out of it. If there is more than one sink, the stationary distribution is not unique. Moreover, under any stationary distribution, any vertex in a non-sink has probability mass zero. Hence, Chung's method essentially deletes all non-sinks before considering the expansion properties of the remaining graph.

In this paper, we explore ways to consider expansion properties that involve the non-sinks of a directed graph that is not strongly connected.

1.2   Our Contribution

The contribution of this paper is mainly conceptual, and offers an approach to extend Chung's spectral analysis of transition matrices to the scenarios when the underlying directed graph is not strongly connected, or when the vertex weights do not follow a stationary distribution. On a high level, our technique to handle both issues is to add a new vertex to the graph and define additional transition probabilities involving the new vertex such that the new underlying graph is strongly connected, and the expansion properties for the old vertices are also preserved in the new graph. Therefore, Chung's technique can be applied after the transformation. We outline our approaches and results as follows.

(1) Transition matrix whose directed graph is not necessarily strongly connected. Given a transition matrix $\mathbf{P}$ corresponding to a random walk on a graph $G(V, E)$ and a subset $S \subseteq V$ of vertices, we denote by $\mathbf{P}|_S$ the submatrix defined by restricting $\mathbf{P}$ only to the rows and the columns corresponding to S.

It is known [7] that the eigenvalues of $\mathbf{P}$ are the union of the eigenvalues of $\mathbf{P}|_C$ over all strongly connected components C in directed graph G. An important observation is that as long as the strongly connected components and the transition probabilities within a component remain the same, the eigenvalues of $\mathbf{P}$ are independent of the transition probabilities between different strongly connected components. This suggests that it might be difficult to use spectral properties to analyze expansion properties involving edges between different strongly connected components.

Therefore, we propose that it makes sense to consider the expansion properties for each strongly connected component separately. If C is a sink (i.e., there is no edge leaving C), then $\mathbf{P}|_C$ itself is a probability transition matrix, for which Chung's approach can be applied by using the (unique) stationary distribution on C.

However, if C is a non-sink, then there is no stationary distribution for $\mathbf{P}|_C$, because there is non-zero probability mass leaking out of C in every step of the random walk. For the non-trivial case when $|C| \geq 2$, by the Perron-Frobenius Theorem [10], there exists some maximal eigenvalue $\lambda > 0$ with respect to the complex norm, and unique (left) eigenvector ϕ with strictly positive coordinates such that $\phi^{\mathrm{T}}\mathbf{P}|_C = \lambda\phi^{\mathrm{T}}$. When ϕ is normalized such that all coordinates sum to 1, we say that ϕ is the *diluted stationary distribution* of $\mathbf{P}|_C$. It is stationary in the sense that if we start the random walk with distribution ϕ, then conditioning on the event that the next step remains in C (which has probability λ), we have the same distribution ϕ on C.

Hence, we can define the expansion of a subset S in C with respect to the diluted stationary distribution ϕ. Given a vertex $u \in C$ and a vertex $v \in V$

(that could be outside C), the weight of the edge (u, v) is $\phi(u) \cdot P(u, v)$. Observe that the sum of weights of edges going out of u is $\phi(u)$. Hence, the expansion of a subset S in C are due to edges leaving S that can either stay in or out of the component C.

In order to analyze this notion of expansion using Chung's approach, we construct a strongly connected graph on the component C together with a new vertex v_0, which absorbs all the probabilities leaking out of C, and returns them to C according to the diluted stationary distribution ϕ. This defines a probability transition matrix $\widehat{P}$ on the new graph that is strongly connected with various nice properties. For instance, $\widehat{P}$ has 1 as the maximal eigenvalue with the corresponding left eigenvector formed from the diluted stationary distribution ϕ by appending an extra coordinate corresponding to the new vertex with value $1 - \lambda$.

One interesting technical result (Lemma 1) is that the new transition matrix $\widehat{P}$ preserves the spectral properties of $P|_C$ in the sense that the eigenvalues of $\widehat{P}$ can be obtained by removing λ from the multi-set of eigenvalues of $P|_C$ and including 1 and $\lambda - 1$. In other words, other than the removal of λ and the inclusion of 1 and $\lambda - 1$, all other eigenvalues are preserved, even up to their algebraic and geometric multiplicities.

Hence, we can use Chung's approach to define a symmetric Laplacian for $\widehat{P}$, and use the recent results from Lee et al. [13] on higher-order Cheeger Inequalities to achieve an analogous result for a strongly connected component in a directed graph. In particular, multi-way partition expansion is considered. For a subset C of vertices, we denote:

$$\rho_k(C) := \min\{\max_{i \in [k]} \rho(S_i) : S_1, S_2, \ldots, S_k \text{ are disjoint subsets of C}\}.$$

Theorem 1. *Suppose C is a strongly connected component of size n associated with some probability transition matrix, and the expansion $\rho(S)$ of a subset S of vertices within C is defined with respect to the diluted stationary distribution ϕ as described above.*

Then, one can define a Laplacian matrix with dimension $(n + 1) \times (n + 1)$ having eigenvalues $0 = \lambda_1 \leq \lambda_2 \leq \cdots \leq \lambda_{n+1}$ such that for $1 \leq k \leq n$, we have
$$\frac{\lambda_k}{2} \leq \rho_k(C) \leq O(k^2) \cdot \sqrt{\lambda_{k+1}}.$$

(2) Vertex weights deviate from stationary distribution. Given a transition matrix P, recall that in Chung's approach, the expansion is defined with the careful choice of setting each vertex's weight according to a stationary distribution. We consider the case when the vertex weights $\phi : V \to \mathbb{R}_+$ can deviate from a stationary distribution of P.

Suppose each vertex is assigned a positive weight according to ϕ. Then, the following parameter measures how much ϕ deviates from a stationary distribution:

$$\varepsilon := 1 - \min_{u \in V} \frac{\phi(u)}{\sum_{v \in V} \phi(v) \cdot P(v, u)}.$$

A smaller value of ε means that ϕ is closer to a stationary distribution. In particular, if $\varepsilon = 0$, then ϕ is a stationary distribution.

Our idea is to first scale down all probabilities in P by a factor of $(1 - \varepsilon)$. We add a new vertex v_0 to absorb the extra ε probability from each existing vertex. Then, we define the transition probabilities from v_0 to the original vertices carefully such that each original vertex u receives the same weight $\phi(u)$ after one step. In other words, we can append a new coordinate corresponding to v_0 to ϕ to obtain a stationary distribution $\widehat{\phi}$ for the transition matrix $\widehat{\mathbf{P}}$ of the augmented random walk. Moreover, for each subset $S \subset V$ of the original vertices, the new expansion $\widehat{\rho}(S)$ with respect to $\widehat{\phi}$ and $\widehat{\mathbf{P}}$ can be related to the old expansion $\rho(S)$ as follows: $\widehat{\rho}(S) = (1 - \varepsilon) \cdot \rho(S) + \varepsilon$.

Therefore, we can apply Chung's approach to $\widehat{\mathbf{P}}$ and $\widehat{\phi}$ to construct a symmetric Laplacian matrix, whose eigenvalues are related to the expansion properties using the results by Lee et al. [13].

Theorem 2. *Suppose n vertices have positive weights defined by $\phi : V \rightarrow \mathbb{R}_+$, and $\varepsilon \geq 0$ is the parameter defined above. Then, there exists a symmetric Laplacian matrix with dimension $(n+1) \times (n+1)$ and eigenvalues $0 = \lambda_1 \leq \lambda_2 \leq \cdots \leq \lambda_{n+1}$ such that for $1 \leq k \leq n$, we have $\frac{\lambda_k}{2} \leq (1 - \varepsilon) \cdot \rho_k(V) + \varepsilon \leq O(k^2) \cdot \sqrt{\lambda_{k+1}}$.*

In the full version, we show that if we allow a self-loop at the new vertex v_0 with negative weight, we can slightly improve the left hand side of the inequality in Theorem 2.

1.3 Related Work

Since the Cheeger's Inequality [6] was introduced in the context of Riemannian geometry, analogous results have been achieved by Alon et al. [1,3] to relate the expansion of an undirected graph with the smallest positive eigenvalue of the associated Laplacian matrix. The reader is referred to the standard textbook by Chung [9] on spectral graph theory for a more comprehensive introduction of the subject.

As far as we know, the only previous attempt to apply spectral analysis to directed graphs was by Chung [8], who reduced the special case of directed instances induced by stationary distributions into undirected instances. On a high level, our approach is to reduce general directed instances into instances induced by stationary distributions.

Recently, for undirected instances, Lee et al. [13] extended Cheeger's Inequality to relate higher order eigenvalues with multi-way spectral partition. This result was further improved by Kwok et al. [12]. Since Chung's approach [8] made use of the Laplacian induced by an undirected instance, the higher order Cheeger Inequalities can be directly applied to the cases considered by Chung.

The reader can refer to the survey on spectral partitioning by Shewchuck [15], who also mentioned expansion optimization problems with negative edge weights. Other applications of spectral analysis include graph coloring [2,4], web search [5,11] clustering [14], image segmentation [16,17], etc.

2 Preliminaries

We consider a graph $G(V, E)$ with non-negative edge weights $w : E \rightarrow \mathbb{R}_+$. In most cases, we consider directed graphs, but we will also use results for undirected graphs. Note that a vertex might have a self-loop (even in an undirected graph).

For a subset $S \subseteq V$, $\partial(S)$ is the set of edges leaving S in a directed graph, whereas in an undirected graph, it includes those edges having exactly one endpoint in S (excluding self-loops). Given the edge weights, vertex weights are defined as follows. For each $u \in V$, its weight $w(u)$ is the sum of the weights of its out-going edges (including its self-loop) in a directed graph, whereas in an undirected graph, the edges incident on u are considered.

The *expansion* $\rho(S)$ of a subset S (with respect to w) is defined as:

$$\frac{w(\partial(S))}{w(S)} = \frac{\sum_{e \in \partial(S)} w(e)}{\sum_{u \in S} w(u)}.$$

In this paper, we use bold capital letters (such as $\mathbf{A}$) to denote matrices and bold small letters (such as $\boldsymbol{\phi}$) to denote column vectors. The transpose of a matrix $\mathbf{A}$ is denoted as $\mathbf{A}^{\mathrm{T}}$. For a positive integer n, $\mathbf{I}_n$ is the $n \times n$ identity matrix, and $\mathbf{0}_n$ and $\mathbf{1}_n$ are the all zero's and all one's column vectors, respectively, of dimension n, where the subscript n is omitted if the dimension is clear from context.

Undirected Graphs. Suppose $\mathbf{W}$ is the symmetric matrix indicating the edge weights w of an undirected graph of size n, and $\boldsymbol{\Phi}$ is the diagonal matrix whose diagonal entries correspond to the vertex weights induced by w as described above. Then, the *normalized Laplacian* of $\mathbf{W}$ is $\mathcal{L} := \mathbf{I}_n - \boldsymbol{\Phi}^{-\frac{1}{2}} \mathbf{W} \boldsymbol{\Phi}^{-\frac{1}{2}}$.

Multi-way partition expansion is considered in Lee et al. [13] by considering the following parameter. For $C \subseteq V$ and positive integer k, denote

$$\rho_k(C) := \min\{\max_{i \in [k]} \rho(S_i) : S_1, S_2, \ldots, S_k \text{ are disjoint subsets of C}\}.$$

The following fact relates the eigenvalues of $\mathcal{L}$ with the multi-way partition expansion with respect to $\mathbf{W}$, which may contain self-loops.

Fact 1. *(Higher Order Cheeger's Inequality [13]) Given a symmetric matrix $\mathbf{W}$ indicating the non-negative edge weights of an undirected graph, suppose its normalized Laplacian $\mathcal{L}$ as defined above has eigenvalues $0 = \lambda_1 \leq \lambda_2 \leq \cdots \leq \lambda_n$. Then, for $1 \leq k \leq n$, we have: $\frac{\lambda_k}{2} \leq \rho_k(V) \leq O(k^2) \cdot \sqrt{\lambda_k}$.*

Chung's Approach [8] to Transition Matrices. Given a probability transition matrix $\mathbf{P}$ (which is a square matrix with non-negative entries such that every row sums to 1) corresponding to a random walk on vertex set V, and non-negative vertex weight $\boldsymbol{\phi}$, one can define edge weights $w : V \times V \to \mathbb{R}_+$ as $w(u, v) := \phi(u) \cdot P(u, v)$. (Observe that these edge weights induce vertex weights that are consistent with $\boldsymbol{\phi}$.)

One interpretation of Chung's approach is that the vertex weights $\boldsymbol{\phi}$ are chosen to be a stationary distribution of $\mathbf{P}$, i.e., $\boldsymbol{\phi}^{\mathrm{T}} \mathbf{P} = \boldsymbol{\phi}^{\mathrm{T}}$. Hence, the edge weights w satisfy the following *Eulerian property*: for any subset $S \subseteq V$, we have $w(\partial(S)) = w(\partial(\overline{S}))$, where $\overline{S} := V \setminus S$.

We can define edge weights $\widehat{w}$ for the (complete) undirected graph with vertex set V such that for $u \neq v$, $\widehat{w}(u, v) = \frac{1}{2}(w(u, v) + w(v, u))$, and each self-loop has the same weight in $\widehat{w}$ and w.

Because of the Eulerian property of w, it is immediate that for all $S \subseteq V$, $w(\partial(S)) = \widehat{w}(\partial(S))$, where $\partial(S)$ is interpreted according to the directed case on the left and to the undirected case on the right. Moreover, for all $u \in V$, $w(u) = \widehat{w}(u)$. Hence, as far as expansion is concerned, it is equivalent to consider the undirected graph with edge weights given by the matrix $\widehat{\mathbf{W}}$, for which the (higher-order) Cheeger Inequalities (as in Fact 1) can be readily applied.

Chung's approach can be applied to any stationary distribution ϕ of $\mathbf{P}$, but special attention is paid to the case when $\mathbf{P}$ is irreducible, i.e., the edges corresponding to non-zero transition probabilities form a strongly connected graph on V. The advantages is that in this case, the stationary distribution is unique, and every vertex has non-zero probability.

In Section 3, we consider how to extend Chung's approach to the case where the underlying directed is not strongly connected. In Section 4, we consider the case when the edge weights ϕ deviate (slightly) from a stationary distribution.

3   Directed Graphs with Multiple Strongly Connected Components

In a directed graph, we say that a strongly connected component C is a *sink*, if there is no edge leaving C. Otherwise, we say that it is a non-sink.

Even if the underlying directed graph of a given transition matrix is not strongly connected, Chung's approach [8] can still be applied if one chooses the vertex weights according to some stationary distribution.

However, under any stationary distribution, the weight on any vertex in a non-sink component must be zero. If we consider expansion using weights induced by a stationary distribution, essentially we are considering only the sink components. In this section, we explore if there is any meaningful way to consider expansion properties involving the non-sink components. As we shall see, it makes sense to consider the expansion properties of each strongly connected component separately.

3.1   Motivation for Considering Components Separately

Suppose $\mathbf{P}$ is a probability transition matrix corresponding to a random walk on some direted graph $G(V, E)$. Given a subset $C \subseteq V$, let $\mathbf{P}|_C$ be the square matrix restricting to the columns and the rows corresponding to C.

It is known [7, Theorem 3.22] that the eigenvalues of $\mathbf{P}$ are the union of the eigenvalues of $\mathbf{P}|_C$ over all strongly connected components C in a directed graph G. An important observation is that as long as the strongly connected components and the transition probabilities within a component remain the same, the eigenvalues of $\mathbf{P}$ are independent of the transition probabilities between different strongly connected components. For instance, the figure below depicts a directed graph, where the edges are labeled with the transition probabilities. Observe that each vertex is its own strongly connected component.

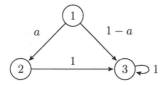

The transition matrix is:

$$\mathbf{P} = \begin{pmatrix} 0 & a & 1-a \\ 0 & 0 & 1 \\ 0 & 0 & 1 \end{pmatrix},$$

whose eigenvalues are $\{0, 0, 1\}$, which are independent of the parameter a. This suggests that it might be difficult to use spectral properties to analyze expansion properties involving edges between different strongly connected components. Hence, we propose that the expansion of each connected component should be analyzed separately.

3.2   Defining Expansion via Diluted Stationary Distribution

Observe that if a strongly connected component C is a sink, then the transition matrix $\mathbf{P}|_C$ has a stationary distribution, and we can apply Chung's approach. However, if C is a non-sink, then not every row of $\mathbf{P}|_C$ sums to 1, and so $\mathbf{P}|_C$ has no stationary distribution.

However, since C is a strongly connected component, by the Perron-Frobenius Theorem [10], $\mathbf{P}|_C$ has a unique maximum eigenvalue $\lambda_{\max} \geq 0$ with algebraic and geometric multiplicity 1 such that every other eigenvalue (which might be complex) has magnitude at most $\lambda_{\max}$. Moreover, the associated eigenvector of $\lambda_{\max}$ has positive coordinates and is unique up to scaling. Suppose ϕ is the (left) eigenvector which is normalized such that the coordinates sum to 1, i.e., $\sum_{u \in V} \phi(u) = 1$ and $\phi^T \mathbf{P}|_C = \lambda_{\max} \phi$. We say that ϕ is the *diluted stationary distribution* of $\mathbf{P}|_C$, because the distribution on vertices in C is diluted by a factor of $\lambda_{\max}$ after one step of the random walk.

We use the diluted stationary distribution ϕ as vertex weights to define expansion $\rho(S)$ for $S \subseteq C$. Observe that the weight of edges leaving component C also contributes to the expansion.

3.3   Augmenting Graph to Achieve Stationary Distribution

Suppose the component C has size n, and has $\lambda_{\max} < 1$. In order to use Chung's approach, we construct an augmented graph $\widehat{G}$ consisting of the component C and an extra vertex v_0. For each $u \in C$, all the original probabilities leaking out of C from u are now directed to v_0. For the new vertex v_0, the transition probabilities from v_0 to vertices in C are given by the diluted stationary distribution ϕ. Hence, the augmented graph $\widehat{G}$ is strongly connected. We write $\mathbf{A} = \mathbf{P}|_C \in \mathbb{R}^{n \times n}$, and $\mu = \mathbf{1}_n - \mathbf{A1}_n \in \mathbb{R}^n$. The new transition matrix is

$$\mathbf{B} = \begin{pmatrix} \mathbf{A} & \mu \\ \phi^T & 0 \end{pmatrix} \in \mathbb{R}^{(n+1) \times (n+1)}.$$

Given a square matrix $\mathbf{M}$, its determinant is denoted as $|\mathbf{M}|$, and $eig(\mathbf{M})$ is the multi-set of its eigenvalues, which are roots of the polynomial $|\lambda \mathbf{I} - \mathbf{M}|$ in λ. We first show that the matrix $\mathbf{B}$ preserves the spectral properties of $\mathbf{A}$.

Lemma 1 (Spectral Preservation). *We have $eig(\mathbf{B}) = eig(\mathbf{A}) - \{\lambda_{max}\} + \{1, \lambda_{max} - 1\}$. Furthermore, if an eigenvalue $\lambda \in eig(\mathbf{A}) - \{\lambda_{max}, \lambda_{max} - 1\}$, it has the same geometric multiplicity in $\mathbf{A}$ and $\mathbf{B}$.*

Proof. To prove the first part, it suffices to show that for all $\lambda \in \mathbb{R}$,

$$\Big|\lambda \mathbf{I}_n - \mathbf{A}\Big| (\lambda - 1)(\lambda - (\lambda_{max} - 1)) = \Big|\lambda \mathbf{I}_{n+1} - \mathbf{B}\Big| (\lambda - \lambda_{max}),$$

because both sides are polynomials in λ of degree $n + 2$. Hence, they must be equivalent polynomials if they are equal for more than $n + 2$ values of λ.

If $\lambda = 1$, then the right hand side is zero because $\mathbf{B}$ has eigenvalue 1; similarly, if $\lambda = \lambda_{max}$, the left hand side is zero because $\mathbf{A}$ has eigenvalue λ_{max}.

For $\lambda \neq 1, \lambda_{max}$, we have:

$$\Big|\lambda \mathbf{I}_{n+1} - \mathbf{B}\Big| = \begin{vmatrix} \lambda \mathbf{I}_n - \mathbf{A} & -\mu \\ -\phi^T & \lambda \end{vmatrix} \cdot \begin{vmatrix} \mathbf{I}_n & \mathbf{1}_n \\ \mathbf{0}_n^T & 1 \end{vmatrix} \tag{1}$$

$$= \begin{vmatrix} \lambda \mathbf{I}_n - \mathbf{A} & (\lambda - 1)\mathbf{1}_n \\ -\phi^T & \lambda - 1 \end{vmatrix} \tag{2}$$

$$= (\lambda - 1)\Big|\lambda \mathbf{I}_n - \mathbf{A} - (\lambda - 1)\mathbf{1}_n(\lambda - 1)^{-1}(-\phi^T)\Big| \tag{3}$$

$$= (\lambda - 1)\Big|\lambda \mathbf{I}_n - \mathbf{A} + \mathbf{1}_n\phi^T\Big|, \tag{4}$$

where (1) follows because the second determinant is 1. Moreover, (2) follows from $|\mathbf{X}| \cdot |\mathbf{Y}| = |\mathbf{XY}|$. Equation (3) follows from the identity that for invertible $\mathbf{H}$,

$$\begin{vmatrix} \mathbf{E} & \mathbf{F} \\ \mathbf{G} & \mathbf{H} \end{vmatrix} = |\mathbf{H}| \cdot |\mathbf{E} - \mathbf{FH}^{-1}\mathbf{G}|.$$

Similarly, using $|\mathbf{X}| \cdot |\mathbf{Y}| = |\mathbf{XY}|$ repeatedly, we have

$$
\left|\lambda \mathbf{I}_n - \mathbf{A} + \mathbf{1}_n \boldsymbol{\phi}^{\mathrm{T}}\right| (\lambda - \lambda_{\max}) = \begin{vmatrix} \mathbf{I}_n & -\mathbf{1}_n \\ \mathbf{0}_n^{\mathrm{T}} & 1 \end{vmatrix} \cdot \begin{vmatrix} \lambda \mathbf{I}_n - \mathbf{A} + \mathbf{1}_n \boldsymbol{\phi}^{\mathrm{T}} & \mathbf{0}_n \\ \boldsymbol{\phi}^{\mathrm{T}} & \lambda - \lambda_{\max} \end{vmatrix}
$$

$$
= \begin{vmatrix} \mathbf{I}_n & \mathbf{0}_n \\ \frac{-\boldsymbol{\phi}^{\mathrm{T}}}{\lambda - \lambda_{\max}} & 1 \end{vmatrix} \cdot \begin{vmatrix} \lambda \mathbf{I}_n - \mathbf{A} & -(\lambda - \lambda_{\max}) \mathbf{1}_n \\ \boldsymbol{\phi}^{\mathrm{T}} & \lambda - \lambda_{\max} \end{vmatrix}
$$

$$
= \begin{vmatrix} \lambda \mathbf{I}_n - \mathbf{A} & -(\lambda - \lambda_{\max}) \mathbf{1}_n \\ \mathbf{0}_n & \lambda - \lambda_{\max} + 1 \end{vmatrix}
$$

$$
= \left|\lambda \mathbf{I}_n - \mathbf{A}\right| (\lambda - \lambda_{\max} + 1).
$$

This completes the proof of the first part. To show that the geometry multiplicities of a common eigenvalue $\lambda \neq \lambda_{\max}, \lambda_{\max} - 1$ are equal, we show that $\mathbf{x} \longleftrightarrow \binom{\mathbf{x}}{0}$ is a bijection between $\mathbf{A}$ and $\mathbf{B}$'s corresponding right eigenvectors.

Since $\lambda \neq \lambda_{\max}$ is an eigenvalue of $\mathbf{A}$, if $\mathbf{x}$ is a corresponding eigenvector, then $\boldsymbol{\phi}^{\mathrm{T}} \mathbf{x} = 0$, because $\lambda_{\max} \boldsymbol{\phi}^{\mathrm{T}} \mathbf{x} = \boldsymbol{\phi}^{\mathrm{T}} \mathbf{A} \mathbf{x} = \lambda \boldsymbol{\phi}^{\mathrm{T}} \mathbf{x}$. Hence, $B\binom{\mathbf{x}}{0} = \lambda \binom{\mathbf{x}}{0}$.

Conversely, suppose $\binom{\mathbf{x}}{y}$ is an eigenvector of $\mathbf{B}$ with eigenvalue λ, where $\mathbf{x} \in \mathbb{R}^n$ and $y \in \mathbb{R}$. We have

$$
\begin{cases} \mathbf{Ax} + \boldsymbol{\mu} y = \lambda \mathbf{x} \\ \boldsymbol{\phi}^{\mathrm{T}} \mathbf{x} = \lambda y \end{cases} \quad \text{or} \quad \begin{cases} (\lambda \mathbf{I} - \mathbf{A}) \mathbf{x} = \boldsymbol{\mu} y \\ \boldsymbol{\phi}^{\mathrm{T}} \mathbf{x} = \lambda y. \end{cases}
$$

Then, $\boldsymbol{\phi}^{\mathrm{T}} (\lambda \mathbf{I} - \mathbf{A}) \mathbf{x} = (\lambda - \lambda_{\max}) \boldsymbol{\phi}^{\mathrm{T}} \mathbf{x} = (\lambda - \lambda_{\max}) \lambda y$.

But we also have $\boldsymbol{\phi}^{\mathrm{T}} \boldsymbol{\mu} y = \boldsymbol{\phi}^{\mathrm{T}} (\mathbf{1} - \mathbf{A} \mathbf{1}) y = (1 - \lambda_{\max}) y$. Since the two quantities are equal, this implies that $\lambda = 1, \lambda_{\max} - 1$ or $y = 0$. By assumption, $\lambda \neq 1, \lambda_{\max} - 1$, and so the only possibility is $y = 0$, then $\mathbf{A} \mathbf{x} = \lambda \mathbf{x}$. This shows that λ has the same geometric multiplicity in $\mathbf{A}$ and $\mathbf{B}$. $\qquad \square$

3.4 Higher-Order Cheeger Inequalities for Component

Given a non-sink component C, we have described how to add an extra vertex v_0 to construct an augmented graph $\widehat{G}$ with transition matrix $\mathbf{B}$. Observe that $\mathbf{B}$ has stationary distribution $\widehat{\boldsymbol{\phi}} = (\boldsymbol{\phi}, 1 - \lambda_{\max})$, where $\boldsymbol{\phi}$ is the diluted stationary distribution of $\mathbf{A}$. Hence, it follows that for all $S \subseteq C$, the old expansion $\rho(S)$ is the same as the new expansion $\widehat{\rho}(S)$ in the augmented graph.

Therefore, we can apply Chung's approach [8] and the spectral analysis by Lee et al. [13] to obtain the following lemma, which is a restatement of Theorem 1.

Lemma 2 (Cheeger Inequalities for Component C). *Suppose $\widehat{\boldsymbol{\Phi}}$ is the diagonal matrix whose diagonal entries are coordinates of the stationary distribution $\widehat{\boldsymbol{\phi}}$ of $\mathbf{B}$. Moreover, suppose the normalized Laplacian $\mathcal{L}$ of the symmetric matrix $\frac{1}{2}(\widehat{\boldsymbol{\Phi}} \mathbf{B} + \mathbf{B}^T \widehat{\boldsymbol{\Phi}})$ has eigenvalues $0 = \lambda_1 \leq \lambda_2 \leq \cdots \leq \lambda_{n+1}$. Then, for all $1 \leq k \leq n$, $\frac{\lambda_k}{2} \leq \rho_k(C) \leq O(k^2) \cdot \sqrt{\lambda_{k+1}}$.*

Proof. We use the following inequality from [13]:

$\frac{\lambda_k}{2} \leq \widehat{\rho}_k(\widehat{G}) \leq O(k^2) \cdot \sqrt{\lambda_k}$.

Observe that if $S \subseteq C$ does not contain the new vertex v_0, then S has the same expansion $\rho(S) = \widehat{\rho}(S)$ in both graphs.

From $k + 1$ disjoint subsets in the augmented graph $\widehat{G}$, we can get at least k subsets of C by removing the one containing v_0. Hence, we have

$\rho_k(C) \leq \widehat{\rho}_{k+1}(\widehat{G}) \leq O((k+1)^2)\sqrt{\lambda_{k+1}}$.

On the other hand, k disjoint subsets in C are also disjoint in $\widehat{G}$. Therefore, we have $\rho_k(C) \geq \rho_k(\widehat{G}) \geq \frac{\lambda_k}{2}$, as required. □

4   Vertex Weights Deviate from Stationary Distribution

In this section, we consider a transition matrix $\mathbf{P}$ whose underlying directed graph $G(V, E)$ (where $n = |V|$) is not necessarily strongly connected. Moreover, each vertex has a positive weight given by a vector $\boldsymbol{\phi} \in \mathbb{R}^n$ that is not necessarily a stationary distribution of $\mathbf{P}$. We wish to analyze the expansion with respect to $\mathbf{P}$ and $\boldsymbol{\phi}$ using spectral techniques. As in Section 3, we shall add an extra vertex v_0 to form an augmented graph $\widehat{G}$.

We measure how much $\boldsymbol{\phi}$ deviates from a stationary distribution by the following parameter:

$$\varepsilon := 1 - \min_{u \in V} \frac{\phi(u)}{\sum_{v \in V} \phi(v) \cdot P(v, u)}.$$

A smaller value of ε means that $\boldsymbol{\phi}$ is closer to a stationary distribution. In particular, if $\varepsilon = 0$, then $\boldsymbol{\phi}$ is a stationary distribution.

Our idea is to first scale down all probabilities in $\mathbf{P}$ by a factor of $(1 - \varepsilon)$. We add a new vertex v_0 to absorb the extra ε probability from each existing vertex. Then, we define the transition probabilities from v_0 to the original vertices carefully such that each original vertex u receives the same weight $\phi(u)$ after one step.

For each vertex u, the weight mass it obtains from vertices V through the scaled-down $\mathbf{P}$ is $(1 - \varepsilon) \sum_{v \in V} \phi(v)P(v, u)$. Hence, the new vertex v_0 needs to return mass weights to vertices in V given by the vector $\mathbf{m} := \boldsymbol{\phi} - (1 - \varepsilon)\mathbf{P}^T\boldsymbol{\phi}$, whose coordinates are non-negative by the choice of ε. Normalizing by $\mathbf{m}^T\mathbf{1}_n = \varepsilon\boldsymbol{\phi}^T\mathbf{1}_n$, we have the vector $\boldsymbol{\mu} := \frac{\mathbf{m}}{\varepsilon\boldsymbol{\phi}^T\mathbf{1}_n}$ of transition probabilities from v_0 to vertices in V.

The transition matrix of the augmented graph $\widehat{G}$ is

$$\widehat{\mathbf{P}} = \begin{pmatrix} (1 - \varepsilon)\mathbf{P} & \varepsilon\mathbf{1}_n \\ \boldsymbol{\mu}^T & 0 \end{pmatrix}.$$

Observe that $\widehat{G}$ is strongly connected, and its stationary distribution can be obtained by normalizing the vector $\widehat{\boldsymbol{\phi}} = (\boldsymbol{\phi}, \varepsilon\boldsymbol{\phi}^T\mathbf{1}_n)$. In other words, we can append a new coordinate corresponding to v_0 to $\boldsymbol{\phi}$ to obtain a left eigenvector $\widehat{\boldsymbol{\phi}}$ with eigenvalue 1 for matrix $\widehat{\mathbf{P}}$.

Moreover, for each subset $S \subseteq V$ of the original vertices, the new expansion $\widehat{\rho}(S)$ with respect to $\widehat{\phi}$ and $\widehat{\mathbf{P}}$ can be related to the old expansion $\rho(S)$ as follows: $\widehat{\rho}(S) = (1 - \varepsilon) \cdot \rho(S) + \varepsilon$.

Hence, we can apply Chung's approach to $\widehat{\mathbf{P}}$ and $\widehat{\phi}$ to construct a symmetric Laplacian matrix, whose eigenvalues are related to the expansion properties using the results by Lee et al. [13]. The following lemma is a restatement of Theorem 2, and its proof uses the same argument as in the proof of Lemma 2.

Lemma 3. *Suppose $\widehat{\Phi}$ is the diagonal matrix whose diagonal entries are coordinates of $\widehat{\phi}$ as defined above. Moreover, suppose the normalized Laplacian $\mathcal{L}$ of the symmetric matrix $\frac{1}{2}(\widehat{\Phi}\widehat{\mathbf{P}} + \widehat{\mathbf{P}}^T\widehat{\Phi})$ has eigenvalues $0 = \lambda_1 \leq \lambda_2 \leq \cdots \leq \lambda_{n+1}$. Then, for all $1 \leq k \leq n$, $\frac{\lambda_k}{2} \leq (1 - \varepsilon) \cdot \rho_k(V) + \varepsilon \leq O(k^2) \cdot \sqrt{\lambda_{k+1}}$.*

References

1. Alon, N.: Eigenvalues and expanders. Combinatorica **6**(2), 83–96 (1986)
2. Alon, N., Kahale, N.: A spectral technique for coloring random 3-colorable graphs. SIAM J. Comput. **26**(6), 1733–1748 (1997)
3. Alon, N., Milman, V.D.: lambda, isoperimetric inequalities for graphs, and super-concentrators. J. Comb. Theory, Ser. B **38**(1), 73–88 (1985)
4. Aspvall, B., Gilbert, J.R.: Graph coloring using eigenvalue decomposition. SIAM Journal on Algebraic Discrete. Methods **5**(4), 526–538 (1984)
5. Brin, S., Page, L.: Reprint of: The anatomy of a large-scale hypertextual web search engine. Computer Networks **56**(18), 3825–3833 (2012)
6. Cheeger, J.: A lower bound for the smallest eigenvalue of the laplacian. Problems in analysis **625**, 195–199 (1970)
7. Chen, W.K., Lauwerier, H.A., Koiter, W.T.: Applied Graph Theory: Graphs and Electrical Networks. North-Holland Series in Applied Mathematics and Mechanics. Elsevier Science (2014)
8. Chung, F.: Laplacians and the cheeger inequality for directed graphs. Annals of Combinatorics **9**(1), 1–19 (2005)
9. Chung, F.R.K.: Spectral graph theory, vol. 92. American Mathematical Soc. (1997)
10. Horn, R.A., Johnson, C.R.: Matrix analysis. Cambridge University Press (1990)
11. Kleinberg, J.M.: Authoritative sources in a hyperlinked environment. J. ACM **46**(5), 604–632 (1999)
12. Kwok, T.C., Lau, L.C., Lee, Y.T., Gharan, S.O., Trevisan, L.: Improved cheeger's inequality: analysis of spectral partitioning algorithms through higher order spectral gap. In: STOC, pp. 11–20 (2013)
13. Lee, J.R., Gharan, S.O., Trevisan, L.: Multiway spectral partitioning and higher-order cheeger inequalities. J. ACM **61**(6), 37 (2014)
14. Ng, A.Y., Jordan, M.I., Weiss, Y.: On spectral clustering: analysis and an algorithm. In: NIPS, pp. 849–856 (2001)
15. Shewchuk, J.R.: Ladies and gentlemen, allow me to introduce spectral and isoperimetric graph partitioning (2011)
16. Shi, J., Malik, J.: Normalized cuts and image segmentation. IEEE Trans. Pattern Anal. Mach. Intell. **22**(8), 888–905 (2000)
17. Tolliver, D., Miller, G.L.: Graph partitioning by spectral rounding: applications in image segmentation and clustering. In: IEEE Computer Society Conference on Computer Vision and Pattern Recognition, pp. 1053–1060 (2006)

Game Theory and Algorithms

Strategy-Proof Mechanism for Obnoxious Facility Location on a Line

Deshi Ye[1], Lili Mei[1], and Yong Zhang[2,3]([✉])

[1] College of Computer Science, Zhejiang University, Hangzhou, China
{yedeshi,meilili}@zju.edu.cn
[2] Shenzhen Institutes of Advanced Technology,
Chinese Academy of Sciences, Beijing, China
zhangyong@siat.ac.cn
[3] Department of Computer Science, The University of Hong Kong, Hong Kong, China

Abstract. In the problem of obnoxious facility location, an obnoxious facility is located in an area. To maximize the social welfare, *e.g.*, the sum of distances from all the agents to the facility, we have to get the true locations of each agent. However, each agent may misreport his/her location to stay far away from the obnoxious facility. In this paper, we design strategy-proof mechanisms on locating an obnoxious facility on a real line. Two objective functions, *i.e.*, maximizing the sum of squares of distances (maxSOS) and maximizing the sum of distances (maxSum), have been considered. For maxSOS, a randomized strategy-proof mechanism with approximation ratio 5/3 is given, meanwhile the lower bound is proved to be at least 1.042. The lower bound of any randomized strategy-proof mechanisms w.r.t. maxSum is proved to be 1.077. Moreover, an extended model that each agent controls multiple locations is considered. For this model, we investigate deterministic and randomized strategy-proof mechanisms w.r.t. maxSum and maxSOS objectives, respectively. The deterministic mechanisms are shown to be tight for both objectives.

1 Introduction

We consider the problem of locating an obnoxious facility on a line, where a set of agents are located. In the view of algorithmic mechanism design, the agent's locations are private information and each agent attempts to maximize his/her utility, *i.e.*, stay far away from the obnoxious facility, by misreporting his/her location. Mechanisms receive the declaration of agents as input and determine the location. Our target is to design strategy-proof (or truthful) mechanisms to maximize social welfare, *i.e.*, the sum of squares of distances (maxSOS) or the sum of distances (maxSum).

In the classical facility location game, the utility of an agent is the distance from his/her true location to the nearest facility if there are multiple facilities,

---

This work is supported by NSFC (No. 61433012, U1435215, 11171086) Natural Science Foundation of Hebei A2013201218, Shenzhen basic research project (JCYJ20120615140531560), HKU small project funding 201309176064.

© Springer International Publishing Switzerland 2015
D. Xu et al. (Eds.): COCOON 2015, LNCS 9198, pp. 45–56, 2015.
DOI: 10.1007/978-3-319-21398-9_4

and each agent attempts to minimize his/her utility. In this work, we consider the obnoxious facility location problem, in which each agent attempts to maximize the utility, *i.e.*, stay as far as possible from the facility. These problems arise in many real applications, such as locating nuclear reactors, garbage dump sites, ammunition dumps, and polluting plants in a community.

To measure the performance of the mechanisms for classical facility location problem, the *approximation ratio* with respect to some given social objectives has been considered. The approximation ratio is defined to be the worst-case ratio between the mechanism's objective value and the optimal solution. There are mainly three objectives: *minSum* (minimizing the sum of agents' utilities), *minMax* (minimizing the maximum utility of agents), and *minSOS* (minimizing the sum of squares of agents' utilities). Feldman and Wilf [2] pointed out that the minSOS function is highly relevant in economic settings, the centroid in geometry, or the center of mass in physics. Accordingly, for obnoxious facility game, the social objectives are *maxSum* (maximizing the sum of agents' utilities), *maxMin* (maximizing the minimum utility of agents), and *maxSOS* (*i.e.*, maximizing the sum of squares of agents's utilities). Han and Du [4] mentioned that for obnoxious facility game, any deterministic mechanisms for maxMin objective are unbounded. Hence, in this paper, we only consider maxSum and maxSOS two objectives. Observe that for obnoxious facility problem, if the number of the facilities is more than one, locating all the facilities the same place is the best choice. Due to the above reasons, in this work, we only consider the single obnoxious facility game for maxSum and maxSOS objectives.

We then extend our model to a setting that each agent can control multiple locations and only one facility can be located. Procaccia and Tennenholtz [5] gave an example of real estate agents for the classical facility location game, which is also suitable for the obnoxious version. There are some other scenarios for the extended model. We can take one community or one company with many branches be an agent. For the extended model, we also consider deterministic and randomized strategy-proof mechanisms for maxSum and maxSOS objectives. And as noted by Schummer and Vohra [16], payment is not available in many real scenarios, especially in the social choice literature. Hence, we focus on the mechanism design without payment.

Related work. The classical setting of the facility location game is to minimize the utility function of each agent, where the utility function is the distance from agent's true location to the facility location. One direction of studying this setting is to investigate the characterizations of strategy-proof mechanisms. Moulin [14] and Schummer et al. [15] provided characterizations of deterministic strategy-proof mechanisms on line, tree, and cycle networks. Recently, Dokow et al. [10] gave a full characterization of strategy-proof mechanisms on line and on sufficiently large cycles on a discrete graph. Fotakis and Tzamos [11] gave a characterization of deterministic strategy-proof mechanisms with a bounded approximation ratio for 2-Facility on the line.

On the algorithmic view of the classical setting, Procaccia and Tennenholtz [5] first studied strategy-proof mechanisms with provable approximation ratios on the line for 1-Facility and 2-Facility for minSum and minMax objectives. Subsequently, Lu et al. [6,7] improved some results for 2-Facility on a line and circle under the minSum objective. Alon et al. [8] provided the approximation ratios achievable by randomized and deterministic mechanisms for 1-Facility in circle and general metrics under the minMax objective. Feldman and Wilf [2] studied the approximation ratios for randomized and deterministic mechanisms under the minSOS objective on a line. The extended model with multiple locations per agent was investigated in [5,7] for locating one facility on a line for minSum and minMax objectives. Thang [12], Fotakis and Tzamos [21] studied the k-Facility location problem.

For the setting of obnoxious facility game, the mechanism design with the social objective maxSum was first studied by Cheng et al. [1]. They presented a 3-approximation group strategy-proof deterministic mechanism and a lower bound of 2. They designed a randomized strategy-proof mechanism with 3/2-approximation. They also extended the work to tree and circle networks. Furthermore, for the general network, they proposed a 4-approximation group strategy-proof (GSP) deterministic mechanism and a 2-approximation group strategy-proof randomized mechanism, respectively.

On the direction of characterizing the mechanisms, Ibara and Nagamochi [3] first studied the characterization of deterministic strategy-proof mechanisms for obnoxious facility location game. In their paper, they showed there is no strategy-proof mechanism such that the number of candidates (locations output by the mechanism for some reported locations) is more than two. Then they characterized (group) strategy-proof mechanisms for two candidates in the general metric, in which a 4-approximation group strategy-proof mechanism in any metric was proposed. Recently, Han and Du [4] also investigated the characterization of deterministic strategy-proof mechanisms for the single-sinked policy domain. Moreover, they showed that any deterministic group strategy-proof mechanisms on the line have approximation ratio at least $(1 + 2^p)^{\frac{1}{p}}$ if the social objective function is L_p-norm, which implies lower bounds of 3 and 5 for maxSum and maxSOS objectives, respectively. Note that the lower bound of 3 is tight with the upper bound given in [1].

Our Contributions. In this paper, we provide approximation results for obnoxious facility game with respect to maxSum and maxSOS objectives on the line. We provide, for any randomized strategy-proof mechanisms, the lower bounds are 1.077 and 1.042 with respect to maxSum and maxSOS objective. We provide a strategy-proof randomized mechanism with approximation ratio 5/3.

We then extend the model such that each agent controls multiple locations. The utility of an agent is the sum of the distances from this agent's locations to the facility. For maxSum objective, we provide a deterministic strategy-proof mechanism with approximation ratio 3, which is tight with the lower bound [4]. In the case of randomized strategy-proof mechanism, we show a lower bound 10/9 and an upper bound 2. For maxSOS objective, the deterministic strategy-proof mechanism is 5 ,which matches the lower bound of 5 in [4]. For the randomized strategy-proof mechanisms, we show a lower bound of 1.13 and an upper bound of 4. A summary of our results are illustrated in Table 1.

Table 1. A summary of our results. UB and LB stands for upper bound and lower bound, respectively. SP and GSP represent strategy-proof and group strategy-proof, respectively. MultiLA stands for the model with multiple locations per agent.

Model	Deterministic		Randomized	
maxSum	UB: 3 GSP [1]	LB: 3 GSP [4]	UB: 3/2 [1]	LB: **1.077** SP
maxSOS	UB: 5 GSP [4]	LB: 5 GSP [4]	UB: **5/3** GSP	LB: **1.042** SP
MultiLA: maxSum	UB: **3** GSP	LB: 3 SP [4]	UB: **2** SP	LB: **10/9** SP
MultiLA: maxSOS	UB: **5** GSP	LB: 5 GSP [4]	UB: **4** SP	LB: **1.13** SP

2   Model and Preliminaries

The possible location area G is represented by a line. For obnoxious facility location problem, locating the facility at the infinity is the optimal location. Hence, we restrict the line to a closed interval. For simplicity, suppose G is within the interval $[0, 2]$, i.e., $G = [0, 2]$. Let $N = \{1, 2, \ldots, n\}$ be a set of agents. Each agent i has an true location $x_i \in G$. We refer to $x = (x_1, x_2, \ldots, x_n) \in G^n$ as the location profile.

The utility of agent i is the distance from x_i to the facility. Each agent attempts to maximize his/her utility. The social welfare is maxSOS (the sum of squares of agents' utilities) or maxSum (the sum of agents' utilities). The goal of a mechanism is to provide a strategy-proof (or truthful) mechanism with the maximum social welfare.

A *deterministic mechanism* is a function $f : G^n \to G$ that maps the reported location profile to the location of a facility (which can be located anywhere in G). When the facility is located at $y \in G$, the utility of agent i is simply the distance between x_i and y, i.e., $D(y, x_i) = |y - x_i|$, for all $i \in N$.

A *randomized mechanism* is a function $f : G^n \to \Delta G$ that maps the reported location profile to a probability distribution over G. If $f(x) = P$, where P is a probability distribution, agent i's utility is defined to be the expected distance between x_i and y, i.e., $D(P, x_i) = E_{y \sim P}[|y - x_i|]$, for all $i \in N$.

A mechanism is *strategy-proof* if no agent can strictly increase his/her utility by misreporting the location, regardless of the declarations of other agents. Formally, given a location profile $x \in G^n$, and for all x_i', we have $D(f(x_i, x_{-i}), x_i) \geq D(f(x_i', x_{-i}), x_i)$, where $x_{-i} = (x_1, \ldots, x_{i-1}, x_{i+1}, \ldots, x_n)$ is the location profile excluding agent i.

A mechanism is said to be *group strategy-proof* (GSP) if no any set of agents can all benefit by misreporting their locations, regardless of the reports of the other agents. That is, given any location profile $x = (x_S, x_{-S})$ for any non-empty subset of agents $S \subseteq N$ and the misreported location x_S', there exists $i \in S$ such that $D(f(x), x_i) \geq D(f(x_S', x_{-S}), x_i)$.

Given a location profile x and the facility location y by a mechanism, the social welfare of the mechanism is defined by $SC(y, x)$. If the social objective function is maxSum, then $SC(y, x) = \sum_{i \in N} D(y, x_i) = \sum_{i \in N} |y - x_i|$. If the social objective function is maxSOS, then $SC(y, x) = \sum_{i \in N} (y - x_i)^2$. Moreover, the social welfare of a distribution P of profile x is $SC(P, x) = E_{y \sim P}[SC(y, x)]$.

We denote $SC(opt, x)$ to be the optimal social welfare with respect to the location profile x, i.e., $SC(opt, x) = \max_y \{SC(y, x)\}$. A mechanism f is said to be ρ-approximation if $SC(opt, x) \leq \rho \cdot SC(f(x), x)$ for any location profile x.

Cheng et al. [1] pointed out that an optimal facility location will be at one of two extreme points 0 or 2 for the maxSum objective. In the following, we show that the property also holds for maxSOS objective.

Proposition 1. *Given a location profile x, at least one of 0 and 2 is an optimal location for maxSOS objective.*

Proof. Given a location profile x, suppose that the optimal facility location is at y, where $x_k \leq y < x_{k+1}$. If necessary, we add x_0 (or x_{n+1}) if y is located between 0 and x_1 (or x_n and 2). In the following, we show that if y is not located at 0 or 2, then we can move the facility to 0 or 2, the social welfare is not decreased, which therefore the proposition follows.

The social welfare of location y given the profile x is

$$SC(opt, x) = \sum_{i=1}^{k}(y - x_i)^2 + \sum_{i=k+1}^{n}(x_i - y)^2.$$

If $k \leq n/2$, let us consider the location at 0, otherwise the location is at 2. Due to the the symmetric, without loss of generality, we only consider the case when $k \leq n/2$.

The social welfare of location 0 is

$$
\begin{aligned}
SC(0, x) &= \sum_{i=1}^{k}(x_i)^2 + \sum_{i=k+1}^{n}(x_i)^2 \\
&= \sum_{i=1}^{k}(y - x_i - y)^2 + \sum_{i=k+1}^{n}(x_i - y + y)^2 \\
&\geq SC(opt, x) - 2\sum_{i=1}^{k}y^2 + \sum_{i=1}^{k}y^2 + \sum_{i=k+1}^{n}y^2 \\
&\geq SC(opt, x).
\end{aligned}
$$

□

To design strategy-proof mechanisms, we need the following notations in the remaining paper. Let $n_1 = |\{i : 0 \leq x_i \leq 1\}|$ be the number of agents whose declarations are in the interval $[0, 1]$. Let $n_2 = |\{i : 1 < x_i \leq 2\}|$ be the number of agents whose declarations are in the interval $(1, 2]$.

A deterministic strategy-proof mechanism was provided in [1] as below.

Mechanism 1: Set $f(x) = 0$ if $n_1 \leq n_2$ and otherwise $f(x) = 2$.

Mechanism 1 was proved to be strategy-proof in [1]. Han and Du [4] showed that Mechanism 1 achieves 5-approximation for maxSOS objective and the approximation ratio is the best possible.

3 Randomized Strategy-Proof Mechanisms

In this section, we study randomized strategy-proof mechanisms with respect to maxSOS objective function. We first show a family of group strategy-proof mechanisms with respect to n_1, the number of agents whose reported locations are within the interval $[0, 1]$.

Mechanism RM: For any given profile x, the mechanism places the facility at 0 with probability α, and places the facility at 2 with probability $1 - \alpha$, where $0 \le \alpha \le 1$ is a number that depends on x.

Lemma 1. *The mechanism RM is group strategy-proof if α is a non-increasing function on n_1. (The proof can be found in the full version of the paper.)*

Theorem 2. *Let $\alpha = \frac{4n_1n_2+n_2^2}{8n_1n_2+n_1^2+n_2^2}$. Mechanism RM is group strategy-proof and its approximation ratio is at most $5/3$ for maxSOS objective.*

Proof. Let $\beta = \frac{n_1}{n_2} = \frac{n_1}{n-n_1}$. Then $\alpha = \frac{1+4\beta}{1+8\beta+\beta^2}$. It is easy to check that α is a non-increasing function on β and then a a non-increasing function on n_1. The group strategy-proof holds due to Lemma 1.

Let us consider the approximation ratio. To state conveniently, we let f denote Mechanism RM. For any given location profile $x = (x_1, x_2, \dots, x_n)$, the social welfare of Mechanism RM is

$$SC(f(x), x) = \alpha \sum_{i=1}^{n} x_i^2 + (1 - \alpha) \sum_{i=1}^{n} (2 - x_i)^2.$$

From Proposition 1, the optimal facility location is either at 0 or 2. We estimate the following upper bounds for optimal social welfare.

$$\sum_{i=1}^{n} x_i^2 = \sum_{x_i \in [0,1)} x_i^2 + \sum_{x_i \in [1,2]} x_i^2 \le n_1 + 4n_2 \tag{1}$$

$$\sum_{i=1}^{n} (2 - x_i)^2 = \sum_{x_i \in [0,1)} (2 - x_i)^2 + \sum_{x_i \in [1,2]} (2 - x_i)^2 \le 4n_1 + n_2 \tag{2}$$

We deal with two cases with respect to the optimal location.

Case 1. The optimal location is at 0, *i.e.*, $SC(opt, x) = \sum_{i=1}^{n} x_i^2$. The social welfare of mechanism RM is

$$SC(f(x), x) = \alpha SC(opt, x) + (1 - \alpha)\left(\sum_{x_i \in [0,1)} (2 - x_i)^2 + \sum_{x_i \in [1,2]} (2 - x_i)^2 \right)$$
$$\ge \alpha SC(opt, x) + (1 - \alpha)n_1$$
$$\ge \alpha SC(opt, x) + (1 - \alpha)\frac{n_1}{n_1 + 4n_2} SC(opt, x) \tag{3}$$

Inequality (3) holds due to the inequality (1).

Case 2. The optimal location is at 2, *i.e.*, $SC(opt, x) = \sum_{i=1}^{n}(2 - x_i)^2$. The social welfare of mechanism RM is

$$
\begin{aligned}
SC(f(x), x) &= \alpha\Big(\sum_{x_i \in [0,1)} x_i^2 + \sum_{x_i \in [1,2]} x_i^2 \Big) + (1 - \alpha)\sum_{i=1}^{n}(2 - x_i)^2 \\
&\geq \alpha n_2 + (1 - \alpha)SC(opt, x) \\
&\geq \alpha \frac{n_2}{n_1 + 4n_2}SC(opt, x) + (1 - \alpha)SC(opt, x), \quad (4)
\end{aligned}
$$

Inequality (4) holds due to the inequality (2).

From Inequality (3) and Inequality (4), we obtain the approximation ratio ρ as below.

$$
\rho \leq \max\Big\{ \frac{1}{\alpha + (1 - \alpha)n_1/(n_1 + 4n_2)}, \frac{1}{\alpha n_2/(4n_1 + n_2) + (1 - \alpha)} \Big\} \quad (5)
$$

The above approximation ρ reaches the maximum when $\alpha = \frac{4n_1 n_2 + n_2^2}{8n_1 n_2 + n_1^2 + n_2^2}$. Let $\beta = n_1/n_2$. Then $\rho = \frac{1 + 8\beta + \beta^2}{1 + 4\beta + \beta^2}$, and ρ reaches its maximum $5/3$ if $\beta = 1$.   □

3.1 Lower Bounds of Randomized Mechanisms

In this subsection, we first show that no randomized strategy-proof mechanism can achieve approximation ratio smaller than 1.077 for maxSum objective. Then, we extend the idea for designing a lower bound 1.04285 for any randomized mechanisms with respect to maxSOS objective.

Theorem 3. *The approximation ratio of any randomized strategy-proof mechanisms for maxSum objective is at least $14/13 \approx 1.077$.*

Proof. Consider two agents with a location profile $x = (1/3, 5/3)$. Then $SC(opt, x) = 2$. Let y be the facility location determined by a randomized mechanism.

We assume there exists a strategy-proof randomized mechanism with approximation ratio $\rho < 1.077$. It is worth to note that $SC(y, x) = E[|y - 1/3| + |y - 5/3|] \leq 2$. Without loss of generality, we assume $E[|y - 5/3|] \leq 1$.

Let us consider a new profile $x' = (1/3, 2)$, where agent 2 misreports its location to 2 other than $5/3$. Let y' be the facility location of profile x'. Due to the strategy-proofness, then we obtain $E[|y' - 5/3|] \leq E[|y - 5/3|] \leq 1$.

Denote $Pr[y' < 1/3] = q$. To state simply, we let $E_{2,l}$ be $E[|y' - 5/3| : y' < 1/3]$ and $E_{2,r}$ be $E[|y' - 5/3| : y' \geq 1/3]$. Since the utility is nonnegative and the expected value is larger than the minimum value, we get that

$$
\frac{4}{3}q \leq E[|y' - 5/3|] = E_{2,l} \cdot q + E_{2,r} \cdot (1 - q) \leq 1,
$$

which therefore, $q \leq 3/4$.

In profile x', the optimal location is at 0. Hence, we have $SC(opt, x') = 7/3$. Note that $E[SC(y', x') : y' \geq 1/3] = 5/3$. Hence, we get an upper bound of any randomized mechanisms,

$$E[SC(y', x')] = E[SC(y', x') : y' < 1/3]q + E[SC(y', x') : y' \geq 1/3](1 - q)$$
$$\leq \frac{7}{3}q + \frac{5}{3}(1 - q).$$

Then, the approximation ratio

$$\rho \geq \frac{7/3}{\frac{7}{3}q + \frac{5}{3}(1 - q)} = \frac{7}{5 + 2q} \geq 14/13 \approx 1.077. \qquad \square$$

Theorem 4. *Any randomized strategy-proof mechanisms have an approximation ratio of at least 1.04285 for maxSOS objective.*

Proof. Consider a location profile with two agents $x = (x_1, x_2)$, where $x_1 = 1 - a, x_2 = 1 + a$ and $0 < a \leq 1$ is a constant that will be specified later. Then $SC(opt, x) = (1 + a)^2 + (1 - a)^2 = 2(1 + a^2)$.

Since $E[(y - x_1)^2] + E[(y - x_2)^2] \leq SC(opt, x)$, w.l.o.g, we assume that $E[|y - x_2|^2] \leq (1 + a^2)$. From Jensen's inequality, we have $(E[|y - x_2|])^2 \leq E[|y - x_2|^2] \leq (1 + a^2)$, which gives $E[|y - x_2|] \leq \sqrt{1 + a^2}$.

Consider a new location profile $x' = (1 - a, 2)$. Let y' be the random variable of the facility location on profile x'. Due to the strategy-proofness, we have

$$E[|y' - x_2|] \leq E[|y - x_2|] \leq \sqrt{1 + a^2}, \qquad (6)$$

otherwise, agent 2 in profile x will lie to 2.

Let $Pr[y' < 1 - a] = q$. Due to space constraints, we let $E_{2,l}$ be $E[|y' - x_2| : y' < 1 - a]$ and $E_{2,r}$ be $E[|y' - x_2| : y' \geq 1 - a]$. From Inequality (6), we have

$$2aq \leq E_{2,l} \cdot q \leq E[|y' - x_2|] = E_{2,l} \cdot q + E_{2,r} \cdot (1 - q) \leq \sqrt{1 + a^2}.$$

Thus, we get that $q \leq \frac{\sqrt{1 + a^2}}{2a}$.

It is worth to note that the optimal location of profile x' is 0, which gives the optimal social welfare $SC(opt, x') = 4 + (1 - a)^2$.

Now let us consider the social welfare for the profile x'. Due to space constraints, we let E_l be $E[SC(y', x') : y' < 1 - a]$ and E_r be $E[SC(y', x') : y' \geq 1 - a]$. We first consider the maximum value of social welfare if a facility $1 - a \leq z \leq 2$ is located. We get that

$$SC(z, x') = (z - (1 - a))^2 + (2 - z)^2 = 4 + (1 - a)^2 + 2z^2 - 2(3 - a)z$$
$$\leq 4 + (1 - a)^2 + 4(a - 1) = SC(opt, x') + 4(a - 1), \qquad (7)$$

where the last inequality holds since $1 - a \leq z \leq 2$ and $0 < a \leq 1$.

Now we turn to the social welfare for the profile x'. The social welfare is

$$E[SC(y', x')] = E_l \cdot q + E_r \cdot (1 - q)$$
$$\leq q \cdot SC(opt, x') + (1 - q) \cdot (SC(opt, x') + 4(a - 1))$$
$$= SC(opt, x') + 4(1 - q)(a - 1),$$

where the inequality holds since $E_l \leq SC(opt, x')$ and Inequality (7).

Due to $0 < a \leq 1$, and $q \leq \frac{\sqrt{1+a^2}}{2a}$, it holds that

$$E[SC(y', x')] \leq 4 + (1-a)^2 + 2(1 - \frac{\sqrt{1+a^2}}{2a})(2a - 2).$$

The approximation ratio of the randomized mechanism is

$$\rho \geq \frac{4 + (1-a)^2}{E[SC(y', x')]} \geq \frac{4 + (1-a)^2}{4 + (1-a)^2 + 2(1 - \frac{\sqrt{1+a^2}}{2a})(2a - 2)}. \tag{8}$$

The right side of Inequality (8) reaches the maximum when $a = 0.758267$, then we finally get $\rho \geq 1.04285$ by setting $a = 0.758267$. $\square$

4   Multiple Locations Per Agent

In this section, we consider an extended model that each agent controls multiple locations. Let w_i be the number of locations controlled by agent $i \in N$. The set of locations by agent i is then $x_i = (x_{i1}, x_{i2}, \ldots, x_{iw_i})$. The location profile is now $x = (x_1, \ldots, x_k)$, and there are total n locations, i.e $\sum_{i=1}^{k} w_i = n$.

As before, the utility of an agent i is the sum of utilities of each location, i.e., $D(f(x), x_i) = \sum_{j=1}^{w_i} |f(x) - x_{ij}|$. For simplicity, let $\sum_i g(x_i)^p = \sum_i \sum_{j=1}^{w_i} g(x_{ij})^p$ for any $p \geq 1$, and any function g. In the remaining paper, the function g usually refers to $g(x_i) = x_i$ or $g(x_i) = 2 - x_i$.

4.1   Deterministic Strategy-Proof Mechanisms for Multiple Locations Per Agent

It is quite natural to consider Mechanism 1 for this model. However, Mechanism 1 is not strategy-proof.

Lemma 2. *Mechanism 1 is not strategy-proof for the multiple locations per agent. (The proof can be found in the full version of the paper.)*

Mechanism 1 is not strategy-proof since n_1 and n_2 only represent the independent locations without considering the agents. The reason motivates us to consider the locations of an agent as an entirety. An agent is said to *prefer* 0 if $D(0, x_i) \geq D(2, x_i)$.

Given profile x, let $N_1(x)$ be the set of agents who prefer 0 and $N_2(x)$ be the set of agents who prefer 0. Hence, $|N_1(x)|$ is the number of agents who prefer 0 and $|N_2(x)|$ is the number of agents who prefer 2.

It is quite natural to consider the mechanism who locates the facility at 0 if more agents prefer 0 than 2.

Mechanism MA:
For given location profile x, if $|N_1(x)| \leq |N_2(x)|$, the mechanism returns location 2, otherwise places the facility at 0.

Lemma 3. *Mechanism MA is a group strategy-proof mechanism. (The proof can be found in the full version of the paper.)*

Theorem 5. *Mechanism MA is a group strategy-proof mechanism, and its approximation ratio is $\Theta(n)$ for maxSOS objective. (The proof can be found in the full version of the paper.)*

Similar as Theorem 5, we obtain that the lower bound of Mechanism MA is $2n - 1$ and the upper bound is at most $2n - 1$. Thus, we have the following Corollary.

Corollary 1. *Mechanism MA is a group strategy-proof mechanism, and its approximation ratio is $\Theta(n)$ for maxSum objective.*

Though Mechanism MA is group strategy-proof, the approximation ratio is not constant. To get better approximation ratio, we provide the following mechanism.

Mechanism TMA:
For given location profile x, the mechanism places the facility at 0 if $\sum_{i \in N_1(x)} w_i \geq \sum_{i \in N_2(x)} w_i$, otherwise the facility is allocated at 2.

Theorem 6. *Mechanism TMA is a strategy-proof mechanism, and its approximation ratio is 3 for maxSum objective and 5 for maxSOS objective. (The proof can be found in the full version of the paper.)*

Remark: Han and Du [4] showed that the lower bounds of the general setting where each agent only has one location are 3 and 5 for maxSum and maxSOS objectives, respectively. The lower bounds are also valid for our models, and thus our mechanism is tight for both objectives.

4.2   Randomized Strategy-Proof Mechanisms for Multiple Locations Per Agent

In this subsection, we study randomized mechanisms for multiple locations per agent model. We first consider the mechanism that choose only location 0 and 2 with positive probabilities. The mechanism is described as below.

Mechanism RMA:
Given a location profile x, the mechanism places the facility at 0 with probability $\sum_{i \in N_1(x)} w_i / n$ and selects the facility at 2 with probability $\sum_{i \in N_2(x)} w_i / n$.

Theorem 7. *Mechanism RMA is a strategy-proof mechanism and its approximation ratio is 4 for maxSOS objective. (The proof can be found in the full version of the paper.)*

Corollary 2. *Mechanism RMA is a strategy-proof mechanism and its approximation ratio is 2 for maxSum objective. (The proof can be found in the full version of the paper.)*

4.3 Lower Bounds of Randomized Mechanisms for Multiple Locations Per Agent

The lower bounds of randomized mechanisms in Section 3 are still valid for this general model. However, we can improve the lower bounds due to multiple locations of each agent. The techniques of designing lower bounds are also different from Section 3.

Theorem 8. *Any randomized strategy-proof mechanisms with multiple locations per agent achieve an approximation ratio at least 10/9 for maxSum objective. (The proof can be found in the full version of the paper.)*

Theorem 9. *The approximation ratio of any randomized strategy-proof mechanisms for multiple locations per agent are at least 1.13 for maxSOS objective. (The proof can be found in the full version of the paper.)*

5 Concluding Remarks

In this paper, we have studied strategy-proof mechanisms for different objective functions of obnoxious facility location game on a real line. Both upper bounds and lower bounds are provided for randomized mechanisms. Besides, we studied an extended model that each agent controls multiple locations. The provided deterministic mechanisms are the best possible.

There exist plentiful researches on facility location and obnoxious facility location (*e.g.* surveys [18,19]). Tamir [20] considered k-maxMin and k-maxSum obnoxious facility location problem on graph, and showed that the problem is strongly NP-hard even the graph is a line. Berman and Drezner [17] studied the obnoxious facility on a network to maximize the minimal distance. This substantial work motivates us to investigate a lot of interesting work for future. First, one may consider to extend our model to different networks, such as a tree or a circle. Second, it is interesting to study two facilities or the general k facilities location games.

References

1. Cheng, Y., Yu, W., Zhang, G.: Strategy-proof approximation mechanisms for an obnoxious facility game on networks. Theoretical Computer Science **35**(3), 513–526 (2011)
2. Feldman, M., Wilf, Y.: Strategyproof facility location and the least squares objective. In: Proceedings of the 14th ACM Conference on Electronic Commerce (EC 2013), pp. 873–890 (2013)
3. Ibara, K., Nagamochi, H.: Characterizing mechanisms in obnoxious facility game. In: Lin, G. (ed.) COCOA 2012. LNCS, vol. 7402, pp. 301–311. Springer, Heidelberg (2012)
4. Han, Q., Du, D.: Moneyless strategy-proof mechanism on single-sinked policy domain: characterization and applications, Issue 2012, Part 8. Working paper series (University of New Brunswick, Faculty of Business Administration)

5. Procaccia, A.D., Tennenholtz, M.: Approximate mechanism design without money. In: Proceedings of the 10th ACM conference on Electronic Commerce (EC 2009), pp. 177–186 (2009)

6. Lu, P., Sun, X., Wang, Y., Zhu, Z.A.: Asymptotically optimal strategy-proof mechanisms for two-facility games. In: Proceedings of the 11th ACM conference on Electronic Eommerce (EC 2010), pp. 315–324 (2010)

7. Lu, P., Wang, Y., Zhou, Y.: Tighter bounds for facility games. In: Leonardi, S. (ed.) WINE 2009. LNCS, vol. 5929, pp. 137–148. Springer, Heidelberg (2009)

8. Alon, N., Feldman, M., Procaccia, A.D., Tennenholtz, M.: Strategyproof approximation of the minimax on networks. Mathematics of Operations Research **35**(5), 513–526 (2010)

9. Nissim, K., Smorodinsky, R., Tennenholtz, M.: Approximately optimal mechanism design via differential privacy. In: Proceedings of the 3rd Innovations in Theoretical Computer Science Conference (ITCS 2012), pp. 203–213 (2012)

10. Dokow, E., Feldman, M., Meir, R., Nehama, I.: Mechanism design on discrete lines and cycles. In: Proceedings of the 13th ACM Conference on Electronic Commerce (EC 2012), pp. 423–440 (2012)

11. Fotakis, D., Tzamos, C.: On the power of deterministic mechanisms for facility location games. In: Fomin, F.V., Freivalds, R., Kwiatkowska, M., Peleg, D. (eds.) ICALP 2013, Part I. LNCS, vol. 7965, pp. 449–460. Springer, Heidelberg (2013)

12. Thang, N.K.: On (group) strategy-proof mechanisms without payment for facility location games. In: Saberi, A. (ed.) WINE 2010. LNCS, vol. 6484, pp. 531–538. Springer, Heidelberg (2010)

13. Gibbard, A.: Manipulation of voting schemes: a general result. Econometrica: Journal of the Econometric Society, 587–601 (1973)

14. Moulin, H.: On strategy-proofness and single peakedness. Public Choice **35**(4), 437–455 (1980)

15. Schummer, J., Vohra, R.V.: Strategy-proof location on a network. Journal of Economic Theory **104**(2), 405–428 (2002)

16. Schummer, J., Vohra, R.V.: Mechanism design without money. In: Algorithmic Game Theory, chap. 10, pp. 243–299, Cambridge (2007)

17. Berman, O., Drezner, Z.: A note on the location of an obnoxious facility on a network. European Journal of Operational Research **120**(1), 215–217 (2000)

18. Erkut, E., Neuman, S.: Analytical models for locating undesirable facilities. European Journal of Operational Research **40**(3), 275–291 (1989)

19. ReVelle, C., Eiselt, H.: Location analysis: A synthesis and survey. European Journal of Operational Research **165**(1), 1–19 (2005)

20. Tamir, A.: Obnoxious facility location on graphs. SIAM Journal on Discrete Mathematics **4**(4), 550–567 (1991)

21. Fotakis, D., Tzamos, C.: Strategyproof facility location for concave cost functions. In: Proceedings of the 14th ACM Conference on Electronic Commerce (EC 2013), pp. 435–452 (2013)

Bin Packing Game with an Interest Matrix

Zhenbo Wang[1], Xin Han[2(✉)], György Dósa[3], and Zsolt Tuza[4,5]

[1] Department of Mathematical Sciences, Tsinghua University, Beijing 100084, China
zwang@math.tsinghua.edu.cn
[2] Software School, Dalian University of Technology, Dalian 116620, China
hanxin.mail@gmail.com
[3] Department of Mathematics, University of Pannonia, Veszprém 8200, Hungary
dosagy@almos.vein.hu
[4] Department of Computer Science and Systems Technology,
University of Pannonia, Veszprém, Hungary
[5] Alfréd Rényi Institute of Mathematics, Budapest, Hungary
tuza@dcs.uni-pannon.hu

Abstract. In this paper we study a game problem, called bin packing game with an interest matrix, which is a generalization of all the currently known bin packing games. In this game, there are some items with positive sizes and identical bins with unit capacity as in the classical bin packing problem; additionally we are given an interest matrix with rational entries, whose element a_{ij} stands for how much item i likes item j. The payoff of item i is the sum of a_{ij} over all items j in the same bin with item i, and each item wants to stay in a bin where it can fit and its payoff is maximized. We find that if the matrix is symmetric, a pure Nash Equilibrium always exists. However the PoA (Price of Anarchy) may be very large, therefore we consider several special cases and give bounds for PoA in them. We present some results for the asymmetric case, too.

1 Introduction

In the bin packing problem, there are n items with positive rational sizes $\{s_1, s_2, ..., s_n\}$, where each item has size at most 1, and infinitely many bins with unit capacity are available. The goal is to pack the items into a minimum number of bins, so that in any bin B_k, the sum of the sizes of the items being packed there (called level of the bin) does not exceed the capacity of the bin; i.e., the quantity $s(B_k) = \sum_{i \in B_k} s_i$ is at most 1. There are many papers on this topic; we refer to [2,3,6,7,9] for details.

The first bin packing game was introduced by Bilò [1]. Later another version was proposed by Ma et al. [10], and recently a general version was developed by

---

Z. Wang—Partially supported by NSFC No. 11371216.

X. Han—Partially supported by NSFC(11101065), LJQ2012003, RGC(HKU716-412E) and "the Fundamental Research Funds for the Central Universities".

G. Dósa, Z. Tuza—Partially supported by TÁMOP-4.2.2.A-11/1/KONV-2012-0072.

© Springer International Publishing Switzerland 2015
D. Xu et al. (Eds.): COCOON 2015, LNCS 9198, pp. 57–69, 2015.
DOI: 10.1007/978-3-319-21398-9_5

Dósa and Epstein [4]. We call these bin packing games (or models) as BPG1, BPG2, and BPG3, respectively.

In case of the BPG1 model, the items in the same bin pay unit cost in total for being in this bin. The items share the cost proportionally to their sizes: a bigger item pays more, a smaller item pays less, i.e. an item with size s_i pays $s_i/s(B_k)$ for being in bin B_k.

In case of the BPG2 model, the cost of any bin is again unit, but the items of any bin pay the same price for being in this bin, i.e. any item pays $1/k$ if there are k items in the bin.

In case of model BPG3, each item i has two parameters s_i and u_i, where s_i is the size of the item (as usually), and a nonnegative weight u_i is also specified for item i. Then, for being in any one bin B_k, the items in B_k pay proportionally to their weights rather than to their sizes or their cardinality; i.e., item i pays u_i/U_k cost for being in B_k, where $U_k = \sum_{i \in B_k} u_i$. This is a common generalization of the two previous models BPG1 and BPG2, since if $u_i = s_i$ for any item, we get model BPG1, or if $u_i = 1$ for any item, we get model BPG2.

Generalized Bin Packing Game. We introduce a new type of bin packing games. This new game is a common generalization of all the above three models, thus we call it Generalized Bin Packing Game and abbreviate it as GBPG for short.

The motivation of this new model is to express that people make their decisions not only considering money or cost, but they often also take into account how much they like a certain situation. Let us consider the next simple example: There is a party where the people sit down at tables (tables = bins). Then a person is interested not only in the cost of sitting at some table (and paying for the food and drinks he will have), but would also like to enjoy the party, and therefore chooses a table where he/she finds the people appealing.

Formally, an instance $\mathcal{I} = (A, S)$ of GBPG is given as follows. There are n items with sizes $S = \{s_1, s_2, ..., s_n\}$, where $0 < s_i \leq 1$, and an $n \times n$ rational matrix $A = [a_{ij}]$, called the interest matrix, is also given. The payoff of item i is $p_i = \sum_{j \in B_k} a_{ij}$ if i is packed into bin B_k. Each item wants to stay in a bin where it can fit and its payoff is maximized. All bins are assumed to be identical with unit capacity. In the discussion below we assume that all $a_{ij} \geq 0$, although some facts remain valid for negative values, too. (Some remarks of this kind will be given.) We note that also a_{ii} is taken into account when defining the payoff p_i of item i.

A packing of the items is called a Nash Equilibrium [11], or NE for short, if no item can improve its payoff by moving to another bin in which it can fit. Moreover, if all the items are packed into the minimum number of bins, we call this packing an optimum packing, and denote it by OPT. Without the danger of confusion, we also use OPT and NE to denote the bins used by an OPT packing and an NE packing, respectively.

Price of Anarchy (PoA). An often used metric in case of bin packing games is the Price of Anarchy (PoA, for short); it measures how large a NE can be compared to OPT, when the value of OPT gets large. More exactly,

$$PoA = \lim_{k \to \infty} \sup \left\{ \frac{NE}{OPT} \mid OPT = k \right\}.$$

There are further metrics, too, such as price of stability (PoS), strong price of anarchy, and so on; in this paper we deal only with PoA.

Previous Results. The bin packing game BPG1 was studied first by Bilò in 2006 [1]. He proved that this game admits a NE, and that the PoA is in the interval $[1.6, 1.667]$. Epstein and Kleiman [8] obtained stronger estimates, proving that the PoA is in $[1.6416, 1.6428]$.

For the BPG2 model it was proven that its PoA is at most 1.7. This result got further improved in [5].

In case of model BPG3 [4], it was shown that many kinds of Nash equlibria (NE, Strong NE, Strongly Pareto Optimal NE and Weakly Pareto Optimal NE) exist. For the case of unit weights (which is equivalent to model BPG2), the PoA is in $[1.6966, 1.6994]$; and for the weighted case, both of the lower and upper bounds are 1.7. For other results, we refer to [5,13].

Our Contribution. When the interest matrix is symmetric, we prove that there exists a NE for any instance. Generally, the value of PoA can be very large, therefore we consider several specific types of the interest matrix $[a_{ij}]$. The results are listed in the following table, where s_i is the size of item i and u_i is the weight of item i. The bounds for $\max\{s_i, s_j\}$ are quoted from [8], and the lower bound in the last two columns is from [4].

$a_{ij} =$	general	$\max\{s_i, s_j\}$	$\min\{s_i, s_j\}$	$\max\{u_i, u_j\}$	$u_i + u_j$
PoA	∞	$[1.6416, 1.6428]$	∞	$[1.6966, 1.723]$	$[1.6966, 2]$

Lastly we investigate the case of asymmetric matrices and find that a NE packing does not exist for some instances. We give a sufficient condition to recognize this situation of non-existing NE. On the other hand we give a sufficient condition which guarantees that an asymmetric interest matrix can be converted to a symmetric one.

2   Preliminaries

In this section, we first recall the definition of a classical algorithm for the bin packing problem, and then prove that the game we study is a generalization of all the known bin packing games.

First Fit for Bin Packing. For an input I of bin packing, let $ALG(I)$ be the number of bins used by algorithm ALG to pack this input, and let $OPT(I)$ denote an optimal solution. Algorithm First Fit (FF) is a classical algorithm, which packs each item into the first bin where it fits. (If the item does not fit into any opened bin, it is packed into a new bin.) First Ullmann [12] proved that

$FF(I) \leq 1.7 \cdot OPT(I) + 3$. Then, after several attempts to decrease the additive constant, finally Dósa and Sgall [6] proved that $FF(I) \leq 1.7 \cdot OPT(I)$, what means that the absolute approximation ratio of FF is at most 1.7. In another work, Dósa and Sgall [7] give a matching lower bound, thus these two papers together prove that the bound 1.7 is tight.

Relation to the Earlier Models. The players in the games introduced earlier wish to minimize their costs, while in our game the players wish to maximize their payoff. In spite of this, the former bin packing game models can be considered as special cases of Model GBPG, in the following way:

- BPG1 model: let $a_{ij} = s_i \cdot s_j$; or $a_{ij} = s_j$.
- BPG2 model: let $a_{ij} = 1$; or $a_{ij} = s_i$.
- BPG3 model: let $a_{ij} = u_i \cdot u_j$; or $a_{ij} = u_j$.

We only prove this claim for the BPG1 model; the assertion for the other two models can be shown in a similar way.

Lemma 1. *If $a_{ij} = s_i \cdot s_j$, or if $a_{ij} = s_j$, for all $1 \leq i, j \leq n$, then our game is equivalent to the BPG1 model in the sense that a NE for GBPG is also a NE for BPG1, and vice versa.*

Proof. Suppose $a_{ij} = s_i \cdot s_j$, and an item i is packed into bin B_k. Then the payoff of this item is $p_i = \sum_{j \in B_k}(s_i \cdot s_j) = s_i \cdot s(B_k)$. This item is intended to go into another bin B_l, if the payoff of item i will be bigger there (and item i fits there). This new payoff is $p_i' = \sum_{j \in B_l}(s_i \cdot s_j) + s_i^2 = s_i \cdot (s(B_l) + s_i)$. Thus the movement is possible if and only if $p_i' > p_i$, i.e. $s(B_l) + s_i > s(B_k)$. This is exactly the same case (when the movement is possible) like the one in model BPG1.   □

3   The Symmetric Case

We prove that for any symmetric matrix A, there always exists a pure NE for our model GBPG. We give the proof by using a potential function.

Theorem 1. *If matrix A is symmetric, then GBPG always has an NE.*

Proof. The high-level idea is to associate each feasible packing with a potential in such a way that the potential function is upper-bounded by a value computable from the input, and to prove that once an item moves from one bin to another, the total potential strictly increases. If this property holds, then no previous state can occur again, thus a NE surely exists because there are only a finite number of different configurations (packings).

Recall that if item i is packed into bin B_k, then its payoff is $p_i = \sum_{j \in B_k} a_{ij}$. Given an input I and a packing for I, we define a potential function as

$$P = \sum_{i \in I} p_i \leq n^2 \max_{i,j}\{a_{ij}\}.$$

Next we prove that if item i moves from bin B_k to bin B_h, then the value of P increases. Before moving from bin B_k to bin B_h, let p_j be the payoff of item j and $P = \sum_{i \in I} p_i$. After the movement, let p'_j be the payoff of item j and $P' = \sum_{i \in I} p'_i$, and name bins B_k and B_h as B'_k and B'_h, respectively.

Observe that for item j which is not packed in bin B_k or B_h, its payoff does not change. Then we have

$$
\begin{aligned}
P' - P &= \sum_{j \in B'_k} p'_j + \sum_{j \in B'_h} p'_j - \sum_{j \in B_k} p_j - \sum_{j \in B_h} p_j \\
&= \left(\sum_{j \in B'_h} p'_j - \sum_{j \in B_h} p_j \right) + \left(\sum_{j \in B'_k} p'_j - \sum_{j \in B_k} p_j \right) \\
&= \left(p'_i + \sum_{j \in B_h} a_{ji} \right) - \left(p_i + \sum_{j \in B'_k} a_{ji} \right) \\
&= (p'_i - p_i) + \left(\sum_{j \in B_h} a_{ij} - \sum_{j \in B'_k} a_{ij} \right) \quad \text{by } a_{ij} = a_{ji} \\
&= (p'_i - p_i) + \left(a_{ii} + \sum_{j \in B_h} a_{ij} - a_{ii} - \sum_{j \in B'_k} a_{ij} \right) \\
&= (p'_i - p_i) + (p'_i - p_i) = 2(p'_i - p_i) > 0.
\end{aligned}
$$

For the convergence steps, if we set A a rational matrix, let $\Delta > 0$ be the minimal integer such that Δa_{ij} is integer for all components of A, and therefore $\Delta(p'_i - p_i) \geq 1$, and the potential function will increase at least $2/\Delta$ after a selfish movement. After at most $\Delta n^2 \max_{i,j}\{a_{ij}/2\}$ steps, we will have a NE. □

Remark 1. One can observe that the above proof works even when matrix A has zero or negative entries. A natural interpretation of this extension is that a_{ij} is positive if person i likes person j, and is negative if i dislikes j.

Observation 1. *For arbitrarily large real k there exists an interest matrix A, for which the PoA is bigger than k, even if $a_{ij} \in \{0,1\}$ is required for all $1 \leq i, j \leq n$.*

Proof. Given any real k, let n be an integer such that $n > k$. There are n items $\{1, 2, ..., n\}$, each with size $\epsilon \leq 1/n$. We pack the items into mutually distinct bins. Let $a_{ii} = 1$ for every i, and $a_{ij} = 0$ for all $i \neq j$. The actual packing is a NE, whereas the optimal solution uses only one bin. □

Recall that if all entries in the interest matrix A have the same value $a_{ij} = 1$, the PoA is upper-bounded by 1.6994 [4]. However, we find that the PoA can be very large even if almost all entries are $a_{ij} = 1$ and all the other elements satisfy $a_{ij} = 1 - \epsilon$ where $\epsilon > 0$ could be arbitrarily small.

Proposition 1. *Given $0 < \delta < 1/2$, let k be an integer for which $k\delta > 1$. Then there is a matrix A with size $n \times n$, where $n = k^4$, in which $a_{ij} = 1$ for at least $(1 - 1/k) \cdot n^2$ different pairs (i, j), moreover $a_{ij} = 1 - \delta$ for all the other entries, and the PoA is at least k^2.*

The proof is left in the Appendix due to the page limitation.

3.1   Special Models

Now we give lower and upper estimates on PoA for several special cases of GBPG: $a_{ij} = \max\{s_i, s_j\}$, $a_{ij} = \min\{s_i, s_j\}$, $a_{ij} = \max\{u_i, u_j\}$, and $a_{ij} = u_i + u_j$, where u_i is the weight of item i, which may be different from s_i.

In the model $a_{ij} = \max\{u_i, u_j\}$, where u_i means something that how much person or item i is important. So a special (and natural) case is, where the items correspond to some persons, some of them are famous while the other are not famous. If we assume that $u_i = 1$ if any person i is famous, and $u_i = p < 1$ otherwise, we model the situation that people like to be in the presence of famous, or important persons. Or, more generally, any person gets an "importance" index, this is the u_i value. Then, the happiness between two persons, i and j is defined as $a_{ij} = \max\{u_i, u_j\}$. Further more in the special model $a_{ij} = \max\{s_i, s_j\}$, the bigger item can enforce his want to the smaller item. This is a typical situation in many cases.

In order to explore the relationship between a matrix A and the corresponding value of PoA, we begin with the settings $a_{ij} = max\{s_i, s_j\}$ and $a_{ij} = \min\{s_i, s_j\}$. For the former we prove that the PoA is at most 1.7, and find that some earlier results also remain valid for this model. But the latter is substantially different as the PoA can be arbitrarily large.

Proposition 2. *If $a_{ij} = \max\{s_i, s_j\}$, then PoA is at most 1.7.*

Proof. Our key observation is as follows: For any bin in a given packing, the payoff of the smallest item is the total size of items in the bin, and the payoff of any other item in the bin is at least this value. Consider a NE with bins $B_1, B_2, ..., B_m$. Assume that the bins are sorted such that $s(B_1) \geq s(B_2) \geq ... \geq s(B_m)$. We claim that no item in B_k fits into B_h for any $h < k$, i.e. the packing can be viewed as a result of FF packing. Let i be the smallest item in B_k, and suppose for a contradiction that it fits into B_h. In B_k, the payoff of item i is $p_i = s(B_k)$, whereas the payoff of this item is at least $s(B_h) + s_i > s(B_k)$ if it moves to B_h; this contradicts the assumption that we are in a NE state. Thus the claim follows. As we know that the asymptotic approximation ratio of FF (and even the absolute approximation ratio of FF) is 1.7, we obtain that the PoA in the current model is at most 1.7.   □

Remark 2. We find that using the methods in [1], one can get $1.6 \leq PoA \leq 1.667$; and using the methods in [8], one can further get $1.6416 \leq PoA \leq 1.6428$.

Proposition 3. *If $a_{ij} = \min\{s_i, s_j\}$, then PoA can be arbitrarily large.*

The proof is left in the appendix.

Theorem 2. *Assume that each item i is associated with two parameters, the size s_i and the weight $u_i > 0$. If $a_{ij} = \max\{u_i, u_j\}$, then PoA is at most $\frac{31}{18} < 1.723$.*

Proof. Let us consider an NE with bins $B_1, B_2, ..., B_m$. Given a bin B_j, let the total weight and total size of the items in B_j be $u(B_j)$ and $s(B_j)$, respectively; and let the number of items in B_j be $|B_j|$. Suppose that there are two bins, say B_1 and B_2, such that $s(B_1) + s(B_2) \leq 1$. Assume without loss of generality that $u(B_1) \geq u(B_2)$. Let i be the item with the smallest weight in B_2. The payoff of i is exactly $u(B_2)$, and it becomes at least $u(B_1) + u_i > u(B_2)$ if the item moves to B_1, contradicting the assumption that the packing is an NE. Consequently, $s(B_1) + s(B_2) > 1$ holds for any two bins. Moreover, we have the following properties.

1. If item i is packed into B_j, its payoff satisfies $p_i \geq u(B_j)$ and equality holds only if i has the minimum weight in B_j.
2. If item k has the smallest or second smallest weight in B_j, then $p_k < u(B_j) + u_k$.

Now we divide the bins into four groups:

$$G_1 = \{B_j \mid 0 < s(B_j) \leq 1/2\}, \quad G_2 = \{B_j \mid 1/2 < s(B_j) \leq 2/3\},$$
$$G_3 = \{B_j \mid 2/3 < s(B_j) \leq 3/4\}, \quad G_4 = \{B_j \mid s(B_j) > 3/4\}.$$

Claim 1. $|G_1| \leq 1$.

Proof: This follows from the fact that, as we have shown the beginning of the argument, any two bins satisfy $s(B_1) + s(B_2) > 1$ in any NE. $\square$

Define $\quad G_2^1 = \{B_j \mid B_j \in G_2, |B_j| = 1\}, \quad G_3^1 = \{B_j \mid B_j \in G_3, |B_j| = 1\},$
$$G_2^{2+} = \{B_j \mid B_j \in G_2, |B_j| \geq 2\}, \quad G_3^{2+} = \{B_j \mid B_j \in G_3, |B_j| \geq 2\}.$$

Claim 2. The sole item in each bin of $G_2^1 \cup G_3^1$ has a size larger than $1/2$ (by definition).

Claim 3. $|G_2^{2+}| \leq 1$.

Proof: Suppose for a contradiction that at least two bins, say B_1 and B_2 belong to G_2^{2+}; assume without loss of generality that $u(B_1) \geq u(B_2)$. From the definition of G_2^{2+}, we see that the item with the smallest weight or the second smallest weight has a size at most $\frac{1}{3}$. Let item k be the item. If item k moves to bin B_1, then its payoff is at least $u(B_1) + u_k \geq u(B_2) + u_k > p_k$, where p_k is the payoff of item k in bin B_2. So it is not difficult to see that the item in B_2 has an incentive to move to bin B_1. Hence the assumption is false and $|G_2^{2+}| \leq 1$. $\square$

Claim 4. With the exception of at most one bin, in each bin of G_3^{2+}, both the item with the minimum weight and the second minimum weight have size larger than $1/4$.

Proof: Consider any two bins of G_3^{2+}, say B_1 and B_2. Assume without loss of generality that $u(B_1) \geq u(B_2)$. In B_2, if any item i with the smallest or the second smallest weight has a size at most $\frac{1}{4}$, then its payoff will get improved from at most $u(B_2) + u_i$ to at least $u(B_1) + u_i$, by Properties 1 and 2. Therefore each of them has a size larger than $1/4$. $\square$

Now we can proceed further with the proof of the theorem. We have an upper bound

$$NE = |G_1| + |G_2^1| + |G_3^1| + |G_2^{2+}| + |G_3^{2+}| + |G_4|$$
$$\leq 2 + |G_2^1| + |G_3^1| + |G_3^{2+}| + |G_4|,$$

and a lower bound

$$OPT \geq \frac{|G_2^1| + |G_3^1|}{2} + \frac{2|G_3^{2+}|}{3} + \frac{3|G_4|}{4}. \tag{1}$$

The size of each item in the bins of $G_2^1 \cup G_3^1$ is larger than $1/2$; let us call these items big. Hence, there are $|G_2^1| + |G_3^1|$ big items. In each bin of G_3^{2+}, except at most one bin, there are at least two items with a size larger than $1/4$ each; let us call these items medium-sized items. Hence, there are at least $2(|G_3^{2+}| - 1)$ medium-sized items. Note that no two big items can be packed into the same bin, therefore

$$OPT \geq |G_2^1| + |G_3^1|. \tag{2}$$

Case 1. $|G_2^1| + |G_3^1| \geq 2(|G_3^{2+}| - 1)$. Then we also have

$$OPT \geq 2|G_3^{2+}| - 2. \tag{3}$$

We multiply the inequalities (1), (2), (3) by $\frac{24}{18}$, $\frac{6}{18}$, and $\frac{1}{18}$, respectively. Adding them we obtain

$$\frac{31}{18}OPT \geq |G_2^1| + |G_3^1| + |G_3^{2+}| + |G_4| - 1/9,$$

thus

$$NE \leq |G_2^1| + |G_3^1| + |G_3^{2+}| + |G_4| + 2 \leq \frac{31}{18}OPT + \frac{19}{9} < 1.723 \cdot OPT + 2.12.$$

Case 2. $|G_2^1| + |G_3^1| < 2(|G_3^{2+}| - 1)$. Now, in any feasible packing, at most one medium-sized item can be packed with a big item into the same bin, and the remaining medium-sized items need at least $\frac{2(|G_3^{2+}|-1)-|G_2^1|-|G_3^1|}{3}$ bins. Therefore,

$$OPT \geq |G_2^1| + |G_3^1| + \frac{2(|G_3^{2+}| - 1) - |G_2^1| - |G_3^1|}{3}$$
$$= \frac{2(|G_2^1| + |G_3^1|)}{3} + \frac{2|G_3^{2+}| - 2}{3}. \tag{4}$$

Now we multiply the inequalities (1), (2), and (4) by $\frac{24}{18}$, $\frac{4}{18}$, and $\frac{3}{18}$, respectively. Adding them we obtain

$$\frac{31}{18}OPT \geq |G_2^1| + |G_3^1| + |G_3^{2+}| + |G_4| - 1/9,$$

therefore we also have $NE \leq \frac{31}{18}OPT + \frac{19}{9} < 1.723OPT + 2.12.$ □

Proposition 4. *Assume that $a_{ij} = u_i + u_j$, where $u_i > 0$ is the weight of item i. Then PoA is at most 2.*

Proof. It suffices to show that the average level of bins in any NE is larger than $1/2$. In fact, the average is this large already for any two bins. Indeed, consider an NE, and assume for a contradiction that there are two bins, say B_1 and B_2, such that their total level is at most 1. Let $u(B_1)$ and $u(B_2)$ be defined as $u(B_1) = \sum_{j \in B_1} u_j$ and $u(B_2) = \sum_{j \in B_2} u_j$, respectively; we assume without loss of generality that $u(B_1) \geq u(B_2)$. Let l and k be the numbers of items in B_1 and B_2, respectively. Suppose first that $l \geq k$, and let i be an arbitrary item in B_2. We claim that i would like to move to B_1. The actual payoff for item i is

$$p_i = \sum_{j \in B_2} a_{ij} = \sum_{j \in B_2} (u_i + u_j) = k \cdot u_i + u(B_2).$$

If i moves to bin B_1, its payoff will be there

$$p_i' = \sum_{j \in B_1} a_{ij} + a_{ii} = \sum_{j \in B_1} (u_i + u_j) + 2u_i = (l + 2) \cdot u_i + u(B_1),$$

which is bigger than p_i, thus the claim is verified. This contradicts the assumption that the packing is a NE, therefore we must have $l < k$. Let i be the item in B_1 for which u_i is the biggest, and j be the item in B_2, for which u_j is the smallest. Then

$$u_i \geq u(B_1)/l \geq u(B_2)/l > u(B_2)/k \geq u_j. \qquad (5)$$

If we move i from B_1 to B_2, its payoff changes by

$$c_1 = p_i' - p_i = (k \cdot u_i + u(B_2) + 2u_i) - (l \cdot u_i + u(B_1))$$
$$= (k + 2 - l)u_i + u(B_2) - u(B_1);$$

and if we move j from B_2 to B_1, its payoff changes by

$$c_2 = p_j' - p_j = (l \cdot u_j + u(B_1) + 2u_j) - (k \cdot u_j + u(B_2))$$
$$= (l + 2 - k)u_j + u(B_1) - u(B_2).$$

Moreover, from (5) we have

$$c_1 + c_2 = 2(u_i + u_j) + (k - l)(u_i - u_j) > 0.$$

This means that at least one of items i and j will improve its payoff by moving to the other bin, a contradiction. □

Remark 3. In the two latter models, if we set $u_i = 1$ for all i, then $PoA \geq 1.6966$ can be proved by the approach given in [4].

4 Asymmetric Case

In this section we deal with the case where the matrix A is not symmetric. First we observe by giving an example that NE may not always exist; more precisely, from a suitably chosen initial packing, NE is not reached after any infinite sequence of selfish steps. Then we describe a sufficient condition for problem inputs, which ensures that there exists an initial packing and an infinite sequence of feasible steps which never lead to an NE. However we find by another example that the condition is not necessary. Finally we also give a sufficient condition, where NE always exist.

Example 1. The following instance admits an initial packing which never terminates with an NE, independently of the value of the parameter $p < 1$. Take three items 1, 2, and 3, each with size 0.5. Let $a_{i,i} = 1$ for $i = 1, 2, 3$, $a_{12} = a_{23} = a_{31} = 1$, and $a_{21} = a_{32} = a_{13} = p$.

Proof. Assume that the three items are packed in three distinct bins. Then first item 1 moves to share a bin with item 2. Then item 2 leaves item 1 alone and moves to share a bin with item 3. Then item 3 leaves item 2 alone and moves to share a bin with item 1. Then item 1 moves and again shares a bin with item 2, and the movement can be continued to infinity. Then there is no NE for the above instance.   □

In the following we give a generalization of the instance in Example 1, where NE does not exits after any sequence of selfish steps. This part is related to line graphs.

Line Graph. According to standard terminology, the vertices in the line graph $L(G)$ of G represent the edges of G, and two vertices of $L(G)$ are adjacent if and only if the corresponding edges of G share a vertex. So, each edge in $L(G)$ identifies three vertices, say v_i, v_j, v_k in G, and two edges $v_i v_j$, $v_i v_k$ on them, sharing one vertex v_i. Originally in G the edges are undirected; but now their common vertex v_i specifies the *ordered* pairs $(i, j), (i, k)$. In case if $a_{ij} = a_{ik}$, we remove the corresponding edge from $L(G)$; and if equality does not hold, then we orient the edge of $L(G)$ from the smaller to the larger a-value; see Fig. 1 for an illustration. We denote by $H = (X, F)$ the oriented graph obtained in this way. (This H, obviously, does not contain cycles of length 2.)

Compatibility Graph. Let $\mathcal{I} = (A, S)$ be an instance of GBPG. We define the *compatibility graph* to represent the pairs of items which can occur together in a bin. This undirected graph, which we denote by $G = (V, E)$, is described by the following rules:

- the vertices are $v_1, v_2, ..., v_n$, indexed according to the items;
- an unordered vertex pair $v_i v_j$ is an edge if and only if $s_i + s_j \leq 1$.

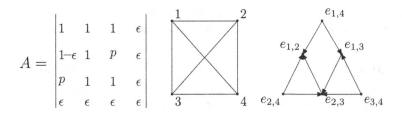

Fig. 1. Matrix, Compatibility Graph and Line Graph

Theorem 3. *Let $\mathcal{I} = (A, S)$ be an instance of GBPG, and let $H = (X, F)$ be the oriented partial line graph as described above. If H contains a directed (cyclically oriented) cycle, then there exists an initial packing of the items and an infinite sequence of feasible steps along which NE is never reached.*

Proof. Let $C = x_1 x_2 \ldots x_\ell$ be a directed cycle in H. We assume, without loss of generality, that C is a *shortest* cycle in H. Each x_k ($1 \le k \le \ell$) corresponds to some edge $e_k := v_{i_k} v_{j_k}$ of G, and we have $e_k \cap e_{k+1} \neq \emptyset$ for all $1 \le k \le \ell$ (subscript addition is taken modulo ℓ throughout the proof). We further observe:

Claim. For all k we have $e_k \cap e_{k+1} \neq e_{k+1} \cap e_{k+2}$.

Proof. Suppose for a contradiction that $e_k \cap e_{k+1} = e_{k+1} \cap e_{k+2} = v_k$, and assume that $e_i = v_k v_{j_i}$ for $i = k, k+1, k+2$. Then, by the construction of H, we have

$$a_{k,j_k} < a_{k,j_{k+1}} < a_{k,j_{k+2}}.$$

As a consequence, also $x_k x_{k+2}$ is an arc in H, because e_k and e_{k+2} share v_k and the corresponding s-values satisfy the required inequality. This contradicts the assumption that C is a shortest cycle in H, and hence the claim follows. $\square$

To prove the theorem we start with the initial packing where the two items corresponding to the vertices of e_1 are in the same bin, and all the other items are in mutually distinct bins. The claim implies that moving the item of $e_1 \cap e_2$ from its bin to the bin of $e_2 \setminus e_1$ is feasible. More generally, from a bin whose contents are the two items belonging to e_k, it is feasible to move $e_k \cap e_{k+1}$ to the bin of $e_{k+1} \setminus e_k$, for any $1 \le k \le \ell$. Consequently, in the first $\ell - 1$ bins the first ℓ items can circulate forever, without reaching NE at any time. $\square$

Remark 4. The line graph of the input can be constructed in linear time in terms of the input size, and it can also be tested in polynomial time whether the line graph contains a directed cycle.

Lemma 2. *Even in a line graph of an input instance there is no directed cycle, NE may not occurs after finite steps of selfish improvement.*

The proof is left in the appendix.

Next we give a sufficient condition, which ensures that the game in an asymmetric case can be converted to a symmetric game. The proof is left in the Appendix.

Proposition 5. *Assume that $a_{ij} > 0$ for any pair (i,j). If there is a univariate function $g : N \to R^+$ such that $\frac{a_{ij}}{a_{ji}} = \frac{g(i)}{g(j)}$, then the asymmetric case can be transformed to the symmetric case, and hence NE exists.*

5   Conclusions and Further Research

In this paper we introduced a new type of bin packing game, which is a common generalization of all the previously considered bin packing games. Then we studied several special types of it. There are several open questions left, let us mention two of them.

1. What are the tight bounds on *PoA* for the models studied?
2. Determine the algorithmic complexity of the following decision problems.
   Given $\mathcal{I} = (A, S)$, an instance of GBPG, together with an initial packing,
   (a) is it true that NE is reached after a suitable sequence of steps?
   (b) is it true that NE is reached after a finite number of steps, no matter which feasible step is chosen at any time?
   Is any of these problems polynomial-time solvable?

References

1. Bilò, V.: On the packing of selfish items. In: Proc. of the 20th International Parallel and Distributed Processing Symposium (IPDPS 2006), 9 p. IEEE (2006)
2. Coffman Jr., E.G., Csirik, J., Galambos, G., Martello, S., Vigo, D.: Bin packing approximation algorithms: survey and classification. In: Pardalos, P.M., Du, D.-Z., Graham, R.L. (eds.) Handbook of Combinatorial Optimization, pp. 455–531. Springer, New York (2013)
3. Coffman, E.G., Galambos, G., Martello, S., Vigo, D.: Bin packing approximation algorithms: combinatorial analysis. In: Du, D.-Z., Pardalos, P.M. (eds.) Handbook of Combinatorial Optimization, pp. 151–208. Kluwer, Dordrecht (1999)
4. Dosa, G., Epstein, L.: Generalized selfish bin packing, arXiv:1202.4080, 1–43 (2012)
5. Dósa, G., Epstein, L.: The convergence time for selfish bin packing. In: Lavi, R. (ed.) SAGT 2014. LNCS, vol. 8768, pp. 37–48. Springer, Heidelberg (2014)
6. Dósa, G., Sgall, J.: First fit bin packing: a tight analysis. In: Portier, N., Wilke, T. (eds) Proceedings of the 30th Symposium on the Theoretical Aspects of Computer Science (STACS 2013), pp. 538–549, Kiel, Germany (2013)
7. Dósa, G., Sgall, J.: Optimal analysis of best fit bin packing. In: Esparza, J., Fraigniaud, P., Husfeldt, T., Koutsoupias, E. (eds.) ICALP 2014. LNCS, vol. 8572, pp. 429–441. Springer, Heidelberg (2014)
8. Epstein, L., Kleiman, E.: Selfish Bin Packing. Algorithmica **60**(2), 368–394 (2011)
9. Garey, M.R., Johnson, D.S.: Computer and Intractability: A Guide to the Theory of NP-Completeness. Freeman, New York (1979)

10. Ma, R., Dósa, G., Han, X., Ting, H.-F., Ye, D., Zhang, Y.: A note on a selfish bin packing problem. Journal of Global Optimization **56**(4), 1457–1462 (2013)
11. Nash, J.: Non-cooperative games. Annals of Mathematics **54**(2), 286–295 (1951)
12. Ullman, J.D.: The performance of a memory allocation algorithm. Technical Report 100, Princeton Univ., Princeton, NJ (1971)
13. Yu, G., Zhang, G.: Bin packing of selfish items. In: Papadimitriou, C., Zhang, S. (eds.) WINE 2008. LNCS, vol. 5385, pp. 446–453. Springer, Heidelberg (2008)

The Least-Core and Nucleolus
of Path Cooperative Games

Qizhi Fang[1], Bo Li[1]([✉]), Xiaohan Shan[2], and Xiaoming Sun[2]

[1] School of Mathematical Sciences, Ocean University of China, Qingdao, China
qfang@ouc.edu.cn, boli198907@gmail.com
[2] Institute of Computing Technology, Chinese Academy of Sciences, Beijing, China
{shanxiaohan,sunxiaoming}@ict.ac.cn

Abstract. Cooperative games provide an appropriate framework for fair and stable profit distribution in multiagent systems. In this paper, we study the algorithmic issues on path cooperative games that arise from the situations where some commodity flows through a network. In these games, a coalition of edges or vertices is successful if they establish a path from the source to the sink in the network, and lose otherwise. Based on dual theory of linear programming and the relationship with flow games, we provide the characterizations on the CS-core, least-core and nucleolus of path cooperative games. Furthermore, we show that the least-core and nucleolus are polynomially solvable for path cooperative games defined on both directed and undirected network.

1 Introduction

A central question in cooperative game theory is how to distribute a certain amount of profit generated by a group of agents N, denoted by a function $\gamma(N)$, to each individual. It is often assumed that the grand coalition N is formed, since in many games the total profit or costs are optimized if all agents work together. To achieve this goal, the collective profit should be distributed properly so as to minimize the incentive of subgroups of agents to deviate and form coalitions of their own. A number of solution concepts have been proposed to capture this intuition, such as the core, the least-core, and the nucleolus, which will be the focus of this paper.

In this paper, we consider a kind of cooperative game models, *path cooperative games* (PC-games), arising from the situations where some commodity (traffic, liquid or information) flows through a network. In these games, each player

---

Q. Fang—The first author is supported by the National Natural Science Foundation of China (NSFC) (NO. 11271341). The fourth author is supported in part by the National Natural Science Foundation of China Grant 61170062, 61222202, 61433014 and the China National Program for support of Top-notch Young Professionals. This work is also partially supported by the National Natural Science Foundation of China (NSFC) (NO. 61173009.) Finally, we would like to acknowledge our editors and a superb set of anonymous referees for their excellent suggestions.

© Springer International Publishing Switzerland 2015
D. Xu et al. (Eds.): COCOON 2015, LNCS 9198, pp. 70–82, 2015.
DOI: 10.1007/978-3-319-21398-9_6

controls an edge or a vertex of the network (called edge path cooperative games or vertex edge path cooperative games, respectively), a coalition of players wins if they establish a path from the source to the sink, and lose otherwise. We will focus on the algorithmic problems on game solutions of path cooperative games, especially core related solutions.

Path cooperative games have a natural correspondence with flow games. Flow games were first introduced by Kalai and Zemel [13] and studied extensively by many researchers. When there are public arcs in the network, the core of the flow game is nonempty if and only if there is a minimum (s, t)-cut containing no public arcs. And in this case, the core can be characterized by the minimum (s, t)-cuts[13,18], and the nucleolus can also be computed efficiently[5,17]. Recently, Aziz et al. [1] introduced the threshold versions of monotone games, including PC-games as a special case. Yoram [3] showed that computing ε-core for threshold network flow games is polynomial time solvable for unit capacity networks, and NP-hard for networks with general capacities. For PC-games defined on series-parallel graphs, Aziz et al.[1] showed that the nucleolus can be computed in polynomial time. However, the complexity of computing the nucleolus for general PC-games remains open, from the algorithmic point of view, the solution concepts of general PC-games have not been systematically discussed.

The algorithmic problems in cooperative games are especially interesting, since except for the fairness and rationality requirements in the solution definitions, computational complexity is suggested be taken into consideration as another measure of rationality for evaluating and comparing different solution concepts (Deng and Papadimitriou [6]). The computational complexity of classical solution concepts has therefore been studied with growing interest during the last decades. On the positive side, efficient algorithms have been proposed for computing the core, the least-core and the nucleolus for, such as, assignment games [20], cardinality matching games [14], unit flow games [5] and weighted voting games [8]. On the negative side, the problems of computing the nucleolus and testing whether a given distribution belongs to the core or the nucleolus are proved to be NP-hard for minimum spanning tree games [9,10], flow games and linear production games [5,11].

The main contribution of this work is the efficient characterizations of the CS-core, least-core and the nucleolus of PC-games, based on linear programming technique and the relationship with flow games. These characterizations yield directly to efficient algorithms for the related solutions. The organization of the paper is as follows.

In Section 2, the relevant definitions in cooperative game are introduced. In Section 3, we first define PC-games (edge path cooperative game and vertex path cooperative game), and then give the the characterizations of the core and CS-core. Section 4 is dedicated to the efficient description of the least-core for PC-games. In Section 5, we prove that the nucleolus is polynomially solvable for both edge and vertex path cooperative games.

2   Preliminaries

A cooperative game $\Gamma = (N, \gamma)$ consists of a player set $N = \{1, 2, \cdots, n\}$ and a characteristic function $\gamma : 2^N \to R$ with $\gamma(\emptyset) = 0$. For each coalition $S \subseteq N$, $\gamma(S)$ represents the profit obtained by S without help of other players. The set N is called the grand coalition. In what follows, we assume that $\gamma(S) \geq 0$ for all $S \subseteq N$, and $\gamma(\emptyset) = 0$.

An imputation of Γ is a payoff vector $x = (x_1, ...x_n)$ such that $\sum_{i \in N} x_i = \gamma(N)$ and $x_i \geq \gamma(\{i\})$, $\forall i \in N$. The set of imputations is denoted by $\mathcal{I}(\Gamma)$. Throughout this paper, we use the shorthand notation $x(S) = \sum_{i \in S} x_i$. Given a payoff vector $x \in \mathcal{I}(\Gamma)$, the excess of coalition $S \subseteq N$ with respect to x is defined as: $e(x, S) = x(S) - \gamma(S)$. This value measures the degree of S's satisfaction with the payoff x.

Core. The *core* of a game Γ, denoted by $\mathcal{C}(\Gamma)$, is the set of payoff vectors satisfying that, $x \in \mathcal{C}(\Gamma)$ if and only if $e(x, S) \geq 0$ for all $S \subseteq N$. These constraints, called group rationality, ensure that no coalition would have an incentive to split from the grand coalition N, and do better on its own.

Least-core. When $\mathcal{C}(\Gamma)$ is empty, it is meaningful to relax the group rationality constraints by $e(x, S) \geq \varepsilon$ for all $S \subseteq N$ ($\varepsilon < 0$). We shall find the maximum value ε^* such that the set $\{x \in \mathcal{I}(\Gamma) : e(x, S) \geq \varepsilon^*, \forall S \subseteq N\}$ is nonempty. This set of imputations is called the *least-core*, denoted by $\mathcal{LC}(\Gamma)$, and ε^* is called the value of $\mathcal{LC}(\Gamma)$ or $\mathcal{LC}$-value.

Nucleolus. Now we turn to the concept of the *nucleolus*. A payoff vector x generates a 2^n-dimensional excess vector $\theta(x) = (e(x, S_1), \cdots, e(x, S_{2^n}))$, whose components are arranged in a non-decreasing order. That is, $e(x, S_i) \leq e(x, S_j)$ for $1 \leq i < j \leq 2^n$. The nucleolus, denoted by $\eta(\Gamma)$, is defined to be a payoff vector that lexicographically maximizes the excess vector $\theta(x)$ over the set of imputations $\mathcal{I}(\Gamma)$. It was proved by Schmeidler [19] that the nucleolus of a game with the nonempty imputation set contains exactly one element.

Monotone Games and Simple Games. A game $\Gamma = (N, \gamma)$ is *monotone* if $\gamma(S') \leq \gamma(S)$ whenever $S' \subseteq S$. A game is called a simple game if it is a monotonic game with $\gamma : 2^N \to \{0, 1\}$ such that $\gamma(\emptyset) = 0$ and $\gamma(N) = 1$. Simple games can be usually used to model situations where there is a task to be completed, a coalition is labeled as winning if and only if it can complete the task. Formally, coalition $S \subseteq N$ is *winning* if $\gamma(S) = 1$, and *losing* if $\gamma(S) = 0$. A player i is called a *veto* player if he or she belongs to all winning coalitions. It is easy to see that, in a simple game, i is a veto player if and only if $\gamma(N) = 1$ but $\gamma(N \setminus \{i\}) = 0$.

For simple games, Osborne[16] and Elkind *et al.* [7] gave the following result on the core and the nucleolus.

Lemma 1. *A simple game* $\Gamma = (N, \gamma)$ *has a nonempty core if and only if there exists a veto player. Moreover,*

1. $x \in \mathcal{C}(\Gamma)$ if and only if $x_i = 0$ for each $i \in N$ who is not a veto player;
2. when $\mathcal{C}(\Gamma) \neq \emptyset$, the nucleolus of Γ is given by $x_i = \frac{1}{k}$ if i is a veto player and $x_i = 0$ otherwise, where k is the number of veto players.

CS-core. Taking coalition structure into consideration, we can arrive at another solution concept, *CS-core*. Given a cooperative game $\Gamma = (N, \gamma)$, a coalition structure over N is a partition of N, *i.e.*, a collection of subsets $CS = \{C^1, \cdots, C^k\}$ with $\cup_{j=1}^k C^j = N$ and $C^i \cap C^j = \emptyset$ for $i \neq j$ and $i, j \in \{1, \cdots, k\}$. A vector $x = (x_1, \cdots, x_n)$ is a payoff vector for a coalition structure $CS = \{C^1, \cdots, C^k\}$ if $x_i \geq 0$ for all $i \in N$, and $x(C^j) = \gamma(C^j)$ for each $j \in \{1, \cdots, k\}$.

In general, an outcome of the game Γ is a pair (CS, x), where CS is a coalition structure and x is a corresponding payoff vector. The *CS-core* of the game $\Gamma = (N, \gamma)$, denoted by $\mathcal{C}_{cs}(\Gamma)$, is the set of outcomes (CS, x) satisfying the constraints of "group rationality". That is,

$$\mathcal{C}_{cs}(\Gamma) = \{(CS, x) : \forall C \in CS, x(C) = \gamma(C) \text{ and } \forall S \subseteq N, x(S) \geq \gamma(S)\}.$$

A stronger property that is also enjoyed by many practically useful games is superadditivity. The game $\Gamma = (N, \gamma)$ is *superadditive* if it satisfies $\gamma(S_1 \cup S_2) \geq \gamma(S_1) + \gamma(S_2)$ for every pair of disjoint coalitions $S_1, S_2 \subseteq N$. This implies that the agents can earn at least as much profit by working together within the grand coalition. Therefore, for superadditive games, it is always assumed that the agents form the grand coalition. For a (non-superadditive) game $\Gamma = (N, \gamma)$, we can define a new game $\Gamma^* = (N, \gamma^*)$ by setting

$$\gamma^*(S) = \max_{CS \in \mathcal{CS}_S} \gamma(CS), \ \forall S \subseteq N$$

where $\mathcal{CS}_S$ denotes the space of all coalition structures over S and $\gamma(CS) = \sum_{C \in CS} \gamma(C)$. It is easy to verify that the game Γ^* is superadditive, and it is called the *superadditive cover* of Γ. The relationship between the CS-core of Γ and the core of its superadditive cover Γ^* is presented in the following lemma [4,12].

Lemma 2. *A cooperative game $\Gamma = (N, \gamma)$ has nonempty CS-core if and only if its superadditive cover $\Gamma^* = (N, \gamma^*)$ has a non-empty core. Moreover, if $\mathcal{C}(\Gamma^*) \neq \emptyset$, then $\mathcal{C}_{cs}(\Gamma) = \mathcal{C}(\Gamma^*)$.*

3   Path Cooperative Game and Its Core

Let $D = (V, E; s, t)$ be a connected flow network with unit arc capacity (called unit flow network), where V is the vertex set, E is the arc set, $s, t \in V$ are the source and the sink of the network respectively. In this paper, an (s, t)-*path* is referred to as a *directed* path from s to t that visits each vertex in V at most once.

Let $U, W \subseteq V$ be a partition of the vertex set V such that $s \in U$ and $t \in W$, then the set of arcs with tails in U and heads in W is called an (s, t)-*edge-cut*, denoted by $\bar{E} \subseteq E$. An (s, t)-*vertex-cut* is a vertex subset $\bar{V} \subseteq V \setminus \{s, t\}$ such that $D \setminus \bar{V}$ is disconnected. An (s, t)-edge(vertex)-cut is minimum if its cardinality is minimum. In the remainder of the paper, (s, t)-edge(vertex)-cuts will be abbreviated as edge(vertex)-cut S for short. Given an edge-cut $\bar{E}$, we denote its indicator vector by $\mathcal{H}_{\bar{E}} \in \{0, 1\}^{|E|}$, where $\mathcal{H}_{\bar{E}}(e) = 1$ if $e \in \bar{E}$, and 0 otherwise. The indicator vector of a vertex-cut is defined analogously.

Now we introduce two kinds of path cooperative games (PC-games), *edge path cooperative games* and *vertex path cooperative games*.

Definition 1 (Path cooperative game, PC-game). *Let $D = (V, E; s, t)$ be a unit flow network.*

1. *The associated edge path cooperative game (EPC-game) $\Gamma_E = (E, \gamma_E)$ is:*
   - *The player set is E;*
   - $\forall S \subseteq E,$ $\begin{cases} \gamma_E(S) = 1 & \text{if } D[S] \text{ admits an } (s, t)\text{-path}; \\ \gamma_E(S) = 0 & \text{otherwise.} \end{cases}$
   Here, $D[S]$ denotes the induced subgraph with vertex set V and edge set S.

2. *The associated vertex path cooperative game(VPC-game) $\Gamma_V = (V, \gamma_V)$ is:*
   - *The player set is $V \setminus \{s, t\}$;*
   - $\forall T \subseteq V,$ $\begin{cases} \gamma_V(T) = 1 & \text{if induced subgraph } D[T] \text{ admits an } (s, t)\text{-path}; \\ \gamma_V(T) = 0 & \text{otherwise.} \end{cases}$

Clearly, PC-games fall into the class of simple games. Therefore, we can get the necessary and sufficient condition of the non-emptiness of the core directly from Lemma 1.

Proposition 1. *Given an EPC-game Γ_E and a VPC-game Γ_V associated with network $D = (V, E; s, t)$, then*

1. *$\mathcal{C}(\Gamma_E) \neq \emptyset$ if and only if the size of the minimum edge-cut of D is 1;*
2. *$\mathcal{C}(\Gamma_V) \neq \emptyset$ if and only if the size of the minimum vertex-cut of D is 1.*

Moreover, when the core of a PC-game is nonempty, the only edge (vertex) in the edge(vertex)-cut is a veto player, both the core and the nucleolus can be given directly. In the following two sections, we only consider PC-games with empty core.

We note that PC-games also have a natural correspondence with *flow games* and in what follows, we will reveal the close relationship between flow games and PC-games. Let $D = (V, E; s, t)$ be a unit flow network. Given $N \subseteq E$, each edge $e \in N$ is controlled by one player, *i.e.*, we can identify the set of edges N with the set of players. Edges not under control of any players, in $E \setminus N$, are called public arcs; they can be used freely by any coalition. Thus, a unit flow network with player set N is denoted as $D\langle N \rangle = (V, E; s, t)$

Definition 2 (Simple flow game). *The simple flow game $\Gamma_f \langle N \rangle = (N, \gamma)$ associated with the unit network $D\langle N \rangle$ is defined as:*

1. *The player set is N;*
2. *$\forall S \subseteq N$, $\gamma(S)$ is the value of the max-flow from s to t in $D[S \cup (E \setminus N)]$ (using only the edges in S and public edges).*

Flow game is a classical combinatorial optimization game, which has been extensively studied. The core of the flow game $\Gamma_f\langle N\rangle$ is nonempty if and only if there is a minimum edge-cut without public edges [18]. In this case, the core is exactly the convex hull of the indicator vectors of minimum edge-cuts without public edges in D [13,18], and the nucleolus can also be computed in polynomial time [5,17].

Now we turn to discuss the *CS*-core of PC-games. It is easy to see that for the network D without public edges, the associated flow game is the superadditive cover of the corresponding EPC-game. Thus, the nonemptiness of *CS*-core of EPC-game follows directly from Lemma 2.

Proposition 2. *Given an EPC-game Γ_E associated with network $D = (V, E; s, t)$, then the CS-core of Γ_E is nonempty and it is exactly the convex hull of the indicator vectors of minimum edge-cuts of D.*

For a VPC-game, we can also establish some relationship with a flow game. Given a network $D = (V, E; s, t)$, we transform it into a new network D_V in the following way.

(1) For each $v \in V \setminus \{s, t\}$, split it into two distinct vertices v' and v'';

(2) Connect v' and v'' by a new directed edge $e_v = (v', v'')$. The set of all such edges is denoted by E_V;

(3) For original edge $e = (u, v) \in E$, transform it into a new edge $e = (u'', v')$ in D_V ($s = s' = s''$ and $t = t' = t''$).

In the new constructed network D_V, the player set is just the set E_V and all the other edges are viewed as public edges. It is easy to show that in the new network D_V, there must be a minimum edge-cut containing only edges in E_V. Hence, we can verify that the flow game associated with the network $D_V\langle E_V\rangle$ is the superadditive cover of the corresponding VPC-game defined on D. Similarly, the nonemptiness of *CS*-core of VPC-game follows from Lemma 2 and the results of core nonemptiness of flow games.

Proposition 3. *Given an VPC-game Γ_V associated with network $D = (V, E; s, t)$, then the CS-core of Γ_V is nonempty and it is exactly the convex hull of the indicator vectors of minimum vertex-cuts of D.*

4   Least-Core of PC-Games

In this section, we first discuss the least-core of EPC-games. Throughout this section, Γ_E is an EPC-game associated with the network $D = (V, E; s, t)$ with $|E| = n$. Denote by $\mathcal{P}$ the set of all (s,t)-path in D, and $|\mathcal{P}| = m$. According to the definitions of EPC-game and the least-core, it is shown that $\mathcal{LC}(\Gamma_E)$ can be formulated as the following linear program:

$$\max \; \varepsilon$$
$$\text{s.t.} \begin{cases} x(E) = 1 \\ x(P) \geq 1 + \varepsilon & \forall P \in \mathcal{P} \\ x_i \geq 0 & \forall \, i \in E \end{cases} \tag{1}$$

In spite that the number of the constraints in (1) may be exponential in $|E|$, the $\mathcal{LC}$-value and a least-core imputation can be found efficiently by ellipsoid algorithm with a polynomial-time separation oracle: Let (x, ε) be a candidate solution for $\text{LP}(\mathcal{LC}_E)$. We first check whether constraints $x(E) = 1$ and $x(e) \geq 0$ ($\forall e \in E$) are satisfied. Then, checking whether $x(P) \geq 1 + \varepsilon$ ($\forall P \in \mathcal{P}$) are satisfied is transformed to solving the shortest (s,t)-path in D with respect to the edge length $x(e)$ ($\forall e \in E$), and this can aslo be done in polynomial time.

In what follows, we aim at giving a succinct characterization of the least-core for EPC-games. We first give the linear program model of the max-flow problem on D and its dual:

$$\text{LP(flow):} \quad \max \; \sum_{j=1}^{m} y_j$$
$$\text{s.t.} \begin{cases} \displaystyle\sum_{P_j : e_i \in P_j} y_j \leq 1 & i = 1, 2, ..., n \\ y_j \geq 0 & j = 1, 2, ..., m \end{cases} \tag{2}$$

$$\text{DLP(flow):} \quad \min \; \sum_{i=1}^{n} x_i$$
$$\text{s.t.} \begin{cases} \displaystyle\sum_{e_i : e_i \in P_j} x_i \geq 1 & j = 1, 2, ..., m \\ x_i \geq 0 & i = 1, ..., n \end{cases} \tag{3}$$

Due to *max-flow and min-cut* theorem, the optimum value of (2) and (3) are equal, and the set of optimal solutions of (3) is exactly the convex hull of the indicator vectors of the minimum edge-cut of D, which is denoted by $\mathbb{C}_E$. On the other hand, it is known that the core of the flow game Γ_f defined on $D\langle E\rangle$ is also the convex hull of the indicator vectors of the minimum edge-cut of D. Hence, we have

Theorem 1. *Let Γ_E and Γ_f be an EPC-game and a flow game defined on $D = (V, E; s, t)$, respectively, f^* be the value of the max-flow of D. Then,*

$$x \in \mathcal{LC}(\Gamma_E) \text{ if and only if } x = z/f^* \text{ for some } z \in \mathbb{C}_E.$$

Proof. Let $x = (1 + \varepsilon)z$ be a transformation, then (1) can be rewritten as

$$\max \varepsilon$$
$$\text{s.t.} \begin{cases} z(E) = 1/(1 + \varepsilon) \\ z(P) \geq 1 & \forall P \in \mathcal{P} \\ z_i \geq 0 & \forall e_i \in E \end{cases} \tag{4}$$

Combining the first constraint $z(E) = 1/(1 + \varepsilon)$ and the objective function, it is easy to see that linear program (4) is the same as DLP(flow) (3). Since the optimal value of (3) is also f^*, Theorem 1 thus follows. □

Based on the relationship between a VPC-game and the corresponding flow game discussed in Section 3, we can obtain a similar result on the least-core for VPC-games (The proof is omitted).

Theorem 2. *Let $\Gamma_V = (E, \gamma_V)$ be a VPC-game defined on $D = (V, E; s, t)$, f^* be the value of the max-flow of D, then*

$$x \in \mathcal{LC}(\Gamma_V) \text{ if and only if } x = z/f^* \text{ for some } z \in \mathbb{C}_V.$$

Here $\mathbb{C}_V$ is the convex hull of the indicator vectors of minimum vertex-cuts in D.

Theorem 1 and 2 show that for the unit flow network, the least-core of the PC-game is equivalent to the core of the corresponding flow game in the sense of scaling down by $1/f^*$. Hence, all the following problems for PC-games can be solved efficiently:

- Computing the $\mathcal{LC}$-value;
- Finding an imputation in $\mathcal{LC}(\Gamma_E)$ and $\mathcal{LC}(\Gamma_V)$;
- Checking whether a given imputation is in $\mathcal{LC}(\Gamma_E)$ or $\mathcal{LC}(\Gamma_V)$.

Remark. Path cooperative games have close relationship with a noncooperative two-person zero-sum game, called *path intercept game* [21]. In this model, an "evader" attempts to select a path P from the source to the sink through a given network. At the same time, an "interdictor" attempts to select an edge e in this network to detect the evader. If the evader traverses through arc e, he is detected; otherwise, he goes undetected. The interdictor aims to find a probabilistic "edge-inspection" strategy to maximize the average probability of detecting the evader. While for the evader, he wants to find a "path-selection strategy" to minimize the interdiction probability. Aziz *et al.*[2] observed that the mixed Nash Equilibrium of path intercept games is the same as the least-core of EPC-games. With max-min theorem in matrix game theory, the same result can be obtained based on the similar analysis as in the proof of Theorem 1.

5   Nucleolus of PC-games

In this section, we aim at showing that the nucleolus of PC-games can be computed in polynomial time. Given a game $\Gamma = (N, \gamma)$, Kopelowitz [15] showed that the nucleolus $\eta(\Gamma)$ can be obtained by recursively solving the following standard sequence of linear programs $SLP(\eta(\Gamma))$:

$$
\begin{array}{ll}
& \max \varepsilon \\
LP_k & \\
(k = 1, 2, \cdots) \quad \text{s.t.} &
\begin{cases}
x(S) = \gamma(S) + \varepsilon_r, & \forall S \in \mathcal{J}_r \quad r = 0, 1, \cdots, k-1 \\
x(S) \geq \gamma(S) + \varepsilon, & \forall \emptyset \neq S \subset N \setminus \cup_{r=0}^{k-1} \mathcal{J}_r \\
x \in \mathcal{I}(\Gamma).
\end{cases}
\end{array}
$$

Initially, set $\mathcal{J}_0 = \{\emptyset, N\}$ and $\varepsilon_0 = 0$. The number ε_r is the optimal value of the r-th program LP_r, and $J_r = \{S \subseteq N : x(S) = \gamma(S) + \varepsilon_r, \forall x \in X_r\}$, where $X_r = \{x \in R^n : (x, \varepsilon_r) \text{ is an optimal solution of } LP_r\}$.

As in the last section, we first discuss the nucleolus of EPC-games. Let Γ_E be the EPC-game associated with network $D = (V, E; s, t)$ with $|E| = n$, $\mathcal{P}$ be the set of all (s, t)-paths and f^* be the value of the max-flow of D. Denote $\mathcal{E}_\Gamma$ be the set of coalitions consisting of one-edge coalitions and path coalitions, $i.e.$,

$$\mathcal{E}_\Gamma = \{\{e\} : e \in E\} \cup \{P \subseteq E : P \in \mathcal{P} \text{ is an } (s, t)\text{-path}\}.$$

We show that the sequential linear programs $SLP(\eta(\Gamma_E))$ of EPC-game Γ_E can be simplified as follows.

$$LP_k' : \quad \begin{array}{ll} \max & \varepsilon \\ \text{s.t.} & \begin{cases} x(e) = \varepsilon_r, & \forall e \in E_r, r = 0, 1, ..., k-1 \\ x(e) \geq \varepsilon, & \forall e \in E \backslash \bigcup_{r=0}^{k-1} E_r \\ x(P) = 1/f^* + \varepsilon_r, & \forall P \in \mathcal{P}_r, r = 0, 1, ..., k-1 \\ x(P) \geq 1/f^* + \varepsilon, & \forall P \in \mathcal{P} \backslash \bigcup_{r=0}^{k-1} \mathcal{P}_r \\ x(e) \geq 0, & \forall e \in E \\ x(E) = 1. \end{cases} \end{array} \qquad (5)$$

where ε_r is the optimum value of LP_r, $X_r = \{x \in R^n : (x, \varepsilon_r) \text{ is an optimal solution of } LP_r\}$, $\mathcal{P}_r = \{P \in \mathcal{P} : x(P) = 1 + \varepsilon_r, \forall x \in X_r\}$ and $E_r = \{e \in E : x(e) = \varepsilon_r, \forall x \in X_r\}$. Initially, $\varepsilon_0 = 0$, $\mathcal{P}_0 = \emptyset$ and $E_0 = \emptyset$.

Proposition 4. *The nucleolus $\eta(\Gamma_E)$ of EPC-game Γ_E defined on the network $D = (V, E; s, t)$ can be obtained by computing the linear programs LP_k' in (5).*

Proof. Firstly, we show that in sequential linear programs $SLP(\eta(\Gamma))$, only the constraints corresponding to the the coalitions in $\mathcal{E}_\Gamma$ ($i.e.$, the one-edge coalitions and path coalitions) are necessary in determining the nucleolus $\eta(\Gamma_E)$.

In fact, for any winning coalition $S \subseteq N$ (not a path), S can be decomposed into a path P and some edges $E' = S \backslash E(P)$. Then,

$$x(S) - \gamma(S) = x(P) - 1 + \sum_{e \in E'} x(e) \geq x(P) - 1.$$

Since $x(e) \geq 0$ for all $e \in E'$, S cannot be fixed before P or any $e \in E'$. After P and all $e \in E'$ are fixed, S is also fixed, $i.e.$, S is redundant. If S is a losing coalition, then S is a set of edges with $\gamma(S) = 0$ and $x(S) - \gamma(S) = \sum_{e \in S} x(e) \geq x(e), \forall e \in S$. That is to say, S cannot be fixed before any $e \in S$. When all edges in S are fixed, S is fixed accordingly, $i.e.$ S is also redundant in this case. Therefore, deleting all the constraints corresponding to the coalitions not in $\mathcal{E}_\Gamma$ will not change the result of $SLP(\eta(\Gamma))$.

The key point in remainder of the proof is the correctness of the third and the forth constraints in (5), where we replace the original constraints $x(P) = 1 + \varepsilon_r$ and $x(P) \geq 1 + \varepsilon$ in $SLP(\eta(\Gamma))$ with new constraints $x(P) = 1/f^* + \varepsilon_r$ and $x(P) \geq 1/f^* + \varepsilon$, respectively.

In the process of solving the sequential linear programs, the optimal values increase with k. Since $\mathcal{C}(\Gamma_E) = \emptyset$, we know $\varepsilon_1 < 0$. Note that we can always find an optimal solution such that $\varepsilon_1 > -1$ (for example $x(e) = \frac{1}{n}, \forall e \in E$ is a feasible solution of the linear programming of $\mathcal{LC}(\Gamma_E)$).

We can divide the process into two stages. The first stage is the programs with $-1 < \varepsilon_r < 0$. In this case, the constraints $x(e) \geq \varepsilon, \forall e \in E$ cannot affect the optimal solutions of the current programs, because $x(e) \geq 0$. Ignoring the invalid constraints we can get (5) directly.

The second stage is the programs with $\varepsilon_r \geq 0$. When the programs arrive at this stage, we can claim that all paths have been fixed. Otherwise, if there is a path satisfying $x(p) = 1 + \varepsilon_r \geq 1$, then we have $x(p) = 1$ (note $x(E) = 1$), contradicting with the precondition that the value of maximum flow $f^* \geq 2$. We can omit the path constraints in this stage and then this implies (5).

This completes the proof of Proposition 4.   □

In the following, we shall show that the nucleolus of PC-games can be solved in polynomial time. Let $\Gamma_f = (E, \gamma)$ be the flow game defined on the unit flow network $D = (V, E; s, t)$. It is easy to show that the sequential linear programs $LP(\eta(\Gamma_f))$ can be simplified as $\widehat{LP}_k, (k = 1, 2, ...)$:

$$
\widehat{LP}_k : \quad
\begin{aligned}
\max \quad & \varepsilon \\
\text{s.t.} \quad &
\begin{cases}
x(e) = \varepsilon_r & \forall e \in E_r, r = 0, 1, ..., k-1 \\
x(P) = 1 + \varepsilon_r & \forall P \in \mathcal{P}_r, r = 0, 1, ..., k-1 \\
x(e) \geq \varepsilon & \forall e \in E \backslash \bigcup_{r=0}^{k-1} E_r \\
x(P) \geq 1 + \varepsilon & \forall P \in \mathcal{P} \backslash \bigcup_{r=0}^{k-1} \mathcal{P}_r \\
x(E) = f^*,
\end{cases}
\end{aligned}
\tag{6}
$$

where ε_r is the optimum value of $\widehat{LP}_r$, $X_r = \{x \in R^n : (x, \varepsilon_r)$ is an optimal solution of $\widehat{LP}_r\}$, $\mathcal{P}_r = \{P \in \mathcal{P} : x(P) = 1 + \varepsilon_r, \forall x \in X_r\}$ and $E_r = \{e \in E : x(e) = \varepsilon_r, \forall x \in X_r\}$. Initially, $\varepsilon_0 = 0$, $\mathcal{P}_0 = \emptyset$ and $E_0 = \emptyset$.

Deng et al. [5] proved that the sequential linear programs $\widehat{LP}_k, (k = 1, 2...)$ can be transformed to another sequential linear programs with only polynomial number of constraints, and it follows that the nucleolus of flow game $\eta(\Gamma_f)$ can be found efficiently. Futhermore, Potters et al. [17], show that the nucleolus of flow games with public edges can also be found in polynomial time when the core is nonempty. Based on these known results, we discuss the algorithmic problem on the nucleolus of PC-games in the following theorems.

Theorem 3. Let Γ_E and Γ_f be the EPC-game and flow game defined on a unit flow network $D = (V, E; s, t)$, respectively. The nucleolus of Γ_E can be computed in polynomial time. Furthermore,

$$x \in \eta(\Gamma_E) \text{ if and only if } z = x \cdot f^* \in \eta(\Gamma_f),$$

where f^* is the value of the max-flow of D.

Proof. Notice that the dimension of the feasible regions of $LP'_k (k = 1, 2...)$ decreases in each step, so we can complete the process within at most $|N|$ steps.

The key point here is to show that there is a one-to-one correspondence between the optimal solutions of $\widetilde{LP}_k$ (6) and that of LP'_k (5) ($\forall k = 1, 2, \cdots$).

We first prove that if $(z^*, \tilde{\varepsilon}^*)$ is an optimal solution of $\widetilde{LP}_k$ (6), then $(x^*, \varepsilon^*) = (z^*/f^*, \tilde{\varepsilon}^*/f^*)$ is an optimal solution of LP'_k (5).

When $k = 1$, we have $E_0 = \emptyset$, $\mathcal{P}_0 = \emptyset$ in LP'_1. And it is easy to check the feasibility and the optimality of (z^*, ε^*) in LP'_1. To continue the proof recursively, we need to explain $E_1 = \tilde{E}_1$ and $\mathcal{P}_1 = \tilde{\mathcal{P}}_1$, i.e., the constraints which become tight in every iteration are exactly the same in the two linear programs. For each $e \in E$, if $z^*(e) = \tilde{\varepsilon}^*$, then $x^*(e) = z^*(e)/f^* = \tilde{\varepsilon}^*/f^* = \varepsilon^*$. And if $z^*(e) > \tilde{\varepsilon}^*$, then we have $x^*(e) > \varepsilon^*$. Thus, $E_1 = \tilde{E}_1$. $\mathcal{P}_1 = \tilde{\mathcal{P}}_1$ can be shown analogously. The other direction of the result can be shown similarly. That is, the conclusion holds for $k = 1$.

For the rest iterations $k = 2, 3, \cdots$, the proof can be carried out in a same way. Here we omit the detail of the proof. Since the nucleolus of flow game can be found in polynomial time, it follows that the nucleolus of EPC-game is also efficiently solvable.   □

As for the nucleolus of VPC-games, we also show that it is polynomially solvable based on the relationship between a VPC-game and the corresponding flow game demonstrated in Section 3. Due to the space limitation, the proof of the following theroem is omitted.

Theorem 4. *The nucleolus of VPC-games can be solved in polynomial time.*

PC-games on Undirected Networks. Given an undirected network $D = (V, E; s, t)$, we construct a directed network $\overrightarrow{D} = (V, \overrightarrow{E}; s, t)$ derived from D as follows (see the following figure):

1. For edge $e \in E$ with end vertices v_1 and v_2, transform it into two directed edges $\overrightarrow{e}_{v_1} = (v_{11}, v_{12})$ and $\overrightarrow{e}_{v_2} = (v_{21}, v_{22})$;
2. Connect the two directed edges into a directed cycle via two supplemental directed edges $\overrightarrow{e}_1$ and $\overrightarrow{e}_2$.

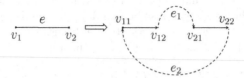

Thus, the EPC-game defined on undirected network $D = (V, E; s, t)$ is transformed to an EPC-game defined on the constructed directed network $\overrightarrow{D} = (V, \overrightarrow{E}; s, t)$. Furthermore, it is easy to check that there exists one-to-one correspondence for the game solution (such as, the core, the least-core and the nucleolus) between the two games. As for a VPC-game defined on an undirected network, we first transform it into EPC-game on an undirected network as demonstrated in Section 3, and then transform it to EPC-game on a directed network in the same way as above. Henceforth, the algorithmic results for PC-games can be generalized from directed networks to undirected networks.

Theorem 5. *Computing the least-core and the nucleolus can be done in polynomial time for both EPC-games and VPC-games defined on undirected networks.*

References

1. Aziz, H., Brandt, F., Harrenstein, P.: Monotone cooperative games and their threshold versions. In: Proceedings of the 9th International Conference on Autonomous Agents and Multiagent Systems, vol. 1, pp. 1107–1114 (2010)
2. Aziz, H., Sørensen, T.B.: Path coalitional games (2011). arXiv preprint arXiv:1103.3310
3. Bachrach, Y.: The least-core of threshold network flow games. In: Murlak, F., Sankowski, P. (eds.) MFCS 2011. LNCS, vol. 6907, pp. 36–47. Springer, Heidelberg (2011)
4. Chalkiadakis, G., Elkind, E., Wooldridge, M.: Computational aspects of cooperative game theory. Synthesis Lectures on Artificial Intelligence and Machine Learning 5(6), 1–168 (2011)
5. Deng, X., Fang, Q., Sun, X.: Finding nucleolus of flow game. Journal of combinatorial optimization 18(1), 64–86 (2009)
6. Deng, X., Papadimitriou, C.H.: On the complexity of cooperative solution concepts. Mathematics of Operations Research 19(2), 257–266 (1994)
7. Elkind, E., Goldberg, L.A., Goldberg, P.W., Wooldridge, M.: Computational complexity of weighted threshold games. In: Proceedings of the National Conference on Artificial Intelligence, vol. 22, p. 718 (2007)
8. Elkind, E., Pasechnik, D.: Computing the nucleolus of weighted voting games. In: Proceedings of the 12th Annual ACM-SIAM Symposium on Discrete Algorithms, pp. 327–335 (2009)
9. Faigle, U., Kern, W., Fekete, S.P., Hochstättler, W.: On the complexity of testing membership in the core of min-cost spanning tree games. International Journal of Game Theory 26(3), 361–366 (1997)
10. Faigle, U., Kern, W., Kuipers, J.: Note computing the nucleolus of min-cost spanning tree games is np-hard. International Journal of Game Theory 27(3), 443–450 (1998)
11. Fang, Q., Zhu, S., Cai, M., Deng, X.: On computational complexity of membership test in flow games and linear production games. International Journal of Game Theory 31(1), 39–45 (2002)
12. Greco, G., Malizia, E., Palopoli, L., Scarcello, F.: On the complexity of the core over coalition structures. In: IJCAI, vol. 11, pp. 216–221. Citeseer (2011)
13. Kalai, E., Zemel, E.: Generalized network problems yielding totally balanced games. Operations Research 30(5), 998–1008 (1982)
14. Kern, W., Paulusma, D.: Matching games: the least-core and the nucleolus. Mathematics of Operations Research 28(2), 294–308 (2003)
15. Kopelowitz, A.: Computation of the kernels of simple games and the nucleolus of n-person games. Technical report, DTIC Document (1967)

16. Osborne, M.J., Rubinstein, A.: A course in game theory. Cambridge, Massachusetts (1994)
17. Potters, J., Reijnierse, H., Biswas, A.: The nucleolus of balanced simple flow networks. Games and Economic Behavior **54**(1), 205–225 (2006)
18. Reijnierse, H., Maschler, M., Potters, J., Tijs, S.: Simple flow games. Games and Economic Behavior **16**(2), 238–260 (1996)
19. Schmeidler, D.: The nucleolus of a characteristic function game. SIAM Journal on applied mathematics **17**(6), 1163–1170 (1969)
20. Solymosi, T., Raghavan, T.E.S.: An algorithm for finding the nucleolus of assignment games. International Journal of Game Theory **23**(2), 119–143 (1994)
21. Washburn, A., Wood, K.: Two-person zero-sum games for network interdiction. Operations Research **43**(2), 243–251 (1995)

Reversible Pebble Game on Trees

Balagopal Komarath[✉], Jayalal Sarma, and Saurabh Sawlani

Department of Computer Science and Engineering,
Indian Institute of Technology Madras, Chennai, India
baluks@gmail.com

Abstract. A surprising equivalence between different forms of pebble games on graphs - Dymond-Tompa pebble game (studied in [4]), Raz-McKenzie pebble game (studied in [10]) and reversible pebbling (studied in [1]) - was established recently by Chan[2]. Motivated by this equivalence, we study the reversible pebble game and establish the following results.

- We give a polynomial time algorithm for computing reversible pebbling number of trees. As our main technical contribution, we show that the reversible pebbling number of any tree is exactly one more than the edge rank colouring of the underlying undirected tree.
- By exploiting the connection with the Dymond-Tompa pebble game, we show that complete binary trees have optimal pebblings that take at most $n^{O(\log \log(n))}$ steps. This substantially improves the previous bound of $n^{O(\log(n))}$ steps.
- Furthermore, we show that *almost optimal* (within $(1 + \epsilon)$ factor for any constant $\epsilon > 0$) pebblings of complete binary trees can be done in polynomial number of steps.
- We also show a time-space tradeoff for reversible pebbling for families of bounded degree trees: for any constant $\epsilon > 0$, such families can be pebbled using $O(n^\epsilon)$ pebbles in $O(n)$ steps. This generalizes a result of Královic[7] who showed the same for chains.

1 Introduction

Pebbling games on graphs of various forms abstracts out resources in different combinatorial models of computation (see [3]). The most obvious connection is to the space used by the computation process. A pebble placed on a vertex in a graph corresponds to storing the value at that vertex and an edge (a, b) in the graph would represent a data-dependency - namely, value at b can be computed only if the value at a is known (or stored). Devising the rules of the pebbling game to capture the moves in the computation, and establishing bounds for the total number of pebbles used at any point in time, give rise to a combinatorial approach to proving bounds on the space used by the computation. The Dymond-Tompa pebble game and the Raz-Mckenzie pebble games depict some of the combinatorial barriers in proving bounds for depth (or parallel time) of Boolean circuits (or parallel algorithms).

© Springer International Publishing Switzerland 2015
D. Xu et al. (Eds.): COCOON 2015, LNCS 9198, pp. 83–94, 2015.
DOI: 10.1007/978-3-319-21398-9_7

Motivated by applications in the context of reversible computation (for example, quantum computation), Bennett[1] introduced the reversible pebbling game. Given any DAG G with a unique sink vertex r, the reversible pebbling game starts with no pebbles on G and ends with a pebble (only) on r. Pebbles can be placed or removed from any vertex according to the following two rules.

1. To pebble v, all in-neighbours of v must be pebbled.
2. To unpebble v, all in-neighbours of v must be pebbled.

The goal of the game is to pebble the DAG G using the minimum number of pebbles (also using the minimum number of steps).

Recently, Chan[2] showed that for any DAG G the number of pebbles required for the reversible pebbling game is exactly the same as the number of pebbles required for the Dymond-Tompa pebble game and the Raz-Mckenzie pebble game. Chan[2] also studied the complexity of the following problem – Given a DAG $G = (V, E)$ with sink r and an integer $1 \leq k \leq |V|$, check if G can be pebbled using at most k pebbles. He showed that this problem is PSPACE-complete.

The irreversible black and black-white pebble games are known to be PSPACE-complete on DAGs (see [5], [6]). When we restrict the irreversible black pebbling game to be read-once (each vertex is pebbled only once), then the problem becomes NP-complete (see [11]). However, if we restrict the DAG to a tree, the irreversible black pebble game[9] and black-white pebble game[13] are solvable in polynomial time. The key insight is that optimal *irreversible* (black or black-white) pebbling number of trees can be achieved by read-once pebblings of trees. This fact simplifies many arguments for irreversible pebblings of trees. For example, deciding whether the pebbling number is at most k is in NP since the optimal pebbling can be used as the certificate. We cannot show that reversible pebbling is in NP using the same argument as we do not know whether the optimal value can always be achieved using pebblings taking only polynomially many steps.

Our Results: In this paper, we study reversible pebblings on trees. We show that the reversible pebbling number of trees along with strategies achieving the optimal value can be computed in polynomial time. Our main technical result is that the reversible pebbling number of any tree is exactly one more than the edge rank colouring of the underlying undirected tree. We then use the linear-time algorithm given by Lam and Yue [8] for finding an optimal edge rank coloring of the underlying undirected tree and show how to convert an optimal edge rank coloring into an optimal reversible pebbling.

Chan[2] also raised the question whether we can find connections between other parameters of different pebbling games. Although, we do not answer this question, we show that the connection with Dymond-Tompa pebble game can be exploited to show that complete binary trees have optimal pebblings that take at most $n^{O(\log \log(n))}$ steps. This is a significant improvement over the trivial $n^{O(\log(n))}$ steps.

Furthermore, we show that "almost" (within $(1 + \epsilon)$ factor for any constant $\epsilon > 0$) optimal pebblings of complete binary trees can be done in polynomial number of steps. We also generalize a time-space tradeoff result given for chains by Královic [7] to families of bounded degree trees showing that for any constant $\epsilon > 0$, such families can be pebbled using $O(n^\epsilon)$ pebbles in $O(n)$ steps.

2   Preliminaries

We assume familiarity with basic definitions in graph theory, such as those found in [12]. A directed tree $T = (V, E)$ is called a *rooted directed tree* if there is an $r \in V$ such that r is reachable from every vertex in T. The vertex r is called the root of the tree.

An *edge rank coloring* of an undirected tree T with k colours $\{1, \ldots, k\}$ labels each edge of T with a colour such that if two edges have the same colour i, then the path between these two edges consists of an edge with some colour $j > i$. The minimum number of colours required for an edge rank colouring of T is denoted by $\chi'_e(T)$.

Definition 1. *(Reversible Pebbling[1]) Let G be a rooted DAG with root r. A reversible pebbling configuration of G is a subset of V which denotes the set of pebbled vertices). A reversible pebbling of G is a sequence of reversible pebbling configurations $\mathcal{P} = (P_1, \ldots, P_m)$ such that $P_1 = \phi$ and $P_m = \{r\}$ and for every $i, 2 \leq i \leq m$, we have*

1. *$P_i = P_{i-1} \cup \{v\}$ or $P_{i-1} = P_i \cup \{v\}$ and $P_i \neq P_{i-1}$ (Exactly one vertex is pebbled/unpebbled at each step).*
2. *All in-neighbours of v are in P_{i-1}.*

The number m is called the time taken by the pebbling $\mathcal{P}$. The number of pebbles or space used in a reversible pebbling of G is the maximum number of pebbles on G at any time during the pebbling. The persistent reversible pebbling number of G, denoted by $R^\bullet(G)$, is the minimum number of pebbles required to pebble G.

A closely related notion is that of visiting reversible pebbling, where the pebbling $\mathcal{P}$ satisfies (1) $P_1 = P_m = \phi$ and (2) there exists a j such that $r \in P_j$. The minimum number of pebbles required for a visiting pebbling of G is denoted by $R^\phi(T)$.

It is easy to see that $R^\phi(G) \leq R^\bullet(G) \leq R^\phi(G) + 1$ for any DAG G.

Definition 2. *(Dymond-Tompa Pebble Game [4]) Let G be a DAG with root r. A Dymond-Tompa pebble game is a two-player game on G where the two players, the pebbler and the challenger take turns. In the first round, the pebbler pebbles the root vertex and the challenger challenges the root vertex. In each subsequent round, the pebbler pebbles a (unpebbled) vertex in G and the challenger either challenges the vertex just pebbled or re-challenges the vertex challenged in the* *previous round. The pebbler wins when the challenger challenges a vertex v and all in-neighbours of v are pebbled.*

The Dymond-Tompa pebble number of G, denoted $DT(G)$, is the minimum number of pebbles required by the pebbler to win against an optimal challenger play.

The Raz-Mckenzie pebble game is also a two-player pebble game played on DAGs. The optimal value is denoted by $RM(G)$. A definition for the Raz-Mckenzie pebble game can be found in [10]. Although the Dymond-Tompa game and the reversible pebbling game look quite different. The following theorem reveals a surprising connection between them.

Theorem 1. *(Theorems 6 and 7, [2]) For any rooted DAG G, we have $DT(G) = R^\bullet(G) = RM(G)$.*

Definition 3. *(Effective Predecessor [2]) Given a pebbling configuration P of a DAG G with root r, a vertex v in G is called an* effective predecessor *of r if there exists a path from v to r with no pebbles on the vertices in the path (except at r).*

Lemma 1. *(Claim 3.11, [2]) Let G be any rooted DAG. There exists an optimal pebbler strategy for the Dymond-Tompa pebble game on G such that the pebbler always pebbles an effective predecessor of the currently challenged vertex.*

We call the above pebbling strategy (resp. pebbler) as an upstream pebbling strategy(resp. upstream pebbler). The height or depth of a tree is defined as the maximum number of vertices in any root to leaf path. We denote by Ch_n the rooted directed path on n vertices with a leaf as the root. We denote by Bt_h the the complete binary tree of height h. We use $root(Bt_h)$ to refer to the root of Bt_h. If v is any vertex in Bt_h, we use $left(v)$ $(right(v))$ to refer to the left (right) child of v. We use $right^i$ and $left^i$ to refer to iterated application of these functions. We use the notation $Ch_i + Bt_h$ to refer to a tree that is a chain of i vertices where the source vertex is the root of a Bt_h.

Definition 4. *We define the language* TREE-PEBBLE *(*TREE-VISITING-PEBBLE*) as the set of all tuples (T, k), where T is a rooted directed tree and k is an integer satisfying $1 \le k \le n$, such that $R^\bullet(T) \le k$ $(R^\phi(T) \le k)$.*

In the rest of the paper, we use the term pebbling to refer to *persistent reversible pebbling* unless explicitly stated otherwise.

3   Main Theorem

Definition 5. *(Strategy Tree) Let T be a rooted directed tree. If T only has a single vertex v, then any strategy tree for T only has a single vertex labelled v. Otherwise, we define a strategy tree for T as any tree satisfying*

1. *The root vertex is labelled with some edge $e = (u, v)$ in T.*

2. *The left subtree of root is a strategy tree for T_u and the right subtree is a strategy tree for $T \setminus T_u$.*

The following properties are satisfied by any strategy tree S of $T = (V, E)$.

1. Each vertex has 0 or 2 children.
2. There are bijections from E to internal vertices of S & from V to leaves of S.
3. Let v be any vertex in S. Then the subtree S_v corresponds to the subtree of T spanned by the vertices labelling the leaves of S_v. If u and v are two vertices in S such that one is not an ancestor of the other, then the subtrees in T corresponding to u and v are vertex-disjoint.

Lemma 2. *Let T be a rooted directed tree. Then $R^\bullet(T) \le k$ if and only if there exists a strategy tree for T of depth at most k.*

Proof. We prove both directions by induction on $|T|$. If T is a single vertex tree, then the statement is trivial.

(if) Assume that the root of a strategy tree for T of depth k is labelled by an edge (u, v) in T. The pebbler then pebbles the vertex u. If the challenger challenges u, the pebbler follows the strategy for T_u given by the left subtree of root. If the challenger rechallenges, the pebbler follows the strategy for $T \setminus T_u$ given by the right subtree of the root. The remaining game takes at most $k - 1$ pebbles by the inductive hypothesis. Therefore, the total number of pebbles used is at most k.

(only if) Consider an upstream pebbler that uses at most k pebbles. We are going to construct a strategy tree of depth at most k. Assume that the pebbler pebbles u in the first move where $e = (u, v)$ is an edge in T. Then the root vertex of S is labelled e. Now we have $R^\bullet(T_u), R^\bullet(T \setminus T_u) \le k - 1$. Let the left (right) subtree be the strategy tree obtained inductively for T_u ($T \setminus T_u$). Since the pebbler is upstream, the pebbler never places a pebble outside T_u ($T \setminus T_u$) once the challenger has challenged u (the root). $\square$

Definition 6. *(Matching Game) Let U be an undirected tree. Let $T_1 = U$. At each step of the matching game, we pick a matching M_i from T_i and contract all the edges in M_i to obtain the tree T_{i+1}. The game ends when T_i is a single vertex tree. We define the* contraction number *of U, denoted $c(U)$, as the minimum number of matchings in the matching sequence required to contract U to the single vertex tree.*

Lemma 3. *Let T be a rooted directed tree and let U be the underlying undirected tree for T. Then $R^\bullet(T) = k + 1$ if and only if $c(U) = k$.*

Proof. First, we describe how to construct a matching sequence of length k from a strategy tree S of depth $k + 1$. Let the leaves of S be the level 0 vertices. For $i \ge 1$, we define the level i vertices to be the set of all vertices v in S such that one child of v has level $i - 1$ and the other child of v has level at most $i - 1$.

Define M_i to be the set of all edges in U corresponding to level i vertices in S. We claim that $M_1, \ldots, M_k$ is a matching sequence for U. Define S_i as the set of all vertices v in S such that the parent of v has level at least $i + 1$ (S_k contains only the root vertex). Let $Q(i)$ be the statement "T_{i+1} is obtained from T_1 by contracting all subtrees corresponding to vertices (see Property 3) in S_i". Let $P(i)$ be the statement "M_{i+1} is a matching in T_{i+1}". We will prove $Q(0)$ and $Q(i) \implies P(i)$ and $(Q(i) \wedge P(i)) \implies Q(i+1)$. Indeed for $i = 0$, we have $Q(0)$ because $T_1 = U$ and S_0 is the set of all leaves in S or vertices in T (Property 2). To prove $Q(i) \implies P(i)$, observe that the edges of M_{i+1} correspond to vertices in S where both children are in S_i. So these edges correspond to edges in T_{i+1} (by $Q(i)$) and these edges are pairwise disjoint since no two vertices in S have a common child.

To prove that $(Q(i) \wedge P(i)) \implies Q(i+1)$, consider the tree T_{i+2} obtained by contracting M_{i+1} from T_{i+1}. Since $Q(i)$ is true, this is equivalent to contracting all subtrees corresponding to S_i and then contracting the edges in M_{i+1} from T_1. The set S_{i+1} can be obtained from S_i by adding all vertices in S corresponding to edges in M_{i+1} and then removing both children (of these newly added vertices) from S_i. This is equivalent to combining the subtrees removed from S_i using the edge joining them. This is because M_{i+1} is a matching by $P(i)$ and hence one subtree in S_i will never be combined with two other subtrees in S_i. But then contracting subtrees in S_{i+1} from T_1 is equivalent to contracting S_i followed by contracting M_{i+1}.

We now show that a matching sequence of length at most k can be converted to a strategy tree of depth at most $k + 1$. We use proof by induction. If the tree T is a single vertex tree, then the statement is trivial. Otherwise, let e be the edge in the last matching M_k in the sequence and let (u, v) be the corresponding edge in T. Label the root of S by e and let the left (right) subtree of root of S be obtained from the matching sequence $M_1, \ldots, M_{k-1}$ restricted to T_u ($T \setminus T_u$). By the inductive hypothesis, these subtrees have height at most $k - 1$.   □

Lemma 4. *For any undirected tree U, we have $c(U) = \chi'_e(U)$.*

Proof. Consider an optimal matching sequence for U. If the edge e is contracted in M_i, then label e with the color i. This is an edge rank coloring. Suppose for contradiction that there exists two edges e_1 and e_2 with label i such that there is no edge labelled some $j \geq i$ between them. We can assume without loss of generality that there is no edge labelled i between e_1 and e_2 since if there is one such edge, we can let e_2 to be that edge. Then e_1 and e_2 are adjacent in T_i and hence cannot belong to the same matching.

Consider an optimal edge rank coloring for U. Then in the ith step all edges labelled i are contracted. This forms a matching since in between any two edges labelled i, there is an edge labelled $j > i$ and hence they are not adjacent in T_i.   □

The theorems in this section are summarized in Fig. 1

Theorem 2. *Let T be a rooted directed tree and let U be the underlying undirected tree for T. Then we have $R^{\bullet}(T) = \chi'_e(U) + 1$.*

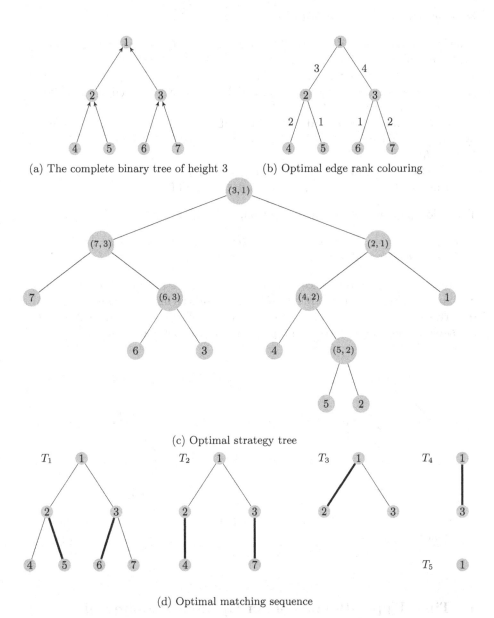

(a) The complete binary tree of height 3   (b) Optimal edge rank colouring

(c) Optimal strategy tree

(d) Optimal matching sequence

Fig. 1. This figure illustrates the equivalence between persistent reversible pebbling, matching game and edge rank coloring on trees by showing an optimal strategy tree and the corresponding matching sequence and edge rank colouring for height 3 complete binary tree

Corollary 1. $R^{\phi}(T)$ and $R^{\bullet}(T)$ along with strategy trees achieving the optimal pebbling value can be computed in polynomial time for trees.

Proof. We show that TREE-PEBBLE and TREE-VISITING-PEBBLE are polynomial time equivalent. Let T be an instance of TREE-PEBBLE. Pick an arbitrary leaf v of T and root the tree at v. By Theorem 2, the reversible pebbling number of this tree is the same as that of T. Let T' be the subtree rooted at the child of v. Then we have $R^{\bullet}(T) \leq k \iff R^{\phi}(T') \leq k - 1$.

Let T be an instance of TREE-VISITING-PEBBLE. Let T' be the tree obtained by adding the edge (r, r') to T where r is the root of T. Then we have $R^{\phi}(T) \leq k \iff R^{\bullet}(T') \leq k + 1$.

The statement of the theorem follows from Theorem 2 and the linear-time algorithm for finding an optimal edge rank coloring of trees[8]. $\square$

The following corollary is immediate from Theorem 1.

Corollary 2. For any rooted directed tree T, we can compute $DT(T)$ and $RM(T)$ in polynomial time.

An interesting consequence of Theorem 2 is that the persistent reversible pebbling number of a tree depends only on its underlying undirected graph. We remark that this does not generalize to DAGs. Below we show two DAGs with the same underlying undirected graph and different pebbling numbers.

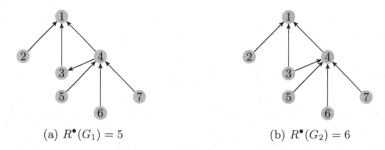

(a) $R^{\bullet}(G_1) = 5$ (b) $R^{\bullet}(G_2) = 6$

Fig. 2. DAGs G_1 and G_2 have the same underlying undirected graph and different persistent pebbling numbers

4   Time Upper-Bound for an Optimal Pebbling of Complete Binary Trees

Proposition 1. The following statements hold.

1. $R^{\bullet}(Bt_h) \geq R^{\bullet}(Bt_{h-1}) + 1$
2. $R^{\bullet}(Bt_h) \geq h + 2$ for $h \geq 3$
3. ([1]) $R^{\bullet}(Ch_n) \leq \lceil \log_2(n) \rceil + 1$ for all n

Proof. (1) In any persistent pebbling of Bt_h, consider the earliest time after pebbling the root at which one of the subtrees of the root vertex has $R^\phi(Bt_{h-1})$ pebbles. At this time, there is a pebble on the root and there is at least one pebble on the other subtree of the root vertex. So, in total, there are at least $R^\phi(Bt_{h-1}) + 2 \geq R^\bullet(Bt_{h-1}) + 1$ pebbles on the tree.

(2) Item (1) and the fact that $R^\bullet(Bt_3) = 5$. □

Theorem 3. *There exists an optimal pebbling of Bt_h that takes at most $n^{O(\log \log(n))}$ steps.*

Proof. We will describe an optimal upstream pebbler in a pebbler-challenger game who pebbles $root(Bt_h)$, $left(root(Bt_h))$, $left(right(root(Bt_h)))$ and so on. In general, the pebbler pebbles $left(right^{i-1}(root(Bt_h)))$ in the ith step for $1 \leq i < h - \log(h)$. An upper bound on the number of steps taken (denoted by $t(h)$) by the reversible pebbling obtained from this game (which is, recursively pebble $left(right^{i-1}(root(Bt_h)))$ for $0 \leq i < h - \log(h)$ and optimally pebble the remaining tree $Ch_{h-\log(h)} + Bt_{\log(h)}$ using any algorithm) is given below. Here the term $(2h - \log(h) + 1)^{3\log(h)}$ is an upper bound on the number of different pebbling configurations with $3\log(h)$ pebbles, and therefore an upper bound for time taken for optimally pebbling the tree $Ch_{h-\log(h)} + Bt_{\log(h)}$.

$$
\begin{aligned}
t(h) &\leq 2\left[t(h-1) + t(h-2) + \ldots + t(\log(h)+1)\right] + (2h - \log(h) + 1)^{3\log(h)} \\
&\leq 2ht(h-1) + (2h - \log(h) + 1)^{3\log(h)} \\
&= O\left((2h)^h(2h)^{3\log(h)}\right) \\
&= (\log(n))^{O(\log(n))} = n^{O(\log \log(n))}
\end{aligned}
$$

In the first step, the pebbler will place a pebble on $left(root(Bt_h))$ and the challenger will re-challenge the root vertex. These moves are optimal. Before the ith step, the tree has pebbles on the root and $left(right^j(root(Bt_h)))$ for $0 \leq j < i - 1$. We argue that if $i < h - \log(h)$, placing a pebble on $left(right^{i-1}(root(Bt_h)))$ is an optimal move. If the pebbler makes this move, then the cost of the game is $\max(R^\bullet(Bt_{h_1-1}), R^\bullet(Ch_i + Bt_{h_1-1})) = R^\bullet(Ch_i + Bt_{h_1-1}) \leq R^\bullet(Bt_{h_1-1}) + 1 = p$, where $h_1 = h - i + 1$. Note that the inequality here is true when $i < h - \log(h)$ by Prop 1. We consider all other possible pebble placements on i^{th} step and prove that all of them are inferior.

- A pebble is placed on the path from the root to $right^{i-1}(root(Bt_h))$ *(inclusive)*: The challenger will challenge the vertex on which this pebble is placed. The cost of this game is then at least $R^\bullet(Bt_{h_1}) \geq p$.
- A pebble is placed on a vertex with height less than $h_1 - 1$: The challenger will re-challenge the root vertex and the cost of the game is at least $R^\bullet(Ch_i + Bt_{h_1-1})$.

The theorem follows. □

5   Almost Optimal Pebblings of Complete Binary Trees

In this section, we show that we can get arbitrarily close to optimal pebblings for complete binary trees using a polynomial number of steps.

Theorem 4. *For any constant $\epsilon > 0$, we can pebble Bt_h using at most $(1 + \epsilon)h$ pebbles and $n^{O(\log(1/\epsilon))}$ steps for sufficiently large h.*

Proof. Let $k \geq 1$ be an integer. Then consider the following pebbling strategy parameterized by k.

1. Recursively pebble the subtrees rooted at $left(right^i(root(Bt_h)))$ for $0 \leq i \leq k - 1$ and $right^k(root(Bt_h))$.
2. Leaving the $(k + 1)$ pebbles on the tree (from the previous step), pebble the root vertex using an additional k pebbles in $2k - 1$ steps.
3. Retaining the pebble on the root, reverse step (1) to remove every other pebble from the tree.

The number of pebbles and the number of steps used by the above strategy on Bt_h for sufficiently large h is given by the following recurrences.

$$S(h) \leq S(h - k) + (k + 1) \leq \frac{(k + 1)}{k} h$$

$$T(h) \leq 2 \left[\sum_{i=1}^{k} T(h - i) \right] + (2k + 2) \leq (2k)^h (2k + 2) \leq n^{\log(k)+1}(2k + 2)$$

where n is the number of vertices in Bt_h.

If we choose $k > 1/\epsilon$, then the theorem follows.   □

6   Time-Space Trade-Offs for Bounded-Degree Trees

In this section, we study time-space trade-offs for bounded-degree trees.

Theorem 5. *For any constant positive integer k, a bounded-degree tree T consisting of n vertices can be pebbled using at most $O\left(n^{1/k}\right)$ pebbles and $O\left(2^k n\right)$ pebbling moves.*

Proof. Let us prove this by induction on the value of k. In the base case ($k = 1$), we are allowed to use $O(n)$ pebbles. So, the best strategy would to place a pebble on every vertex of T in bottom-up fashion, starting from the leaf vertices. After the root is pebbled, we unpebble each vertex in exactly the reverse order, while leaving the root pebbled.

In this strategy, clearly, each vertex is pebbled and unpebbled at most once. Hence the number of pebbling moves must be bounded by $2n$. Hence, a tree can be pebbled using $O(n)$ pebbles in $O(2n)$ moves.

Now consider that for $k \leq k_0 - 1$, where k_0 is an integer ≥ 2, any bounded-degree tree T with n vertices can be pebbled using $O\left(n^{1/k}\right)$ pebbles in $O\left(2^k n\right)$ moves. Assume that we are allowed $O\left(n^{1/k_0}\right)$ pebbles. To apply induction, we will be decomposing the tree into smaller components. We prove the following.

Claim. Let T' be any bounded-degree tree with $n' > n^{(k_0-1)/k_0}$ vertices and maximum degree Δ. There exists a subtree T'' of T' such that the number of vertices in T'' is at least $\lfloor n^{(k_0-1)/k_0}/2 \rfloor$ and at most $\lceil n^{(k_0-1)/k_0} \rceil$.

Proof. From the classical tree-separator theorem, we know that T' can be divided into two subtrees, where the larger subtree has between $\lfloor n'/2 \rfloor$ and $\left\lceil n' \cdot \dfrac{\Delta}{\Delta+1} \right\rceil$ vertices. The key is to recursively subdivide the tree in this way and continually choose the larger subtree. However, we need to show that in doing this we will definitely strike upon a subtree with the number of vertices within the required range. Let $T_1', T_2', \ldots$ be the sequence of subtrees we obtain in these iterations. Also let n_i be the number of vertices in T_i' for every i. Note that $\forall i, \lfloor n_i/2 \rfloor \leq n_{i+1} \leq \left\lceil v_i \cdot \dfrac{\Delta}{\Delta+1} \right\rceil$. Assume that j is the last iteration where $n_j > \lceil n^{(k_0-1)/k_0} \rceil$. Clearly $n_{j+1} \geq \lfloor n^{(k_0-1)/k_0}/2 \rfloor$. Also, by the definition of j, $n_{j+1} \leq \lceil n^{(k_0-1)/k_0} \rceil$. Hence the proof. □

The final strategy will be as follows:

1. Separate the tree into $\theta(n^{1/k_0})$ connected subtrees, each containing $\theta(n^{(k_0-1)/k_0})$ vertices. Claim 6 shows that this can always be done.
2. Let us number these subtrees in the following inductive fashion: denote by T_1, the 'lowermost' subtree, i.e. every path to the root of T_1 must originate from a leaf of T. Denote by T_i, the subtree for which every path to the root originates from either a leaf of T or the root of some T_j for $j < i$. Also, let n_i denote the number of vertices in T_i.
3. Pebble T_1 using $O\left(n_1^{1/(k_0-1)}\right) = O\left(n^{1/k_0}\right)$ pebbles. From the induction hypothesis, we know that this can be done using $O\left(2^{k_0-1}n_1\right)$ pebbling moves.
4. Retaining the pebble on the root vertex of T_1, proceed to pebble T_2 in the same way as above. Continue this procedure till the root vertex of T is pebbled. Then proceed to unpebble every vertex other than the root of T by executing every pebble move upto this instant in reverse order.

Now we argue the bounds on the number of pebbles and pebbling moves of the algorithm. Recall that the number of these subtrees is $O\left(n^{1/k_0}\right)$. Therefore, the number of intermediate pebbles at the root vertices of these subtrees is $O\left(n^{1/k_0}\right)$. Additionally, while pebbling the last subtree, $O\left(n^{1/k_0}\right)$ pebbles are used. Therefore, the total number of pebbles at any time remains $O\left(n^{1/k_0}\right)$. Each of the subtrees are pebbled and unpebbled once (effectively pebbled twice). Thus, the total number of pebbling moves is at most $\sum_i 2O\left(2^{k_0-1}n_i\right) = O\left(2^{k_0}n\right)$. □

7 Discussion and Open Problems

We studied reversible pebbling on trees. Although there are polynomial time algorithms for computing black and black-white pebbling numbers for trees, it was unclear, prior to our work, whether the reversible pebbling number for trees could be computed in polynomial time. We also established that almost optimal pebbling can be done in polynomial time. .

We conclude with the following open problems.

- Prove or disprove that there is an optimal pebbling for complete binary trees that takes at most $O\left(n^k\right)$ steps for a fixed k.
- Prove or disprove that the there is a constant k such that optimal pebbling for any tree takes at most $O\left(n^k\right)$ (for black and black-white pebble games, this statement is true with $k = 1$).
- Give a polynomial time algorithm for computing optimal pebblings of trees that take the smallest number of steps.

References

1. Bennett, C.H.: Time/space trade-offs for reversible computation. SIAM Journal of Computing **18**(4), 766–776 (1989)
2. Chan, S.M.: Just a pebble game. In: Proceedings of the 28th Conference on Computational Complexity (CCC), pp. 133–143 (2013)
3. Chan, S.M.: Pebble Games and Complexity. PhD thesis, EECS Department, University of California, Berkeley, August 2013
4. Dymond, P.W., Tompa, M.: Speedups of deterministic machines by synchronous parallel machines. Journal of Computer and System Sciences **30**(2), 149–161 (1985)
5. Gilbert, J.R., Lengauer, T., Tarjan, R.E.: The pebbling problem is complete in polynomial space. SIAM Journal on Computing **9**(3), 513–524 (1980)
6. Hertel, P., Pitassi, T.: The pspace-completeness of black-white pebbling. SIAM J. Comput. **39**(6), 2622–2682 (2010)
7. Král'ovic, R.: Time and space complexity of reversible pebbling. In: Pacholski, L., Ružička, P. (eds.) SOFSEM 2001. LNCS, vol. 2234, p. 292. Springer, Heidelberg (2001)
8. Lam, T.W., Yue, F.L.: Optimal edge ranking of trees in linear time. In: Proc. of the 9th Annual ACM-SIAM Symposium on Discrete Algorithms, pp. 436–445 (1998)
9. Loui, M.C.: The space complexity of two pebbles games on trees. Technical Report MIT/LCS/TM-133, Massachusetts Institute of Technology (1979)
10. Raz, R., McKenzie, P.: Separation of the monotone NC hierarchy. Combinatorica **19**(3), 403–435 (1999). Conference version appeared in proceedings of 38th Annual Symposium on Foundations of Computer Science (FOCS 1997, Pages 234–243)
11. Sethi, R.: Complete register allocation problems. SIAM Journal on Computing, pp. 226–248(1975)
12. West, D.B.: Introduction to Graph Theory, 2nd edn. Prentice Hall, September 2000
13. Yannakakis, M.: A polynomial algorithm for the min-cut linear arrangement of trees. Journal of the ACM **32**(4), 950–988 (1985)

Computational Complexity

Combinations of Some Shop Scheduling Problems and the Shortest Path Problem: Complexity and Approximation Algorithms

Kameng Nip, Zhenbo Wang$^{(\boxtimes)}$, and Wenxun Xing

Department of Mathematical Sciences, Tsinghua University, Beijing, China
zwang@math.tsinghua.edu.cn

Abstract. We study several combinatorial optimization problems which combine the classic shop scheduling problems (open shop scheduling or job shop scheduling) and the shortest path problem. The objective of the considered problem is to select a subset of jobs that forms a feasible solution of the shortest path problem, and to execute the selected jobs on the open shop or job shop machines such that the makespan is minimized. We show that these problems are NP-hard even if the number of machines is two, and they cannot be approximated within a factor of less than 2 if the number of machines is an input unless P = NP. We present several approximation algorithms for these problems.

Keywords: Approximation algorithm · Combination of optimization problems · Job shop · Open shop · Scheduling · Shortest path

1 Introduction

Combinatorial optimization involves many active subfields, e.g. network flows, scheduling, bin packing. Usually these subfields are motivated by various applications or theoretical interests, and separately developed. The development of science and technology makes it possible to integrate manufacturing, service and management. At the same time, the decision-makers always need to deal with problems involving more than one combinatorial optimization problems. For instance, the network monitoring scenario described in [17] and the railway manufacturing scenario [12].

Wang and Cui [17] introduced a problem combining two classic combinatorial optimization problems, namely parallel machine scheduling and the vertex cover problem. The combination problem is to select a subset of jobs that forms a vertex cover, and to schedule it on some identical parallel machines such that the makespan is minimized. This work also inspired the study of the combination of different combinatorial optimization problems.

Flow shop, open shop and job shop are three basic models of multi-stage scheduling problems. Nip and Wang [12] studied a combination problem that combines two-machine flow shop scheduling and the shortest path problem.

© Springer International Publishing Switzerland 2015
D. Xu et al. (Eds.): COCOON 2015, LNCS 9198, pp. 97–108, 2015.
DOI: 10.1007/978-3-319-21398-9_8

They argued that this problem is NP-hard, and proposed two approximation algorithms with worst-case ratio 2 and $\frac{3}{2}$ respectively. Recently, Nip et al. [13] extended the results to the case that the number of flow shop machines is arbitrary. One motivation of this problem is manufacturing rail racks. We plan to build a railway between two cities. How should we choose a feasible path in a map, such that the corresponding rail tracks (jobs) can be manufactured on some shop machines as early as possible? Similar scenarios can be found in telecommunications and other transportation industries. It connects two classic combinatorial optimization problems, say shop scheduling and the shortest path problem. An intuitive question is what will happen if the shop environment is one of the other two well-known shop environments, i.e. open shop and job shop. This is the core motivation for this current work. In this paper, we mainly study two problems: the combination of open shop scheduling and the shortest path problem, and the combination of job shop scheduling and the shortest path problem.

The contributions of this paper are described as follows: (1) we argue that these combination problems are NP-hard even if the number of machines is two, and if the number of machines is an input, these problems cannot be approximated within a factor less than 2 unless P = NP; (2) we present several approximation algorithms with performance ratio summarized as follows in which $\epsilon > 0$ is any constant and μ is the maximum operations per job in job shop scheduling.

Table 1. Performance of our algorithms

Number of Machines	Open Shop	Job Shop
2	FPTAS	$\frac{3}{2} + \epsilon^*$
m (fixed)	PTAS**	$O\left(\frac{\log^2(m\mu)}{\log\log(m\mu)}\right)$
m (input)	m	m

* Assume that each job has at most 2 operations.
** A $(2 + \epsilon)$-approximation algorithm is also proposed.

The rest of the paper is organized as follows. In Section 2, we give a formal definition of the combination problems stated above, and briefly review some related problems and algorithms that will be used subsequently. In Section 3, we study the computational complexity of these combination problems and give an inapproximability result when the number of machines is an input. Section 4 provides several approximation algorithms for these problems. Some concluding remarks are provided in Section 5.

2   Preliminaries

2.1   Problem Description

We first recall the definitions of open shop and the job shop scheduling problems in the literatures.

Given a set of n jobs $J = \{J_1, \cdots, J_n\}$ and m machines $M = \{M_1, \cdots, M_m\}$, each job has several operations. At the same time, each machine can process at most one job and each job can be processed on one machine. In the open shop scheduling problem ($Om||C_{\max}$), each job must be processed on each machine exactly once, but the processing order can be arbitrary (in other words, the sequence of machines through which job passes can differ between jobs). In the job shop scheduling ($Jm||C_{\max}$), the processing order of each job is given in advance, and may differ between jobs. Furthermore, each job is allowed to be processed on the same machine more than once but consecutive operations of the same job must be processed on different machines, and is not necessary to go through all machines in the job shop. The goal of $Om||C_{\max}$ or $Jm||C_{\max}$ is to find a feasible schedule such that the makespan, that is, the completion time of the last stage among all the jobs is minimum.

Now we define the combination problems considered in this paper.

Definition 1 ($Om|$shortest path$|C_{\max}$). *Given a directed graph $G = (V, A)$ with two distinguished vertices $s, t \in V$, and m machines. Each arc $a_j \in A$ corresponds to a job $J_j \in J$. The $Om|$shortest path$|C_{\max}$ problem is to find an $s - t$ directed path P of G, and to schedule the jobs of J_P on the open (job) shop machines such that the minimum makespan over all P, where J_P denotes the set of jobs corresponding to the arcs in P.*

Definition 2 ($Jm|$shortest path$|C_{\max}$). *Given a directed graph $G = (V, A)$ with two distinguished vertices $s, t \in V$, and m machines. Each arc $a_j \in A$ corresponds to a job $J_j \in J$. The $Jm|$shortest path$|C_{\max}$ problem is to find an $s - t$ directed path P of G, and to schedule the jobs of J_P on the job shop machines such that the minimum makespan over all P, where J_P denotes the set of jobs corresponding to the arcs in P.*

Let the number of jobs (arcs) be n, i.e. $|A| = |J| = n$. Let p_{ij} be the processing times for J_j on machine M_i, and μ_{ij} be the frequency of J_j processed on M_i. Notice $\mu_{ij} = 1$ in the open shop.

It is not difficult to see that the shop scheduling problem (open shop or job shop) and the classic shortest path problem are special cases of our problems. For example, consider the following instances with $m = 2$ ([13]). If there is a unique path from s to t in G, as shown in the left of Fig. 1, our problem is the two-machine shop scheduling problem (open shop or job shop). If all the processing times on the second machine are zero, as shown in the right of Fig. 1, then our problem is equivalent to the classic shortest path with respect to the processing times on the first machine. Therefore we say the considered problems are the combinations of the shop scheduling problems and the shortest path problem.

In this paper, we will use the results of some optimization problems that have a similar structure to the classic shortest path problem. We introduce the generalized shortest path problem defined in [13].

Definition 3. *Given a weighted directed graph $G = (V, A, w^1, \cdots, w^K)$ and two distinguished vertices $s, t \in V$ with $|A| = n$, each arc $a_j \in A, j = 1, \cdots, n$ is*

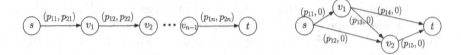

Fig. 1. Special cases of our problems

associated with K weights $w_j^1, \cdots, w_j^K$, and we define vector $w^k = (w_1^k, w_2^k, \cdots, w_n^k)$ for $k = 1, 2, \cdots, K$. The goal of our shortest path problem $SP(G, s, t, f)$ is to find an $s - t$ directed path P that minimizes $f(w^1, w^2, \cdots, w^K; x)$, in which f is a given objective function and $x \in \{0, 1\}^n$ contains the decision variables such that $x_j = 1$ if and only if $a_j \in P$.

For simplicity of notation, we denote SP instead of $SP(G, s, t, f)$ in the rest of the paper. Notice SP is a generalization of various shortest path problems. For example, if we set $K = 1$ and $f(w^1, x) = w^1 \cdot x$, where $\cdot$ is the dot product, it is the classic shortest path problem. If $f(w^1, w^2, \cdots, w^K; x) = \max\{w^1 \cdot x, w^2 \cdot x, \cdots, w^K \cdot x\}$, it is the min-max shortest path problem [1].

2.2   Review of Open Shop and Job Shop Scheduling

Gonzalez and Sahni [5] first gave a linear time optimal algorithm for $O2||C_{\max}$. They also proved that $Om||C_{\max}$ is NP-hard for $m \geq 3$, however whether it is strongly NP-hard is still an outstanding open problem. A feasible shop schedule is called dense when any machine is idle if and only if there is no job that could be processed on it. Rácsmány (see Bárány and Fiala [2]) observed that for any dense schedule, the makespan is at most twice that of the optimal solution, which leads to a greedy algorithm. Sevastianov and Woeginger [15] presented a PTAS for fixed m, which is obtained by dividing jobs into large jobs and small jobs. Their algorithm first optimally schedules the large jobs, then fills the operations of the small jobs into the 'gaps'. In this paper, we will use these algorithms, and refer to them as the GS algorithm, Rácsmány algorithm and the SW algorithm respectively. We present the main results of these algorithms as follows.

Theorem 1 ([5]). *The GS algorithm returns an optimal schedule for $O2||C_{\max}$ in linear time such that* $C_{\max} = \max\left\{\max_{J_j \in J}(p_{1j} + p_{2j}), \sum_{J_j \in J} p_{1j}, \sum_{J_j \in J} p_{2j}\right\}$.

Theorem 2 ([2,16]). *Rácsmány algorithm returns a 2-approximation algorithm for $Om||C_{\max}$ such that $C_{\max} \leq \sum_{J_j \in J} p_{lj} + \sum_{i=1}^{m} p_{ik} \leq 2C_{\max}^*$, where J_k is the last completed job and it is processed on M_l, and $C_{\max}^*$ denotes the optimal makespan.*

Theorem 3 ([15]). *The SW algorithm is a PTAS for $Om||C_{\max}$.*

For job shop scheduling problems, few polynomially solvable cases are known. One is $J2|op \leq 2|C_{\max}$, which can be solved by Jackson's rule [6] that is an extension of Johnson's rule for $F2||C_{\max}$ (flow shop scheduling problem with two machines [8]), where $op \leq 2$ means there are at most 2 operations per job.

In fact, a slight change may lead to NP-hard problems. For instance, $J2|op \leq 3|C_{\max}$ and $J3|op \leq 2|C_{\max}$ are NP-hard [9], $J2|p_{ij} \in \{1,2\}|C_{\max}$ and $J3|p_{ij} = 1|C_{\max}$ are strongly NP-hard [10]. For the general case $J||C_{\max}$, Shmoys, Stein and Wein [16] constructed a randomized approximation algorithm with worst-case ratio $O\left(\frac{\log^2(m\mu)}{\log\log(m\mu)}\right)$, where μ is the maximum number of operations per job. Schmidt, Siegel and Srinivasan [14] obtained a deterministic algorithm with the same bound by derandomizing. We refer to it as the SSW-SSS algorithm. Moreover, for fixed m, the best known approximation algorithm is also proposed in [16] with an approximation factor $2 + \epsilon$, where $\epsilon > 0$ is an arbitrary constant. If μ is a constant, the problem is denoted as $Jm|op \leq \mu|C_{\max}$ and admits a PTAS [7]. We list the main results mentioned above as follows.

Theorem 4 ([6]). *Jackson's rule solves $J2|op \leq 2|C_{\max}$ in $O(n \log n)$ time.*

Theorem 5 ([14,16]). *The SSW-SSS algorithm solves $Jm||C_{\max}$ in polynomial time, and returns a schedule with makespan*

$$O\left(\frac{\log^2(m\mu)}{\log\log(m\mu)}\left(\max_{i \in \{1,\cdots,m\}} \sum_{J_j \in J} \mu_{ij}p_{ij} + \max_{J_j \in J} \sum_{i=1}^{m} \mu_{ij}p_{ij}\right)\right).$$

Furthermore, a well-known inapproximability result is that $O||C_{\max}$, $F||C_{\max}$ and $J||C_{\max}$ cannot be approximated within $\frac{5}{4}$ unless P = NP [18]. Recently, Mastrolilli and Svensson [11] showed that $J||C_{\max}$ cannot be approximated within $O(\log(m\mu)^{1-\epsilon})$ for $\epsilon > 0$ based on a stronger assumption than P $\neq$ NP.

To conclude this subsection, we list some trivial bounds for a dense shop schedule. Denote by $C_{\max}$ the makespan of an arbitrary dense shop schedule with job set J, and we have

$$C_{\max} \geq \max_{i \in \{1,\cdots,m\}} \left\{ \sum_{J_j \in J} \mu_{ij}p_{ij} \right\}, \tag{1}$$

and

$$C_{\max} \leq \sum_{J_j \in J} \sum_{i=1}^{m} \mu_{ij}p_{ij}. \tag{2}$$

For each job, we have

$$C_{\max} \geq \sum_{i=1}^{m} \mu_{ij}p_{ij}, \qquad \forall J_j \in J. \tag{3}$$

2.3 Review of Shortest Path Problems

It is well-known that Dijkstra algorithm solves the classic shortest path problem with nonnegative edge weights in $O(|V|^2)$ time [3]. We have mentioned the min-max shortest path problem, that is NP-hard even for $K = 2$, and Aissi, Bazgan and Vanderpooten proposed an FPTAS if K is a fixed number [1]. We refer to their algorithm as the ABV algorithm, which has the following result.

Theorem 6 ([1]). *Given $\epsilon > 0$, in a directed graph with K nonnegative weights on each arc, where K is a fixed number, the ABV algorithm finds a path P between two specific vertices satisfying* $\max_{i \in \{1,2,\cdots,K\}} \left\{ \sum_{a_j \in P} w_j^i \right\} \leq (1+\epsilon) \max_{i \in \{1,2,\cdots,K\}} \left\{ \sum_{a_j \in P'} w_j^i \right\}$ *for any path P' between the two specified vertices, and the running time is $O(|A||V|^{K+1}/\epsilon^K)$.*

In this paper, sometimes we need to find the min-max shortest path among all the paths visiting some specified arcs if such a path exists. We propose a modified ABV algorithm for this problem, which will be involved in the complete version.

3 Computational Complexity

First, notice that $Om||C_{\max}$ and $Jm||C_{\max}$ are special cases of the corresponding combination problems, thus the combination problem is at least as hard as its component optimization problems. On the other hand, we know that $O2||C_{\max}$ and $J2|op \leq 2|C_{\max}$ are polynomially solvable. However, we can simply verify that the corresponding combination problems, say $O2|$shortest path$|C_{\max}$ and $J2|op \leq 2,$shortest path$|C_{\max}$, are NP-hard by adopting the same reduction proposed in Theorem 2 of [12] for the NP-hardness of $F2|$shortest path$|C_{\max}$. We summarize the results as Theorem 7.

Theorem 7. *$J2|$shortest path$|C_{\max}$ is strongly NP-hard; $O2|$shortest path$|C_{\max}$ and $J2|op \leq 2,$shortest path$|C_{\max}$ are NP-hard.*

Now we consider the case where the number of machines m is part of the input. Williamson et al.[18] showed that it is NP-hard to approximate $O||C_{\max}$, $F||C_{\max}$ or $J||C_{\max}$ within a factor less than $\frac{5}{4}$ by a reduction from the restricted versions of 3-SAT. They also showed that deciding if there is a scheduling of length at most 3 is in P. We show that for these problems combining with shortest path problem, deciding if there is a scheduling of length at most 1 is still NP-hard. Our proof is established by constructing a reduction from 3-Dimensional Matching (3DM) that is NP-complete [4].

Theorem 8. *For $O|$shortest path$|C_{\max}$, deciding if there is a scheduling of length at most 1 is NP-hard.*

Notice that the reduction in Theorem 8 is also valid for $F|$shortest path$|C_{\max}$ and $J|$shortest path$|C_{\max}$, since each job in the reduction has only one nonzero processing time. Therefore we have the following result.

Corollary 1. *The problems $O|$shortest path$|C_{\max}$, $F|$shortest path$|C_{\max}$ and $J|$shortest path$|C_{\max}$ do not admit an approximation algorithm with worst-case ratio less than 2, unless* P $=$ NP.

To our knowledge, the best known inapproximability results based on P $\neq$ NP for $F||C_{\max}$, $O||C_{\max}$ and $J||C_{\max}$ are still $\frac{5}{4}$. The corollary implies that the combination problems of the three shop scheduling problems and the shortest path problem may have stronger inapproximability results than the original problems.

4   Approximation Algorithms

4.1   An Intuitive Algorithm for Arbitrary m

An intuitive algorithm was proposed for $F2|$shortest path$|C_{\max}$ in [12]. The idea is to find the classic shortest path by setting the weight of an arc to be the sum of processing times of its corresponding job, and then schedule the returned jobs by Johnson's rule. This simple idea can be extended to the combination problems we considered, even if the number of machines is an input.

Algorithm 1. The SD algorithm for $O|$shortest path$|C_{\max}$ ($J|$shortest path$|C_{\max}$)

1: Find the shortest path in G with weights $w_j^1 := \sum_{i=1}^m \mu_{ij} p_{ij}$ by Dijkstra algorithm. For the returned path P, construct the job set J_P.

2: Obtain a dense schedule for the jobs of J_P by an arbitrary open (job) shop scheduling algorithm. Let σ be the returned job schedule and $C_{\max}$ the returned makespan, and denote the job set J_P by S.

3: **return** S, σ and $C_{\max}$.

Theorem 9. *For $O|$shortest path$|C_{\max}$ and $J|$shortest path$|C_{\max}$, the SD algorithm is m-approximated, and this bound is tight.*

4.2   A Unified Algorithms for Fixed m

In [12], a $\frac{3}{2}$-approximation algorithm was proposed for $F2|$shortest path$|C_{\max}$. The idea is to iteratively find a feasible path by the ABV algorithm [1] with two weights for each arc, and schedule the corresponding jobs by Johnson's rule, then adaptively modify the weights of arcs and repeat the procedures until we obtain a feasible schedule with good guarantee. We generalize this idea to solve the combination problems considered in this paper. We first propose a unified framework which is denoted as UAR(Alg, ρ, m), where Alg is a polynomial time algorithm used for shop scheduling, ρ is a control parameter to decide the termination rule of the iterations and the jobs to be modified, and m is the number of machines. The pseudocode of the UAR(Alg, ρ, m)algorithm is described by Algorithm 2.

By setting the appropriate scheduling algorithms and control parameters, we can derive algorithms for different combination problems. Notice that at most n jobs are modified in the UAR(Alg, ρ, m) algorithm, therefore the iterations

---

Algorithm 2. Algorithm UAR(Alg, ρ, m)

1: Initially,$(w_j^1, w_j^2, \cdots, w_j^m) := (\mu_{1j}p_{1j}, \mu_{2j}p_{2j}, \cdots, \mu_{mj}p_{mj})$, for $a_j \in A$ corresponding to J_j.

2: Given $\epsilon > 0$, use the ABV algorithm [1] to obtain a feasible path P to SP, and construct the corresponding job set as J_P.

3: Schedule the jobs of J_P by the algorithm Alg, denote the returned makespan as $C'_{\max}$, and the job schedule as σ'.

4: $S := J_P$, $\sigma := \sigma'$, $C_{\max} := C'_{\max}$, $D := \emptyset$, $M := (1 + \epsilon) \sum\limits_{J_j \in J} \sum\limits_{i=1}^{m} \mu_{ij}p_{ij} + 1$.

5: **while** $J_P \cap D = \emptyset$ **and** there exists J_j in J_P satisfying $\sum\limits_{i=1}^{m} \mu_{ij}p_{ij} \geq \rho C'_{\max}$ **do**

6:    **for** all jobs satisfy $\sum\limits_{i=1}^{m} \mu_{ij}p_{ij} \geq \rho C'_{\max}$ in $J \backslash D$ **do**

7:        $(w_j^1, w_j^2, \cdots, w_j^m) := (M, M, \cdots, M)$, $D := D \cup \{J_j\}$.

8:    **end for**

9:    Use the ABV algorithm [1] to obtain a feasible path P to SP, and construct the corresponding job set as J_P.

10:    Schedule the jobs of J_P by the algorithm Alg, denote the returned makespan as $C'_{\max}$, and the job schedule as σ'.

11:    **if** $C'_{\max} < C_{\max}$ **then**

12:        $S := J_P$, $\sigma := \sigma'$, $C_{\max} := C'_{\max}$.

13:    **end if**

14: **end while**

15: **return** S, σ and $C_{\max}$.

---

execute at most n times. Since the scheduling algorithms for shop scheduling and the ABV algorithm [1] are all polynomial time algorithms (for fixed m and ϵ), we claim that the following algorithms based on UAR(Alg, ρ, m) are polynomial-time algorithms. We present the algorithms and prove their performance as follows.

We first apply the UAR(Alg, ρ, m) algorithm to $O2|$shortest path$|C_{\max}$ by setting Alg be the GS algorithm [5] and $\rho = 1$. We refer to this algorithm as the GAR algorithm.

Algorithm 3. The GAR algorithm for $O2|$shortest path$|C_{\max}$

1: Let $m = 2$, Alg be the GS algorithm [5] for $O2||C_{\max}$ and $\rho = 1$.

2: Solve the problem by using UAR(Alg, ρ, m).

Theorem 10. *The GAR algorithm is an FPTAS for $O2|$shortest path$|C_{\max}$.*

We point out that the proofs of the worst-case performance of algorithms based on UAR(Alg, ρ, m) are quite similar. In the following proofs of this subsection, we will only describe the key ideas and main steps since the results can be obtained by analogous arguments. We will adopt the same notations as in the proof of Theorem 10, and also analyze the same two cases.

For $Om|$shortest path$|C_{\max}$ where m is fixed, we obtain the following RAR algorithm based on UAR(Alg, ρ, m) and Rácsmány algorithm [2,16].

Algorithm 4. The RAR algorithm for $Om|$shortest path$|C_{\max}$

1: Let Alg be Rácsmány algorithm [2,16] for $Om||C_{\max}$ and $\rho = \frac{1}{2}$.
2: Solve the problem by using UAR(Alg, ρ, m).

Theorem 11. *Given $\epsilon > 0$, the RAR algorithm is a $(2+\epsilon)$-approximation algorithm for $Om|$shortest path$|C_{\max}$.*

The framework can also be applied to the combination problem of job shop scheduling and the shortest path problem. For the combination of $J2|op \leq 2|C_{\max}$ and the shortest path problem, we obtain a $(\frac{3}{2} + \epsilon)$-approximation algorithm by using Jackson's rule and setting $\rho = \frac{2}{3}$ in the UAR(Alg, ρ, m) algorithm. We refer to this algorithm as the JJAR algorithm, and describe it in Algorithm 5. Recall that all $\mu_{ij} = 1$ in $J2|op \leq 2|C_{\max}$.

Algorithm 5. The JJAR algorithm for $J2|op \leq 2,$ shortest path$|C_{\max}$

1: Let $m = 2$, Alg be Jackson's rule for $J2|op \leq 2|C_{\max}$ and $\rho = \frac{2}{3}$.
2: Solve the problem by using UAR(Alg, ρ, m).

Before studying the worst-case performance of the JJAR algorithm, we establish the following lemma. Let $(1 \to 2)$ $((2 \to 1))$ indicate the order that a job needs to be processed on M_1 (M_2) first and then on M_2 (M_1).

Lemma 1. *For $J2|op \leq 2|C_{\max}$, let $C_{\max}^J$ be the makespan returned by Jackson's rule. Suppose we change the processing order of all jobs to be $(1 \to 2)$ $((2 \to 1))$, and the processing times keep unchanged. Then schedule the jobs by Johnson's rule for $F2||C_{\max}$, and denote the makespan as $C_{\max}^1$ $(C_{\max}^2)$. We have $C_{\max}^J \leq \max\{C_{\max}^1, C_{\max}^2\}$.*

Now we can study the performance of the JJAR algorithm for $J2|op \leq 2,$ shortest path$|C_{\max}$.

Theorem 12. *Given $\epsilon > 0$, the JJAR algorithm is a $(\frac{3}{2} + \epsilon)$-approximation algorithm for $J2|op \leq 2,$ shortest path$|C_{\max}$.*

Finally, we study the general case $Jm|$shortest path$|C_{\max}$, where m is fixed. By Theorem 5, we know that there exists $\alpha > 0$, such that the SSW-SSS algorithm [14,16] returns a schedule satisfying

$$C'_{\max} \leq \alpha \frac{\log^2(m\mu)}{\log\log(m\mu)} \left(\max_{i \in \{1,\cdots,m\}} \sum_{J_j \in J'} \mu_{ij} p_{ij} + \max_{j \in J'} \sum_{i=1}^{m} \mu_{ij} p_{ij} \right). \quad (4)$$

The factor α is decided by choosing the probability of the randomized steps and the subsequent operations in the SSW-SSS algorithm [14,16] [14,16], and its value can be obtained by complicated calculation. Assume we determine such value of α. We can design an approximation algorithm with worst-case ratio $O\left(\frac{\log^2(m\mu)}{\log\log(m\mu)}\right)$ for $Jm|$shortest path$|C_{\max}$. We refer to this algorithm as the SAR algorithm, and describe it in Algorithm 6.

Algorithm 6. The SAR algorithm for $Jm|$shortest path$|C_{\max}$

1: Let Alg be the SSW-SSS algorithm [14,16] for $Jm||C_{\max}$ and $\rho = \frac{\log\log(m\mu)}{2\alpha\log^2(m\mu)}$.
2: Solve the problem by using UAR(Alg, ρ, m).

Theorem 13. *The SAR algorithm is an* $O\left(\frac{\log^2(m\mu)}{\log\log(m\mu)}\right)$*-approximation algorithm for* $Jm|$shortest path$|C_{\max}$.

Remind that the SAR algorithm relies on the assumption, that we can determine the constant α for the SSW-SSS algorithm [14,16]. We can calculate it by following the details of the SSW-SSS algorithm [14,16], and in fact we can choose α large enough to guarantee the performance ratio of our algorithm.

4.3   A PTAS for $Om|$shortest path$|C_{\max}$

In the previous subsection, we introduced a $(2+\epsilon)$-approximation algorithm for $Om|$shortest path$|C_{\max}$ based on the UAR(Alg, ρ, m) algorithm. By a different approach, we propose a $(1+\epsilon)$-approximation algorithm for any $\epsilon > 0$, i.e. a PTAS. We also iteratively find feasible solutions, but guarantee that one of the returned solutions has the same first N-th largest jobs with an optimal solution where N is a given constant. Precisely speaking, we say job J_j is larger than job J_k if $\max\limits_{i\in\{1,\cdots,m\}} p_{ij} > \max\limits_{i\in\{1,\cdots,m\}} p_{ik}$. To do this, we enumerate all size N subsets J^N of J, and then iteratively modify the weights of the graph such that the jobs larger than any job in J^N will not be chosen. Then find a feasible solution which contains all the jobs in J^N corresponding to the modified graph, i.e., the corresponding path is constrained to visit all the arcs corresponding to J^N if such a path exists.

To find a feasible solution in each iteration, we adopt the modified ABV algorithm to obtain a near optimal min-max shortest path among all the paths visiting the arcs corresponding to J^N if such a path exists. Then we schedule the selected jobs by [15] which is denoted as the SW algorithm [15] for $Om||C_{\max}$. We refer to our algorithm as the SAE algorithm, and describe it in Algorithm 7.

There are $\binom{n}{N}$ distinct subsets J^N, thus the iterations between line 4 - line 12 run at most $O(n^N)$ times, that is a polynomial of n since N is a constant when m and ϵ are fixed. Since the modified ABV algorithm is an FPTAS and the SW algorithm [15] is a PTAS, the running time of each iteration is also bounded by the polynomial of n if m and ϵ are fixed. It suffices to show that the SAE algorithm terminates in polynomial time. The following theorem indicates the SAE algorithm is a PTAS.

Algorithm 7. The SAE algorithm for $Om|$shortest path$|C_{\max}$

1: Given $0 < \epsilon < 1$, set $N = m \left(\frac{m(3+\epsilon)}{\epsilon} \right)^{2^{\frac{m(3+\epsilon)}{\epsilon}}}$.

2: Let $D := \emptyset$, $M := (1 + \frac{\epsilon}{3}) \sum_{J_j \in J} \sum_{i=1}^{m} p_{ij} + 1$, $C_{\max} := \sum_{J_j \in J} \sum_{i=1}^{m} p_{ij}$.

3: Initially, $(w_j^1, w_j^2, \cdots, w_j^m) := (p_{1j}, p_{2j}, \cdots, p_{mj})$, for $a_j \in A$ corresponding to J_j.

4: **for** all $J^N \subset J$, with $|J^N| = N$ **do**

5:   $(w_j^1, w_j^2, \cdots, w_j^m) := (p_{1j}, p_{2j}, \cdots, p_{mj})$, $D := \emptyset$.

6:   For jobs $J_k \in J \setminus J^N$ with $\max_{i \in \{1, \cdots, m\}} p_{ik} > \min_{J_j \in J^N} \max_{i \in \{1, \cdots, m\}} p_{ij}$, set
     $(w_k^1, w_k^2, \cdots, w_k^m) := (M, M, \cdots, M)$, $D := D \cup \{J_k\}$.

7:   Use the modified ABV algorithm to obtain a feasible path P of SP such that
     the returned path visits all the arcs corresponding to J^N if such a path exists.
     Construct the corresponding job set as J_P.

8:   Schedule the jobs of J_P by the SW algorithm [15], denote the returned makespan
     as $C'_{\max}$, and the job schedule as σ'.

9:   **if** $C'_{\max} < C_{\max}$ **then**

10:    $S := J_P$, $\sigma := \sigma'$, $C_{\max} := C'_{\max}$.

11:   **end if**

12: **end for**

13: **return** S, σ, $C_{\max}$.

Theorem 14. *The SAE algorithm is a PTAS for $Jm|$shortest path$|C_{\max}$.*

5 Conclusions

This paper studies several problems combining two well-known combinatorial optimization problems. We show the hardness of the problems, and present some approximation algorithms. It is interesting to find approximation algorithms with better worst-case ratios for $J2|op \leq 2$, shortest path$|C_{\max}$ and $Jm|$shortest path$|C_{\max}$. Moreover, it needs further study to close the gap between the 2-inapproximability results and the m-approximation algorithms for $O|$shortest path$|C_{\max}$ and $J|$shortest path$|C_{\max}$. We can also consider other interesting combinations of combinatorial optimization problems.

Acknowledgments. Wang's research has been supported by NSFC No. 11371216 and Bilateral Scientific Cooperation Project between Tsinghua University and KU Leuven. Xing's research has been supported by NSFC No. 11171177.

References

1. Aissi, H., Bazgan, C., Vanderpooten, D.: Approximating min-max (regret) versions of some polynomial problems. In: Chen, D.Z., Lee, D.T. (eds.) COCOON 2006. LNCS, vol. 4112, pp. 428–438. Springer, Heidelberg (2006)

2. Bárány, I., Fiala, T.: Többgépes ütemezési problémák közel optimális megoldása (in Hungarian). Szigma - Matematikai - Közgazdasági Folyóirat **15**, 177–191 (1982)

3. Dijkstra, E.W.: A note on two problems in connexion with graphs. Numerische Mathematik **1**, 269–271 (1959)
4. Garey, M.R., Johnson, D.S.: Computers and Intractability: A Guide to the Theory of NP-completeness. Freeman, San Francisco (1979)
5. Gonzalez, T., Sahni, S.: Open shop scheduling to minimize finish time. Journal of the Association for Computing Machinery **23**, 665–679 (1976)
6. Jackson, J.R.: An extension of Johnson's results on job-lot scheduling. Naval Research Logistics Quarterly **3**, 201–203 (1956)
7. Jansen, K., Solis-Oba, R., Sviridenko, M.: Makespan minimization in job shops: A linear time approximation scheme. SIAM Journal on Discrete Mathematics **16**, 288–300 (2003)
8. Johnson, S.M.: Optimal two- and three-stage production schedules with setup times included. Naval Research Logistics Quarterly **1**, 61–68 (1954)
9. Lenstra, J.K., Kan, A.R., Brucker, P.: Complexity of machine scheduling problems. Annals of Operations Research **1**, 343–362 (1977)
10. Lenstra, J.K., Rinnooy Kan, A.: Computational complexity of discrete optimization. Annals of Operations Research **4**, 121–140 (1979)
11. Mastrolilli, M., Svensson, O.: Hardness of approximating flow and job shop scheduling problems. Journal of the Association for Computing Machinery **58**, 20:1–20:32 (2011)
12. Nip, K., Wang, Z.: Combination of two-machine flow shop scheduling and shortest path problems. In: Du, D.-Z., Zhang, G. (eds.) COCOON 2013. LNCS, vol. 7936, pp. 680–687. Springer, Heidelberg (2013)
13. Nip, K., Wang, Z., Talla Nobibon, F., Leus, R.: A Combination of Flow Shop Scheduling and the Shortest Path Problem. Journal of Combinatorial Optimization **29**, 36–52 (2015)
14. Schmidt, J.P., Siegel, A., Srinivasan, A.: Chernoff-hoeffding bounds for applications with limited independence. SIAM Journal on Discrete Mathematics **8**, 223–250 (1995)
15. Sevastianov, S.V., Woeginger, G.J.: Makespan minimization in open shops: A polynomial time approximation scheme. Mathematical Programming **82**, 191–198 (1998)
16. Shmoys, D.B., Stein, C., Wein, J.: Improved approximation algorithms for shop scheduling problems. SIAM Journal on Computing **23**, 617–632 (1994)
17. Wang, Z., Cui, Z.: Combination of parallel machine scheduling and vertex cover. Theoretical Computer Science **460**, 10–15 (2012)
18. Williamson, D.P., Hall, L.A., Hoogeveen, J.A., Hurkens, C.A.J., Lenstra, J.K., Sevast'janov, S.V., Shmoys, D.B.: Short shop schedules. Operations Research **45**, 288–294 (1997)

Complexity of Grundy Coloring and Its Variants

Édouard Bonnet[1,2], Florent Foucaud[3], Eun Jung Kim[1], and Florian Sikora[1](✉)

[1] PSL, LAMSADE - CNRS UMR 7243, Université Paris-Dauphine, Paris, France
{edouard.bonnet,eunjung.kim,florian.sikora}@dauphine.fr
[2] Hungarian Academy of Sciences, Budapest, Hungary
[3] LIMOS - CNRS UMR 6158, Université Blaise Pascal, Clermont-ferrand, France
florent.foucaud@gmail.com

Abstract. The Grundy number of a graph is the maximum number of colors used by the greedy coloring algorithm over all vertex orderings. In this paper, we study the computational complexity of GRUNDY COLORING, the problem of determining whether a given graph has Grundy number at least k. We show that GRUNDY COLORING can be solved in time $O^*(2.443^n)$ on graphs of order n. While the problem is known to be solvable in time $f(k, w) \cdot n$ for graphs of treewidth w, we prove that under the Exponential Time Hypothesis, it cannot be computed in time $O^*(c^w)$, for any constant c. We also study the parameterized complexity of GRUNDY COLORING parameterized by the number of colors, showing that it is in FPT for graphs including chordal graphs, claw-free graphs, and graphs excluding a fixed minor.

Finally, we consider two previously studied variants of GRUNDY COLORING, namely WEAK GRUNDY COLORING and CONNECTED GRUNDY COLORING. We show that WEAK GRUNDY COLORING is fixed-parameter tractable with respect to the weak Grundy number. In stark contrast, it turns out that checking whether a given graph has connected Grundy number at least k is NP-complete already for $k = 7$.

1 Introduction

A k-*coloring* of a graph G is a surjective mapping $\varphi : V(G) \to \{1, \ldots, k\}$ and we say v is colored with $\varphi(v)$. A k-coloring φ is *proper* if any two adjacent vertices receive different colors in φ. The *chromatic number* $\chi(G)$ of G is the smallest k such that G has a k-coloring. Determining the chromatic number of a graph is the most fundamental problem in graph theory. Given a graph G and an ordering $\sigma = v_1, \ldots, v_n$ of $V(G)$, the *first-fit algorithm* colors vertex v_i with the smallest color that is not present among the set of its neighbors within $\{v_1, \ldots, v_{i-1}\}$. The *Grundy number* $\Gamma(G)$ is the largest k such that G admits a vertex ordering on which the first-fit algorithm yields a proper k-coloring. First-fit is presumably the simplest heuristic to compute a proper coloring of

---

F. Foucaud—This research was done while this author was a postdoctoral fellow at the Department of Mathematics of University of Johannesburg (South Africa) and at LAMSADE.

© Springer International Publishing Switzerland 2015
D. Xu et al. (Eds.): COCOON 2015, LNCS 9198, pp. 109–120, 2015.
DOI: 10.1007/978-3-319-21398-9_9

a graph. In this sense, the Grundy number gives an algorithmic upper bound on the performance of any heuristic for the chromatic number. This notion was first studied by Grundy in 1939 in the context of digraphs and games [11], and formally introduced 40 years later by Christen and Selkow [8]. Many works have studied the first-fit algorithm in connection with on-line coloring algorithms, see e.g. [21]. A natural relaxation of this concept is the *weak Grundy number*, introduced by Kierstead and Saoub [17], where the obtained coloring is not asked to be proper. A more restricted concept is the one of *connected Grundy number*, introduced by Benevides et al. [3], where the algorithm is given an additional "local" restriction: at each step, the subgraph induced by the colored vertices must be connected.

The goal of this paper is to advance the study of the computational complexity of determining the Grundy number, the weak Grundy number and the connected Grundy number of a graph.

Let us introduce the problems formally. Let G be a graph and let $\sigma = v_1, \ldots, v_n$ be an ordering of $V(G)$. A (not necessarily proper) k-coloring $\varphi : V(G) \to \{1, \ldots, k\}$ of G is a *first-fit coloring with respect to* σ if for every vertex v_i and every color c with $c < \varphi(v_i)$, v_i has a neighbor v_j with $\varphi(v_j) = c$ for some $j < i$. In particular, $\varphi(v_1) = 1$. A vertex ordering $\sigma = v_1, \ldots, v_n$ is *connected* if for every i, $1 \leqslant i \leqslant n$, the subgraph induced by $\{v_1, \ldots, v_i\}$ is connected. A k-coloring $\varphi : V(G) \to \{1, \ldots, k\}$ is called the (i) *weak Grundy*, (ii) *Grundy*, (iii) *connected Grundy coloring* of G, respectively, if it is a first-fit coloring with respect to some vertex ordering σ such that (i) φ and σ has no restriction, (ii) φ is proper, (iii) φ is proper and σ is connected, respectively.

The maximum number of colors used in a (weak, connected, respectively) Grundy coloring is called the (*weak, connected*, respectively) Grundy number and is denoted $\Gamma(G)$ ($\Gamma'(G)$ and $\Gamma_c(G)$, respectively). In this paper, we study the complexity of computing these invariants.

GRUNDY COLORING
Input: A graph G, an integer k.
Question: Do we have $\Gamma(G) \geqslant k$?

The problems WEAK GRUNDY COLORING and CONNECTED GRUNDY COLORING are defined analogously.

Note that $\chi(G) \leqslant \Gamma(G) \leqslant \Delta(G) + 1$, where $\chi(G)$ is the chromatic number and $\Delta(G)$ is the maximum degree of G. However, the difference $\Gamma(G) - \chi(G)$ can be (arbitrarily) large, even for bipartite graphs. For example, the Grundy number of the tree of Figure 1 is 4, whereas its chromatic number is 2. Note that this is not the case for Γ_c for bipartite graphs, since $\Gamma_c(G) \leqslant 2$ for any bipartite graph G [3]. However, the difference $\Gamma_c(G) - \chi(G)$ can be (arbitrarily) large even for planar graphs [3].

Previous Results. GRUNDY COLORING remains NP-complete on bipartite graphs [14] and their complements [25] (and hence claw-free graphs and P_5-free graphs), on chordal graphs [23], and on line graphs [13]. Certain graph classes admit polynomial-time algorithms. There is a linear-time algorithm for GRUNDY COLORING on trees [15]. This result was extended to graphs of bounded treewidth by Telle and Proskurowski [24], which proposed a dynamic programming algorithm running in time $k^{O(w)}2^{O(wk)}n = O(n^{3w^2})$ for graphs of treewidth w (in other words, their algorithm is in FPT for parameter $k + w$ and in XP for parameter w).[1] A polynomial-time algorithm for P_4-laden graphs, which contains all cographs as a subfamily, was given in [2].

Note that GRUNDY COLORING admits a polynomial-time algorithm when the number k of colors is fixed [26], in other words, it is in XP for parameter k.

GRUNDY COLORING has polynomial-time constant-factor approximation algorithms for inputs that are interval graphs [12,21], complements of chordal graphs [12], complements of bipartite graphs [12] and bounded tolerance graphs [17]. In general, however, there is a constant $c > 1$ s.t. approximating GRUNDY COLORING within c is impossible unless NP $\subseteq$ RP [18]. It is not known if a polynomial-time $o(n)$-factor approximation algorithm exists.

When parameterized by the graph's order minus the number of colors, GRUNDY COLORING was shown to be in FPT by Havet and Sempaio [14].

CONNECTED GRUNDY COLORING was introduced by Benevides *et al.* [3], who proved it to be NP-complete, even for chordal graphs and for co-bipartite graphs. WEAK GRUNDY COLORING is NP-complete [10].

Our Results. As pointed out in [24], no (extended) monadic second order expression is known for the property "$\Gamma(G) \geqslant k$". Therefore it is not clear whether the algorithm of [24] can be improved, e.g. to an algorithm of running time $f(w) \cdot poly(n)$. Nevertheless, on general graphs, we show that GRUNDY COLORING can be solved in time $O^*(2.443^n)$.

As a lower bound to the positive algorithmic bounds, we show that under the Exponential Time Hypothesis (ETH) [16], an $O(c^w \cdot poly(n))$-time algorithm for GRUNDY COLORING does not exist (for any fixed constant c). Hence the exponent n cannot be replaced by the treewidth in our $O^*(2.443^n)$-time algorithm.

We also study the parameterized complexity of GRUNDY COLORING parameterized by the number of colors, showing that it is in FPT for graphs including chordal graphs, claw-free graphs, and graphs excluding a fixed minor.

Finally, we show that WEAK GRUNDY COLORING and CONNECTED GRUNDY COLORING exhibit opposite computational behavior when viewed through the lense of parameterized complexity (for the parameter "number of colors"). While WEAK GRUNDY COLORING is shown to be FPT on general graphs, CONNECTED GRUNDY COLORING is NP-complete even when $k = 7$, i.e. does not belong to XP (it is the only of the three studied problems to be in this case). Note that the known NP-hardness proof for CONNECTED GRUNDY COLORING was only for an unbounded number of colors [3].

---

[1] The first running time is not explicitly stated in [24] but follows from their meta-theorem. The second one is deduced by the authors of [24] from the first one by bounding k by $w \log_2 n + 1$.

Due to space constraints, some proofs are deferred to the full version of the paper [6].

2 Preliminaries

We defer many (classic) technical definitions to the full version [6], and only give the ones related to Grundy colorings. Given a graph G, a *colored witness* of height ℓ, or simply called an ℓ-witness, is a subgraph G' of G, which comes with a partition $\mathcal{W} = W_1 \uplus \cdots \uplus W_\ell$ of $V(G')$ such that for every i in $1, \ldots, \ell$ (1) $W_i \neq \emptyset$, and (2) W_i is an independent dominating set of $G[W_i \cup \cdots \cup W_\ell]$. The cell W_i under $\mathcal{W}$ is called the *color class* of color i. A witness G' of height ℓ is said to be *minimal* if for every $u \in V(G')$, $G' - u$ with the partition $\mathcal{W}|_{V(G')-\{u\}}$ is not an ℓ-witness.

Observation 1. *For any graph G, $\Gamma(G) \geqslant k$ if and only if G allows a minimal k-witness.*

Observation 2. *A minimal k-witness has a vertex of degree $k - 1$ (the root), order at most 2^{k-1}, and is included in the distance-k neighborhood of the root.*

By these observations, k-GRUNDY COLORING can be solved by checking, for every subset of 2^{k-1} vertices, if it contains a k-witness as an induced subgraph:

Corollary 3 ([26]). GRUNDY COLORING *can be solved in time $f(k)n^{2^{k-1}}$, i.e.* GRUNDY COLORING *parameterized by the number k of colors is in* XP.

Observation 4. *In any Grundy coloring of G, a vertex with degree d cannot be colored with color $d + 2$ or larger.*

Proposition 5. *Let G be a graph with a minimal Grundy coloring achieving color k and let W be the corresponding minimal witness. Then, if a vertex u of W is colored with $k' < k$, u has a neighbor colored with some color $k'', k'' > k'$.*

Proof. If not, one could remove u from the witness, a contradiction. ∎

Lemma 6. *Let G be a graph and let G' be the corresponding minimal ℓ-witness with the partition $\mathcal{W} := W_1 \uplus \cdots \uplus W_\ell$. Then, W_i is an independent set which dominates the set $\bigcup_{j \in [i+1,\ell]} W_j$ (and no proper subset of W_i has this property). In particular, W_1 is a minimal independent dominating set of $V(G')$.*

For each $i \in [l]$, let t_i be a rooted tree. We define $v[t_1, t_2, \ldots, t_l]$ as the tree rooted at node v where v is linked to the root of each tree t_i. The set $(T_k)_{k \geqslant 1}$ is a family of rooted trees (known as *binomial trees*) defined as follows (see Figure 1 for an illustration):

- T_1 consist only of one node (incidentally the root), and
- $\forall k \geqslant 1, T_{k+1} = v[T_1, T_2, \ldots, T_k]$.

Fig. 1. The binomial tree T_4, where numbers denote the color of each vertex in a first-fit proper coloring with largest number of colors

In a tree T_k with root v, for each $i \in [k]$, $v(i)$ denotes the root of T_i (i.e. the i-th child of v).

We now show a useful lemma about Grundy colorings of the tree T_k.

Lemma 7. *The Grundy number of T_k is k. Moreover, there are exactly two Grundy colorings achieving color k, and a unique coloring if we impose that the root is colored k.*

The following result of Chang and Hsu [7] will prove useful:

Theorem 8 ([7]). *Let G be a graph on n vertices for which every subgraph H has at most $d|V(H)|$ edges. Then $\Gamma(G) \leqslant \log_{d+1/d}(n) + 2$.*

3   Grundy Coloring: Algorithms and Complexity

3.1   An Exact Algorithm

A straightforward way to solve GRUNDY COLORING is to enumerate all possible orderings of the vertex set and to check whether the greedy algorithm uses at least k colors. This is a $\Theta(n!)$-time algorithm. A natural question is whether there is a faster exact algorithm. We now give such an algorithm.

We rely on two observations: (a) in a colored witness, every color class W_i is an independent dominating set in $G[\bigcup_{j \geqslant i} W_j]$ (Lemma 6), and (b) any independent dominating set is a maximal independent set (and vice versa). The algorithm is obtained by dynamic programming over subsets, and uses an algorithm which enumerates all maximal independent sets.

Theorem 9. GRUNDY COLORING *can be solved in time $O^*(2.44225^n)$.*

Proof. Let $G = (V, E)$ be a graph. We present a dynamic programming algorithm to compute $\Gamma(G)$. For simplicity, given $S \subseteq V$, we denote the Grundy number of the induced subgraph $G[S]$ by $\Gamma(S)$. We recursively fill a table $\Gamma^*(S)$ over the subset lattice $(2^V, \subseteq)$ of V in a bottom-up manner starting from $S = \emptyset$. The base case of the recursion is $\Gamma^*(\emptyset) = 0$. The recursive formula is given as

$$\Gamma^*(S) = \max\{\Gamma^*(S \setminus X) + 1 \mid X \subseteq S \text{ is an independent dominating set of } G[S]\}.$$

Now let us show by induction on $|S|$ that $\Gamma^*(S) = \Gamma(S)$ for all $S \subseteq V$. The assertion trivially holds for the base case. Consider a nonempty subset $S \subseteq V$;

by induction hypothesis, $\Gamma^*(S') = \Gamma(S')$ for all $S' \subset S$. Let X be a subset of S achieving $\Gamma^*(S) = \Gamma^*(S \setminus X) + 1$ and X' be the set of the color class 1 in the ordering achieving the Grundy number $\Gamma(S)$.

Let us first see that $\Gamma^*(S) \leqslant \Gamma(S)$. By induction hypothesis we have $\Gamma^*(S \setminus X) = \Gamma(S \setminus X)$. Consider a vertex ordering σ on $S \setminus X$ achieving $\Gamma(S \setminus X)$. Augmenting σ by placing all vertices of X at the beginning of the sequence yields a (set of) vertex ordering(s). Since X is an independent set, the first-fit algorithm gives color 1 to all vertices in X, and since X is also a dominating set for $S \setminus X$, no vertex of $S \setminus X$ receives color 1. Therefore, the first-fit algorithm on such ordering uses $\Gamma(S \setminus X) + 1$ colors. We deduce that $\Gamma(S) \geqslant \Gamma(S \setminus X) + 1 = \Gamma^*(S \setminus X) + 1 = \Gamma^*(S)$.

To see that $\Gamma^*(S) \geqslant \Gamma(S)$, we first observe that $\Gamma(S \setminus X') \geqslant \Gamma(S) - 1$. Indeed, the use of the optimal ordering of S ignoring vertices of X' on $S \setminus X'$ yields the color $\Gamma(S) - 1$. We deduce that $\Gamma(S) \leqslant \Gamma(S \setminus X') + 1 = \Gamma^*(S \setminus X') + 1 \leqslant \Gamma^*(S \setminus X) + 1 = \Gamma^*(S)$.

As a minimal independent dominating set is a maximal independent set, we can estimate the computation of the table by restricting X to the family of maximal independent sets of $G[S]$. On an n-vertex graph, one can enumerate all maximal independent sets in time $O(1.44225^n)$ [20]. Checking whether a given set is a minimal independent set is polynomial and thus, the number of execution steps is dominated (up to a polynomial factor) by the number of recursion steps taken. This is

$$\sum_{i=0}^{n} \binom{n}{i} \cdot 1.44225^i = (1 + 1.44225)^n. \square$$

We leave as an open question to improve this running time. However, we note that the *fast subset convolution* technique [4] does not seem to be directly applicable.

3.2   Lower Bound on the Treewidth Dependency

Let us recall that GRUNDY COLORING is known to be in XP for the parameter treewidth, but its membership in FPT remains open.

The following result is inspired by ideas in [19] for proving near-optimality of known algorithm on bounded treewidth graphs. Unlike [19] which is based on the *Strong* ETH, our result is based on the ETH.

Theorem 10. *Under the ETH, for any constant c, GRUNDY COLORING is not solvable in time $O^*(c^w)$ on graphs with feedback vertex set number (and hence treewidth) at most w.*

3.3   Grundy Coloring on Special Graph Classes

For each fixed k, GRUNDY COLORING can be solved in polynomial time [26] and thus GRUNDY COLORING parameterized by the number of colors is in XP.

However, it is unknown whether it is in FPT for this parameter. We will next show several positive results for H-minor-free, chordal and claw-free graphs. Note that GRUNDY COLORING is NP-complete on chordal graphs [23] and on claw-free graphs [25].

We first observe that the XP algorithm of [24] implies a pseudo-polynomial-time algorithm on apex-minor-free graphs (such as planar graphs).

Proposition 11. GRUNDY COLORING *is* $n^{O(\log^2 n)}$-*time solvable on apex-minor-free graphs.*

Proposition 12. GRUNDY COLORING *parameterized by the number of colors is in* FPT *for the class of graphs excluding a fixed graph H as a minor.*

Proof. Notice that G contains a k-witness H as an induced subgraph if and only if $\Gamma(G) \geqslant k$. We can check, for every k-witness H, whether the input graph G contains H as an induced subgraph. By Observation 1, it suffices to test only the minimal k-witnesses. The number of minimal k-witnesses is bounded by some function of k and H-INDUCED SUBGRAPH ISOMORPHISM is in FPT when parameterized by $|V(H)|$ on graphs excluding H as a minor [9]. Therefore, one can check if $\Gamma(G) \geqslant k$ by solving H-INDUCED SUBGRAPH ISOMORPHISM for all minimal k-witnesses H.   □

Proposition 13. *Let $\mathcal{C}$ be a graph class for which every member G satisfies $tw(G) \leqslant f(\Gamma(G))$ for some function f. Then, GRUNDY COLORING parameterized by the number of colors is in* FPT *on $\mathcal{C}$. In particular, GRUNDY COLORING is in* FPT *on chordal graphs.*

Proof. Since GRUNDY COLORING is in FPT for parameter combination of the number of colors and the treewidth [24], the first claim is immediate. Moreover $\omega(G) \leqslant \Gamma(G)$, hence if $tw(G) \leqslant f(\omega(G))$ we have $tw(G) \leqslant f(\Gamma(G))$. For any chordal graph G, $tw(G) = \omega(G) - 1$ [5].   □

Proposition 14. GRUNDY COLORING *can be solved in time* $O\left(nk^{\Delta^{k+1}}\right) = n\Delta^{\Delta^{O(\Delta)}}$ *for graphs of maximum degree Δ.*

Proof. Observation 2 implies that one can enumerate every distance-k-neighbourhood of each vertex, test every k-coloring of this neighborhood, and check if it is a valid Grundy k-coloring. Every such neighborhood has size at most $\Delta^{k+1} \leqslant \Delta^{\Delta+3}$ since by Observation 4, $k \leqslant \Delta + 2$. There are at most k^x k-colorings of a set of x elements.   □

Corollary 15. *Let $\mathcal{C}$ be a graph class for which every member G satisfies $\Delta(G) \leqslant f(\Gamma(G))$ for some function f. Then, GRUNDY COLORING parameterized by the number of colors is in* FPT *for graphs in $\mathcal{C}$. In particular, this holds for the class of claw-free graphs.*

Proof. Straightforward by Proposition 14. Moreover, let G be a claw-free graph, and consider a vertex v of degree $\Delta(G)$. Since G is claw-free, the subgraph induced by the neighbors of v has independence number at most 2, and hence $\Gamma(G) \geqslant \chi(G) \geqslant \chi(N(v)) \geqslant \frac{\Delta(G)}{2}$.   □

4   Weak and Connected Grundy Coloring

Among the three versions of GRUNDY COLORING we consider in this paper, WEAK GRUNDY COLORING is the least constrained while CONNECTED GRUNDY COLORING appears to be the most constrained one. This intuition turns out to be true when it comes to their parameterized complexity. When parameterized by the number of colors, WEAK GRUNDY COLORING is in FPT while CONNECTED GRUNDY COLORING does not belong to XP.

We recall that WEAK GRUNDY COLORING is NP-complete [10].

Theorem 16. WEAK GRUNDY COLORING *parameterized by number of colors is in* FPT.

The FPT-algorithm is based on the idea of *color-coding* by Alon et al. [1]. The height of a minimal witness for $\Gamma' \geqslant k$ is bounded by a function of k. Since those vertices of the same color do not need to induce an independent set, a random coloring will identify a *colorful* minimal witness with a good probability.

We also remark that the approach used to prove Theorem 16 does not work for GRUNDY COLORING because there is no control on the fact that a color class is an independent set.

Minimal connected Grundy k-witnesses, contrary to minimal Grundy k-witnesses (Observation 2), have arbitrarily large order: for instance, the cycle C_n of order n ($n > 4$, n odd) has a Grundy 3-witness of order 4, but its unique *connected* Grundy 3-witness is of order n: the whole cycle.

Observe that $\Gamma_c(G) \leqslant 2$ if and only if G is bipartite. Hence, CONNECTED GRUNDY COLORING is polynomial-time solvable for any $k \leqslant 3$. However, we will now show that this is not the case for larger values of k, contrary to GRUNDY COLORING (Corollary 3). Hence, the parameterized version of the problem does not belong to XP.

Theorem 17. CONNECTED GRUNDY COLORING *is* NP*-hard even for* $k = 7$.

Proof. We give a reduction from 3-SAT 3-OCC, an NP-complete restriction of 3-SAT where each variable appears in at most three clauses [22], to CONNECTED GRUNDY COLORING with $k = 7$. We first give the intuition of the reduction. The construction consists of a tree-like graph of constant order (resembling binomial tree T_6) whose root is adjacent to two vertices of a K_6 (this constitutes W) and contains three special vertices a_4, a_{21}, and a_{24} (which will have to be colored with colors 1, 3, and 2 respectively), a connected graph P_1 which encodes the variables and a path P_2 which encodes the clauses. One in every three vertices of P_2 is adjacent to a_4, a_{21} and a_{24}. To achieve color 7, we will need to color those vertices with color strictly greater than 3. This will be possible if and only if the assignment corresponding to the coloring of P_1 satisfies all the clauses.

We now formally describe the construction. Let $\phi = (X = \{x_1, \ldots, x_n\}, \mathcal{C} = \{C_1, \ldots, C_m\})$ be an instance of 3-SAT 3-OCC where no variable appears always as the same literal. $P_1 = (\{i_1, i_2, v\} \cup \{v_i, \overline{v_i} \mid i \in [n]\}, \{\{i_1, i_2\}, \{i_2, v\}\} \cup \{\{v, v_i\} \cup \{v, \overline{v_i}\} \cup \{v_i, \overline{v_i}\} \mid i \in [n]\})$ consists of n triangles sharing the vertex v.

$P_2 = (\{p_j \mid j \in [3m-1]\}, \{\{p_j, p_{j+1}\} \mid j \in [3m-2]\})$ consists of a path of length $3m-1$. For each $j \in [m]$ and $i \in [n]$, $c_j \overset{def}{=} p_{3j-1}$ is adjacent to v_i if x_i appears positively in C_j, and is adjacent to $\overline{v_i}$ if x_i appears negatively in C_j. For each $j \in [m]$, c_j is adjacent to a_4, a_{21}, and a_{24}.

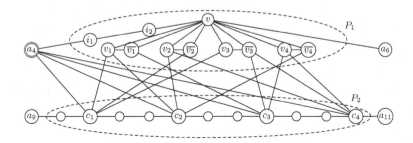

Fig. 2. P_1 and P_2 for the instance $\{x_1 \vee \neg x_2 \vee x_3\}, \{x_1 \vee x_2 \vee \neg x_4\}, \{\neg x_1 \vee x_3 \vee x_4\}, \{x_2 \vee \neg x_3 \vee x_4\}$

Intuitively, setting a literal to true consists of coloring the corresponding vertices with 3. Therefore, a clause C_j is satisfied if c_j has a 3 among its neighbors. To actually satisfy a clause, one has to color c_j with 4 or higher. Thus, c_j must also see a 2 in its neighborhood. We will show that the unique way of doing so is to color p_{3j-2} with 2, so all the clauses have to be checked along the path P_2.

We give, in Figure 3, a coloring of P_1 corresponding to a truth assignment of the instance SAT formula. One can check that when going along P_2 all the c_j's are colored with color 4.

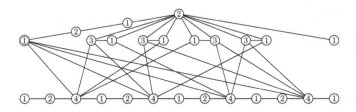

Fig. 3. A connected Grundy coloring such that all the c_j's are colored with color at least 4

The constant gadget W is depicted in Figure 4. The waves between a_4 and a_6 and between a_9 and a_{11} correspond, respectively, to the gadgets encoding the variables (P_1) and the clauses (P_2) described above and drawn in Figure 2. A connected Grundy coloring achieving color 7 is given in Figure 5 provided that going from a_9 to a_{11} can be done without coloring any vertex c_j with color 2 or less.

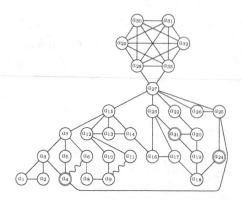

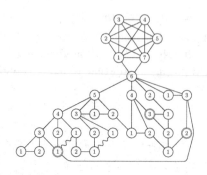

Fig. 4. The constant gadget. The doubly-circled vertices are adjacent to all the c_j's $(j \in [m])$.

Fig. 5. A connected Grundy coloring of the constant gadget achieving color 7. The order is given by the sequence $(a_i)_{1 \leqslant i \leqslant 33}$.

In the following claims, we use extensively Observation 1 which states that a vertex with degree d gets color at most $d + 1$. We observe that coloring a vertex of degree d with color $d + 1$ is useful only if we want to achieve color $d + 1$. Indeed, otherwise, the vertex has all its neighbors already colored and cannot be used in the sequel. Moreover, if one wants to color a neighbor y of a vertex x in order to color x with a higher color, y cannot receive a color greater than its degree $d(y)$. Hence, the only vertices that could achieve color k are vertices of degree at least $k - 1$ having at least one neighbor of degree at least $k - 1$.

In the sequel, we call *doubly-circled vertices* the special vertices a_4, a_{21} and a_{24}, as they are doubly-circled in our figures.

Claim 17.A. *To achieve color 7, a_{27} needs to be colored with color 6 (while for all $i \in [28, 33]$, a_i is still uncolored).*

Claim 17.B. *Vertices a_{26}, a_{22}, a_{25}, a_{23}, a_{15} must receive color 1, 2, 3, 4, 5 respectively.*

Claim 17.C. *Vertex a_7 must receive color 4.*

Claim 17.D. *Vertex a_3 must receive color 3.*

Claim 17.D has further consequences: we must start the connected Grundy coloring by giving colors 1 and 2 to a_1 and a_2. The only follow-up, for connectivity reasons, is then to color a_3 with color 3 and a_4 with color 1. Thus, vertices a_5 and a_6 has to be colored with colors 2 and 1 respectively (so that a_7 can be colored 4). As, by Claim 17.B, a_{25} must receive color 3, a_{24} must receive color 2 (since a_4 has already color 1), so a_{18} must be colored 1.

Claim 17.E. *Vertex a_{21} must receive color 3.*

Claim 17.F. *The unique way of coloring a_{11} with color 1 without coloring any vertex c_j with color 1, 2, or 3 is to color all the c_j's for each $j \in [m]$.*

We remark that opposite literals are adjacent, so for each $i \in [n]$, only one of v_i and $\overline{v_i}$ can be colored with color 3. We interpret coloring v_i with 3 as setting x_i to true and coloring $\overline{v_i}$ with 3 as setting x_i to false.

Claim 17.G. *To color each c_j ($j \in [m]$) of the path P_2 with a color at least 4, the SAT formula must be satisfiable.*

So, to achieve color 7 in a connected Grundy coloring, the SAT formula must be satisfiable. The reverse direction consists of completing the coloring by giving a_{13} color 1 and a_{14} color 2, as shown in Figure 3 and Figure 5.

5   Concluding Remarks and Questions

We presented several positive and negative results concerning GRUNDY COLOR-ING and two of its variants. To conclude this article, we suggest some questions which might be useful as a guide for further studies.

There is a gap between the $f(k, w) \cdot n$ (and XP) algorithm of [24] and the lower bound of Theorem 10. Is GRUNDY COLORING in FPT when parameterized by treewidth? Two simpler questions are whether there is a better $f(k, w)poly(n)$ algorithm (for example with $f(k, w) = k^{O(w)}$), and whether GRUNDY COLORING is in FPT when parameterized by the feedback vertex set number (it is easy to see that it is the case when parameterized by the vertex cover number).

GRUNDY COLORING (parameterized by the number of colors) is in XP, and we showed it to be in FPT on many important graph classes. Yet, the question whether it is in FPT or W[1]-hard remains unsolved. A perhaps more accessible research direction is to settle this question on bipartite graphs.

It would also be interesting to determine the (classic) complexity of GRUNDY COLORING on interval graphs. Also, we saw that the algorithm of [24] implies a pseudo-polynomial algorithm for planar (even apex-minor-free) graphs, making it unlikely to be NP-complete on this class. Is there a polynomial-time algorithm?

Concerning CONNECTED GRUNDY COLORING, we showed that it becomes NP-complete for $k = 7$. As CONNECTED GRUNDY COLORING is polynomial-time solvable for $k \leqslant 3$, its complexity status for $4 \leqslant k \leqslant 6$ and/or on restricted graph classes remains open.

References

1. Alon, N., Yuster, R., Zwick, U.: Color-coding. Journal of the ACM **42**(4), 844–856 (1995)
2. Araujo, J., Sales, C.L.: On the Grundy number of graphs with few P_4's. Discrete Applied Mathematics **160**(18), 2514–2522 (2012)
3. Benevides, F., Campos, V., Dourado, M., Griffiths, S., Morris, R., Sampaio, L., Silva, A.: Connected greedy colourings. In: Pardo, A., Viola, A. (eds.) LATIN 2014. LNCS, vol. 8392, pp. 433–441. Springer, Heidelberg (2014)

4. Björklund, A., Husfeldt, T., Kaski, P., Koivisto, M.: Fourier meets Möbius: fast subset convolution. In: Proc. 39th Annual ACM Symposium on Theory of Computing, STOC 2007, pp. 67–74. ACM (2007)
5. Bodlaender, H.L.: A tourist guide through treewidth. Acta Cybernetica **11**(1–2), 1–21 (1993)
6. Bonnet, E., Foucaud, F., Kim, E.J., Sikora, F.: Complexity of grundy coloring and its variants (2014). CoRR, abs/1407.5336
7. Chang, G.J., Hsu, H.-C.: First-fit chromatic numbers of d-degenerate graphs. Discrete Mathematics **312**(12–13), 2088–2090 (2012)
8. Christen, C.A., Selkow, S.M.: Some perfect coloring properties of graphs. Journal of Combinatorial Theory, Series B **27**(1), 49–59 (1979)
9. Flum, J., Grohe, M.: Fixed-parameter tractability, definability, and model-checking. SIAM Journal on Computing **31**(1), 113–145 (2001)
10. Goyal, N., Vishwanathan, S.: NP-completeness of undirected Grundy numbering and related problems. Manuscript (1997)
11. Grundy, P.M.: Mathematics and games. Eureka **2**, 6–8 (1939)
12. Gyárfás, A., Lehel, J.: On-line and first fit colorings of graphs. Journal of Graph Theory **12**(2), 217–227 (1988)
13. Havet, F., Maia, A.K., Yu, M.-L.: Complexity of greedy edge-colouring. Research report RR-8171, INRIA, December 2012
14. Havet, F., Sampaio, L.: On the Grundy and b-chromatic numbers of a graph. Algorithmica **65**(4), 885–899 (2013)
15. Hedetniemi, S.M., Hedetniemi, S.T., Beyer, T.: A linear algorithm for the Grundy (coloring) number of a tree. Congressus Numerantium **36**, 351–363 (1982)
16. Impagliazzo, R., Paturi, R., Zane, F.: Which problems have strongly exponential complexity? Journal of Computer and System Sciences **63**(4), 512–530 (2001)
17. Kierstead, H.A., Saoub, K.R.: First-fit coloring of bounded tolerance graphs. Discrete Applied Mathematics **159**(7), 605–611 (2011)
18. Kortsarz, G.: A lower bound for approximating Grundy numbering. Discrete Mathematics & Theoretical Computer Science **9**(1) (2007)
19. Lokshtanov, D., Marx, D., Saurabh, S.: Known algorithms on graphs of bounded treewidth are probably optimal. In: Proc. 22nd Annual ACM-SIAM Symposium on Discrete Algorithms, SODA 2011, pp. 777–789. SIAM (2011)
20. Moon, J., Moser, L.: On cliques in graphs. Israel Journal of Mathematics **3**(1), 23–28 (1965)
21. Narayanaswamy, N.S., Babu, R.S.: A note on first-fit coloring of interval graphs. Order **25**(1), 49–53 (2008)
22. Papadimitriou, C.H.: Computational complexity. Addison-Wesley (1994)
23. Sampaio, L.: Algorithmic aspects of graph colouring heuristics. PhD thesis, University of Nice-Sophia Antipolis, November 2012
24. Telle, J.A., Proskurowski, A.: Algorithms for vertex partitioning problems on partial k-trees. SIAM Journal on Discrete Mathematics **10**(4), 529–550 (1997)
25. Zaker, M.: The Grundy chromatic number of the complement of bipartite graphs. Australasian Journal of Combinatorics **31**, 325–329 (2005)
26. Zaker, M.: Results on the Grundy chromatic number of graphs. Discrete Mathematics **306**(23), 3166–3173 (2006)

On the Complexity of the Minimum Independent Set Partition Problem

T.-H. Hubert Chan[1], Charalampos Papamanthou[2], and Zhichao Zhao[1(✉)]

[1] Department of Computer Science, The University of Hong Kong, Hong Kong, China
{hubert,zczhao}@cs.hku.hk
[2] Department of Electrical and Computer Engineering,
UMIACS University of Maryland, College Park, USA
cpap@umd.edu

Abstract. We consider the Minimum Independent Set Partition Problem (MISP) and its dual (MISPDual). The input is a multi-set of N vectors from $\{0,1\}^n$, where $U := \{1,\ldots,n\}$ is the index set. In MISP, a threshold k is given and the goal is to partition U into a minimum number of subsets such that the projected vectors on each subset of indices have multiplicity at least k, where the multiplicity is the number of times a vector repeats in the (projected) multi-set. In MISPDual, a target number χ is given instead of k, and the goal is to partition U into χ subsets to maximize k such that each projected vector appears at least k times.

The problem is inspired from applications in private voting verification. Each of the N vectors corresponds to a voter's preference for n contests. The n contests are partitioned into χ subsets such that each voter receives a verifiable tracking number for each subset. For each subset of contests, each voter's tracking number together with the votes for that subset is released in some public bulletin, which can be verified by each voter. The multiplicity k of the vectors' projection onto each subset of indices ensures that the bulletin for each subset of contests satisfies the standard privacy notion of k-anonymity.

In this paper, we show strong inapproximability results for both problems. For MISP, we show the problem is hard to approximate to within a factor of $n^{1-\epsilon}$. For MISPDual, we show the problem is hard to approximate to within a factor of $N^{1-\epsilon}$. Here, ϵ can be any small constant. Note that factors n and N approximation are trivial for MISP and MISPDual respectively. Hence, our results imply that any polynomial-time algorithm can almost do no better than the trivial one.

1 Introduction

We study the Minimum Independent Set Partition problem (MISP) and its dual problem (MISPDual). This problem was raised by Wagner on cstheory. stackexchange [12] in the context of data privacy [6]. We first describe the problem and an application scenario.

This research is partially funded by a grant from Hong Kong RGC under the contract HKU719312E.

© Springer International Publishing Switzerland 2015
D. Xu et al. (Eds.): COCOON 2015, LNCS 9198, pp. 121–132, 2015.
DOI: 10.1007/978-3-319-21398-9_10

In MISP, a multi-set Y of N vectors in $\{0,1\}^n$ is given together with a multiplicity threshold k. Our goal is to partition the indices $[n]$ into minimum number χ of subsets such that the projection of Y on each subset has multiplicity at least k.

The dual problem MISPDual is also of interest, in which a multi-set Y of vectors is also given. However, the target number χ of parts is given, and the goal is to return a χ-partition of the indices $[n]$ such that the minimum multiplicity k of the projected vectors is maximized.

Application Scenario. The problem is motivated by privacy in voting verification. We have N voters, each of whom is voting for n contests (with $\{0,1\}$ voting). To verify that all votes have been counted, each voter gets assigned a verifiable tracking number during voting. Then, there is a public bulletin board where all pairs of tracking numbers and votes are posted (where names of voters are withheld) such that each voter can verify that his votes are correct using his own tracking number. This could provide verifiability, but it is well-known in the privacy community that simply replacing a user's name with a random id cannot achieve privacy [6], since a voter might be uniquely identified by the way he votes in the n contests.

An expensive solution would be for each voter to get a separate tracking number for each contest, but this would increase the space complexity to store n numbers for each voter. Observe that if k is the minimum of the number of minority votes over all n contests, this expensive solution achieves the standard notion of k-anonymity [11].

To obtain a tradeoff between the space complexity of each voter and the anonymity parameter, one solution is: after receiving all votes, partition the n contests into some small number χ of subsets such that within each subset of contests, each voter has at least $k - 1$ other voters who vote in exactly the same way in that subset of contests, for some parameter k. In the public bulletin board, the χ subsets of contests are released independently. Each voter needs to store only χ tracking numbers (one for each subset of contests), and k-anonymity is achieved.

The case for MISP corresponds to the scenario when a parameter k is given, and the goal is the minimize the number χ of subsets to achieve k-anonymity. For the dual problem MISPDual, the number χ of subsets is given, and the goal is to partition the contests into χ subsets such that the anonymity parameter k is maximized. Hence, it is of interest to investigate the complexity and hardness of approximation for these problems.

Our Results and Techniques. We prove strong inapproximability results for both problems MISP and MISPDual. We first give a reduction from graph coloring, which is NP-hard; in graph coloring, each vertex is assigned a color such that no two adjacent vertices receive the same color. Intuitively, each index in $[n]$ stands for a vertex, while the vectors capture the properties of the graph coloring problem. In our construction, a valid coloring corresponds to a partition with multiplicity k, while an invalid coloring corresponds to one with multiplicity 1.

The inapproximability of graph coloring implies that the approximation hardness of MISP and MISPDual with $\chi \geq 3$ is at least $n^{1-\epsilon}$ and $N^{1-\epsilon}$ respectively.

However, we show that MISPDual with $\chi = 2$ is much harder than graph coloring with $\chi = 2$. Observe that deciding if a graph is 2-colorable can be solved in polynomial time. Hence, to show the hardness of MISPDual with $\chi = 2$, we need new reduction techniques. We give a novel reduction from the NP-hard problem 3-SAT. Similar to graph coloring, some indices stand for variables and their negations. Intuitively, one subset stands for "true" and the other stands for "false". In order to show approximation hardness, for any threshold k, our reduction is carefully constructed such that a satisfiable assignment corresponds to a partition with multiplicity at least k, while an unsatisfiable formula corresponds to an instance such that any 2-partition has multiplicity only 1. This gap property allows us to prove that it is NP-hard to approximate MISPDual within factor $N^{1-\epsilon}$.

Our strong inapproximability results imply that there can be no efficient approximation algorithms for the problems MISP and MISPDual in their most general form. However, in real-world applications, the instances might have special structures that facilitate useful heuristic algorithms, which we leave as future research directions.

1.1   Historical Overview on Inapproximability

NP-Completeness has been developed in the 1970s [2,9]. Its success motivated the study of approximation algorithms. The first such paper was by Johnson [8]. He considered the problems Max SAT, Independent Set, Coloring and Set Cover. Several approximation algorithms have been proposed for these problems in this paper.

The design and analysis of approximation algorithms have grown since then. Several problems are shown to admit polynomial time approximation schemes (PTAS), meaning that they can be approximated as close to the optimum as possible. It was known from the very beginning of approximation algorithms that some problems do not admit PTAS. For instance, coloring can not be approximated within $\frac{4}{3} - \epsilon$, since 3-coloring is NP-hard. However, the inapproximabilities for many hard problems remains unknown.

Modern theory of inapproximability starts from the development of PCP systems, which are proved in [1]. Unlike conventional NP-hardness reduction, PCP systems can be used more readily to achieve inapproximability hardness. Based on PCP systems, several strong inapproximability have been proved since then, e.g., MAX 3SAT [7], Set Cover [5] and Coloring [4,13]. In particular, one of our reductions is based on the hardness of graph coloring [13].

Other Vector Partition Problems. Onn and Schulman [10] have also considered vector partition problems in which the input is also a collection of vectors. However, the goal is to partition the vectors (as opposed the coordinate index set) to maximize some convex objective function on the sum of vectors in each part. They showed that if both the dimension and the number of parts are fixed, the problem can be solved in strongly polynomial time.

2   Problem Definition

We give the formal definition of the Minimum Independent Set Partition Problem (MISP). The input is a positive integer n, a multi-set $Y := \{y_1, y_2, \ldots, y_N\}$ of N vectors in $\{0,1\}^n$, and a multiplicity threshold k. We use $U := [n] = \{1, 2, \ldots, n\}$ to denote the set of indices.

Given a vector y and subset $I \subseteq U$ of indices, we use $y|_X$ to denote the projection of vector y on I. For instance, for $y = (0, 1, 0, 1, 0, 1)$ and $I = \{2, 4, 5\}$, $y|_I = (1, 1, 0)$. Given a multi-set Y, the projection of Y on I is a multi-set defined similarly $Y|_I := \{y|_I : y \in Y\}$.

A subset $I \subseteq U$ of indices is k-*independent* (with respect to Y) if each vector in the multi-set $Y|_I$ has multiplicity at least k, where multiplicity denotes the number of times a vector in $Y|_I$ repeats. A partition $\{I_1, I_2, \ldots, I_\chi\}$ of U is k-independent if each part I_i is k-independent.

The goal is to find the smallest integer χ and partition U into χ subsets $I_1, \ldots, I_\chi$ such that each partition I_i is k-independent.

Dual Problem. We also describe a dual version of the problem that we call MISPDual. Similarly, a multi-set Y of vectors are given, and a target number χ of partitions is given instead of k. The goal is to maximize k and partition the indices U into χ subsets $I_1, \ldots, I_\chi$ such that each I_i is k-independent.

3   General Reduction Schema

In this section, we reduce from the problem of graph coloring to MISP (and MISPDual with $\chi \geq 3$); in a *valid* coloring of an undirected graph, each vertex is assigned a color such that no two adjacent vertices receive the same color. We convert from an undirected graph $G = (V, E)$ to a multi-set of vectors such that a valid coloring corresponds to satisfying some fixed multiplicity threshold k while an invalid coloring leads to multiplicity 1. The use of this "k vs 1"-gap will be clear in the proof of the hardness of MISPDual. Because graph coloring is hard to approximate [13], our reduction readily implies the approximation hardness of MISP (and MISPDual with $\chi \geq 3$).

Our reduction depends on an arbitrarily chosen parameter $k > 1$ that is the same as the given threshold in MISP or may depend on the graph size $n = |V|$ in MISPDual. The index set is $U := [n]$. The multi-set Y consists of $N = k(n + 1) + \binom{n}{2} + (k - 1)|\overline{E}|$ vectors in $\{0, 1\}^n$, where $\overline{E}$ is the edges in the complement graph of G. We also use $u \in V$ to denote an index of a vector. Let MISP(G, k) be the instance reduced from graph G with parameter k. The vectors in MISP(G, k) are defined as follows.

(I) An all-0's vector, and the n vectors in the standard basis (each having exactly one non-zero coordinate). Each of these vectors are repeated k times. There are $k(n + 1)$ such vectors.

(II) Vectors of exactly two non-zero coordinates. There are $\binom{n}{2}$ such vectors.

(III) For each $(u,v) \notin E$, the vector with exactly two non-zero entries at indices u and v. Each of these vectors are repeated $(k-1)$ times. There are $(k-1)|\overline{E}|$ such vectors.

Figure 1 contains an example of the vectors for graph $G = (V = \{a,b,c,d\}, E = \{\{a,b\},\{a,c\},\{a,d\}\})$ and $k = 3$. Observe that parts (I) and (II) only depend on the size of graph G and k.

Note that a coloring of the graph gives a partition on U (and vice versa) in a natural way, where vertices having the same color corresponds to a subset of indices. Next we prove the relationship between colorings and partitions.

Theorem 1. *For any $k > 1$ and graph G, G has a valid χ-coloring iff* $\mathsf{MISP}(G,k)$ *has a k-independent χ-partition. If G does not have any valid χ-coloring, then any χ-partition of $\mathsf{MISP}(G,k)$ is not 2-independent.*

Proof. When G has a valid χ-coloring, we can induce a χ-partition from the coloring. We prove it is k-independent. Given a subset I of indices, consider the projected vectors in each part of the reduction.

Each projected vector in part (I) appears at least k times by the construction. Each projected vector in (III) appears at least k times since it repeats $k-1$ times in (III) and we can find a same one in (II).

For a vector in part (II), it depends on the indices u and v at which the entries are non-zero. If at most one of them is included in I, then the projected vector already appears k times in (I); otherwise, both u and v are included in I.

Two vertices u and v can be included in the same part I only if they are not neighbors in G; hence, the projected vector appears once from (II) and $k-1$ times from (III). This proves the "only if" part.

On the other hand, if a χ-partition is 2-independent, then we induce a χ-coloring for G from the partition. We claim the coloring is valid. For any vertices u, v with the same color, we have a vector in (II) with u, v in the same part I (the part corresponding to their color). Such vector appears only once in (II). It appears in (III) at least once, since the partition is 2-independent. Hence u, v cannot be neighbours in G, thus it is a valid coloring. Notice k-independent implies 2-independent. This proves the "if" part and the contrapositive proves the second statement.

Theorem 2. *The inapproximability of MISP is $n^{1-\epsilon}$ for arbitrarily small $\epsilon > 0$, unless $\mathsf{P} = \mathsf{NP}$; this means that if a k-independent partition has minimum number of parts χ, it is NP-hard to return a k-independent partition with at most $n^{1-\epsilon} \cdot \chi$ parts. Moreover, the result holds for any constant $k \geq 2$.*

Proof. We want to show a reduction from a coloring instance G to an instance of MISP. Use the "reduction schema" in Theorem 1 with $k \geq 2$ to get a multi-set Y, which is a MISP instance with threshold k.

It is immediate from Theorem 1 that the minimum χ such that there is a k-independent partition with χ parts in the MISP instance is the same as the chromatic number of G (the minimum number of colors needed to color G).

Thus, the inapproximability of graph coloring can be applied to MISP. The inapproximability of chromatic number is $n^{1-\epsilon}$, by [13], meaning that it is NP-hard to approximate chromatic number within a factor of $n^{1-\epsilon}$. Hence, it is also NP-hard to approximate MISP within a factor of $n^{1-\epsilon}$.

4   Approximation Hardness of MISPDual

In this section, we show that the dual problem of maximizing the multiplicity of the projections into partitions with $\chi \geq 3$ is hard to approximate. In Section 5, we show that even for $\chi = 2$, the problem is hard.

Theorem 3. *For arbitrarily small constant $\epsilon > 0$, there is no polynomial time algorithm that approximates* MISPDual *within a factor of $N^{1-\epsilon}$, where N is the number of vectors in the given multi-set Y; moreover this result holds for any constant $\chi \geq 3$, unless* P=NP.

Remark 1. We comment on choosing "n vs N" as the parameter to express approximation hardness. In MISP, a trivial solution is to partition U into n singletons, and hence, it is natural to compare with the trivial solution with approximation ratio n. Hence, inapproximability within factor $n^{1-\epsilon}$ is a strong indicator that no efficient algorithm would exist.

In MISPDual, since any partition would give multiplicity 1, and the maximum possible multiplicity is the number N of vectors, inapproximability within factor $N^{1-\epsilon}$ indicates that there is no efficient algorithm. Observe that we can also derive n^C hardness for MISPDual for any constant C.

Proof. We use the fact [3] that the problem of deciding whether a graph is χ-colorable is NP-complete for any $\chi \geq 3$. We reduce the problem of deciding whether a graph G is χ-colorable to MISPDual, such that for a "YES" instance, the multiplicity of MISPDual solution is at least k, otherwise the multiplicity is at most 1. Later we will set $k = n^C$ for some large enough constant $C = \Omega(\frac{1}{\epsilon})$.

Given a graph G, we use the "reduction schema" in Theorem 1 with $k = n^C$ to get a multi-set Y, which is a MISPDual instance with the same χ (target number of parts). Suppose the graph is χ-colorable. From Theorem 1, we know that the MISPDual has a solution with multiplicity at least k. On the other hand, if the graph is not χ-colorable, then MISPDual only has solutions with multiplicity 1, since otherwise it will contradict Theorem 1.

Note that the gap between "NO" and "YES" instances is 1 vs k.

We next prove no polynomial algorithm can approximate MISPDual within a factor better (smaller) than $k = n^C$. Note that the size N of Y is at most $k(n+1) + \binom{n}{2} + (k-1)|\overline{E}| \leq n^{C+10}$, hence this will imply no polynomial algorithm can approximate MISPDual within a factor better than $N^{\frac{C}{C+10}}$.

Suppose there is an algorithm $\mathcal{A}$ that can approximate MISPDual within a factor better than k. Then, we can decide whether a graph is χ-colorable by examining if the multiplicity is greater than 1. Hence, it is NP-hard to approximate MISPDual within a factor better than $k = n^C > N^{\frac{C}{C+10}}$. Note that for constant C, this is a polynomial-time reduction.

Setting C large enough such that $\frac{C}{C+10} > 1 - \epsilon$ gives the result.

5   Improved Approximation Hardness of **MISPDual**

This is the most technical part of the paper. In view of Section 4, it is natural to ask whether MISPDual with $\chi = 2$ is polynomial-time solvable, as deciding if a graph is 2-colorable has an easy solution.

In this section, we answer this question negatively. We show strong inapproximability result for MISPDual with $\chi = 2$. Observe that the reduction from graph coloring no longer works. To derive such a result, we need some problem with binary choice to tackle 2-partition. It turns out that 3-SAT does the job. In our construction the two parts correspond to "true" and "false" literals. At the same time, "true" and "false" literals are distinguishable via additional indices. The inapproximability comes from the fact that any satisfiable assignment corresponds to a 2-partition with high multiplicity, while any non-satisfiable assignment corresponds to a 2-partition with low multiplicity. In particular, we prove the following result.

Theorem 4. *For arbitrarily small constant $\epsilon > 0$, there is no polynomial algorithm that approximates MISPDual with $\chi = 2$ within a factor of $N^{1-\epsilon}$, unless P=NP.*

Proof. We use the fact that 3-SAT is NP-hard [9]. We construct a reduction from 3-SAT to MISPDual with $\chi = 2$. Consider an instance of 3-SAT: $C = \wedge_{i=1}^{l} C_i = \wedge_{i=1}^{l}(c_{i,1} \vee c_{i,2} \vee c_{i,3})$, with l clauses and p distinct variables.

Here $c_{i,j}$ can be x or $\neg x$. Without loss of generality, we assume that x and $\neg x$ do not appear in the same clause. It is obvious that $p \leq 3l$, and we further assume that $p, l \geq 2$ to avoid trivial cases. The property of our reduction is that a satisfiable 3-SAT instance corresponds a MISPDual solution with multiplicity at least k (later fixed to be $l^{\Omega(\frac{1}{\epsilon})}$), while a non-satisfiable 3-SAT corresponds to a MISPDual solution with multiplicity at most 1. Notice that the gap "1 vs k" is used to derive the inapproximability result.

We next give the construction for the reduction from 3-SAT to MISPDual with $\chi = 2$. We need a parameter $k \geq 2$ to be fixed later, which will be polynomially related to l. We denote the resulting MISPDual instance by MISPDual(C, k), where C is the 3-SAT instance and k is the parameter.

Our reduction will generate a multi-set Y of vectors from $\{0,1\}^{(1+l+2p)}$, with index set $U := [l + 1 + 2p]$. The first l indices are identification indices and are denoted by $[1..l]$. The $(l + 1)$-th index is the separation index and is denoted by $(l + 1)$. The last $2p$ indices correspond to literals (and their negations) and are denoted by the literals, e.g., x or $\neg x$. The use of identification and separation indices will become clear in the proof.

NOTATION. To simplify description, coordinates not mentioned are 0.

There are four parts of vectors as below:

(I) There are $2k$ vectors:

The 1st vector is the vector with the first l coordinates being 1.

The 2nd to the k-th vectors are the vectors with the first $l+1$ coordinates being 1.

The $(k+1)$-st vector is the vector with the $(l+1)$-st coordinate being 1.

The remaining $k-1$ vectors are all zero vectors.

The use of (I) is to force the identification indices $1..l$ to be in different part from the separation index $l+1$ in a "good" partition. Notice that some $(0,1)$ will appear only once otherwise.

(II) There are $(2k+1)p$ vectors. For each variable x, we have $(2k+1)$ vectors described as below:

(II.x) The first k vectors are the vectors with coordinates $(x, \neg x)$ being $(0,1)$.

The next k vectors are the vectors with coordinates $(x, \neg x)$ being $(1,0)$.

The last vector is a vector with indices $(x, \neg x)$ being $(1,1)$.

The use of (II) is to force x and $\neg x$ to be apart. Since there will be only one $(1,1)$ if the two indices are put together. In a "good" partition, literals setting to be "true" are supposed to be within the identification indices' (the first l indices) partition, while the "false" are in the separation index's (the $l+1$-st index) partition.

(III) There are $(3k+1)l$ vectors. For each clause $C_i = x \vee y \vee z$ (with literals x, y and z), we have $3k+1$ vectors:

(III.i) The first k vectors are the vectors with the i-th coordinate set to 1 and coordinates $(\neg y, \neg z)$ set to 1.

The next k vectors are the vectors with the i-th coordinate set to 1 and the coordinates $(\neg x, \neg z)$ set to 1.

The next k vectors are the vectors with the i-th coordinate set to 1 and the coordinates $(\neg x, \neg y)$ set to 1.

The last vector is the vector with the i-th coordinate set to 1 and the coordinates $(\neg x, \neg y, \neg z)$ set to 1.

Note that for all the $(3k+1)$ vectors, the i-th coordinate is set to 1.

The use of (III) is to force the variables to satisfy the constraints. Notice that if a clause is not satisfied, then all the indices $\neg x, \neg y, \neg z$ are on the "true" side (together with the first l indices), causing $(1,1,1)$ to appear only once in the projection onto the coordinates $(\neg x, \neg y, \neg z)$. On the other hand, as long as the not all indices $\neg x, \neg y, \neg z$ are included on the "true" side, any vector will appear at least k times. Notice the use of identification indices (the first l indices) here. With different identification indices, clauses will not affect each other.

(IV) There are lk vectors. For each clause $C_i = x \vee y \vee z$ there are k vectors as follows.

(IV.i) The k vectors are the same with the coordinates $(\neg x, \neg y, \neg z)$ set to 1.

Notice that in (IV) the identifier columns are set to 0, which is different from (III). The idea is to handle the situation when in (III.i) all $\neg x, \neg y, \neg z$ are partitioned into the "false" side. If this happens, the vector (projected on the "false" side with the $(l+1)$-st index) will repeat at least k times. Figure 2 (in appendix) gives an example for $(x \vee y \vee z) \wedge (\neg y \vee \neg x \vee w)$ with $k=2$.

It remains to show that a satisfiable assignment corresponds to an k-independent partition, while a non-satisfiable assignment corresponds to a MISPDual instance such that any 2-partition is not 2-independent.

Lemma 1. *For all $k > 1$ and 3-SAT instance C, if C has a satisfiable assignment, then* MISPDual(C, k) *has a k-independent 2-partition.*

Proof. Give a satisfiable assignment, we partition the indices set U into 2 subsets T and F as follows. The first l indices $[1..l]$ are included in T, and the $(l + 1)$-st index is in F. For each literal x, if $x = true$ then the index x is included in T and the index $\neg x$ is included in F; otherwise, the index $\neg x$ is in T and the index x is in F.

We next consider the vectors in each part projected on T and F.

Claim. In (I), each vector appears at least k times on both T and F.

Proof. First we consider the each vector in (I) projected on T. By construction, in the first l coordinates, each of the all 1's and all 0's vectors is repeated k times, and other coordinates are all set to 0.

For the projections on F, only the $(l + 1)$-st index has non-zero values and it contains exactly k 1's and k 0's. Hence, in (I), each projected vector repeats at least k times.

Claim. In (II.x), each vector appears at least k times on both T and F.

Proof. It can be seen that the only non-zero values are at indices x and $\neg x$. At both x and $\neg x$, we have more than k 0's and k 1's.

By the construction we know that x and $\neg x$ are assigned to different parts. In each part, the only non-zero coordinate is repeated at least k times, for each of the two values 0 and 1.

Claim. In (III.i), each vector appears at least k times on both T and F.

Proof. We denote the i-th clause by $C_i = x \vee y \vee z$, where x, y, z can be a variable or its negation. By construction, at least 1 of $\neg x, \neg y, \neg z$ is in F, since it is a satisfiable assignment. For instance, suppose $\neg z$ is in F; other situations follow the same argument.

Consider projections on F. Since the first l indices are not in F, we can find at least k same vectors in (III.i) and (IV.i) (in case $\neg x, \neg y, \neg z \in F$).

Now consider the projections on T. Vectors in (III.i) projected on T only differ at indices $\neg x$ and $\neg y$. It can be seen from the construction that no matter which part each of the indices $\neg x$ and $\neg y$ goes, each projected vector still appears at least k times. Hence, result of Claim 5 follows.

Claim. In (IV), each vector appears at least k times on both T and F.

Proof. This follows immediately from the construction.

The result of Lemma 1 follows, since each projected vector repeats at least k times.

Lemma 2. *For all $k > 1$ and 3-SAT instance C, if* MISPDual(C, k) *has a 2-independent 2-partition, then C has a satisfiable assignment.*

Proof. We first argue that if the 2-partition is 2-independent, then the identification (first l) indices and the separation ($l + 1$-st) index should be in different subsets. Similarly, x and $\neg x$ should be in different subsets. Then, an assignment is derived (such that literals on the same side as the identification indices are set to true) and analyzed.

Claim. Each of the indices $[1..l]$ is in the subset different from the subset containing the index $l + 1$.

Proof. Note that the only 1's at index $l+1$ happens in vectors 2 to $k+1$. Suppose on the contrary that some index in $j \in [1..l]$ is in the same subset at index $l + 1$. Then, at the coordinates $(j, l + 1)$, the projection $(0, 1)$ will appear only once due to vector $k + 1$. This contradicts 2-independence.

We denote T as the subset containing $[1..l]$, and F as the other subset F.

Claim. For each literal x, x and $\neg x$ are in different subsets.

Proof. Notice that we assume that no x and $\neg x$ appear in the same clause. As a result, there will be no vector with coordinates $(x, \neg x)$ being $(1, 1)$ in (I,III,IV). Such a vector appears only once in (II). The result follows as the partition is 2-independent.

From this point it is obvious that we should assign *true* to the literals in T and *false* to the literals in F. Next we prove that this is indeed a satisfying assignment.

Claim. Every clause C_i is satisfied by the above assignment.

Proof. Suppose $C_i = x \vee y \vee z$ is not satisfied. Then, it must be the case that $\neg x, \neg y, \neg z \in T$. We consider the vectors in (III.i) projected on T. From the construction, in (III.i) there will be exactly one vector with coordinates $(\neg x, \neg y, \neg z)$ being $(1, 1, 1)$.

We argue that this vector projected on T does not appear anywhere else. To see this, note that the identification indices are included in T, which is different from all other parts except (III.i). In (III.i), such vector (projected on T) only appears once, and hence the result follows.

This completes the proof of Lemma 2.

The following corollary is the contrapositive of Lemma 2.

Corollary 1. *For all $k > 1$ and 3-SAT instance C, if C does not have any satisfiable assignment, then any 2-partition for* MISPDual(C, k) *is not 2-independent.*

At this point, we can see that there is a gap of 1 vs k, meaning that to distinguish satisfiable 3-SAT from unsatisfiable ones, we only need to distinguish between multiplicity 1 and k. Hence, any polynomial algorithm that approximates MISPDual within a factor better than k will imply P=NP.

We can set $k = l^C$ for some large enough constant C, and observing that $N \le l^{C+10}$, we conclude that there is no polynomial algorithm with approximation ratio better than $N^{\frac{C}{C+10}}$.

Choosing C large enough (depending on ϵ) completes the proof of Theorem 4.

Appendix

Fig. 1 — $G = (V = \{a,b,c,d\}, E = \{\{a,b\},\{a,c\},\{a,d\}\})$ with $k = 3$

	a	b	c	d
(I)	0	0	0	0
	0	0	0	0
	0	0	0	0
	1	0	0	0
	1	0	0	0
	1	0	0	0
	0	1	0	0
	0	1	0	0
	0	1	0	0
	0	0	1	0
	0	0	1	0
	0	0	1	0
	0	0	0	1
	0	0	0	1
	0	0	0	1
(II)	1	1	0	0
	1	0	1	0
	1	0	0	1
	0	1	1	0
	0	1	0	1
	0	0	1	1
(III). (b, c)	0	1	1	0
	0	1	1	0
(III). (b, d)	0	1	0	1
	0	1	0	1
(III). (c, d)	0	0	1	1
	0	0	1	1

Fig. 2 — $(x \lor y \lor z) \land (\neg y \lor \neg x \lor w)$ with $k = 2$; unspecified entries are 0

	Identifiers		Separator	x	¬x	y	¬y	z	¬z	w	¬w
(I)	1	1	0								
	1	1	1								
	0	0	1								
	0	0	0								
(II. x)				0	1						
				0	1						
				1	0						
				1	0						
				1	1						
(II. y)						0	1				
						0	1				
						1	0				
						1	0				
						1	1				
(II. z)								0	1		
								0	1		
								1	0		
								1	0		
								1	1		
(II. w)										0	1
										0	1
										1	0
										1	0
										1	1
(III. 1)	1	0		0		1		1			
	1	0		0		1		1			
	1	0		1		0		1			
	1	0		1		0		1			
	1	0		1		1		0			
	1	0		1		1		0			
(III. 2)	0	1			1		0			1	
	0	1			1		0			1	
	0	1			0		1			1	
	0	1			0		1			0	
	0	1			1		1			0	
	0	1			1		1			1	
(IV. 1)	0	0				1		1	1		
	0	0				1		1	1		
(IV. 2)	0	0		1	1					1	
	0	0		1	1					1	

Acknowledgments. We would like to thank David Wagner for posting the problem online [12] and for useful discussions.

References

1. Arora, S., Safra, S.: Probabilistic checking of proofs: A new characterization of np. J. ACM **45**(1), 70–122 (1998)
2. Cook, S.A.: The complexity of theorem-proving procedures. In: Proceedings of the Third Annual ACM Symposium on Theory of Computing, STOC 1971, pp. 151–158. ACM, New York (1971)
3. Dailey, D.P.: Uniqueness of colorability and colorability of planar 4-regular graphs are np-complete. Discrete Mathematics **30**(3), 289–293 (1980)
4. Feige, U., Kilian, J.: Zero knowledge and the chromatic number. In: Proceedings of the Eleventh Annual IEEE Conference on Computational Complexity, pp. 278–287 (1996)
5. Feige, U.: A threshold of ln n for approximating set cover. J. ACM **45**(4), 634–652 (1998)
6. Ganta, S.R., Kasiviswanathan, S.P., Smith, A.: Composition attacks and auxiliary information in data privacy. In: Proceedings of the 14th ACM SIGKDD International Conference on Knowledge Discovery and Data Mining, KDD, pp. 265–273. ACM, New York (2008)
7. Håstad, J.: Some optimal inapproximability results. J. ACM **48**(4), 798–859 (2001)
8. Johnson, D.S.: Approximation algorithms for combinatorial problems. In: Proceedings of the Fifth Annual ACM Symposium on Theory of Computing, STOC 1973, pp. 38–49. ACM, New York (1973)
9. Karp, R.M.: Reducibility Among Combinatorial Problems. In: Miller, R.E., Thatcher, J.W. (eds.) Complexity of Computer Computations, pp. 85–103. Plenum Press (1972)
10. Onn, S., Schulman, L.J.: The vector partition problem for convex objective functions. Math. Oper. Res. **26**(3), 583–590 (2001)
11. Sweeney, L.: k-anonymity: A model for protecting privacy. International Journal of Uncertainty, Fuzziness and Knowledge-Based Systems **10**(5), 557–570 (2002)
12. Wagner, D.: Find index set partition that has large projections (2013). http://cstheory.stackexchange.com/questions/17562/find-index-set-partition-that-has-large-projections
13. Zuckerman, D.: Linear degree extractors and the inapproximability of max clique and chromatic number. Theory of Computing **3**(6), 103–128 (2007)

Bivariate Complexity Analysis of Almost Forest Deletion

Ashutosh Rai[✉] and Saket Saurabh

The Institute of Mathematical Sciences, Chennai, India
{ashutosh,saket}@imsc.res.in

Abstract. In this paper we study a generalization of classic FEEDBACK VERTEX SET problem in the realm of multivariate complexity analysis. We say that a graph F is an l-forest if we can delete at most l edges from F to get a forest. That is, F is at most l edges away from being a forest. In this paper we introduce the ALMOST FOREST DELETION problem, where given a graph G and integers k and l, the question is whether there exists a subset of at most k vertices such that its deletion leaves us an l-forest. We show that this problem admits an algorithm with running time $2^{\mathcal{O}(l+k)}n^{\mathcal{O}(1)}$ and a kernel of size $\mathcal{O}(kl(k+l))$. We also show that the problem admits a $c^{\mathbf{tw}}n^{\mathcal{O}(1)}$ algorithm on bounded treewidth graphs, using which we design a subexponential algorithm for the problem on planar graphs.

1 Introduction

In the field of graph algorithms, vertex deletion problems constitute a considerable fraction. In these problems we need to delete a small number of vertices such that the resulting graph satisfies certain properties. Many well known problems like VERTEX COVER and FEEDBACK VERTEX SET fall under this category. Most of these problems are NP-complete due to a classic result by Lewis and Yannakakis [19]. The field of parameterized complexity tries to provide efficient algorithms for these NP-complete problems by going from the classical view of single-variate measure of the running time to a multi-variate one. It aims at getting algorithms of running time $f(k)n^{\mathcal{O}(1)}$, where k is an integer measuring some aspect of the problem. The integer k is called the *parameter*. In most of the cases, the solution size is taken to be the parameter, which means that this approach gives faster algorithms when the solution is of small size. For more background, the reader is referred to the monographs [4,7,20].

Recently, there has been a trend of exploiting other structural properties of the graph other than the solution size [2,3,6,14,18]. For further reading, reader may refer to the recent survey by Fellows et al. [5]. In an earlier work, Guo et al. [13] introduced the notion of *"distance from triviality"* which looked at structural parameterization as a natural way to deal with problems which are polynomial time solvable on some graph classes. They argued that we could ask the same problems on some other (bigger) graph class which is close to the graph class on which the problem is polynomial time solvable, but the parameter is the

© Springer International Publishing Switzerland 2015
D. Xu et al. (Eds.): COCOON 2015, LNCS 9198, pp. 133–144, 2015.
DOI: 10.1007/978-3-319-21398-9_11

closeness or the *distance* from the original graph class instead of the solution size.

In the same spirit, we introduce the notion of *distance from tractability*. We know that vertex deletion problems deal with deletion of vertices to get to some graph class. For example, the VERTEX COVER problem deals with deleting vertices such that the resulting graph does not have any edge. Similarly, the well known FEEDBACK VERTEX SET problem talks about deleting vertices such that the resulting graph is a forest. What if we want to delete vertices so that the resulting graph class is *close* to the earlier graph class, while taking the *measure of closeness* as a parameter? This approach takes us from single variate parameterized algorithms to multivariate ones, which throws some light on the interplay between the parameters concerned. More precisely, they tell us about the trade-offs if we want to go away from the tractable version of a problem, and hence the term "distance from tractability". We also point out that the term can be a bit misleading, since in the case of this paper, we do find out tractable (FPT) algorithms for these problems when we consider both the parameters. But for want of a better term, we use it here.

There is already some work done which can be seen as examples of the notion of parameterizing by distance from tractability. For example, the PARTIAL VERTEX COVER and other related partial cover problems [1,10] come to mind, where after the deletion of a small set of vertices, the resulting graph is close to an edgeless graph. In these problems, the measure of closeness is the number of edges. Similarly, work has been done on vertex deletion to get a graph of certain treewidth, which can be looked as deletion of vertices to get a graph close to a forest, where the measure of the closeness is the treewidth of the graph.

The algorithms of the above mentioned kind show the correlation between the solution size and the distance from tractability. Let the distance from tractability be the parameter ℓ, and k be the number of vertices to be deleted. Suppose we have an algorithm with running time say $f(\ell)^{g(k)}n^{\mathcal{O}(1)}$, then is it possible to obtain an algorithm with running time $h_1(\ell)h_2(k)n^{\mathcal{O}(1)}$? That is, could we disentangle the function depending on both ℓ and k to a product of functions where each function depends only on one of ℓ and k. Answer to this question is *yes* if we take $f(\ell)^{g(\ell)}f(k)^{g(k)}n^{\mathcal{O}(1)}$. However, if we ask for an algorithm with running time $h(\ell)2^{\mathcal{O}(g(k))}n^{\mathcal{O}(1)}$ then it becomes interesting. This kind of question can be asked for several problems. For an example, it is known that the TREEWIDTH-η-DELETION problem, where the objective is to test whether there exists a vertex subset of size at most k such that its deletion leaves a graph of treewidth at most η. For $\eta = 0$ and 1 this correspond to the VERTEX COVER and FEEDBACK VERTEX SET problems, respectively. It is known that TREEWIDTH-η-DELETION admits an algorithm with running time $f(\eta)^k n^{\mathcal{O}(1)}$ [9,15]. However, it is not known whether there exists an algorithm with running time $h(\eta)2^{\mathcal{O}(k)}n^{\mathcal{O}(1)}$. Clearly, algorithms with running time $h(\ell)2^{\mathcal{O}(g(k))}n^{\mathcal{O}(1)}$ are more desirable.

The FEEDBACK VERTEX SET problem has been widely studied in the field of parameterized algorithms. A series of results have improved the running times to $\mathcal{O}^*(3.619^k)$ in deterministic setting [17] and $\mathcal{O}^*(3^k)$ in randomized setting [3], where the $\mathcal{O}^*$ notation hides the polynomial factors. Looking at this problem in the notion of distance from tractability, the number of edges comes to mind as a natural measure of distance. More precisely, we try to address the question of deleting vertices such that the resulting graph is l edges away from being a forest. We call such forests l-forests, and the problem ALMOST FOREST DELETION. The main focus of this paper is to design algorithm for this problem with parameters both l and k.

Our Results. We show that ALMOST FOREST DELETION can be solved in time $\mathcal{O}^*(5.0024^{(k+l)})$. We arrive at the result using the iterative compression technique which was introduced in [21] and a non-trivial measure which helps us in getting the desired running time. Then we explore the kernelization complexity of the problem, and show that ALMOST FOREST DELETION admits a polynomial kernel with $\mathcal{O}(kl(k + l))$ edges. For arriving at the result, we first make use of the Expansion Lemma and Gallai's theorem for reducing the maximum degree of the graph, and then we bound the size of the graph. It is easy to see that for a YES instances (G, k, l) of ALMOST FOREST DELETION, the treewidth of G is bounded by $k + l$. Since we have an algorithm of the form $\mathcal{O}^*(c^{(k+l)})$ on general graphs, the question of finding an $\mathcal{O}^*(c^{\mathbf{tw}})$ algorithm becomes interesting for bounded treewidth graphs. We answer this question affirmatively by giving an $\mathcal{O}^*(17^{\mathbf{tw}})$ algorithm for graphs which come with a tree decomposition of width $\mathbf{tw}$. This algorithm, along with the notion of bidimensionality gives rise to an algorithm for ALMOST FOREST DELETION on planar graphs running in time $2^{\mathcal{O}(\sqrt{l+k})}n^{\mathcal{O}(1)}$. Our methods are based on the known methods to solve the FEEDBACK VERTEX SET problem.

2 Preliminaries

For a graph G, we denote the set of vertices of the graph by $V(G)$ and the set of edges of the graph by $E(G)$. For a set $S \subseteq V(G)$, the *subgraph of G induced by S* is denoted by $G[S]$ and it is defined as the subgraph of G with vertex set S and edge set $\{(u, v) \in E(G) : u, v \in S\}$, and the subgraph obtained after deleting S is denoted as $G - S$. If H is a subgraph of G, we write $H \subseteq G$ and for two graphs $G_1 = (V_1, E_1)$ and $G_2 = (V_2, E_2)$, by $G_1 \cup G_2$, we denote the graph $(V_1 \cup V_2, E_1 \cup E_2)$. All vertices adjacent to a vertex v are called neighbours of v and the set of all such vertices is called the neighbourhood of v. A *k-flower* in a graph is a set of k cycles which are vertex disjoint except for one vertex v, which is shared by all the cycles in the set. The vertex v is called *center* of the flower and the cycles are called the *petals* of the flower. A *forest* is a graph which does not contain any cycle. An *l-forest* is a graph which is at most l edges away from being a forest, i.e. the graph can be transformed into a forest by deleting at most l edges. For a connected component C of a graph, we call the quantity $|E(G[C])| - |C| + 1$ the *excess of C* and denote it by $\mathsf{ex}(C)$. It can also be equivalently defined as the minimum number of edges we need to delete from the connected component to get to a tree. For a graph G, let $\mathcal{C}$ be the set of its connected components. We define the excess of the graph, $\mathsf{ex}(G)$ as follows.

$$\mathsf{ex}(G) = \sum_{C \in \mathcal{C}} \mathsf{ex}(C)$$

As in the case of components, this measure can be equivalently defined as the minimum number of edges we need to delete from G to to get to a forest. It is easy to see that a graph G is an l-forest if and only if $\mathsf{ex}(G) \leq l$. For $X \subseteq V(G)$ such that $G - X$ is an l-forest, we call X an l-forest deletion set of G. We denote $\{1, \ldots, n\}$ by $[n]$. We define the ALMOST FOREST DELETION problem as follows.

ALMOST FOREST DELETION
Input: A graph G, integers k and l.
Parameter(s): l, k
Question: Does there exist $X \subseteq V(G)$ such that $|X| \leq k$ and $G - X$ is an l-forest?

Observation 1. *Let G' be a subgraph of G. If G is an l-forest, then so is G'.*

Observation 2. *If G is an l-forest, it has at most $V(G) - 1 + l$ edges.*

Lemma 3. *Let G be a graph. If there exists a vertex v such that v is not part of any cycle in G, then $\mathsf{ex}(G - \{v\}) = \mathsf{ex}(G)$. Furthermore, if v is part of a cycle in G, then $\mathsf{ex}(G - \{v\}) \leq \mathsf{ex}(G) - 1$.*

Lemma 4. *Let $X \subseteq V(G)$ be a set of vertices of G which do not belong to any cycle. Then, G is an l-forest if and only if $G - X$ is an l-forest.*

Lemma 5. *Any l-forest can have at most l edge disjoint cycles.*

Kernelization. A *kernelization* algorithm for a parameterized language L is a polynomial time procedure which takes as input an instance $(x, k_1, \ldots, k_l)$, where k_i's are the parameters and returns an instance $(x', k'_1, \ldots, k'_l)$ such that $(x, k_1, \ldots, k_l) \in L$ if and only if $(x', k'_1, \ldots, k'_l) \in L$ and $|x'| \leq h(k_1, \ldots, k_l)$ and $k'_i \leq g(k_1, \ldots, k_l)$ for all $i \in [l]$, for some computable functions h, g. The returned instance is said to be a *kernel* for L and the function h is said to be the *size of the kernel*.

Treewidth. Let G be a graph. A *tree-decomposition* of a graph G is a pair $(\mathbb{T}, \mathcal{X} = \{X_t\}_{t \in V(\mathbb{T})})$ such that

- $\cup_{t \in V(\mathbb{T})} X_t = V(G)$,
- for every edge $xy \in E(G)$ there is a $t \in V(\mathbb{T})$ such that $\{x, y\} \subseteq X_t$, and
- for every vertex $v \in V(G)$ the subgraph of $\mathbb{T}$ induced by the set $\{t \mid v \in X_t\}$ is connected.

The *width* of a tree decomposition is $\max_{t \in V(\mathbb{T})} |X_t| - 1$ and the *treewidth* of G is the minimum width over all tree decompositions of G and is denoted by $\mathbf{tw}(G)$.

A tree decomposition $(\mathbb{T}, \mathcal{X})$ is called a *nice tree decomposition* if $\mathbb{T}$ is a tree rooted at some node r where $X_r = \emptyset$, each node of $\mathbb{T}$ has at most two children, and each node is of one of the following kinds:

1. **Introduce node:** a node t that has only one child t' where $X_t \supset X_{t'}$ and $|X_t| = |X_{t'}| + 1$.
2. **Forget node:** a node t that has only one child t' where $X_t \subset X_{t'}$ and $|X_t| = |X_{t'}| - 1$.
3. **Join node:** a node t with two children t_1 and t_2 such that $X_t = X_{t_1} = X_{t_2}$.
4. **Base node:** a node t that is a leaf of $\mathbb{T}$, is different than the root, and $X_t = \emptyset$.

Notice that, according to the above definition, the root r of $\mathbb{T}$ is either a forget node or a join node. It is well known that any tree decomposition of G can be transformed into a nice tree decomposition maintaining the same width in linear time [16]. We use G_t to denote the graph induced by the vertex set $\cup_{t'} X_{t'}$, where t' ranges over all descendants of t, including t. By $E(X_t)$ we denote the edges present in $G[X_t]$.

3   An $O^*(c^{(l+k)})$ Algorithm for ALMOST FOREST DELETION

In this section we will present a $c^{(l+k)} n^{\mathcal{O}(1)}$ algorithm for ALMOST FOREST DELETION. We use the well known technique of iterative compression and arrive at the desired running time after defining a non-trivial measure.

Given an instance (G, k, l) of ALMOST FOREST DELETION, let $V(G) = \{v_1, \ldots, v_n\}$ and define vertex sets $V_i = \{v_1, \ldots, v_i\}$, and let the graph $G_i = G[V_i]$. We iterate through the instances (G_i, k, l) starting from $i = k + 1$. For the ith instance, we try to find an l-forest deletion set $\hat{S}_i$ of size at most k, with the help of a *known* l-forest deletion set S_i of size at most $k + 1$. Formally, the compression problem we address is the following.

ALMOST FOREST DELETION COMPRESSION
Input: A graph G, an l-forest deletion set S of G of size at most $k+1$, integers k and l.
Parameter(s): k, l
Question: Does there exist $X \subseteq V(G)$ such that $|X| \leq k$ and $G - X$ is an l-forest?

Lemma 6. *If* ALMOST FOREST DELETION COMPRESSION *can be solved in* $f(k, l) n^c$ *time, then* ALMOST FOREST DELETION *can be solved in* $f(k, l) n^{c+1}$ *time.*

For designing an algorithm for ALMOST FOREST DELETION COMPRESSION, let the input instance be (G, S, k, l). We guess a subset $Y \subseteq S$ with the intention of picking these vertices in our hypothetical solution for this instance and not picking the rest of the vertices of S in the solution. We delete the set Y from the graph and decrease k by $|Y|$. We then check if the graph $G[S \setminus Y]$ is an l-forest and if it is not, then reject this guess of Y as a spurious guess. Suppose that $G[S \setminus Y]$ is indeed an l-forest. Then, it remains only to check if there is an l-forest deletion set S' of the size $k' = k - |Y|$ which is disjoint from $S \setminus Y$, and $G - (Y \cup S')$ is an l-forest. More precisely, we have an instance of AFDDC, which is defined as follows.

ALMOST FOREST DELETION DISJOINT COMPRESSION (AFDDC)
Input: A graph G, an l-forest deletion set S of G, integers k and l.
Parameter(s): k, l
Question: Does there exist $X \subseteq V(G)$ such that $X \cap S = \emptyset$, $|X| \leq k$ and $G - X$ is an l-forest?

To solve the problem, we first design a set of reduction rules.

Reduction Rule 1. *If there exists a vertex v of degree at most 1 in the graph, delete it.*

Reduction Rule 2. *If there exists $v \in V(G) \setminus S$ such that $G[S \cup \{v\}]$ is not an l-forest, delete v and decrease k by 1.*

Reduction Rule 3. *If there exists a vertex $v \in V(G) \setminus S$ of degree two, such that at least one of its neighbours is in $V(G) \setminus S$, then delete v and put a new edge between its neighbours (even if they were already adjacent). If both of v's edges are to the same vertex, delete v and put a new self loop on the adjacent vertex (even if it has self loop(s) already).*

It is easy to see that after the exhaustive applications of reduction rules 1-3, if there exists a vertex of degree at most 1 in $G - S$, then it has at least 2 neighbours in S.

Now we are ready to describe our algorithm for AFDDC. Given an input instance (G, S, k, l) of AFDDC, we first apply reduction rules 1, 2 and 3 exhaustively. If $k < 0$, then we return that the given instance is a NO instance.

Now, we look for a vertex v of degree at most 1 in $G - S$ and we branch by either including v in our solution or excluding it. More precisely, we call the algorithm recursively on $(G - \{v\}, S, k - 1, l)$ and $(G, S \cup \{v\}, k, l)$. If one of the recursive call returns YES, then we say that the instance was a YES instance. If there does not exist a vertex of degree at most 1 in $G - S$, then there must be a vertex v which is part of a cycle. In this case we branch on this vertex, and call the algorithm recursively on $(G - \{v\}, S, k - 1, l)$ and $(G, S \cup \{v\}, k, l)$ as we did in the previous case. This concludes the description of the algorithm. The correctness of the algorithm follows from the correctness of reduction rules and the fact that the branching is exhaustive.

To analyze the running time of the algorithm, we define a measure $\phi(I)$ for the input instance $I = (G, S, k, l)$ as follows.

$$\phi(I) = \alpha k + \beta \mathsf{cc}(S) + \gamma(l - \mathsf{ex}(G[S])) + \delta(\mathsf{ex}(G - S))$$

Here, $\mathsf{cc}(S)$ denotes the number of connected components of $G[S]$ and α, β, γ, δ are positive constants such that $\delta > \beta$. We will assume these properties for now, and will fix the values of these constants later.

Lemma 7. *None of the reduction rules 1-3 increases the measure $\phi(I)$.*

Lemma 8. *AFDDC can be solved in time $\mathcal{O}^*((4.0024)^k(5.0018)^l)$.*

Proof. When be branch on a vertex v of degree at most 1 in $G - S$, the measure drops by α in the first branch. In the other branch, depending on whether v's neighbours in S belong to different components or to the same component, the measure drops by at least β or γ. This gives us branching factors of (α, β) and (α, γ). Branching on a vertex v, which is part of a cycle in $G - S$, gives us a branching factor of $(\alpha + \delta, \delta - \beta)$. We ran a numerical program to find values of α, β, γ and δ, which optimize the running time of the algorithm. Putting $\alpha = 1.45$, $\beta = 1.35$, $\gamma = 1.35$ and $\delta = 1.9$ gives us the worst case running time of $(4.0024)^k(5.0018)^l n^{\mathcal{O}(1)}$. □

Given Lemma 8, the algorithm for ALMOST FOREST DELETION COMPRESSION runs in time $\mathcal{O}(\sum_{i=0}^{k} \binom{k+1}{i} \cdot (4.0024)^i(5.0018)^l \cdot n^{\mathcal{O}(1)}) = \mathcal{O}^*(5.0024^{(k+l)})$. Here, the factor of $\binom{k+1}{i}$ is for the guesses we make for the set S. Finally applying Lemma 6, we get the following theorem.

Theorem 9. ALMOST FOREST DELETION *can be solved in $\mathcal{O}^*(5.0024^{(k+l)})$ time.*

4 $\mathcal{O}(kl(k + l))$ Kernel for ALMOST FOREST DELETION

In this section, we give the kernelization algorithm for ALMOST FOREST DELETION. First we give a set of reduction rules which help us bound the size of the output instance. Throughout the section, we apply the reduction rules in order, that is, while applying a reduction rule we assume that all the reduction rules stated previously in the section have been applied exhaustively.

Reduction Rule 4. *If there exists a vertex v of degree at most 1 in the graph, delete it.*

Reduction Rule 5. *If there exists a vertex $v \in V(G)$ of degree two then delete v and put a new edge between its neighbours (even if they were already adjacent). If both of v's edges are to the same vertex, delete v and put a new self loop on the adjacent vertex (even if it has self loop(s) already).*

Reduction Rule 6. *If any edge has multiplicity more that $l+2$, then delete all but $l+2$ copies of that edge.*

Given an instance (G, k, l) of ALMOST FOREST DELETION, we apply reduction rules 4-6 exhaustively. Observe that after the application of these reduction rules, the graph has degree at least 3, as all the vertices of degrees 1 and 2 are taken care of by Reduction Rule 4 and Reduction Rule 5 respectively.

Lemma 10. *If a graph G has minimum degree at least 3, maximum degree at most d, and an l-forest deletion set of size at most k, then it has less than $2l + k(d+1)$ vertices and less than $2kd + 3l$ edges.*

Lemma 10 gives rise to the following reduction rule immediately.

Reduction Rule 7. *After the application of reduction rules 4, 5 and 6 exhaustively, if either $|V(G)| \geq 2l + k(d+1)$ or $|E(G)| \geq 2kd + 3l$, where d is the maximum degree of the graph, return that the given instance is a No instance.*

After this, all that is left is to reduce the maximum degree of the graph. After the exhaustive application of reduction rules 4, 5 and 6, if the maximum degree of the graph is already bounded by $(k + l)(3l + 8)$ then we already have a kernel with $\mathcal{O}(kl(k+l))$ vertices and $\mathcal{O}(kl(k+l))$ edges. Hence we assume, for the rest of the section, that after the exhaustive application of reduction rules 4- 7, there exists a vertex v with degree greater than $(k + l)(3l + 8)$. We need one more reduction rule before we proceed further.

Reduction Rule 8. *If there is a vertex v with more than l self loops, delete v and decrease k by 1.*

We now try to reduce the high degree vertices. The idea is that either a high degree vertex participates in many cycles (and contributes many excess edges) and hence should be part of the solution, or only a small part of its neighbourhood is relevant for the solution. We formalize these notions by use of Gallai's theorem to find flowers and applying a set of reduction rules. Given a set $T \subseteq V(G)$, by T-path we mean set of paths of positive length with both endpoints in T.

Theorem 11 (Gallai, [11]). *Given a simple graph G, a set $T \subseteq V(G)$ and an integer s, one can in polynomial time find either*

- *a family of $s + 1$ pairwise vertex-disjoint T-paths, or*
- *a set B of at most $2s$ vertices, such that in $G - B$ no connected component contains more than one vertex of T.*

We would want to have the neighborhood of a high degree vertex as the set T for applying Gallai's theorem and for detecting flowers. But we need to be careful, as the graph in its current form contains multiple edges and self loops. Let v be a vertex with high degree. The vertices in $N(v)$ which have at least two parallel edges to v can be greedily picked to form a petal of the flower. Let L be the set of vertices in $N(v)$ which have at least two parallel edges to v.

Reduction Rule 9. *If $|L| > k + l$, delete v and decrease k by 1.*

Let $\widehat{G}$ be the graph $G - L$ with all parallel edges replaced with single edges, and all self loops removed. It is not hard to show that finding an f-flower in G centered at v is equivalent to finding an $f - |L|$ flower in $\widehat{G}$ centered at v for any $f \geq |L|$. Now we apply Gallai's theorem on $\widehat{G}$ with $T = N(v)$ and $s = k+l-|L|$. If the theorem returns a collection of vertex disjoint T-paths, then it is easy to see that they are in one to one correspondence with cycles including v, and hence can be considered petals of the flower centered at v.

Reduction Rule 10. *If the application of Gallai's theorem returns a flower with more than s petals, then delete v and decrease k by 1.*

We now deal with the case when the application of Gallai's theorem returns a set B of at most $2(k+l-|L|)$ vertices, such that in $\widehat{G}-B$ no connected component contains more than one vertex of T. Let $Z = B \cup L$. Clearly, $|Z| \leq 2(k+l) - |L|$. Now we look at the set of connected components of $\widehat{G} - (Z \cup \{v\})$. Let us call this set C.

Reduction Rule 11. *If more than $k+l$ components of C contain a cycle, then return that the instance is a NO instance.*

Lemma 12. *After applying reduction rules $4-11$ exhaustively, there are at least $2(l+2)(k+l)$ components in C which are trees and connected to v with exactly one edge.*

Proof. The number of self loops on v is bounded by l due to Reduction Rule 8. Number of edges from v to Z is bounded by $|B| + (l+2)|L| \leq 2(k+l-|L|)+(l+2)|L| = 2(k+l)+l|L| \leq (k+l)(l+2)$. As degree of v is greater than $(k+l)(3l+8)$, at least $(k+l)(3l+8) - (k+l)(l+2) - l \geq (k+l)(2l+5)$ connected components in C have exactly one vertex which is is neighbour of v. Out of these, the number of connected components containing cycles is bounded by $k + l$ by Reduction Rule 11. Hence, at least $2(l+2)(k+l)$ connected components are trees and are connected to v by exactly one edge. $\qquad\square$

Before we proceed further, we state the Expansion Lemma. Let G be a bipartite graph with vertex bipartition (A, B). For a positive integer q, a set of edges $M \subseteq E(G)$ is called a q-expansion of A into B if every vertex of A is incident with exactly q edges of M, and exactly $q|A|$ vertices in B are incident to M.

Lemma 13 (Expansion Lemma, [8]). *Let $q \geq 1$ be a positive integer and G be a bipartite graph with vertex bipartition (A, B) such that $|B| \geq q|A|$ and there are no isolated vertices in B. Then there exist nonempty vertex sets $X \subseteq A$ and $Y \subseteq B$ such that there is a q-expansion of X into Y and no vertex in Y has a neighbor outside X, that is, $N(Y) \subseteq X$. Furthermore, the sets X and Y can be found in time polynomial in the size of G.*

Let D the set of connected components of C which are trees and connected to v with exactly one edge. We have shown that $|D| \geq 2(l+2)(k+l)$. Now we construct an auxiliary bipartite graph H as follows. In one partition of H, we have a vertex for every connected component in D, and the other partition is Z. We put an edge between $A \in D$ and $v \in Z$ if some vertex of A is adjacent to v. Since every connected component in D is a tree and has only one edge to v, some vertex in it has to have a neighbour in Z, otherwise Reduction Rule 1 would apply. Now we have that $|Z| \leq 2(k+l)$ and every vertex in D is adjacent to some vertex in Z, we may apply Expansion Lemma with $q = l+2$. This means, that in polynomial time, we can compute a nonempty set $\widehat{Z} \subseteq Z$ and a set of connected components $\widehat{D} \subseteq D$ such that:

1. $N_G(\bigcup_{D \in \widehat{D}} D) = \widehat{Z} \cup \{v\}$, and
2. Each $z \in \widehat{Z}$ will have $l+2$ private components $A_z^1, A_z^2, \ldots A_z^{l+2} \in \widehat{D}$ such that $z \in N_G(A_z^i)$ for all $i \in [l+2]$. By private we mean that the components $A_z^1, A_z^2, \ldots A_z^{l+2}$ are all different for different $z \in \widehat{Z}$.

Lemma 14. *For any l-forest deletion set X of G that does not contain v, there exists an l-forest deletion set X' in G such that $|X'| \leq |X|$, $X' \cap (\bigcup_{A \in \widehat{D}} A) = \emptyset$ and $\widehat{Z} \subseteq X'$.*

Now we are ready to give the final reduction rule.

Reduction Rule 12. *Delete all edges between v and $\bigcup_{A \in \widehat{D}} A$ and put $l+2$ parallel edges between v and z for all $z \in \widehat{Z}$.*

Theorem 15. ALMOST FOREST DELETION *admits a kernel with $\mathcal{O}(kl(k+l))$ vertices and $\mathcal{O}(kl(k+l))$ edges.*

Proof. First we show that either we have a kernel of the desired size or one of the reduction rules 4-12 apply. So, we only need to define a measure which is polynomial in the size of the graph and show that each of the reduction rules decrease the measure by a constant. We define the measure of a graph G to be $\phi(G) = 2|V(G)| + |E_{\leq l+2}|$, where $E_{\leq l+2}$ is set of edges with multiplicity at most $l+2$. Then we show that each of the reduction rules either terminate the algorithm or decrease the measure by a constant. $\square$

5 An $O^*(c^{\mathbf{tw}})$ Algorithm for ALMOST FOREST DELETION

In this section, we first design an algorithm, which given an instance (G, k) of ALMOST FOREST DELETION along with a tree decomposition of G of width at most $\mathbf{tw}$, solves it in time $O^*(c^{\mathbf{tw}})$. Then, using that algorithm, we give a subexponential time algorithm for ALMOST FOREST DELETION on planar graphs.

Theorem 16. *Given an instance (G, k, l) of ALMOST FOREST DELETION along with tree decomposition of G of width at most $\mathbf{tw}$, it can be solved in $\mathcal{O}^*(c^{\mathbf{tw}})$ time.*

Proof. For solving the problem in desired running time, we do dynamic programming on the tree decomposition of G in a bottom-up manner, while storing partial solution for each node of the tree. We use representative sets to store the information about connectivity of the partial solutions. This can be done using graphic matroid of a clique on vertices which are mapped to the tree node. We also need to store the information about the extra edges. But since in an l-forest it does not matter where exactly the extra edges are, we can make use of this fact and solve the problem efficiently. □

Theorem 17 (Planar Extended Grid Theorem [12, 22]). *Let t be a nonnegative integer. Then every planar graph G of treewidth at least $\frac{9}{2}t$ contains $\boxplus_t$ as a minor. Furthermore, for every $\epsilon > 0$ there exists an $\mathcal{O}(n^2)$ algorithm that, for a given n-vertex planar graph G and integer t, either outputs a tree decomposition of G of width at most $(\frac{9}{2} + \epsilon)t$, or returns that $\boxplus_t$ is a minor of G, where $\boxplus_t$ denotes a grid of dimension $t \times t$.*

Lemma 18. *Let X be an l-forest deletion set of $\boxplus_t$ of size at most k, then $t \le \sqrt{l + 3k} + 1$.*

Theorems 16 and 17, along with Lemma 18 are combined to get the subexponential time algorithm on planar graphs.

Theorem 19. ALMOST FOREST DELETION *can be solved in $2^{\mathcal{O}(\sqrt{l+k})}n^{\mathcal{O}(1)}$ time on planar graphs.*

6 Conclusions

In this paper we studied ALMOST FOREST DELETION and obtained a polynomial kernel as well as a single exponential time algorithm for the problem. It would be interesting to study other classical problems from this view-point of distance from tractability using a suitable measure of distance.

References

1. Amini, O., Fomin, F.V., Saurabh, S.: Implicit branching and parameterized partial cover problems. J. Comput. Syst. Sci. **77**(6), 1159–1171 (2011)
2. Bodlaender, H.L., Koster, A.M.C.A.: Combinatorial optimization on graphs of bounded treewidth. Comput. J. **51**(3), 255–269 (2008)
3. Cygan, M., Nederlof, J., Pilipczuk, M., Pilipczuk, M., van Rooij, J.M.M., Wojtaszczyk, J.O.: Solving connectivity problems parameterized by treewidth in single exponential time. In: IEEE 52nd Annual Symposium on Foundations of Computer Science, FOCS 2011, Palm Springs, CA, USA, October 22–25, 2011, pp. 150–159 (2011)
4. Downey, R.G., Fellows, M.R.: Fundamentals of Parameterized Complexity. Texts in Computer Science. Springer (2013)

5. Fellows, M.R., Jansen, B.M.P., Rosamond, F.A.: Towards fully multivariate algorithmics: Parameter ecology and the deconstruction of computational complexity. Eur. J. Comb. **34**(3), 541–566 (2013)
6. Fellows, M.R., Lokshtanov, D., Misra, N., Mnich, M., Rosamond, F.A., Saurabh, S.: The complexity ecology of parameters: An illustration using bounded max leaf number. Theory Comput. Syst. **45**(4), 822–848 (2009)
7. Flum, J., Grohe, M.: Parameterized Complexity Theory (Texts in Theoretical Computer Science. An EATCS Series). Springer-Verlag New York Inc., Secaucus (2006)
8. Fomin, F.V., Lokshtanov, D., Misra, N., Philip, G., Saurabh, S.: Hitting forbidden minors: approximation and kernelization. In: 28th International Symposium on Theoretical Aspects of Computer Science, STACS 2011, March 10–12, 2011, Dortmund, Germany, pp. 189–200 (2011)
9. Fomin, F.V., Lokshtanov, D., Misra, N., Saurabh, S.: Planar f-deletion: approximation, kernelization and optimal FPT algorithms. In: 53rd Annual IEEE Symposium on Foundations of Computer Science, FOCS 2012, New Brunswick, NJ, USA, October 20–23, 2012, pp. 470–479 (2012)
10. Fomin, F.V., Lokshtanov, D., Raman, V., Saurabh, S.: Subexponential algorithms for partial cover problems. Inf. Process. Lett. **111**(16), 814–818 (2011)
11. Gallai, T.: Maximum-minimum stze und verallgemeinerte faktoren von graphen. Acta Mathematica Academiae Scientiarum Hungarica **12**(1–2), 131–173 (1964)
12. Gu, Q., Tamaki, H.: Improved bounds on the planar branchwidth with respect to the largest grid minor size. Algorithmica **64**(3), 416–453 (2012)
13. Guo, J., Hüffner, F., Niedermeier, R.: A structural view on parameterizing problems: distance from triviality. In: Downey, R.G., Fellows, M.R., Dehne, F. (eds.) IWPEC 2004. LNCS, vol. 3162, pp. 162–173. Springer, Heidelberg (2004)
14. Kawarabayashi, K., Mohar, B., Reed, B.A.: A simpler linear time algorithm for embedding graphs into an arbitrary surface and the genus of graphs of bounded tree-width. In 49th Annual IEEE Symposium on Foundations of Computer Science, FOCS 2008, October 25–28, 2008, Philadelphia, PA, USA, pp. 771–780 (2008)
15. Kim, E.J., Langer, A., Paul, C., Reidl, F., Rossmanith, P., Sau, I., Sikdar, S.: Linear kernels and single-exponential algorithms via protrusion decompositions. In: Fomin, F.V., Freivalds, R., Kwiatkowska, M., Peleg, D. (eds.) ICALP 2013, Part I. LNCS, vol. 7965, pp. 613–624. Springer, Heidelberg (2013)
16. Kloks, T.: Treewidth, Computations and Approximations. LNCS, vol. 842. Springer, Heidelberg (1994)
17. Kociumaka, T., Pilipczuk, M.: Faster deterministic feedback vertex set. Inf. Process. Lett. **114**(10), 556–560 (2014)
18. Komusiewicz, C., Niedermeier, R., Uhlmann, J.: Deconstructing intractability - A multivariate complexity analysis of interval constrained coloring. J. Discrete Algorithms **9**(1), 137–151 (2011)
19. Lewis, J.M., Yannakakis, M.: The node-deletion problem for hereditary properties is NP-complete. J. Comput. Syst. Sci. **20**(2), 219–230 (1980)
20. Niedermeier, R.: Invitation to Fixed Parameter Algorithms (Oxford Lecture Series in Mathematics and Its Applications). Oxford University Press, USA (2006)
21. Reed, B.A., Smith, K., Vetta, A.: Finding odd cycle transversals. Oper. Res. Lett. **32**(4), 299–301 (2004)
22. Robertson, N., Seymour, P.D., Thomas, R.: Quickly excluding a planar graph. J. Comb. Theory, Ser. B **62**(2), 323–348 (1994)

Approximation Algorithms

Improved Approximation Algorithms for Min-Max and Minimum Vehicle Routing Problems

Wei Yu and Zhaohui Liu[✉]

Department of Mathematics, East China University of Science and Technology,
Shanghai 200237, China
{yuwei,zhliu}@ecust.edu.cn

Abstract. Given an undirected weighted graph $G = (V, E)$, a set $C_1, C_2, \ldots, C_k$ of cycles is called a *cycle cover* of V' if $V' \subset \cup_{i=1}^{k} V(C_i)$ and its cost is the maximum weight of the cycles. The Min-Max Cycle Cover Problem(MMCCP) is to find a minimum cost cycle cover of V with at most k cycles. The Rooted Min-Max Cycle Cover Problem(RMMCCP) is to find a minimum cost cycle cover of $V \backslash D$ with at most k cycles and each cycle contains one vertex in D. The Minimum Cycle Cover Problem(MCCP) aims to find a cycle cover of V of cost at most λ with minimum number of cycles. We propose approximation algorithms for the MMCCP, RMMCCP and MCCP with ratios 5, 6 and 24/5, respectively. Our results improve the previous algorithms in term of both approximation ratios and running times. Moreover, we transform a ρ-approximation algorithm for the TSP into approximation algorithms for the MMCCP, RMMCCP and MCCP with ratios 4ρ, $4\rho + 1$ and 4ρ, respectively.

Keywords: Vehicle routing · Cycle cover · Traveling salesman problem · Approximation algorithm

1 Introduction

In the last two decades, considerable research attention has been devoted to the following fundamental vehicle routing problem. Given a fleet of k vehicles and a general network, there is exactly one customer located at each vertex. Each vehicle has to start from some vertex to visit some customers and return to the same vertex. There is a travel cost for each pair of vertices that obeys the triangle inequality. The goal is to find a routing for the vehicles to collectively visit all the customers such that the maximum traveling cost of the vehicles is minimum. If described by graph theoretic language, the above problem is to cover all the vertices of an undirected weighted graph with at most k cycles such that the maximum weight of the cycles is minimum. It is called the Min-Max Cycle Cover Problem(MMCCP) in the literature(see [15]). In the rooted version, called the Rooted Min-Max Cycle Cover Problem(RMMCCP), the objective is to use at most k rooted cycles, i.e., cycles contain one vertex of a given depot set

© Springer International Publishing Switzerland 2015
D. Xu et al. (Eds.): COCOON 2015, LNCS 9198, pp. 147–158, 2015.
DOI: 10.1007/978-3-319-21398-9_12

of vertices, to cover the non-depot vertices such that the maximum weight of the cycles is minimum. In the MMCCP and RMMCCP, if an upper bound $\lambda > 0$ is given for the weight of each cycle and the goal is to minimize the number of cycles used to cover the vertices, we obtain the Minimum Cycle Cover Problem(MCCP) and the Rooted Minimum Cycle Cover Problem(RMCCP), respectively.

The above-mentioned vehicle routing problems and their variants find numerous applications in operations research and computer science. RMMCCP and MMCCP were introduced by Even et al. [6] to model "Nurse Station Location Problem". Campbell et al. [4] illustrated how disaster relief efforts can be improved by efficient algorithms for min-max cycle/path cover problems. Xu et al. [15] described some applications of cycle cover problems in wireless sensor networks. For more practical examples involving min-max and minimum vehicle routing problems we refer to [1,14,16–18] and the references therein.

Unfortunately, all the problems RMMCCP, MMCCP, RMCCP, MCCP are NP-hard since they are extensions of the well-know Traveling Salesman Problem. Therefore, previous results mainly focus on devising approximation algorithms with good performance ratios.

1.1 Previous Works

Xu et al. [18] showed that both the MMCCP and the RMMCCP cannot be approximated within ratio 4/3, unless P=NP. Xu and Wen [16] gave an inapproximability bound of 20/17 for the single-depot RMMCCP. By the NP-completeness of the well-known Hamiltonian Cycle Problem, both the MCCP and the RMCCP can not be approximated within ratio 2.

For the MMCCP, a closely related problem, called the Min-Max Tree Cover Problem(MMTCP), can be obtained by replacing cycles with trees. On the one hand, the optimal value of the MMTCP can not be greater than that of the MMCCP. On the other hand, by duplicating each edge of a feasible solution of the MMTCP we obtain a feasible solution of MMCCP with the objective value doubled. Therefore, any α-approximation algorithm for the MMTCP implies a 2α-approximation algorithm for the MMCCP. Even et al. [6] and Arkin et al. [1] developed independently 4-approximation algorithms for the MMTCP by different algorithmic techniques. Khani and Salavatipour [10] give an improved 3-approximation algorithm, which implies a 6-approximation algorithm for the MMCCP. Xu et al. [18] also derived an approximation algorithm with the same ratio 6. These algorithms were improved to a 16/3-approximation algorithm by Xu et al. [15].

For the RMMCCP, Xu et al. [18] proposed a 7-approximation algorithm. Later, Xu et al. [15] improved the approximation ratio to 19/3. For the single-depot RMMCCP, Frederickson et al. [7] achieved a better ratio of $\rho+1$, where ρ is the approximation ratio of the best available algorithm for the Traveling Salesman Problem. By using the well-known Christofides' Algorithm[5] this implies a 5/2-approximation algorithm. Moreover, Nagamochi [11], Nagamochi and Okada [11,13] obtained better results on a special case of the RMMCCP where the graph is the metric closure of a tree.

In the MCCP, by replacing cycles with trees we derive the Minimum Tree Cover Problem(MTCP), which is also named as Bounded Tree Cover Problem in [10]. Since a cycle of weight at most λ can be splitted into two paths(which are also trees) of weight at most $\frac{\lambda}{2}$ by removing properly two edges, two times the optimal value of the MCCP can not be less than the optimal value of the corresponding MTCP with the upper bound on the weight of the trees reset to $\frac{\lambda}{2}$. On the other hand, by doubling the edges of a feasible solution of the MTCP with the revised upper bound we obtain a feasible solution of the MCCP. Therefore, any α-approximation algorithm for the MTCP implies a 2α-approximation algorithm for the MCCP. Arkin et al. [1] developed a 6-approximation algorithm for the MCCP. This result is also implied by the 3-approximation algorithm for the MTCP in the same paper. Khani and Salavatipour [10] gave an improved 5/2-approximation algorithm for the MTCP, which indicates a 5-approximation algorithm for the MCCP.

The approximability of the RMCCP is far less understood. All the existing results focus on the case of one single depot vertex, which are called the Distance Constrained Vehicle Routing Problem by Nagarajan and Ravi [14]. The authors proposed a $\min\{\log n, \log \lambda\}$-approximation algorithm for the general problem and a 2-approximation algorithm for the problem defined on the metric closure of a tree. Recently, Friggstad and Swamy [8] obtained an improved $\frac{\log \lambda}{\log \log \lambda}$-approximation algorithm for the general problem.

1.2   Our Results and Techniques

In this paper we focus on the MMCCP, RMMCCP and MCCP. Our main contributions are fourfold. Firstly, we obtain a 5-approximation algorithm for the MMCCP, which improves the previous best 16/3-approximation algorithm by Xu et al. [15]. Meanwhile, this algorithm also improves the running time from $O(n^5 \log \sum_{e \in E} w(e))$ to $O(n^3 \log \sum_{e \in E} w(e))$, where n is the number of vertices, E is the edge set of the graph and $w(e)$ is the weight of edge e. Moreover, we transform a ρ-approximation algorithm for the TSP into a 4ρ-approximation algorithm for the MMCCP, which implies further improvement on the performance ratio for the problem defined on some special metrics(e.g. Euclidean metric). Secondly, we show that any α-approximation algorithm for the MMCCP implies an $(\alpha + 1)$-approximation algorithm for the RMMCCP. This indicates a 6-approximation algorithm for the RMMCCP, beating the 19/3-approximation algorithm in [15] in term of both performance ratio and running time. Thirdly, we devise a 24/5-approximation algorithm for the MCCP with running time $O(n^4)$. In contrast, the previous best 5-approximation algorithm by Khani and Salavatipour [10] runs in $O(n^5)$ time. Lastly, we introduce a new matching-based upper bound analysis for the MMCCP which proves to be more efficient than the strategy of doubling tree edges in the literature.

The rest of the paper is organized as follows. We formally state the problems and give some preliminary results in Section 2. We deal with the MMCCP and RMMCCP in Section 3 and Section 4, respectively. The MCCP is treated in Section 5.

2   Preliminaries

Given an undirected weighted graph $G = (V, E)$ with vertex set V and edge set E, $w(e)$ denotes the weight or length of edge e. If $e = (u, v)$, we also use $w(u, v)$ to denote the weight of e. For $B > 0$, $G[B]$ is the subgraph of G obtained by removing all the edges in E with weight greater than B. For a subgraph H(e.g. tree, cycle, path, matching) of G, let $V(H), E(H)$ be the vertex set and edge set of H, respectively. The weight of H is defined as $w(H) = \sum_{e \in E(H)} w(e)$. If H is connected, let $MST(H)$ be the minimum spanning tree on $V(H)$ and its weight $w(MST(H))$ is simplified to $w_T(H)$. A cycle C is also called a tour on $V(C)$. The weight of a path or cycle is also called its length. A cycle(path, tree) containing only one vertex and no edges is a trivial cycle(path, tree) and its weight is defined as zero.

For a subset V' of V, a set $C_1, C_2, \ldots, C_k$ of cycles(some of them may be trivial cycles) is called a *cycle cover* of V' if $V' \subset \cup_{i=1}^k V(C_i)$. And the cost of this cycle cover is defined as $\max_{1 \leq i \leq k} w(C_i)$, i.e., the maximum weight of the cycles. Particularly, a cycle cover of V is simply called a cycle cover. By replacing cycles with trees we can define *tree cover* and its cost similarly.

Now we formally state the problems.

Definition 1. *In the Min-Max Cycle Cover Problem(MMCCP), we are given an undirected complete graph $G = (V, E)$ with a metric nonnegative weight function w on E and a positive integer k, the goal is to find a minimum cost cycle cover with at most k cycles.*

Definition 2. *In the Rooted Min-Max Cycle Cover Problem(RMMCCP), we are given an undirected complete graph $G = (V, E)$ with a metric nonnegative weight function w on E, a depot set $D \subset V$, and a positive integer k, the objective is to find a minimum cost cycle cover of $V \backslash D$ with at most k cycles such that each cycle contains exactly one vertex of D.*

Definition 3. *In the Minimum Cycle Cover Problem(MCCP), we are given an undirected complete graph $G = (V, E)$ with a metric nonnegative weight function w on E and a positive λ, the aim is to find a cycle cover of cost at most λ such that the number of cycles in the cycle cover is minimum.*

Note that in the above problem definitions we assume that the graph is complete. This involves no loss of generality, since we can take the metric closure of a connected graph G if it is not complete. When G is not connected, we simply consider the corresponding problems defined on each connected component. We also suppose w.l.o.g. that the weight of the edges and λ are integers.

Given an instance of the MMCCP(RMMCCP, MCCP), OPT indicates the optimal value and each cycle in the optimal solution is called an optimum cycle. By the triangle inequality, we can assume w.l.o.g that any two optimum cycles are vertex-disjoint. We use n to denote the number of vertices of G.

The following cycle-splitting and tree-decomposition results are very useful in the design and analysis of our algorithms.

Lemma 1. *[1, 7, 18] Given a tour C on V' and $B > 0$, we can split the tour into $\lceil \frac{w(C)}{B} \rceil$ paths of length at most B such that each vertex is located at exactly one path in $O(|V'|)$ time.*

Lemma 2. *[6, 10] Given $B > 0$ and a tree T with $\max_{e \in E(T)} w(e) \le B$, we can decompose T into $k \le \max\{\lfloor \frac{w(T)}{B} \rfloor, 1\}$ edge-disjoint trees $T_1, T_2, \ldots, T_k$ with $w(T_i) \le 2B$ for each $i = 1, 2, \ldots, k$ in $O(|V(T)|)$ time.*

3   Min-Max Cycle Cover

In this section we first show how to transform a ρ-approximation algorithm for the TSP into a 4ρ-approximation algorithm for the MMCCP. Next we give a 5-approximation algorithm for the MMCCP with running time $O(n^3 \log \sum_{e \in E} w(e))$.

As in the previous results, an α-approximation algorithm can be derived in two phases. First, we guess a objective value λ. If $OPT \le \lambda$, we succeed in constructing a cycle cover with at most k cycles of cost no more than $\alpha\lambda$. Second, by a binary search in $[0, \sum_{e \in E} w(e)]$ we find the minimum value λ^* such that a cycle cover with at most k cycles whose cost is at most $\alpha\lambda^*$ can be constructed. This cycle cover is an α-approximate solution since $\lambda^* \le OPT$ by definition. Since the second phase is standard, we focus on the first phase.

Given a ρ-approximation algorithm for the TSP, our first algorithm is described below.

Algorithm $MMCCP(\rho, \lambda)$

Step 1. Delete all the edges with weight greater than $\frac{\lambda}{2}$ in G. The resulted graph $G[\frac{\lambda}{2}]$ has p connected components $F_1, F_2, \ldots, F_p$.

Step 2. For each $i = 1, 2, \ldots, p$, find a ρ-approximate tour TSP_i on $V(F_i)$ and split it into $k_i = \max\{\lceil \frac{w(TSP_i)}{2\rho\lambda} \rceil, 1\}$ paths of length no more than $2\rho\lambda$ by Lemma 1.

Step 3. Connect the two end vertices of each path constructed in Step 2 to obtain $\sum_{i=1}^{p} k_i$ cycles which constitute a cycle cover. If $\sum_{i=1}^{p} k_i \le k$, return this cycle cover; otherwise, return failure.

Let $C_1^*, C_2^*, \ldots, C_{k'}^*$ with $k' \le k$ be all the vertex-disjoint optimum cycles. First, we give two observations that are also noted in [18] and [15]. For any $e = (u, v) \in C_i^*$, the weight of C_i^* consists of $w(e)$ and the weight of a path P from u to v. By the triangle inequality, $w(P) \ge w(e)$. This implies $OPT \ge w(C_i^*) \ge 2w(e)$. So we have

Observation 1. *If $OPT \le \lambda$, then $w(e) \le \frac{\lambda}{2}$ for each $e \in \cup_{i=1}^{k'} E(C_i^*)$.*

By this observation the vertex set of each optimum cycle is contained entirely in exactly one of $V(F_1), V(F_2), \ldots, V(F_p)$. As a consequence, the optimum cycles whose vertex sets are contained in $V(F_i)$ constitute a cycle cover of $V(F_i)$. Moreover, the cost of this cycle cover of $V(F_i)$ is at most OPT since the length of each optimum cycle is no more than OPT.

Observation 2. *If $OPT \leq \lambda$, the optimum cycles can be partitioned into p groups such that the $i^{th}(i = 1, 2, \ldots, p)$ group consisting of $k_i^* \geq 1$ optimum cycles is a cycle cover of $V(F_i)$ with cost at most λ.*

This observation leads to an upper bound on the length of the tour on $V(F_i)$.

Lemma 3. *If $OPT \leq \lambda$, $w(TSP_i) \leq \rho(2k_i^* - 1)\lambda$ for $i = 1, 2, \ldots, p$.*

Proof. For each $i = 1, 2, \ldots, p$, by Observation 2 we have $k_i^* \geq 1$ optimum cycles that constitute a cycle cover of $V(F_i)$. Since F_i is a connected component we can add a set $E_i \subset E(F_i)$ of $k_i^* - 1$ edges to these cycles to obtain a connected subgraph H_i. After that we double all the edges of E_i in H_i to obtain a Eulerian graph H_i' on $V(F_i)$. By shortcutting the repeated vertices of the Eulerian tour of H_i' we generate a tour T_i on $V(F_i)$. Due to $E_i \subset E(F_i)$ and Step 1, $w(e) \leq \frac{\lambda}{2}$ for all $e \in E_i$, which implies $w(E_i) \leq (k_i^* - 1) \cdot \frac{\lambda}{2}$. So

$$w(T_i) \leq w(H_i') \leq k_i^*\lambda + 2w(E_i) \leq k_i^*\lambda + 2(k_i^* - 1) \cdot \frac{\lambda}{2} = (2k_i^* - 1)\lambda, \qquad (1)$$

where the second inequality follows from $OPT \leq \lambda$.

Since TSP_i is a ρ-approximate solution, we have $w(TSP_i) \leq \rho w(T_i)$. Combining this inequality with (1) proves the lemma. $\qquad \square$

Now we are ready to prove the following lemma.

Lemma 4. *If $OPT \leq \lambda$, Algorithm $MMCCP(\rho, \lambda)$ returns a cycle cover with at most k cycles whose cost is at most $4\rho\lambda$ in polynomial time.*

Proof. For each $i = 1, 2, \ldots, p$, TSP_i is a tour on $V(F_i)$. By Lemma 1, each vertex of $V(F_i)$ is contained in some path splitted from TSP_i and hence included in some cycle constructed in Step 3. Consequently, the set of cycles in Step 3 constitute a cycle cover. Since $OPT \leq \lambda$, by Lemma 3 we have

$$\sum_{i=1}^{p} k_i = \sum_{i=1}^{p} \max\{\lceil \frac{w(TSP_i)}{2\rho\lambda} \rceil, 1\} \leq \sum_{i=1}^{p} \max\{\lceil k_i^* - \frac{1}{2} \rceil, 1\} \leq \sum_{i=1}^{p} k_i^* = k' \leq k,$$

where the second inequality follows from the integrality of $k_i^* \geq 1$ and the last inequality holds since k' is the number of cycles used by the optimal solution.

Therefore, the algorithm returns the cycle cover generated in Step 3. To see that the cost of this cycle cover is at most $4\rho\lambda$, it is sufficient to note that all the paths derived in Step 2 have a length of at most $2\rho\lambda$ and the weight of the edge connecting the two end vertices of each path cannot exceed the length of the path due to the triangle inequality.

As for the time complexity, Step 1 takes $O(n^2)$ time. In Step 2, finding the approximate tour can be done in polynomial time since we run a polynomial time approximation algorithm for the TSP, and the cycle-splitting procedure takes $O(n)$ time by Lemma 1. Step 3 can also be completed in $O(n)$ time. To sum up, Algorithm $MMCCP(\rho, \lambda)$ runs in polynomial time. $\qquad \square$

Using Lemma 4 at most $\log \sum_{e \in E} w(e)$ times to conduct a binary search we obtain the following theorem.

Theorem 1. *Given a ρ-approximation algorithm for the TSP, there is a 4ρ-approximation algorithm for the MMCCP.*

This theorem implies good approximation algorithms for the MMCCP defined on some special metrics. Particularly, by the PTAS for the Euclidean TSP given by Arora [2] we have the following result.

Corollary 1. *For any $\epsilon > 0$, there is a $(4 + \epsilon)$-approximation algorithm for the MMCCP defined on any fixed d-dimensional Euclidean space.*

Remark 1. Karakawa et al. [9] proved a stronger version of Lemma 2 for trees on a Euclidean space, which can derive by a similar approach in [15] approximation algorithms for MMCCP defined on a d-dimensional Euclidean space with ratios $5.208(d = 2)$ and $5.237(d \geq 3)$, respectively.

In what follows, we plug in Christofides' Algorithm in Step 2 of Algorithm $MMCCP(\rho, \lambda)$ and make a refined analysis to obtain a 5-approximation algorithm. The modified algorithm is described as follows:

Algorithm $MMCCP(\lambda)$

Step 1. Delete all the edges with weight greater than $\frac{\lambda}{2}$ in G. The resulted graph $G[\frac{\lambda}{2}]$ has p connected components $F_1, F_2, \ldots, F_p$.

Step 2. For each $i = 1, 2, \ldots, p$, compute $MST(F_i)$ and determine the set S_i of vertices in $V(F_i)$ that are of odd degree in $MST(F_i)$. Find a minimum weight perfect matching M_i on S_i and add M_i to $MST(F_i)$ to obtain an Eulerian graph G_i on $V(F_i)$. Shortcut the repeated vertices of the Eulerian tour of G_i to obtain a tour TSP_i on $V(F_i)$. Split the tour into at most $k_i = \max\{\lceil \frac{w(TSP_i)}{\frac{5}{2}\lambda} \rceil, 1\}$ paths of length no more than $\frac{5}{2}\lambda$ by Lemma 1.

Step 3. Connect the two end vertices of each path constructed in Step 2 to obtain $\sum_{i=1}^{p} k_i$ cycles which constitute a cycle cover. If $\sum_{i=1}^{p} k_i \leq k$, return this cycle cover; otherwise, return failure.

Lemma 5. *If $OPT \leq \lambda$, $w(TSP_i) \leq (\frac{5}{2}k_i^* - 1)\lambda$ for $i = 1, 2, \ldots, p$.*

Proof. For each $i = 1, 2, \ldots, p$, we construct subgraphs H_i, H_i' and tour T_i in exactly the same way as in the proof of Lemma 3. Since H_i is a connected subgraph of F_i we have

$$w_T(F_i) \leq w(H_i) \leq k_i^* \lambda + w(E_i) \leq k_i^* \lambda + (k_i^* - 1) \cdot \frac{\lambda}{2} = \left(\frac{3}{2}k_i^* - \frac{1}{2}\right)\lambda, \quad (2)$$

where the second inequality follows from $OPT \leq \lambda$ and the third inequality holds by $w(e) \leq \frac{\lambda}{2}$ for all $e \in E_i$.

By shortcutting we can transform tour T_i into a tour T_i' on $V(M_i)$ with $w(T_i') \leq w(T_i)$ due to the triangle inequality. It is well known that T_i' can be decomposed into two edge-disjoint perfect matching on $V(M_i)$. Thus by the optimality of M_i we have

$$w(M_i) \le \frac{1}{2}w(T_i') \le \frac{1}{2}w(T_i) \le \left(k_i^* - \frac{1}{2}\right)\lambda, \tag{3}$$

where the last inequality follows from (1). Therefore, by (2) and (3) we obtain

$$w(TSP_i) \le w_T(F_i) + w(M_i) \le \left(\frac{3}{2}k_i^* - \frac{1}{2}\right)\lambda + \left(k_i^* - \frac{1}{2}\right)\lambda = \left(\frac{5}{2}k_i^* - 1\right)\lambda. \ \square$$

It can be seen that finding TSP_i for each $i = 1, 2, \ldots, p$ takes $O(|V(F_i)|^3)$, which dominates the time complexity of Algorithm $MMCCP(\lambda)$. Consequently, the algorithm runs in $\sum_{i=1}^{p} O(|V(F_i)|^3)) = O(\sum_{i=1}^{p} |V(F_i)|^3) = O(n^3)$ time.

By Lemma 5 and a similar proof to Lemma 4 we derive the following lemma.

Lemma 6. *If $OPT \le \lambda$, Algorithm $MMCCP(\lambda)$ returns a cycle cover with at most k cycles whose cost is at most 5λ in $O(n^3)$ time.*

Using this lemma to perform a binary search we obtain

Theorem 2. *There is a 5-approximation algorithm for the MMCCP that runs in $O(n^3 \log \sum_{e \in E} w(e))$ time.*

Combining Theorem 1 and Theorem 2 we have

Theorem 3. *Given a ρ-approximation algorithm for the TSP, there exists a $\min\{4\rho, 5\}$-approximation algorithm for the MMCCP.*

4   Rooted Min-Max Cycle Cover

One can transform an α-approximation algorithm for the MMCCP into an $(\alpha + 1)$-approximation algorithm for the RMMCCP as follows. First, by ignoring the depot set D we obtain an instance of the MMCCP. Then we run the α-approximation algorithm for this instance to obtain a cycle cover $C_1, C_2, \ldots, C_{\bar{k}}$ of $V \backslash D$ with $\bar{k} \le k$. Next for each $i = 1, 2, \ldots, \bar{k}$ we choose an arbitrary vertex v_i from C_i with $(u_i', v_i), (u_i'', v_i) \in E(C_i)$, determine $d_i \in D$ such that $w(v_i, d_i) = \min_{d \in D} w(v_i, d)$ and derive a cycle C_i' with $E(C_i') = (E(C_i) \backslash \{(u_i', v_i), (u_i'', v_i)\}) \cup \{(u_i', d_i), (u_i'', d_i)\}$. Then $C_1', C_2', \ldots, C_{\bar{k}}'$ is a feasible cycle cover for the RMMCCP. To show this is indeed an $(\alpha + 1)$-approximation algorithm, we only need two facts: (i)the optimal value of the instance of the MMCCP can not exceed OPT, i.e., the optimal value of the original instance of the RMMCCP; (ii)$w(C_i') \le w(C_i) + 2w(v_i, d_i)$ and $w(v_i, d_i) \le OPT/2$ for each i. The first inequality follows from the triangle inequality. The second one holds because each v_i must locate in the same optimum cycle with some depot $d_i' \in D$ of weight no more than OPT, one of the two paths along the optimum cycle between v and d_i' is of length at most $OPT/2$. By the triangle inequality, $w(v_i, d_i') \le OPT/2$ and hence $w(v_i, d_i) \le w(v_i, d_i') \le OPT/2$.

Therefore, all the results on the MMCCP can be applied to the RMMCCP with a loss of 1 in the approximation ratio.

Theorem 4. *Given a ρ-approximation algorithm for the TSP, there exists a* $\min\{4\rho + 1, 6\}$-*approximation algorithm for the RMMCCP. Particularly, there is an $O(n^3 \log \sum_{e \in E} w(e))$ time 6-approximation algorithm for the RMMCCP.*

5   Minimum Cycle Cover

In this section we give approximation algorithms for the MCCP. First, we show how to apply the results for the MMCCP in Section 3 to obtain approximation algorithms with the same ratio. After that we propose an algorithm with better performance ratio for the MCCP.

Recall that in the MCCP, $\lambda > 0$ is given in advance and the aim is to find a cycle cover of cost at most λ such that the number of cycles is minimum. To turn Algorithm $MMCCP(\rho, \lambda)$ into an approximation algorithm for the MCCP, we need only split the tour TSP_i into paths of length at most $\frac{\lambda}{2}$ instead of $2\rho\lambda$ in Step 2. Moreover, in Step 3 we always return the cycle cover of cost at most λ.

Algorithm $MCCP(\rho)$

Step 1. Delete all the edges with weight greater than $\frac{\lambda}{2}$ in G. The resulted graph $G[\frac{\lambda}{2}]$ has p connected components $F_1, F_2, \ldots, F_p$.

Step 2. For each $i = 1, 2, \ldots, p$, find a ρ-approximate tour TSP_i on $V(F_i)$ and split it into $k_i = \max\{\lceil \frac{w(TSP_i)}{\frac{\lambda}{2}} \rceil, 1\}$ paths of length at most $\frac{\lambda}{2}$ by Lemma 1.

Step 3. Connect the two end vertices of each path constructed in Step 2 to obtain $\sum_{i=1}^{p} k_i$ cycles which constitute a cycle cover. Return this cycle cover.

By a similar analysis to Algorithm $MMCCP(\rho, \lambda)$ one can show the above algorithm is a 4ρ-approximation algorithm for the MCCP. A counterpart to Theorem 2 can also be established. So we have

Theorem 5. *Given a ρ-approximation algorithm for the TSP, there exists a* $\min\{4\rho, 5\}$-*approximation algorithm for the MCCP. Particularly, there exists a 5-approximation algorithm for the MCCP that runs in $O(n^3)$ time.*

Next we present a 24/5-approximation algorithm for the MCCP that runs in $O(n^4)$ time. In contrast, Khani and Salavatipour [10] gave a 5/2-approximation algorithm for the Minimum Tree Cover Problem that runs in $O(n^5)$ time, which implies a 5-approximation for the MCCP with the same running time. Our algorithm adopts a similar approach to the algorithm for the Min-Max Tree Cover Problem also proposed in [10]. However, we make a refined analysis on cycles instead of trees which leads to an improved approximation ratio and simplify the algorithm to obtain a better running time.

The basic idea of the algorithm is as follows. First, we delete all the edges with weight greater than $\frac{\lambda}{5}$ to obtain the graph $G[\frac{\lambda}{5}]$. Let $F_1, F_2, \ldots, F_l$ be the connected components of $G[\frac{\lambda}{5}]$ with $w_T(F_i) \leq \frac{\lambda}{2}(i = 1, 2, \ldots, l)$, called *light components*. The rest of connected components $F_{l+1}, F_{l+2}, \ldots, F_{l+h}$ of $G[\frac{\lambda}{5}]$ with $w_T(F_i) > \frac{\lambda}{2}(i = l+1, l+2, \ldots, l+h)$ are called *heavy components*. Next we construct a tree cover of cost at most $\frac{\lambda}{2}$. Since $w_T(F_i) \leq \frac{\lambda}{2}$ for $i = 1, 2, \ldots, l$ we choose the minimum spanning trees of some light components as the trees in the final tree cover. For the minimum spanning trees of the other light components we connect them properly to the heavy components, which results in h *modified heavy components* $F'_{l+1}, F'_{l+2}, \ldots, F'_{l+h}$. And for each modified heavy component we decompose its minimum spanning tree into a set of trees of weight at most $\frac{\lambda}{2}$ by Lemma 2 and put them into the final tree cover. Lastly, for each tree of the tree cover we double all the edges to obtain a Eulerian graph and shortcut the repeated vertices of the Eulerian tour to derive a cycle cover of cost at most λ.

To guide the choice of the minimum spanning trees of light components to be connected to heavy components, we define $w_{\min}(F_i)$ with $1 \leq i \leq l$ as the minimum weight of edges in G with one vertex in $V(F_i)$ and the other vertex in $\cup_{s=l+1}^{l+h} V(F_s)$ and construct a bipartite graph.

Definition 4. *Given an integer a with $0 \leq a \leq l$, the bipartite graph H_a has l light vertices $u_1, u_2, \ldots, u_l$, a null vertices $x_1, x_2, \ldots, x_a$ and $l - a$ heavy vertices $y_1, y_2, \ldots, y_{l-a}$. For all $i = 1, 2, \ldots, l$ and $j = 1, 2, \ldots, a$, there is an edge (u_i, x_j) of weight 0. For $i = 1, 2, \ldots, l$, if $w_{\min}(F_i) \leq \frac{\lambda}{2}$ we add an edge (u_i, y_j) of weight $w_T(F_i) + w_{\min}(F_i)$ to H_a for each $j = 1, 2, \ldots, l - a$. There are no other edges in H_a.*

Now we formally describe our algorithm below.

Algorithm $MCCP$

Step 1. Delete all the edges with weight greater than $\frac{\lambda}{5}$ in G to obtain $G[\frac{\lambda}{5}]$ with light components $F_1, F_2, \ldots, F_l$ and heavy components $F_{l+1}, F_{l+2}, \ldots, F_{l+h}$.

Step 2. For $a = 0, 1, \ldots, l$, set $T_a := \emptyset$ and $S_a := \emptyset$.

(i) Find a minimum weight perfect matching M_a in H_a (if there is no perfect matching in H_a set $a := a + 1$ and go to Step 2);

(ii) If $(u_i, x_j) \in M_a$, put $MST(F_i)$ into T_a;

(iii) If $(u_i, y_j) \in M_a$, connect $MST(F_i)$ to some heavy component by the edge corresponding to $w_{\min}(F_i)$. This results in h modified heavy components $F'_{l+1}, F'_{l+2}, \ldots, F'_{l+h}$. For each $s = l+1, l+2, \ldots, l+h$, decompose $MST(F'_s)$ by Lemma 2 into a set of trees of weight at most $\frac{\lambda}{2}$ and put them into T_a.

(iv) For each tree in T_a, double all the edges to obtain an Eulerian graph and shortcut the repeated vertices of the Eulerian tour of this graph to obtain a cycle. Put this cycle into S_a.

Step 3. Among all the nonempty S_a, return the one contains the minimum number of cycles.

By construction it is easy to see that each nonempty S_a is a cycle cover of cost no greater than λ. We proceed to show that for some a the number of cycles in S_a is not greater than $\frac{24}{5}OPT$.

Let $OPT = k$ and $C_1^*, C_2^*, \ldots, C_k^*$ be the vertex-disjoint optimum cycles. Given a cycle C, if $V(F_i) \cap V(C) \neq \emptyset$ for some i with $1 \leq i \leq l + h$, we say that F_i is *incident* to C or C is *incident* to F_i. Therefore, all the optimum cycles can be classified into three types. The first type of optimum cycles, called *light cycles*, are incident to only light components. The second type of optimum cycles, i.e. *heavy cycles*, are incident to only heavy components. The last type of optimum cycles, known as *bad cycles*, are incident to at least one light component and at least one heavy component. Let k_l, k_h, k_b be the number of light, heavy and bad cycles, respectively. Clearly, $k = k_l + k_h + k_b$.

Next we analyze the execution of the algorithm for $a = a'$ with $0 \leq a' \leq l$, where a' is the number of light components incident to at least one light cycle, and bound the number of cycles in $S_{a'}$ which is identical to the number of trees in $T_{a'}$. If we define F_0 as an empty light component which contains neither a vertex nor an edge, we can assume without loss of generality that $F_0, F_1, \ldots, F_{a'}$ are the light components incident to at least one light cycle. An edge of weight greater than $\frac{\lambda}{5}$ is referred to as a *long edge*. Since an optimum cycle(particularly a light cycle) is of length no more than λ, it cannot contain more than four long edges and hence $a' \leq 4k_l$, which implies

Lemma 7. *For $a = a'$, Step 2(ii) of Algorithm MCCP generates $a' \leq 4k_l$ trees.*

Lemma 8. *For $a = a'$, $\sum_{s=l+1}^{l+h} w_T(F_s') \leq \frac{6}{5}(k_h + k_b)\lambda$.*

Lemma 9. *For $a = a'$, Step 2(iii) of Algorithm MCCP generates at most $\frac{24}{5}(k_h + k_b)$ trees.*

By Lemma 7 and Lemma 9 we deduce

Lemma 10. $|S_{a'}| = |T_{a'}| \leq \frac{24}{5}k$.

By his lemma and a simple analysis of the complexity of Algorithm *MCCP* we have

Theorem 6. *There is a $\frac{24}{5}$-approximation algorithm for the MCCP that runs in $O(n^4)$ time.*

Combining this theorem with Theorem 5 we obtain

Theorem 7. *Given a ρ-approximation algorithm for the TSP, there exists a $\min\{4\rho, \frac{24}{5}\}$-approximation algorithm for the MCCP.*

Acknowledgments. The authors are grateful to the anonymous referees for their helpful comments. This research is supported in part by the National Natural Science Foundation of China under grants number 11171106, 11301184, 11301475.

References

1. Arkin, E.M., Hassin, R., Levin, A.: Approximations for minimum and min-max vehicle routing problems. Journal of Algorithms **59**, 1–18 (2006)
2. Arora, S.: Polynomial time approximation schemes for euclidean traveling salesman and other geometric problems. Journal of the ACM **45**, 753–782 (1998)

3. Bhattacharya, B., Hu, Y.: Approximation algorithms for the multi-vehicle scheduling problem. In: Cheong, O., Chwa, K.-Y., Park, K. (eds.) ISAAC 2010, Part II. LNCS, vol. 6507, pp. 192–205. Springer, Heidelberg (2010)
4. Campbell, A.M., Vandenbussche, D., Hermann, W.: Routing for relief efforts. Transportation Science **42**, 127–145 (2008)
5. Christofides, N.: Worst-case analysis of a new heuristic for the traveling salesman problem. Technical Report, Graduate School of Industrial Administration, Carnegie-Mellon University, Pittsburgh, PA (1976)
6. Even, G., Garg, N., Koemann, J., Ravi, R., Sinha, A.: Min-max tree covers of graphs. Operations Research Letters **32**, 309–315 (2004)
7. Frederickson, G.N., Hecht, M.S., Kim, C.E.: Approximation algorithms for some routing problems. SIAM Journal on Computing **7**(2), 178–193 (1978)
8. Friggstad, Z., Swamy, C.: Approximation algorithms for regret-bounded vehicle routing and applications to distance-constrained vehicle routing. In: the Proceedings of the 46th Annual ACM Symposium on Theory of Computing, pp. 744–753 (2014)
9. Karakawa, S., Morsy, E., Nagamochi, H.: Minmax tree cover in the euclidean space. Journal of Graph Algorithms and Applications **15**, 345–371 (2011)
10. Khani, M.R., Salavatipour, M.R.: Approximation algorithms for min-max tree cover and bounded tree cover problems. Algorithmica **69**, 443–460 (2014)
11. Nagamochi, H.: Approximating the minmax rooted-subtree cover problem. IEICE Transactions on Fundamentals of Electronics **E88–A**, 1335–1338 (2005)
12. Nagamochi, H., Okada, K.: Polynomial time 2-approximation algorithms for the minmax subtree cover problem. In: Ibaraki, T., Katoh, N., Ono, H. (eds.) ISAAC 2003. LNCS, vol. 2906, pp. 138–147. Springer, Heidelberg (2003)
13. Nagamochi, H., Okada, K.: Approximating the minmax rooted-tree cover in a tree. Information Processing Letters **104**, 173–178 (2007)
14. Nagarajan, V., Ravi, R.: Approximation algorithms for distance constrained vehicle routing problems. Networks **59**(2), 209–214 (2012)
15. Xu, W., Liang, W., Lin, X.: Approximation algorithms for Min-max Cycle Cover Problems. IEEE Transactions on Computers (2013). doi:10.1109/TC.2013.2295609
16. Xu, Z., Wen, Q.: Approximation hardness of min-max tree covers. Operations Research Letters **38**, 408–416 (2010)
17. Xu, Z., Xu, L., Li, C.-L.: Approximation results for min-max path cover problems in vehicle routing. Naval Research Logistics **57**, 728–748 (2010)
18. Xu, Z., Xu, L., Zhu, W.: Approximation results for a min-max location-routing problem. Discrete Applied Mathematics **160**, 306–320 (2012)

Improved Approximation Algorithms for the Maximum Happy Vertices and Edges Problems

Peng Zhang[1(✉)], Tao Jiang[2,3], and Angsheng Li[4]

[1] School of Computer Science and Technology,
Shandong University, Jinan 250101, China
algzhang@sdu.edu.cn
[2] Department of Computer Science and Engineering,
University of California, Riverside, CA 92521, USA
jiang@cs.ucr.edu
[3] MOE Key Lab of Bioinformatics and Bioinformatics Division,
TNLIST/Department of Computer Science and Technology,
Tsinghua University, Beijing 100084, China
[4] State Key Laboratory of Computer Science, Institute of Software,
Chinese Academy of Sciences, Beijing 100190, China
angsheng@ios.ac.cn

Abstract. The Maximum Happy Vertices (MHV) problem and the Maximum Happy Edges (MHE) problem are two fundamental problems arising in the study of the homophyly phenomenon in large scale networks. Both of these two problems are NP-hard. Interestingly, the MHE problem is a natural generalization of Multiway Uncut, the complement of the classic Multiway Cut problem. In this paper, we present new approximation algorithms for MHV and MHE based on randomized LP-rounding techniques. Specifically, we show that MHV can be approximated within $\frac{1}{\Delta+1}$, where Δ is the maximum vertex degree, and MHE can be approximated within $\frac{1}{2} + \frac{\sqrt{2}}{4} f(k) \geq 0.8535$, where $f(k) \geq 1$ is a function of the color number k. These results improve on the previous approximation ratios for MHV, MHE as well as Multiway Uncut in the literature.

1 Introduction

Homophyly [5, Chapter4] is one of the basic laws governing the structures of large scale networks, which states that edges in a network tend to connect nodes with the same or similar attributes. For example, in a social network, people are more likely to connect with people they like, as the old proverb says, "birds of a feather flock together".

As another example, Li et al. [9] recently conducted an interesting experiment to predict the keywords of a paper from the citation network in high energy

---

P. Zhang—Work was done while the first author was visiting at the University of California - Riverside, USA.

© Springer International Publishing Switzerland 2015
D. Xu et al. (Eds.): COCOON 2015, LNCS 9198, pp. 159–170, 2015.
DOI: 10.1007/978-3-319-21398-9_13

physics theory [1] by using the homophyly law. The network consists of $27,770$ vertices (*i.e.*, papers) and $352,807$ directed edges (*i.e.*, citations). However, only $1,214$ papers have keywords annotated by their authors. The task was to predict the keywords of the remaining papers. By the homophyly law, papers within a small community of the network should share common keywords. The prediction algorithm used by [9] is as follows. (1) Find a small community for each paper, if any. After this step, there are $20,310$ papers found in communities, and only $1,409$ papers from amongst them have keywords. (2) Extract the most popular 10 keywords from the known keywords in each community as the remarkable common attributes of this community. (3) For every paper in a community, predict that all or some of the 10 remarkable common keywords of this community are the keywords of the paper by checking whether the keywords appear in either the title or the abstract of the paper. Surprisingly, this simple rule successfully finds keywords for $14,123$ (70%) un-annotated papers. The experiment suggests that real networks do satisfy the homophyly law, and that the homophyly law could be used as a principle for predicting common attributes in a large network.

In a network where the homophyly law holds but some vertices have unknown attributes, as in the case of the above example where keywords are considered as attributes, one may consider the natural question of how to assign (or predict) attributes so that the homophyly law is followed to the greatest degree. Following this idea, and identifying attributes with colors, Li and Zhang [10] recently introduced two interesting maximization problems in terms of graph coloring. For simplicity, they focused on the case that each vertex has only one color.

Definition 1. The Maximum Happy Vertices (MHV) problem.

(Instance) We are given an undirected graph $G = (V, E)$ with vertex weights $\{w_v\}$, a color set $C = \{1, 2, \cdots, k\}$, and a partial vertex coloring function $c\colon V \mapsto C$. That is, c assigns colors only to a part of the vertices in V.

(Goal) A vertex is happy *if it shares the same color with all its neighbors. The goal is to color all the uncolored vertices such that the total weight of happy vertices is maximized.*

Definition 2. The Maximum Happy Edges (MHE) problem.

(Instance) We are given an undirected graph $G = (V, E)$ with edge weights $\{w_e\}$, a color set $C = \{1, 2, \cdots, k\}$, and a partial vertex coloring function $c\colon V \mapsto C$.

(Goal) An edge is happy *if its two endpoints have the same color. The goal is to color all the uncolored vertices such that the total weight of happy edges is maximized.*

Remarks. (i) When every vertex (resp., edge) has unit weight, the MHV (resp., MHE) problem is to maximize the total number of happy vertices (resp., edges). (ii) The partial vertex coloring function c is given in the input. When a vertex v has a color specified by function c, we also say that vertex v has a

---

[1] http://snap.stanford.edu/data/cit-HepTh.html.

pre-specified color (that is, $c(v)$). The MHV and MHE problems actually ask for a total vertex coloring. (iii) The coloring in MHV and MHE is completely different from the well-known Graph Coloring problem, which asks to color all vertices with the minimum number of colors such that each edge has its two endpoints with different colors.

The MHV and MHE problems can also be viewed as two classification problems. Given a set of objects to be classified and a set of colors, a classification problem can be depicted as from a very high level assigning a color to each object in a way that is consistent with some observed data or structure [1,7]. In our problems, the observed structure is homophyly.

1.1 Related Work

MHV and MHE are two quite natural and fundamental algorithmic problems. Surprisingly, as introduced before, they arise only very recently from the study of network homophyly [10]. Li and Zhang [10] proved that both MHV and MHE are already NP-hard even if the color number k is fixed. More precisely, when $k \geq 3$, MHV and MHE are NP-hard. When $k = 2$, MHV and MHE are polynomial time solvable. Li and Zhang [10] proposed a $\frac{1}{2}$-approximation algorithm for (the unit weight version of) MHE based on a combinatorial partitioning strategy. For (the unit weight version of) MHV, they gave two approximation algorithms. One algorithm is based on a greedy approach, whose approximation ratio is $\frac{1}{k}$. The other algorithm is based on a subset-growth technique, whose approximation ratio is $\Omega(\Delta^{-3})$, where Δ is the maximum vertex degree in the input graph.

The MHE problem is closely related to the Multiway Uncut problem [8], which is the complement of the classic Multiway Cut problem [2–4,6,11].

Definition 3. The Multiway Uncut problem.
    (Instance) An undirected graph $G = (V, E)$ with edge weights $\{w_e\}$, and a terminal set $S = \{s_1, s_2, \cdots, s_k\}$.
    (Goal) Find a partition $\{V_1, V_2, \cdots, V_k\}$ of V such that for each i, s_i is contained in V_i, and the total weights of edges not cut by the partition is maximized.

The goal of the Multiway Uncut problem is equivalent to coloring all the non-terminal vertices such that the total weight of happy edges is maximized. From the viewpoint of coloring, in the Multiway Uncut problem there is only one vertex s_i that has the pre-specified color i, for each $1 \leq i \leq k$. So, Multiway Uncut is just a special case of the MHE problem. The current best approximation ratio for Multiway Uncut is 0.8535, due to Langberg et al. [8].

Given an undirected graph $G = (V, E)$ with costs defined on edges and a terminal set $S \subseteq V$, the Multiway Cut problem asks for a set of edges with the minimum total cost such that its removal from graph G separates all terminals in S from one another. The Multiway Cut problem is NP-hard even if there are only three terminals and each edge has a unit cost [4]. The current best approximation ratio known for the Multiway Cut problem is 1.2965 [11]. Since the optimization goals of Multiway Cut and MHE are completely different (minimization vs. maximization), the approximation results for Multiway Cut do not *directly* extend to MHE.

1.2 Our Results

In this paper, we give improved approximation algorithms for MHV and MHE based on randomized rounding in linear programming. Specifically, we show that MHV can be approximated within $\frac{1}{\Delta+1}$, and MHE can be approximated within $\frac{1}{2} + \frac{\sqrt{2}}{4} f(k)$, where $f(k) = \frac{(1-1/k)\sqrt{k(k-1)}+1/\sqrt{2}}{k-1+1/2k} \geq 1$. These results significantly improve on the previous approximation ratios for MHV and MHE in [10]. Our randomized rounding approach is motivated by the work of Kleinberg and Tardos [7] on the uniform Metric Labeling problem and the work of Langberg et al. [8] on the Multiway Uncut problem. However, our approximation ratio analyses require nontrivial extension. From a high-level viewpoint, the analyses of the randomized rounding scheme in [7,8] were performed in an edge-by-edge manner. In contrast, our analysis for the MHV problem considers a group of vertices at each time, and this extension essentially requires the structural properties of the MHV problem.

Since Multiway Uncut is a special case of MHE, the above results also means that the Multiway Uncut problem can be approximated within $\frac{1}{2} + \frac{\sqrt{2}}{4} f(k)$. For fixed values of k, this ratio improves upon the result 0.8535 in [8]. For example, when $k = 3, 4, 5$ and 10, our ratios are 0.8818, 0.8739, 0.8694, and 0.8611, respectively. We get this improvement because we unite a simple randomized algorithm and the randomized rounding procedure in [8]. When k approaches infinity, the ratio tends to $\frac{1}{2} + \frac{\sqrt{2}}{4} = 0.8535 \cdots$, coinciding with the ratio given in [8].

Notations. Throughout the paper, we use OPT to denote the optimal value of an optimization problem and OPT_f to denote the optimal value of the corresponding fractional problem (when linear programming is involved).

2 Algorithms for MHV

Let Δ be the maximum vertex degree in the input graph. In this section, we show that the MHV problem can be approximated within $\frac{1}{\Delta+1}$ in polynomial time by randomized LP-rounding. Li and Zhang [10] gave a simple greedy $\frac{1}{k}$-approximation algorithm for the MHV problem, which is briefly shown as Algorithm $\mathcal{G}$ below. Therefore, to achieve the $\frac{1}{\Delta+1}$-approximation, we can safely assume $k \geq \Delta + 1$.

Algorithm $\mathcal{G}$.
1 Just color all the uncolored vertices in the same color. Since there are k colors, we obtain k vertex colorings for graph G.
2 Output the coloring that has the largest total weight of happy vertices.

Lemma 1. *Without loss of generality, we may assume that* $k \geq \Delta + 1$.   □

The linear programming relaxation for the MHV problem is shown as (LP-V) below. To explain that (LP-V) is really an LP-relaxation for MHV, consider its corresponding integer linear program. That is, the variable constraint is replaced by the constraint $x_v, x_v^i, y_v^i \in \{0, 1\}, \forall i, \forall v$.

Here, variable y_v^i indicates whether vertex v is colored in i, x_v^i indicates whether v is happy by color i, and x_v indicates whether v is happy. Constraint (2) is concerned with vertices that have pre-specified colors. Then, constraint (1) says that each vertex must be colored and can have only one color. In constraint (3), the notation $B(v)$ means the *ball* centered at vertex v, i.e., the set of vertex v itself and all its neighbors. By constraint (3), vertex v is happy by color i only when all the vertices in $B(v)$ are colored in i. Note that constraint (3) can be replaced by the linear constraint $x_v^i \leq y_u^i, \forall i, \forall v, \forall u \in B(v)$.

$$\max \quad \sum_{v \in V} w_v x_v \qquad \text{(LP-V)}$$

$$\text{s.t.} \quad \sum_i y_v^i = 1, \qquad \forall v \qquad (1)$$

$$y_v^i = 1, \qquad \forall i, \forall v \text{ s.t. } c(v) = i \qquad (2)$$

$$x_v^i = \min_{u \in B(v)} \{y_u^i\}, \quad \forall i, \forall v \qquad (3)$$

$$x_v = \sum_i x_v^i, \qquad \forall v \qquad (4)$$

$$x_v, x_v^i, y_v^i \geq 0, \qquad \forall i, \forall v$$

We mention that in the integer version of (LP-V), any vertex can be happy in only one color. For otherwise suppose for some vertex v, we have $x_v^{i_1} = 1$ and $x_v^{i_2} = 1$, where $i_1 \neq i_2$. Then by constraint (3), we have $y_u^{i_1} = 1$ and $y_u^{i_2} = 1$ for any vertex $u \in B(v)$, but this immediately contradicts constraint (1). The similar property holds for LP-relaxation (LP-V), as shown in the following Lemma 2.

Lemma 2. *In any feasible solution to (LP-V), we have $0 \leq x_v \leq 1$ for any vertex v.*

Proof. Suppose for some vertex v we have $x_v = \sum_{i=1}^k x_v^i > 1$. Take any vertex $u' \in B(v)$. For this vertex u', by constraint (3), we have $\sum_i y_{u'}^i \geq \sum_i \min_{u \in B(v)} \{y_u^i\} = \sum_i x_v^i > 1$. This is in contradiction with constraint (1). □

The straightforward strategy that colors vertex u by color i with probability y_u^i would yield an integral solution with poor approximation ratio. Instead, we use the rounding technique proposed by Kleinberg and Tardos [7] to round a fractional solution to (LP-V). The algorithm is shown as Algorithm $\mathcal{R}$. In step 3 of the algorithm, the notation $[k]$ denotes the set $\{1, 2, \cdots, k\}$.

Algorithm $\mathcal{R}$.
1 Solve (LP-V) to obtain an optimal solution (x, y).
2 **while** there exists some uncolored vertex **do**

3   Pick a color $i \in [k]$ uniformly at random.
4   Pick a parameter $\rho \in [0,1]$ uniformly at random.
5   For each uncolored vertex v, if $y_v^i \geq \rho$, then color v in i.
6 **endwhile**

Langberg et al. [8] used the same rounding technique as in Algorithm $\mathcal{R}$ for the Multiway Uncut problem. We adopt the high-level idea of the analyses in [7] and [8]. However, both the analyses in [7] and [8] for the randomized rounding scheme were performed in an edge-by-edge manner. We non-trivially extend the analyses of [7,8] to Algorithm $\mathcal{R}$ for the MHV problem below. In contrast, our analysis considers a group of vertices at each time, and this extension essentially requires the structural properties of the MHV problem. We begin with a simple assumption.

Lemma 3. *Without loss of generality, we may assume that $|B(v)| \geq 2$ for every vertex v.*

Proof. By definition, if $|B(v)| = 1$, then v is an isolated vertex. In this case, either v is already happy (It has a pre-specified color), or v can become happy (It is uncolored. Then we just color v in any color). So, all the isolated vertices can be safely removed from the graph without affecting the analysis of approximation ratio of the algorithm $\mathcal{R}$.   □

In Algorithm $\mathcal{R}$, each execution of steps 3 to 5 is called a *round*. The algorithm may iterate steps 3 to 5 for many rounds. If a ball $B(v)$ contains no colored vertices, then $B(v)$ is called a *blank* ball. If a ball $B(v)$ contains only one color, then $B(v)$ is called a *monochrome* ball.

Lemma 4. *Consider a ball $B(v)$ which is blank at the beginning of some round. Then the probability that in this round all (uncolored) vertices in $B(v)$ are colored in the same color is $\frac{x_v}{k}$.*

Proof. Fix some color i. The probability that color i is picked is $\frac{1}{k}$. All vertices in $B(v)$ are colored in i if and only if $\rho \leq \min_{u \in B(v)} \{y_u^i\} = x_v^i$. So, conditioned on the event that color i is picked, the probability that all vertices in $B(v)$ is colored in i is x_v^i.

By the above analysis, the probability that all vertices in $B(v)$ are colored in the same color is $\sum_i \frac{1}{k} x_v^i = \frac{x_v}{k}$.   □

Remarks. Let $B(v)$ be a ball which is blank at the beginning of some round. It is easy to see that even if parts of the vertices in $B(v)$ are colored in this round, it is still possible that all vertices in $B(v)$ are colored in the same color in the course of the algorithm (due to the subsequent round(s)). This probability is beyond Lemma 4 and is omitted in the analysis of the algorithm. See the proof of Theorem 1. Of course, the approximation ratio proved in Theorem 1 holds even with this omission.

Lemma 5. *Consider a ball $B(v)$ which is blank at the beginning of some round. Then the probability that in this round some vertex in $B(v)$ is colored is $\leq \frac{1}{k}(\Delta + 1 - x_v)$.*

Proof. By step 5 of Algorithm $\mathcal{R}$, there is a vertex in $B(v)$ that will be colored in i by the current round, if and only if ρ falls in the interval $[0, \max_{u \in B(v)}\{y_u^i\}]$. So, the probability that in this round some vertex in $B(v)$ is colored is

$$\sum_i \frac{1}{k} \max_{u \in B(v)} \{y_u^i\} = \frac{1}{k} \sum_i \max_{u \in B(v)} \{y_u^i\}. \tag{5}$$

On the other hand, for any color i, we have $\min_{u \in B(v)}\{y_u^i\} + \max_{u \in B(v)}\{y_u^i\} \leq \sum_{u \in B(v)} y_u^i$, since $|B(v)| \geq 2$ by Lemma 3. This implies

$$\sum_i \left(\min_{u \in B(v)} \{y_u^i\} + \max_{u \in B(v)} \{y_u^i\} \right)$$
$$\leq \sum_i \sum_{u \in B(v)} y_u^i = \sum_{u \in B(v)} \sum_i y_u^i \underset{(1)}{=} |B(v)| \leq \Delta + 1.$$

Therefore, we have

$$\sum_i \max_{u \in B(v)} \{y_u^i\} \leq \Delta + 1 - \sum_i \min_{u \in B(v)} \{y_u^i\} \underset{(3),(4)}{=} \Delta + 1 - x_v. \tag{6}$$

By (5) and (6), the probability that in this round some vertex in $B(v)$ is colored is $\leq \frac{1}{k}(\Delta + 1 - x_v)$. □

Lemma 6. *Consider a blank ball $B(v)$. The probability that all vertices in $B(v)$ are colored in the same round (and hence in the same color) is $\geq \frac{x_v}{\Delta + 1 - x_v}$.*

Proof. Let p be the probability defined in the lemma. Fix the r-th round $(r \geq 1)$ in the execution course of the algorithm. Define $N_{<r}$ as the event that all vertices in $B(v)$ are not colored *before* the r-th round, and A_r the event that all vertices in $B(v)$ are colored in the r-th round. Since the random variables i and ρ in Algorithm $\mathcal{R}$ are chosen independently across rounds, we have

$$p = \sum_{r=1}^{\infty} \Pr[N_{<r}] \cdot \Pr[A_r | N_{<r}]. \tag{7}$$

We first calculate the probability $\Pr[N_{<r}]$. Let t be a round $(t \geq 1)$. Define N_t as the event that all vertices in $B(v)$ are not colored *in* the t-th round. Then

$$\Pr[N_{<r}] = \Pr[N_{<r-1}] \Pr[N_{r-1} | N_{<r-1}]$$
$$= (\Pr[N_{<r-2}] \Pr[N_{r-1} | N_{<r-2}]) \Pr[N_{r-1} | N_{<r-1}]$$
$$= \cdots$$
$$= \Pr[N_{<1}] \Pr[N_1 | N_{<1}] \Pr[N_2 | N_{<2}] \Pr[N_3 | N_{<3}] \cdots \Pr[N_{r-1} | N_{<r-1}]$$
$$= \prod_{t=1}^{r-1} \Pr[N_t | N_{<t}], \tag{8}$$

where the last equality holds since, by the given condition in the lemma, $B(v)$ is a blank ball (at the beginning of the algorithm), and hence we have $\Pr[N_{<1}] = 1$. Note that we also have $\Pr[N_1 | N_{<1}] = \Pr[N_1]$.

Define E_t as the event that there exists a vertex in $B(v)$ that is colored in the t-th round. Then we have $\Pr[N_t | N_{<t}] = 1 - \Pr[E_t | N_{<t}]$. By Lemma 5, we obtain $\Pr[N_t | N_{<t}] \geq 1 - \frac{\Delta + 1 - x_v}{k}$. Consequently, we get

$$\Pr[N_{<r}] \underset{(8)}{\geq} \left(1 - \frac{\Delta + 1 - x_v}{k}\right)^{r-1}. \tag{9}$$

By Lemma 4, we know

$$\Pr[A_r | N_{<r}] = \frac{x_v}{k}. \tag{10}$$

Now we can give our estimation of the probability p:

$$p \underset{(7),(9),(10)}{\geq} \sum_{r=1}^{\infty} \left(1 - \frac{\Delta + 1 - x_v}{k}\right)^{r-1} \cdot \frac{x_v}{k} = \frac{x_v}{\Delta + 1 - x_v}. \qquad \square$$

The following Lemma 7 is used to analyze the probability that a vertex v is happy whose ball $B(v)$ is a monochrome ball. While it is the counterpart of Lemma 6 for a vertex whose ball is a blank ball, we have to be careful for the analysis in Lemma 7, since in a monochrome ball, there is already a used color. Although the two proofs are similar, the proof of Lemma 7 is more complicated than that of Lemma 6. The proof of Lemma 7 is omitted here due to space limitation and will be given in the full version.

Lemma 7. *Consider a monochrome ball $B(v)$ with pre-specified color i^*. The probability that all the uncolored vertices in $B(v)$ are colored in i^* in the same round is $\geq \frac{x_v}{\Delta - x_v}$.*

Define $\mathcal{B}_0$ as the set of vertices whose $B(v)$'s are blank at the beginning of Algorithm $\mathcal{R}$, and $\mathcal{B}_1$ the set of vertices whose $B(v)$'s are monochrome at the beginning of Algorithm $\mathcal{R}$.

Lemma 8. *For vertex $v \notin \mathcal{B}_0 \cup \mathcal{B}_1$, we have $x_v = 0$.*

Proof. Since $v \notin \mathcal{B}_0 \cup \mathcal{B}_1$, by definition, there are at least two different pre-specified colors in $B(v)$. Suppose u_1 and u_2 are two vertices in $B(v)$ with two pre-specified colors i_1 and i_2, respectively. Without loss of generality, we can assume that $i_1 = 1$ and $i_2 = 2$.

For vertex u_1, we have $y_{u_1}^1 = 1$ and $y_{u_1}^2 = \cdots = y_{u_1}^k = 0$. So, for each $2 \leq i \leq k$, we have $\min_{u \in B(v)} \{y_u^i\} = 0$. Then consider color 1. Since $y_{u_2}^2 = 1$, we know $y_{u_2}^1 = 0$. Therefore, $\min_{u \in B(v)} \{y_u^1\} = 0$. This means $x_v = \sum_i x_v^i = 0$. $\square$

Theorem 1. *Algorithm $\mathcal{R}$ is a $\frac{1}{\Delta + 1}$-approximation algorithm for MHV.*

Proof. Let random variable SOL represent the total weight of happy vertices found by Algorithm $\mathcal{R}$. Obviously, only vertices in $\mathcal{B}_0$ and $\mathcal{B}_1$ may be happy. So, we have

$$E[SOL] = \sum_{v \in \mathcal{B}_0} w_v \Pr[B(v) \text{ is finally colored in only one color}] +$$
$$\sum_{v \in \mathcal{B}_1} w_v \Pr[B(v) \text{ is finally colored in only one color}]$$
$$\geq \sum_{v \in \mathcal{B}_0} \frac{w_v x_v}{\Delta + 1 - x_v} + \sum_{v \in \mathcal{B}_1} \frac{w_v x_v}{\Delta - x_v} \quad \text{(by Lemmas 6 and 7)}$$
$$\geq \frac{1}{\Delta + 1} \sum_{v \in \mathcal{B}_0 \cup \mathcal{B}_1} w_v x_v \underset{\text{(LM8)}}{=} \frac{1}{\Delta + 1} \sum_v w_v x_v \geq \frac{1}{\Delta + 1} OPT,$$

where the first equality is due to the linearity of expectation. Note that at the first inequality, we omit the probability that all vertices in $B(v)$ are colored in the same color across rounds (see remarks after Lemma 4). □

Algorithm $\mathcal{R}$ for the MHV problem can be derandomized in polynomial time by the standard conditional expectation method. The derandomization details can be found in [7, Section 5]. Hence, the MHV problem actually can be approximated deterministically within $\frac{1}{\Delta+1}$.

3   Algorithms for MHE

The following linear program (LP-E) is an LP-relaxation for the MHE problem. In the corresponding integer program of (LP-E), variable y_v^i indicates whether vertex v is colored in i, x_e^i indicates whether edge e is happy by color i (*i.e.*, its two endpoints are all colored in i), and x_e indicates whether edge e is happy.

$$\max \quad \sum_{e \in E} w_e x_e \qquad\qquad\qquad\qquad \text{(LP-E)}$$

$$\text{s.t.} \quad \sum_i y_v^i = 1, \qquad\qquad \forall v \qquad\qquad\qquad (11)$$

$$y_v^i = 1, \qquad\qquad \forall i, \forall v \text{ s.t. } c(v) = i \qquad (12)$$

$$x_e^i = \min\{y_u^i, y_v^i\}, \quad \forall i, \forall e = (u, v) \qquad (13)$$

$$x_e = \sum_i x_e^i, \qquad\qquad \forall e$$

$$x_e, x_e^i, y_v^i \geq 0, \qquad\qquad \forall i, \forall v, \forall e$$

Constraint (12) describes the vertices that have pre-specified colors. Constraint (11) says that each vertex has exactly one color. Constraint (13) says that edge e is happy by color i only when both its two endpoints are colored in i. Note that constraint (13) is linear since it can be replaced by $x_e^i \leq y_u^i$

and $x_e^i \leq y_v^i$. Furthermore, by constraint (11), it is impossible for an edge to be simultaneously satisfied by two different colors. Therefore, (LP-E) is really an LP-relaxation for the MHE problem.

Let (x, y) be an optimal fractional solution to (LP-E). To obtain an integral solution, a straightforward LP-rounding technique is to color vertex v in i with probability y_v^i. We can show this strategy will generate an integral solution whose value could be as bad as $1/k$ times OPT_f(LP-E) (the fractional optimum of (LP-E)). This approximation is unsatisfactory, and thus we adopt the randomized rounding technique of Kleinberg and Tardos [7] again to round (x, y). The algorithm is essentially the same as Algorithm $\mathcal{R}$, except that in step 1 we solve (LP-E) instead of (LP-V). For simplicity, we still call the randomized rounding algorithm Algorithm $\mathcal{R}$.

Then Consider the following simple randomized algorithm, namely, Algorithm $\mathcal{P}$ for MHE.

Algorithm $\mathcal{P}$.
1 Pick a color $i \in [k]$ uniformly at random.
2 Color all the uncolored vertices in i.

The final algorithm for MHE is shown as Algorithm $\mathcal{A}$.

Algorithm $\mathcal{A}$.
1 With probability λ run Algorithm $\mathcal{R}$, and with probability $1 - \lambda$ run Algorithm $\mathcal{P}$.
2 Return the coloring found (by either $\mathcal{R}$ or $\mathcal{P}$) in step 1.

Let E_2 be the set of edges with both endpoints being given some pre-specified colors, E_1 the set of edges with only one endpoint being given a pre-specified color, and E_0 the set of edges with no endpoints that have pre-specified colors. Then (E_2, E_1, E_0) is a partition of the edge set E. The edges in E_2 can be further classified into two categories: happy edges and unhappy edges. Since Algorithm $\mathcal{A}$ can obtain all happy edges in E_2 (namely, the approximation ratio for this part of edges is 1), for simplicity of analysis, we may assume that there is no edges of E_2 in the input graph.

Let $e = (u, v)$ be an edge. Note that $0 \leq x_e = \sum_i x_e^i \leq \sum_i y_u^i = 1$. The following lemmas are known from [7,8].

Lemma 9 ([7,8]). *Let e be an edge in E_1. Then $\Pr[e$ is happy in $\mathcal{R}] = x_e$.*

Lemma 10 ([8]). *Let e be an edge in E_0. Then $\Pr[e$ is happy in $\mathcal{R}] \geq \frac{x_e}{2 - x_e}$.*

Theorem 2. *The MHE problem can be approximated within $\frac{1}{2} + \frac{\sqrt{2}}{4} f(k)$, where k is the number of colors and $f(k) = \frac{(1-1/k)\sqrt{k(k-1)}+1/\sqrt{2}}{k-1+1/2k} \geq 1$.*

Proof. Let random variables R_1, P_1, A_1 be the total weight of happy edges in E_1 found by Algorithms $\mathcal{R}$, $\mathcal{P}$, and $\mathcal{A}$, respectively. Then, by Lemma 9, we have

$$
\begin{aligned}
E[A_1] &= (1 - \lambda)E[P_1] + \lambda E[R_1] \\
&= (1 - \lambda) \sum_{e \in E_1} w_e \Pr[e \text{ is happy in } \mathcal{P}] + \lambda \sum_{e \in E_1} w_e \Pr[e \text{ is happy in } \mathcal{R}] \\
&= (1 - \lambda) \sum_{e \in E_1} \frac{w_e}{k} + \lambda \sum_{e \in E_1} w_e x_e = \sum_{e \in E_1} \left(\frac{1 - \lambda}{k} w_e + \lambda w_e x_e \right) \\
&\geq \sum_{e \in E_1} \left(\frac{1 - \lambda}{k} + \lambda \right) w_e x_e.
\end{aligned}
$$

Let random variables R_0, P_0, A_0 be the total weight of happy edges in E_0 found by Algorithms $\mathcal{R}$, $\mathcal{P}$, and $\mathcal{A}$, respectively. Then, by Lemma 10, we have

$$
\begin{aligned}
E[A_0] &= (1 - \lambda)E[P_0] + \lambda E[R_0] \\
&= (1 - \lambda) \sum_{e \in E_0} w_e \Pr[e \text{ is happy in } \mathcal{P}] + \lambda \sum_{e \in E_0} w_e \Pr[e \text{ is happy in } \mathcal{R}] \\
&\geq (1 - \lambda) \sum_{e \in E_0} w_e + \lambda \sum_{e \in E_0} \frac{w_e x_e}{2 - x_e} = \sum_{e \in E_0} \left((1 - \lambda) w_e + \frac{\lambda w_e x_e}{2 - x_e} \right) \\
&\geq \sum_{e \in E_0, x_e > 0} \left(\frac{1 - \lambda}{x_e} + \frac{\lambda}{2 - x_e} \right) w_e x_e \geq \sum_{e \in E_0, x_e > 0} \left(\frac{1}{2} + \sqrt{\lambda(1 - \lambda)} \right) w_e x_e \\
&= \sum_{e \in E_0} \left(\frac{1}{2} + \sqrt{\lambda(1 - \lambda)} \right) w_e x_e,
\end{aligned}
$$

where the last inequality holds since $\min_{x_e \in [0,1]} \{ \frac{1-\lambda}{x_e} + \frac{\lambda}{2 - x_e} \} \geq \frac{1}{2} + \sqrt{\lambda(1 - \lambda)}$.

Let random variable A be the total weight of happy edges found by Algorithm $\mathcal{A}$. Therefore, we get

$$
\begin{aligned}
E[A] &= E[A_1] + E[A_0] \\
&\geq \sum_{e \in E_1} \left(\frac{1 - \lambda}{k} + \lambda \right) w_e x_e + \sum_{e \in E_0} \left(\frac{1}{2} + \sqrt{\lambda(1 - \lambda)} \right) w_e x_e \\
&\geq \min \left\{ \frac{1 - \lambda}{k} + \lambda, \frac{1}{2} + \sqrt{\lambda(1 - \lambda)} \right\} \sum_{e \in E} w_e x_e.
\end{aligned}
$$

Solving $\frac{1-\lambda}{k} + \lambda = \frac{1}{2} + \sqrt{\lambda(1 - \lambda)}$, we get $\lambda = \frac{2k^2 + 2 - 3k + \sqrt{2k^4 - 2k^3}}{2(2k^2 - 2k + 1)}$. (The other root of λ cannot lead to good approximation ratio and is omitted.) For our choice of λ, its value is between $0.8227 \cdots$ and $0.8535 \cdots$ when $k \geq 3$. (The problem is polynomial time solvable when $k = 2$). With this value of λ, the approximation ratio is $\frac{1}{2} + \frac{\sqrt{2}}{4} f(k)$, where $f(k) = \frac{(1 - 1/k)\sqrt{k(k-1)} + 1/\sqrt{2}}{k - 1 + 1/2k} \geq 1$. $\qquad \square$

Algorithms $\mathcal{A}$, $\mathcal{P}$ and $\mathcal{R}$ can all be derandomized in polynomial time by the conditional expectation method. So the result in Theorem 2 is a deterministic result.

Since Multiway Uncut is a special case of MHE, Theorem 2 also means that the Multiway Uncut problem can be approximated within $\frac{1}{2} + \frac{\sqrt{2}}{4} f(k)$. For fixed values of k, this ratio improves upon the ratio 0.8535 of [8]. For example, when $k = 3, 4, 5$, and 10, our ratios are 0.8818, 0.8739, 0.8694, and 0.8611, respectively. When k approaches infinity, the ratio tends to $\frac{1}{2} + \frac{\sqrt{2}}{4} = 0.8535 \cdots$, coinciding with the ratio given in [8]. We get this improvement because we make use of the randomized Algorithm $\mathcal{P}$, while [8] does not.

Acknowledgments. Peng Zhang is supported by the State Scholarship Fund of China, Natural Science Foundation of Shandong Province (ZR2012FZ002 and ZR2013FM030), and the Fundamental Research Funds of Shandong University (2015JC006).

Angsheng Li is supported by the hundred talent program of the Chinese Academy of Sciences, and the grand challenge program, *Network Algorithms and Digital Information*, Institute of Software, Chinese Academy of Sciences.

References

1. Breiman, L., Friedman, J.H., Olshen, R.A., Stone, C.J.: Classification and Regression Trees. Wadsworth and Brooks, Monterey, CA, USA (1984)
2. Buchbinder, N., Naor, J., Schwartz, R.: Simplex partitioning via exponential clocks and the multiway cut problem. In: Proc. STOC, pp. 535–544 (2013)
3. Calinescu, G., Karloff, H., Rabani, Y.: An improved approximation algorithm for multiway cut. Journal of Computer and System Sciences **60**(3), 564–574 (2000)
4. Dahlhaus, E., Johnson, D., Papadimitriou, C., Seymour, P., Yannakakis, M.: The complexity of multiterminal cuts. SIAM Journal on Computing **23**, 864–894 (1994)
5. Easley, D., Kleinberg, J.: Networks, Crowds, and Markets: Reasoning About a Highly Connected World. Cambridge University Press (2010)
6. Karger, D., Klein, P., Stein, C., Thorup, M., Young, N.: Rounding algorithms for a geometric embedding of minimum multiway cut. Mathematics of Operations Research **29**(3), 436–461 (2004)
7. Kleinberg, J., Tardos, É.: Approximation algorithms for classification problems with pairwise relationships: metric labeling and markov random fields. Journal of the ACM **49**(5), 616–639 (2002)
8. Langberg, M., Rabani, Y., Swamy, C.: Approximation algorithms for graph homomorphism problems. In: Díaz, J., Jansen, K., Rolim, J.D.P., Zwick, U. (eds.) APPROX 2006 and RANDOM 2006. LNCS, vol. 4110, pp. 176–187. Springer, Heidelberg (2006)
9. Li, A., Li, J., Pan, Y.: Homophyly/kinship hypothesis: natural communities, and predicting in networks. Physica A **420**, 148–163 (2015)
10. Li, A., Zhang, P.: Algorithmic aspects of homophyly of networks. Manuscript (2012). arXiv:1207.0316
11. Sharma, A., Vondrák, J.: Multiway cut, pairwise realizable distributions, and descending thresholds. In: Proc. STOC, pp. 724–733 (2014)

An Approximation Algorithm for the Smallest Color-Spanning Circle Problem

Yin Wang[1,2]([✉]) and Yinfeng Xu[1,2]

[1] School of Management, Xi'an Jiaotong University, Xi'an 710049, China
[2] The State Key Lab for Manufacturing Systems Engineering, Xi'an 710049, China
yinaywang@stu.xjtu.edu.cn, yfxu@mail.xjtu.edu.cn

Abstract. To find a minimum radius circle in which at least one point of each color lies inside, we have researched the smallest enclosing circle problem for n points with m different colors. The former research proposed a π-approximation algorithm with a running time $O(n^2 + nm \log m)$. In this paper, we construct a color-spanning set for each point and find the smallest enclosing circle to cover all points of each color-spanning set. The approach to find each color-spanning set is based on the nearest neighbor points which have different colors. An approximation algorithm to compute the minimum diameter of the enclosing circle is proposed with the time of $O(nm \log n + n \log m)$ at most. The approximation ratio of our algorithm is less than 2. In conclusion, both approximation ratio and complexity are improved by our proposed algorithm.

Keywords: Computational geometry · Colored set · Approximation algorithm · The minimum diameter color-spanning set problem

1 Introduction

The motivation for the geometric facility location problem comes from the base site selection problem. Based on the arrangement of no different points, a couple of related optimization problems for a set s of n points have already been studied in the literature. Dobkin et al.[1] and Overmas et al.[2] treated the problems of finding minimum perimeter convex k-gons. Their $O(k^2 n \log n + k^5 n)$ Alogrithm result was inmproved to $O(n \log n + k^5 n)$ by Agarwal et al. [3]. Under the condition of $k \leq \frac{n}{2}$, Eppstein et al. [4] studied the smallest ploytopes and proposed a $O(n^2 \log n)$ time algorithms using the Farthest color Voronoi diagram.

Depending on the practical requirements of different applications, variations of problems have been proposed. As associated with different categories, all points are marked by different colors. However, sometimes the information like the locations are not known exactly (Beresfor et al. [5], Cheng et al. [6]). A continuous region which can be a disc model is proposed in difference papers (Wenqi Ju et al. [7]). Then, it is always the objective to find the boundary of a geometrical model to cover all properties.

© Springer International Publishing Switzerland 2015
D. Xu et al. (Eds.): COCOON 2015, LNCS 9198, pp. 171–182, 2015.
DOI: 10.1007/978-3-319-21398-9_14

Abellanas et al.[8] showed an algorithm for smallest color-spanning objects of axis-parallel rectangle and the narrowest strip. Two constant factor approximation algorithms were proposed for the minimum perimeter of the convex hull by Wenqi Ju et al. [7]. And their algorithms cost time of $O(n^2 + nm\log m)$ and $O(\min\{n(n-m)^2, nm(n-m)\})$ with the ratio of π and $\sqrt{2}$ respectively. However, as well as some other variants of theses problems are based on the smallest plogons but Morton et al.[9] proved that such k-gons might not exist for $k > 7$.

However, the former research mostly focus on computing the smallest covering polygons with colored vertices. If the colored points are in general position, it is proved to be NP-hard for some problems (Fleischer and Xu [10], Wenqi Ju et al.[7]). The versions of constrained circle were proposed (Kamrmakar et al. [11]) and solved by the polynomial time algorithms. Zhang et al. [12] showed a brute force algorithm with $O(n^k)$ time for the minimum diameter color-spanning set problem(MDCS).

The problems above are often referred to as the color-spanning set problem (CSSP). Each imprecise point is modelled by a set of k points with m kind of given colors. Each point only has one of the m given colors and the possible location of each point is already known. The color-spanning set contain one point in each color set at least and its number of points is not less than m ($n \geq k \geq m$). In this paper, we study the smallest color-spanning circle in general position and without the constrained of center and color numbers.

For multicolored point sets, there are solutions to several problems, such as the bichromatic closest, see e.g. Agarwal et al. [13]. Interestingly, the approach in paper of Eppstein et al. [14] for the smallest polytopes, like the group Steiner tree by Graf et al. [15], is also based on the nearest neighbors. In this paper, for each point, we cosider the nearest neighbor point which has different color firstly. Then we can find a color-spanning set based on these nearest neighbors. For every color-spanning set, we can find a circle to cover all points of the set. Our purpose is to find the minimum diameter of these circles.

2   Problem Statement and Notations

2.1   Problem Statement

In the color-spanning set problem, each point in the set $P = \{p_1, p_2, ..., p_n\}$ is associated with a color from a set of m colors ($m \leq n$). Let S_i be the set of points in the color class i ($i = 1, 2, ..., m$). We use i to denote the color of the point p_{ij} and j to lable the differentiate points in S_i which have different position in the plane. A new point set named the color-spanning set is constituted such that at least one point of each color lies inside it.

Without loss of generality, we make the following assumption on general position: (1) each point has and only has one of the m given colors and there is no horizontal or vertical line passing through two or more points. (2) the number of colors is fixed. (3) considering the concision, the Euclidean distance between two points is supposed to be known as its position information, but the exact position of each point is uncertain.

Definition 1. We call a circle color-spanning if all points in this circle contains at least one of each color set S_i. The objective is to find the minimum diameter circle c_s to cover a color-spanning set.

Meanwhile, this circle should meet the following conditions: (1) covering all the points in color-spanning set, (2) having the minimum diameter, (3) containing all m given colors. From all above, we can define this problem to find the minimum diameter of this circle as the smallest color-spanning circle (SCSC) problem.

2.2 Notations

The two properties of every point are color and position. j is used to mark each point with unique position in the plane, since the exact position is unknown. Let p_{ij} denote the point with the color of i when its position is j, for $1 \le i \le m$ and $1 \le j < n$.

Definition 2. Let V_{ij} denote a point set of all points in each color-spanning circle. For each point p_{ij}, we can constitute a color-spanning set V_{ij} with $m - 1$ different colors from p_{ij}. Then we have the point set

$$V_{ij} = \{p_{ij}, p_1, \ldots, p_k, \ldots, p_{m-2}, p_{m-1}\}.$$

We assume the number of points in each V_{ij} is not less than m. The rank of points in each V_{ij} is denoted as k ($1 \le k \le m$). Let each p_{ij} be the first one of V_{ij} and called as initial point. The distance between p_{ij} and other points of V_{ij} decides their rank k in each V_{ij}.

The points in each color-spanning set V_{ij} should satisfy two properties: different colors and shortest interval with p_{ij}. Satisfied all the property above, the point set can be regard as the color-spanning set V_{ij} of each point p_{ij}.

Definition 3. The intervals of p_{ij} and other points in V_{ij} constitute a sequence I_{ij}. Similarly, let the sequence I_{ij} as $I_{ij} = \{d_1, d_2, \ldots, d_{m-2}, d_{m-1}\}$. We denote a set of the nearest distinct color distance as I_{ij}.

An approach to find the candidates of a color-spanning set is searching the distance sequence I_{ij}. That is to say, the furthest distance of the nearest distinct color neighbors to p_{ij} is the maximum one in the sequence I_{ij} and denoted as follows: $d_{ij \to m-1} = \max\{I_{ij}\}$. With the hypothesis of its uniqueness, $d_{ij \to m-1}$ is accoresponding to the endpoint p_{m-1} in V_{ij}. By the analogy, $d_{m-2 \to ij}$ is the second furthest one. For the further computing, d_{ij} stands for the solution to the diameter of SCSC problem.

3 An Algorithm for the Color-Spanning Sets

Our aim is to cover all m given colors using the smallest circle. We conside one of the n points briefly. By that analogy, the others can be computed by the same way. Finally, the smallest one of these n color-spanning circles is our solution.

According to the definitions above, we can deduce that the color points in each color-spanning set should satisfy two constraints: closest to initial point p_{ij} in each color set s_i and cover all m colors.

As we know, V_{ij} is the color-spanning set of p_{ij}. The purpose of this paper is to find the smallest color-spanning circle. We present an algorithm to find the nearest distinct colored points of each point p_{ij}. As this purpose, the approach is based on the nearest distinct color distance I_{ij} for each p_{ij}. After finding the furthest distinct colored distance for each given point, the color-spanning set V_{ij} can be constituted. The processes are shown in Algorithm 1.

Algorithm 1. Contitute a color-spanning set for each p_{ij}

For each initial point p_{ij}:
1: **for all** color sets except the one p_{ij} belong to **do**
2:    **for all** points in each color set S_x except the one p_{ij} belong to **do** find the nearest point to initial point p_{ij} in each color set
3:        **if** a point is the nearest one to p_{ij} in color set s_x **then**
4:            save this point as p_x and add it to point set V_{ij}.
5:        **end if**
6:    **end for**
7: **end for**

For n points with m colors, the Algorithm 1 above can constitute a color-spanning point set for each p_{ij}.

Definition 4. With each I_{ij}, we can get the rainbow circles of this color-spanning set. The rainbow circle set of each color-spanning set V_{ij} is defined as a group of concentric circles. Seen in Fig.1 . Thus, we have the following result.

The rainbow circles set $C_{ij}(k, d_k) = \{c_1, c_2, \ldots, c_k, \ldots, c_{m-2}, c_{m-1}\}$ is a set of $m - 1$ concentric circles with the center on p_{ij}. The radius of rainbow circle c_k is depended on the nearest distinct color distance d_k in each I_{ij}. For each p_{ij}, every colored circle c_k has two propertie as we proposed by Lemma 1 and Lemma 2 as below.

Lemma 1. There is k colors covered by the colored circle c_k.

Proof. The center of these circles is the initial point p_{ij}. The radiu of c_k is equal to $d_{ij \to m-1}$, which is belong to I_{ij}. If the distance between any point and p_{ij} is not longer than the radiu of c_k, the color of this point is covered by c_k. According to Algorithm 1, there is k points contain this condition. Besides, all these points have different colors as the definition of V_{ij}.   □

We can know the c_{m-1} and c_{m-2} are the largest and second largest circle in each set of rainbow circles from definition 4. According to Lemma 1, the circle c_{m-1} of each p_{ij} contains all given m colors.

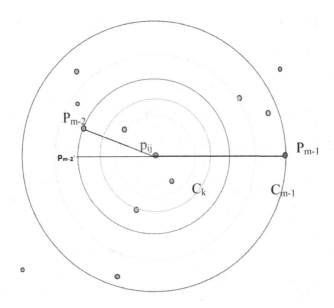

Fig. 1. The rainbow circles for each p_{ij}

Lemma 2. The point p_{m-1} is the only one with the color of circle c_{m-1}.

Proof. Let the color of circle $c_m - 1$ be x. If there is any point with color x inside the rainbow circle c_{m-1}, the point p_{m-1} cannot be the nearest to p_{ij} with color x. $\qquad\square$

Theorem 1. The complexity for constituting each color-spanning set V_{ij} is $O(m \log n)$.

Proof. Each point can be the initial point p_{ij} of n possible color-spanning sets and its distances to other points are supposed to be known. Otherwise, we need time $O(\log n_x)$ to find the nearest point to p_{ij} in each color set S_i , where n_x is the number of points in color set S_x. As we know, there are n points in the plane. Then,

$$\sum_{x=1}^{m-1} n_x + n_i = n$$

the n_i is the number of points in S_i which is the color set the p_{ij} belong to. Therefore, it spends $O(\log n)$ to comupte the nearest distance to p_{ij} in each color set at most. For $m - 1$ color set except the one p_{ij} belong to, it cost time of $O(m \log n)$ to constitute each V_{ij} at most.

$\qquad\square$

4   An Approximation Algorithm for SCSC Problem

4.1   An Approximation Algorithm

For each initial point, according to the Algorithm 1 above, each V_{ij} can be consisted by the nearest neighbor points to p_{ij} with m given different colors. Based on the conclusions above, we come up with an approximation algorithm for the SCSC problem. The first and quite simple idea to do is computing d_{m-1} and d_{m-2} of each V_{ij}. The procedure is shown by Algorithm 2.

Algorithm 2. An Approximation Algorithm

    For given n point
1: **for all** initial point p_{ij} **do**
2:     **for all** points in V_{ij} **do**
3:         save the furthest one to p_{ij} as p_{m-1} and define their interval as d_{m-1} ;
4:     **end for**
5:     **for all** points in V_{ij} except p_{m-1} **do**
6:         save the furthest one to p_{ij} as p_{m-2} and define their interval as d_{m-2} ;
7:     **end for**
8:     **for all** p_{m-1} and p_{m-2} of p_{ij} **do**
9:         compute $d_{ij} = d_{m-1} + d_{m-2}$
10:    **end for**
11: **end for**
12: **for all** d_{ij} **do**
13:     compute the minimal one and save as $d_s = \min\{d_{m-1} + d_{m-2}\}$
14: **end for**

The d_{ij} is the sum of d_{m-1} and d_{m-2} of each V_{ij} in this Approximation Algorithm. We can draw a circle c_s with diameter of the minimum value of d_{ij}. The diameter of c_s is the approximation solution and denoted as d_s. Therefore the following assertion holds.

Definition 5. A union set contains p_{m-1} and all points in circle c_{m-2}. We define it as A for each V_{ij}, where $A \subseteq V_{ij}$.

We draw a circle denoted as c_{ij} with the diameter of d_{ij} for each V_{ij}. Let the line L_{ij} through the initial point p_{ij} and p_{m-1}. For each V_{ij}, the projection of each point p_k on L_{ij} is noted as $p_k{}'$. Find a point On L_{ij} saved as $p_{m-2}{}'$, which is on the left of p_{ij} and has the same interval with p_{m-2} to p_{ij}. Each candidate c_{ij} of the approximation circle should satisfy properties as follows. Firstly, it can cover all given colors. Secondly, the center of this circle is at the middle of $p_{m-2}{}'$ and p_{m-1}. Finally, the diameter d_{ij} of c_{ij} is equal to the sum of d_{m-1} and d_{m-2}.

Lemma 3. Each c_{ij} can cover all given m colors.

Proof. As we know, c_{ij} contains all points in the union set A above. Seen in Fig. 2. Furthermore, the points in A contain all points the circle c_{m-2} for each p_{ij}. Therefore, the $m-1$ colors of points in c_{m-2} are also covered by A. Furthermore, the color of p_{m-1} is unique in V_{ij} and it is contain by A too. As the result,A can cover all m given colors, so does c_{ij}.   □

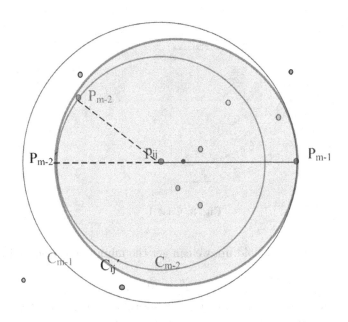

Fig. 2. The color-spanning circle c_{ij} of each p_{ij}

4.2   The Upper Bound and Lower Bound of the SCSC Problem

According to the conclusions above, we know each c_{ij} can contain all given m colors. In each c_{ij}, p_{m-1} and p_{m-2}' is always on the boundary. Shown in the Fig.2, we can find the position of each point in A desides the optimal circle. The exact position is uncertainty, while distances between each two points are certain in this paper. Therefore, three cases of the possible position of all point in A are existed. We donot think about the region outside the rainbow circle c_{m-1} for each V_{ij}.

   Case 1: The projection on L_{ij} of p_{m-2} in on the left of p_{ij}, just like the Fig 2. Let p_{m-2}'' be the projection of p_{m-2}. The radius of c_{ij}, which is saved as d_{ij}, should be equal to the distance of p_{m-2}'' and its center o_{ij}. It can be show as :

$$r_{ij} = d_{ij}/2 = |o_{ij}p_{m-2}|$$

And if the p_{m-2} should not be out of circle c_{ij}, then

$$d_{m-2} = |p_{ij}p_{m-2}| \geq |p_{ij}p_{m-2}''|.$$

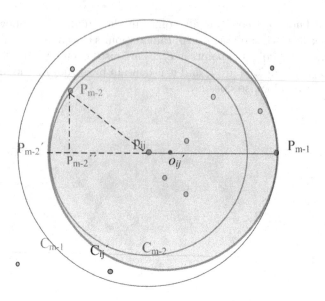

Fig. 3. Case 1

If and only if the equality hold up, we can get the minimum of the radius. That means,

$$r_{ij} = |o_{ij}p_{m-2}''| = |o_{ij}p_{m-2}|$$

Therefore, when the position of p_{m-2} is overlap with the point p_{ij}, we can find the worst case of the circle c_{ij}. The diameter of the smallest c_{ij} is the minimal value under the worst case.

Case 2: Denote the p_k as any point in $A.c$ between p_{ij} and p_{m-1}. As $d_{m-1} \geq d_{m-2}$, the minimum of this radius should not be less than d_{m-1}. The optimal diameter is the minimal value of d_{m-1} and save as d_t. Seen in Fig. 4.

Case 3: When not all points in A satify

$$d_{p_{m-1}p_k}^2 \leq d_{p_{ij}p_{m-1}}^2 - d_{p_{ij}p_k}^2$$

and their projections is not on the right of p_{ij} but the point p_{m-2}. Seen in Fig. 5.

The argument for case 3 is quite simlar to the proof of case 1. If the diameter of circle is less than d_{m-1}, it is certain that the points whose projection is on the right of p_{ij} cannot be covered by this circle. Therefore, the optimal diameter in cases should be more than d_t in case 2 but less than the d_s.

The maximum value of the four cases is the finally worst case of each color-spanning set of p_{ij}. As the result, we can prove the worst case is the situation where p_{m-2}, p_{ij} and p_{m-1} are on one line L_{ij}.

Lemma 4. The lower bound of the SCSC problem is the circle with the diameter d_t, where $d_t = \min\{d_{m-1}\}$.

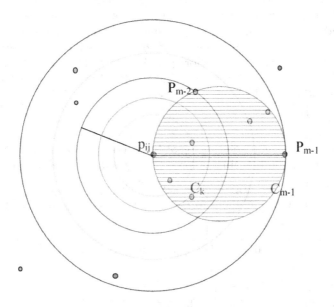

Fig. 4. Case 2

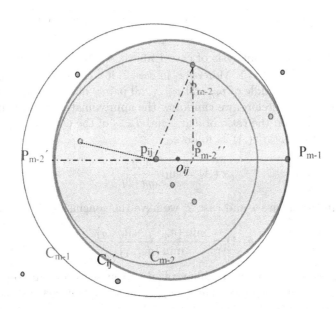

Fig. 5. Case 3

Proof. As the conclusion of the lemma 4, we can find the minimum possible radius in the case 2, which is shown as Fig 4. As the case of $\min\{d_{m-1}\}$, the point p_{ij} and p_{m-1} in c_t cannot be out of this optimal circle. Moreover, $|p_{ij}p_{m-2}|$ is the diameter of the circle. Chords cannot be longer than diameter in one circle. So the diameter d_{opt} of color-spanning circle cannot be less than the interval between p_{ij} and p_{m-1} of c_t. The length of minimal value of d_{m-1} is the lower bound of the color-spanning circle. Denote it as d_t, then $d_t = \min\{d_{m-1}\}$. □

Lemma 5. The optimal solution of SCSC problem cannot be less than the minimum of the furthest distinct colored distances d_{m-1}, as $d_{opt} \geq \min\{d_{m-1}\}$.

Proof. According to the argument of case 1, the upper bound of the optimal solution depends on the minimal worst case. That solution cannot be less than the approximation solution $r_s = \left\{ \frac{d_{m-1}+d_{m-2}}{2} \Big| \min\{d_{m-1}\} \right\}$. □

From the argument above, Theorem 2 can be concluded.

Theorem 2. The upper bound of the SCSC problem is the smallest c_{ij}. The low bound of the SCSC problem is the circle with the diameter of the minimum d_{m-1}.

4.3   The Approximation Ratio

Theorem 3. The approximation ratio $r(A)$ is the value between 1 and 2, $r(A) \in [1, 2]$.

Proof. According to the analysis of case 1 above, when the p_{m-2} and p_{m-2}' in circle c_s is overlap, $r(A) = 1$. Moreover, in case 2, if a circle with radius of $\frac{d_{m-1}}{2}$ is satisfied center at middle of p_{ij} and p_{m-1}, all points of V_{ij} can be covered. As a result, $r(A) = 2$. Therefore, we can know the approximation ratio is between 1 and 2 and depends on the ratio of d_{m-1} and d_{m-2} of the circle r_s. We compute the approximation ratio $r(A)$ as follows.

$$r(A) = \sup_{I} \frac{A(I)}{opt(I)}$$

As the argument of case 1 and case 2, we have the conclusion as below.

$$r(A) = \frac{\min\{d_{m-1} + d_{m-2}\}}{\min\{d_{m-1}\}}$$

Because d_s is the minimum value of d_{ij}, the sum of $\min\{d_{m-1}\}$ and $d_{m-2}\big|_{\min\{d_{m-1}\}}$ can not more than d_s. $d_{m-2}\big|_{\min\{d_{m-1}\}}$ means the d_{m-2} in the circle c_t. Then,

$$r(A) \leq \frac{\min\{d_{m-1}\} + d_{m-2}\big|_{\min\{d_{m-1}\}}}{\min\{d_{m-1}\}} = 1 + \frac{d_{m-1}}{d_{m-1}}\big|_{\min\{d_{m-1}\}}$$

As d_{m-2} is not alway more than d_{m-1} in each V_{ij}, we have the conclusion:

$$1 \leq r(A) \leq 2.$$

The ratio of our approximation algorithm can be proved.

$\square$

4.4   The Complexity of the Approximation Algorithm

Theorem 4. The time complexity for computing the approximation solution of the minimum circle color-spanning set is $O(nm \log n + n \log m)$.

Proof. According to Algorithm 1, each initial point runs less than $O(m \log n)$ time to find the point set V_{ij}. The step 1 and step 2 cost the same time of $O(\log m)$ to find d_{m-1} and d_{m-2} of each V_{ij}. However, the number of initial point p_{ij} is n, such that we can get n color-spanning sets V_{ij}. As a result, traversal of all initial points still need n times. Furthermore, it spend $O(\log n)$ to find the smallest one. Then, we know

$$O(n) = T(n(m \log n + 2 \log m) + \log n) = O(nm \log n + n \log m).$$

As the result, the time complexity of this approximation algorithm is $O(nm \log n + n \log m)$. $\square$

Conclusion

Given a set of n points, which is a unit set of m colored sets, we study the smallest circle to cover at least one of each colored set. Based on the worst case analysis, we present an algorithm to find a color-spanning set for every point. Based on the theory of geometric properties, the upper bound and lower bound of the SCSC problem are computed. The proposed approximation algorithm has an improved ratio (less than) 2. If the distance of each two points is known, the running time of the approximation algorithm is improved to O(nm log n + n log m).

Acknowledgments. The authors would like to acknowledge the financial support of Grants (No.61221063) from NSF of China and (No.IRT1173) from PCSIRT of China.

References

1. Dobkin, D.P., Drysdale, R.L., Guibas, L.J.: Finding smallest polygons. Computational Geometry **1**, 181–214 (1983)
2. Overmars, M.H., Rote, G., Woeginger, G.: Finding minimum area k-gons. Utrecht University, Department of Computer Science (1989)
3. Aggarwal, A., Imai, H., Katoh, N., et al.: Finding k points with minimum spanning trees and related problems. In: Proceedings of the Fifth Annual Symposium on Computational Geometry, pp. 283–291. ACM (1989)

4. Eppstein, D., Overmars, M.H., Rote, G., Woeginger, G.: Finding minimum area k-gons. Discrete Comput. Geom. **7**, 45–58 (1992)
5. Beresford, A.R., Stajano, F.: Location privacy in pervasive computing. IEEE Pervasive computing **2**(1), 46–55 (2003)
6. Cheng, R., Kalashnikov, D.V., Prabhakar, S.: Querying imprecise data in moving object environments. IEEE Transactions on Knowledge and Data Engineering **16**(9), 1112–1127 (2004)
7. Ju, W., Fan, C., Luo, J., et al.: On some geometric problems of color-spanning sets. Journal of Combinatorial Optimization **26**(2), 266–283 (2013)
8. Abellanas, M., Hurtado, F., Icking, C., Klein, R., Langetepe, E., Ma, L., Palop, B., Sacristán, V.: Smallest color-spanning objects. In: Meyer auf der Heide, F. (ed.) ESA 2001. LNCS, vol. 2161, pp. 278–289. Springer, Heidelberg (2001)
9. Morton, J.D., et al.: Sets with no empty convex 7-gons. Canadian Mathematical Bulletin **26**(4), 482 (1983)
10. Fleischer, R., Xu, X.: Computing minimum diameter color-spanning sets. In: Lee, D.-T., Chen, D.Z., Ying, S. (eds.) FAW 2010. LNCS, vol. 6213, pp. 285–292. Springer, Heidelberg (2010)
11. Karmakar, A., Roy, S., Das, S.: Fast computation of smallest enclosing circle with center on a query line segment. Information Processing Letters **108**(6), 343–346 (2008)
12. Zhang, D., et al.: Keyword search in spatial databases: towards searching by document. In: IEEE 25th International Conference on Data Engineering, ICDE 2009, pp. 688–699 (2009)
13. Agarwal, P.K., Edelsbrunner, H., Schwarzkopf, O., et al.: Euclidean minimum spanning trees and bichromatic closest pairs. Discrete and Computational Geometry **6**(1), 407–422 (1991)
14. Eppstein, D., Erickson, J.: Iterated nearest neighbors and finding minimal polytopes. Discrete and Computational Geometry **11**(1), 321–350 (1994)
15. Graf, T., Hinrichs, K.: Algorithms for proximity problems on colored point sets. Universitt Mnster, Angewandte Mathematik und Informatik (1992)

Approximation Algorithms for the Connected Sensor Cover Problem

Lingxiao Huang, Jian Li, and Qicai Shi[✉]

Institute for Interdisciplinary Information Sciences (IIIS),
Tsinghua University, Beijing 100084, China
{huanglx12,sqc12}@mails.tsinghua.edu.cn, lijian83@mail.tsinghua.edu.cn

Abstract. We study the minimum connected sensor cover problem (MIN-CSC) and the budgeted connected sensor cover (Budgeted-CSC) problem, both motivated by important applications in wireless sensor networks. In both problems, we are given a set of sensors and a set of target points in the Euclidean plane. In MIN-CSC, our goal is to find a set of sensors of minimum cardinality, such that all target points are covered, and all sensors can communicate with each other (i.e., the communication graph is connected). We obtain a constant factor approximation algorithm, assuming that the ratio between the sensor radius and communication radius is bounded. In Budgeted-CSC problem, our goal is to choose a set of B sensors, such that the number of targets covered by the chosen sensors is maximized and the communication graph is connected. We also obtain a constant approximation under the same assumption.

1 Introduction

In many applications, we would like to monitor a region or a collection of targets of interests by deploying a set of wireless sensor nodes. A key challenge in such applications is the limited energy supply for each sensor node. Hence, designing efficient algorithms for minimizing energy consumption and maximizing the lifetime of the network is an important problem in wireless sensor networks and many variations have been studied extensively. We refer interested readers to the book by Du and Wan [11] for many algorithmic problems in this domain.

In this paper, we consider two important sensor coverage problems. Now, we introduce some notations and formally define our problem. We are given a set S of n sensors in $\mathbb{R}^d$. All sensors in S have the same communication range R_c and the same sensing range R_s. In other words, two sensors s and s' can communicate with each other if $\mathrm{dist}(s, s') \leq R_c$, and a target point p can be covered by sensor s if $\mathrm{dist}(p, s) \leq R_s$. We use $D(s, R)$ to denote the disk with radius R centered at point s. Let $D_c(s) = D(s, R_c)$ and $D_s(s) = D(s, R_s)$.

Research supported in part by the National Basic Research Program of China Grant 2015CB358700, 2011CBA00300, 2011CBA00301, the National Natural Science Foundation of China Grant 61202009, 61033001, 61361136003.

© Springer International Publishing Switzerland 2015
D. Xu et al. (Eds.): COCOON 2015, LNCS 9198, pp. 183–196, 2015.
DOI: 10.1007/978-3-319-21398-9_15

Assumption 1. *In this paper, we assume that R_s/R_c can be upper bounded by a constant $C = O(1)$ (i.e., $R_s/R_c \leq C$). Note that this assumption holds for most practical applications. Without loss of generality, we can assume that $R_c = 1$. Hence, $R_s = O(1)$.*

The first problem we study is the the *minimum Connected sensor covering* (MIN-CSC) problem. This problem considers the problem of selecting the minimum number of sensors that form a connected network and detect all the targets. It is somewhat similar, but different from, the connected dominating set problem. We will discuss the difference shortly. The formal problem definition is as follows:

Definition 1. MIN-CSC: *Given a set S of sensors and a set $\mathcal{P}$ of target points, find a subset $S' \subseteq S$ of minimum cardinality such that all points in $\mathcal{P}$ are covered by the union of sensor areas in S' and the communication links between sensors in S' form a connected graph.*

In some applications, instead of monitoring a set of discrete target points, we would like to monitor a continuous range R, such as a rectangular area. Such problems can be easily converted into a MIN-CSC with discrete points, by creating a target point (which we need to cover) in each cell of the arrangement of the sensing disks $\{D_s(s)\}_{s \in S}$ restricted in R (see [36] for details).

The second problem studied in this paper is the *Budgeted connected sensor cover* (Budgeted-CSC) problem. The problem setting is the same as MIN-CSC, except that we have an upper bound on the number of sensors we can open, and the goal becomes to maximize the number of covered targets.

Definition 2. Budgeted-CSC: *Given a set S of sensors , a set $\mathcal{P}$ of target points and a positive integer B, find a subset $S' \subseteq S$ such that $|S'| \leq B$ and the number of points in $\mathcal{P}$ covered by the union of sensor areas in S' is maximum and the communication links between sensors in S' form a connected graph.*

1.1   Previous Results and Our Contributions

MIN-CSC. The MIN-CSC problem was first proposed by Gupta et al. [19]. They gave an $O(r \ln n)$-approximation (r is an upper bound of the hop-distance between any two sensors having nonempty sensing intersections). Wu et al. [36] give an $O(r)$-approximation algorithm, which is best approximation ratio known so far (in terms of r). If $R_s \leq R_c/2$, $r = 1$ and the above result implies a constant approximation. However, even R_s is slightly larger than $R_c/2$, r may still be arbitrarily large. We also notice that if $r = O(1)$, we must have $R_s/R_c = O(1)$. So Assumption 1 is a weaker assumption than the assumption that $r = O(1)$.

MIN-CSC is in fact a special case the *group Steiner tree* problem (as also observed in Wu et al [36]). In fact, this can be seen as follows: consider the communication graph (the edges are the communication links). For each target, we create a group which consists for all sensor nodes that can cover the target. The goal is to find a minimum cost tree spanning all groups.[1] Garg et al [16], combined with the optimal probabilistic tree embedding [13], obtained an $O(\log^3 n)$ factor approximation algorithm the group Steiner tree problem via LP rounding. Chekuri et al. [6] obtained nearly the same approximation ratio using pure combinatorial method.

Our first main contribution is a constant factor approximation algorithm for MIN-CSC under Assumption 1, improving on the aforementioned results. Our improvement heavily rely on the geometry of the problem (which the group Steiner tree approach ignores).

Theorem 1. *There is a polynomial time approximation algorithm which can achieve an approximation factor $O(C^2)$ for* MIN-CSC. *Under Assumption 1, the approximation factor is a constant.*

Budgeted-CSC. Recall in Budgeted-CSC, we have a budget B, which is the upper bound of the number of sensors we can use and our goal is to maximize the number of covered target points. Kuo et al.[25] study this problem under the assumption that the communication and the sensing radius of sensors are the same (i.e., $R_s = R_c$). They obtained an $O(\sqrt{B})$-approximation by transforming the problem to a more general connected submodular function maximization problem.

Recently, Khuller et al. [23] obtained a constant approximation for the *budgeted generalized connected dominating set problem*, defined as follows: Given an undirected graph $G(V, E)$ and budget B, and a monotone *special submodular function* [2] $f : 2^V \to \mathbb{Z}^+$, find a subset $S \subseteq V$ such that $|S| \le B$, S induces a connected subgraph and $f(S)$ is maximized. If $R_s \le R_c/2$ in Budgeted-CSC, the coverage function $f(S)$ (the number of targets covered by sensor set S) is a special submodular function. Hence, we have a constant approximation for Budgeted-CSC when $R_s \le R_c/2$. When $R_s > R_c/2$, $f(S)$ may not be special submodular and the algorithm and analysis in [23] do not provide any approximation guarantee for Budgeted-CSC.

We note that it is also possible to adapt the greed approach developed by group Steiner tree [6] and polymatroid Steiner tree [4] to get polylogarithmic approximation for Budgeted-CSC. However, it is unlike the approach can be made to achieve constant approximation factors, and we omit the details.

In this paper, we improve the above results by presenting the first constant factor approximation algorithm under the more general Assumption 1.

---

[1] Notice that the group Steiner tree is edge-weighted but MIN-CSC is node-weighted. However, since all nodes have the same (unit) weight, the edge-weight and node-weight of a tree differ by at most 1.

[2] f is a special submodular function if (1) f is submodular: $f(A \cup \{v\}) - f(A) \ge f(B \cup \{v\}) - f(B)$ for any $A \subset B \subseteq V$; (2) $f(A \cup X) - f(A) = f(A \cup B \cup X) - f(A \cup B)$ if $N(X) \cup N(B) = \emptyset$ for any $X, A, B \subseteq V$. Here, $N(X)$ denotes the neighborhood of X (including X).

Theorem 2. *There is a polynomial time approximation algorithm which can achieve approximation factor of $\frac{1}{102C^2}$ for* Budgeted-CSC. *Under Assumption 1, the approximation factor is $O(1)$.*

Our algorithm is inspired by, but completely different from [23]. In particular, we make crucial use of the geometry of the problem to get around the issue required by [23] (i.e., the coverage function is required to be special submodular in their work).

1.2 Other Related Work

MIN-CSC is closely related to the the *minimum dominating set* (MIN-DS) and the *minimum connected dominating set* (MIN-CDS) problem. In fact, if the communication radius R_c is the same as the sensing radius R_s, MIN-CSC reduces to MIN-CDS. In general graphs, MIN-CDS inherits the inapproximability of set cover, so it is NP-hard to approximation MIN-CDS within a factor of $\rho \ln n$ for any $\rho < 1$ [10,14]. Improving upon Klein et al. [24], Guha et al.[18] obtained a $1.35 \ln n$-approximation, which is the best result known for general graphs.

Lichtenstein et al. [27] proved that MIN-CDS in unit disk graphs (UDG) is NP-hard (which also implies that MIN-CSC is NP-hard). The first constant approximation algorithm for the unweighted MIN-CDS problem in UDG was obtained by Wan et al.[33]. This was later improved by Cheng et al.[7], who gave the first PTAS. For the weighted (connected) dominating set problem , Ambühl et al. [1] obtained the first constant ratio approximation algorithms for both problems (the constants are 72 and 94 for MIN-DS and MIN-CDSrespectively). The constants were improved in a series of subsequent papers [9,21,35,38]. Very recently, Li and Jin [26] obtained the first PTAS for weighted MIN-DS and an improved constant approximation for weighted MIN-CDS in UDG. Many variants of MIN-DS and MIN-CDS, motivated by various applications in wireless sensor network, have been studied extensively. See [11] for a comprehensive treatment.

Budgeted-CSC is a special case of the submodular function maximization problem subject to a cardinality constraint and a connectivity constraint. Submodular maximization under cardinality constraint, which generalizes the maximum coverage problem, is a classical combinatorial optimization problem and it is known the optimal approximation is $1 - 1/e$ [14,29]. Submodular maximization under various more general combinatorial constraints (in particular, downward monotone set systems) is a vibrant research area in theoretical computer science and there have been a number of exciting new developments in the past few years (see e.g., [3,32] and the references therein). The connectivity constraint has also been considered in some previous work [23,25,37], some of which we mentioned before.

2   Preliminaries

We need the following *maximum coverage* (MaxCov) in our algorithms.

Definition 3. MaxCov: *Given a universe U of elements and a family S of subsets of U, and a positive integer B, find a subset $S' \subseteq S$ such that $|S'| \le B$ and the number of elements covered by $\cup_{S \in S'} S$ is maximized.*

We need to following well known result, by [20,29].

Lemma 1 (Corollary 1.1 of Hochbaum et al. [20]). *The greedy algorithm is a $(1 - \frac{1}{e})$-approximation for MaxCov.*

A closely related problem is the *hitting set* problem.

Definition 4. HitSet: *Given a universe U of weighted elements (with weight function $c : U \to \mathbb{R}^+$) and a family S of subsets of U find a subset $H \subseteq U$ such that $H \cap S \neq \emptyset$ for all $S \in S$ (i.e., H hits every subset in S) and $\sum_{u \in H} c_u$ is minimized.*

The HitSet problem is equivalent to the set cover problem (where the elements and subsets switch roles). It is well known that a simple greedy algorithm can achieve an approximation factor of $\ln n$ for HitSet and the factor is essentially optimal [10,14]. In this paper, we use a geometric version of HitSet in which the set of given elements are points in $\mathbb{R}^2$ and the subsets are induced by given disks (i.e., each $S \in S$ is the subset of points that can be covered by a given disk). Geometric hitting set admits constant factor approximation algorithms (even PTAS) for many geometric objects (including disks) [2,5,8,28,31]. As mentioned in the introduction, MIN-CSC is a special case of the following *group Steiner tree* (GST) problem.

Definition 5. GST: *We are given an undirected graph $G = (V, E, c, \mathcal{F})$ where $c : E \to \mathbb{Z}^+$ is the edge cost function, and $\mathcal{F}$ is a collection of subsets of V. Each subset in $\mathcal{F}$ is called a group. The goal is to find a subtree T, such that $T \cap S \neq \emptyset$ for all $S \in \mathcal{F}$ (i.e., T spans all groups) and the cost of the tree $\sum_{e \in T} c_e$ is minimized.*

Our algorithm for Budgeted-CSC also needs the following *quota Steiner tree* (QST) problem.

Definition 6. QST: *Given an undirected graph $G = (V, E, c, p)$ ($c : E \to \mathbb{Z}^+$ is the edge cost function, $p : V \to \mathbb{Z}^+$ is the vertex profit function) and an integer q, find a subtree $T = \arg\max_{T \subset E, \sum_{e \in T} c(e) \le q} \sum_{v_i \in T} p(v_i)$ of the graph G (T tries to collect as much profit as possible subject to the quota constraint).*

Johnson et al. [22] proposed the QST problem and proved that any α-approximation for the k-MST problem yields an α-approximation for the QST problem. Combining with the 2-approximation for k-MST developed by Garg [15], we can get a 2-approximation for the QST problem.

Lemma 2. *These is an approximation algorithm with approximation factor 2 for QST.*

3   Minimum Connected Sensor Cover

We first construct an edge-weighted graph $\mathcal{G}_c$ as follows: If $\text{dist}(s, s') \leq R_c$, we add an edge between s and s' (It is easy to see that $\mathcal{G}_c$ is in fact a unit disk graph). $\mathcal{G}_c$ is called *the communication graph*. Recall that MIN-CSC requires us to find a set of vertices that induces a connected subgraph in the communication graph $\mathcal{G}_c$.

First, we note that $\mathcal{G}_c$ may have several connected components. We can see any feasible solution must be contained in a single connected component (otherwise, the solution can not induce a connected graph). Our algorithm tries to find a solution in every connected component. Our final solution will be the one with the minimum cost among all connected component. Note that for some connected component, there may not be a feasible solution in that component (some target point can not be covered by any point in that component), and our algorithm ignores such component.

From now on, we fix a connected component $\mathcal{C}$ in $\mathcal{G}_c$. Similar with Wu et al. [36], we formulate the MIN-CSC problem as a group Steiner tree (GST) problem. Each edge $e \in \mathcal{G}[\mathcal{C}]$ is associated with a cost $c_e = 1$. For each target $p \in \mathcal{P}$, we create a group

$$\text{gp}(p) = \mathcal{C} \cap D(p, R_s) = \{s \mid s \in \mathcal{C}, \text{dist}(p, s) \leq R_s\}.$$

The goal is to find a tree $\mathcal{T}$ (in $\mathcal{G}[\mathcal{C}]$) such that $\mathcal{T} \cap \text{gp}(p) \neq \emptyset$ for all $p \in \mathcal{P}$ and the cost is minimized. We can easily see the GST instance constructed above is equivalent to the original MIN-CSC problem (the cost of the tree $\mathcal{T}$ is the number of nodes in $\mathcal{T}$ minus 1). The GST problem can be formulated as the following linear integral program: We pick a root $r \in \mathcal{C}$ for the tree $\mathcal{T}$ (we need to enumerate all possible roots). For each edge $e \in \mathcal{G}[\mathcal{C}]$, we use Boolean variable x_e to denote whether we choose edge e.

$$\text{minimize} \quad \sum_{e \in \mathcal{G}[\mathcal{C}]} x_e \tag{1}$$

$$\text{subject to} \quad \sum_{e \in \partial(S)} x_e \geq 1, \quad \text{for all } S \subset \mathcal{C} \text{ such that } r \in S \text{ and } \exists p, S \cap G_p = \emptyset;$$

$$x_e \in \{0, 1\}, \quad \forall e \in \mathcal{G}[\mathcal{C}].$$

The second constraint says that for any cut $\partial(S)$ that separates the root r from any group, there must be at least one chosen edge. By replacing $x_e \in \{0, 1\}$ with $x \in [0, 1]$, we obtain the linear programming relaxation of (1) (denoted as Lp-GST). By the duality between flow and cut, we can see that the second constraint is equivalent to dictating that we can send at least 1 unit of flow from the root r to nodes in $\text{gp}(p)$, for each p. This flow viewpoint (also observed in the original GST paper [16]) will be particularly useful to us later. So we write down the flow LP explicitly as follows. We first replace every undirected edge $e = (u, v)$ by two directed arcs (u, v) and (v, u). For each $p \in \mathcal{P}$ and each directed arc (u, v), we have a variable x_{uv}^p indicating the flow of commodity p on arc (u, v). We use $y_v^p = \sum_u x_{uv}^p - \sum_w x_{vw}^p$ to denote the net flow of commodity p into node v. Then Lp-GST can be equivalently rewritten as the following linear program (denoted as Lp-flow):

$$\text{minimize} \quad \sum_{(u,v) \in \mathcal{G}[\mathcal{C}]} x_{uv} \quad (2)$$

$$\text{subject to} \quad y_v^p = \sum_u x_{uv}^p - \sum_w x_{vw}^p \quad \text{for all } v \in \mathcal{C}$$

$$y_r^p = -1 \quad \text{for all } p \in \mathcal{P},$$

$$\sum_{v \in \text{gp}(p)} y_v^p \geq 1 \quad \text{for all } p \in \mathcal{P},$$

$$y_u^p = 0 \quad \text{for all } u \notin \text{gp}(p), u \neq r,$$

$$x_{uv}^p \leq x_{uv} \quad \text{for all } p \in \mathcal{P}, u, v \in \mathcal{C},$$

$$x_{uv}^p, y_v^p \in [0, 1], \quad \text{for all } u, v \in \mathcal{G}[\mathcal{C}].$$

Now, we describe our algorithm. Our algorithm mainly consists of two steps. In the first step, we extract a *geometric hitting set* instance from the optimal fractional solution of Lp-flow. We can find an integral solution H for the hitting set problem and we can show its cost is at most $O(C^2\text{OPT})$. Moreover all sensors in H can cover all target points $p \in \mathcal{P}$. In the second step, we extract a Steiner tree instance, again from the optimal fractional solution of Lp-flow. We show it is possible to round the Steiner tree LP to get a constant approximation integral Steiner tree, which can connect all points in H.

Step 1: Constructing the Hitting Set Problem :

We first solve the linear program Lp-flow and obtain the fractional optimal solution (x_{uv}, y_v). Let Opt(Lp-flow) to denote the optimal value of Lp-flow. We place a grid with grid size $l = \frac{\sqrt{2}}{2}$ in the plane (i.e., each cell is a $\frac{\sqrt{2}}{2} \times \frac{\sqrt{2}}{2}$ square). For each $p \in \mathcal{P}$, consider the set of sensors $\text{gp}(p)$, that is the set of sensors which can cover p. Since $\text{gp}(p)$ is contained in a disk of radius $R_s \leq C$, there are at most $\frac{\sqrt{2}}{2}O(C^2) = O(1)$ grid cells that may contain some points in $\text{gp}(p)$. Since $\sum_{v \in \text{gp}(p)} y_v^p \geq 1$, there must be a cell (say $\text{cl}(p)$) such that

$$\sum_{v \in \text{gp}(p) \cap \text{cl}(p)} y_v^p \geq \Omega(1/C^2) = \Omega(1). \quad (3)$$

Now, we construct a geometric hitting set (HitSet) instance $(\mathcal{U}, \mathcal{F})$ as follows: Let the set of points be $\mathcal{U} = \cup_{p \in \mathcal{P}}(\text{gp}(p) \cap \text{cl}(p))$ and the family of subsets be $\mathcal{F} = \{\text{gp}(p)\}_{p \in \mathcal{P}}$. The goal is to choose a subset H of $\mathcal{U}$ such that $\text{gp}(p) \cap H \neq \emptyset$ for all $p \in \mathcal{P}$ (i.e., we want to hit every set in $\mathcal{F}$). Write the linear program relaxation for the HitSet problem (denoted as Lp-HS):

$$\text{minimize} \quad \sum_{u \in \mathcal{U}} z_u \tag{4}$$

$$\text{subject to} \quad \sum_{u \in \text{gp}(p)} z_u \geq 1 \quad \text{for all } p \in \mathcal{P},$$

$$z_u \in [0,1], \quad \text{for all } u \in \mathcal{U}.$$

Let Opt(Lp-HS) to denote the optimal value of Lp-HS. We need the following simple lemma. Due to space constraints, the proof can be found in the full version of this paper.[3]

Lemma 3. Opt(Lp-HS) $\leq O(C^2 \text{Opt(Lp-flow)})$.

Bronnimann et al. [2], combined with the existence of ϵ-net of size $O(1/\epsilon)$ for disks (see e.g., [30]), showed that we can round the above linear program Lp-HS to obtain an integral solution (i.e., an actual hitting set) $H \subset \mathcal{U}$ such that $|H| \leq O(\text{Opt(Lp-HS)})$ (the connection to ϵ-net was made simpler and more explicit in Even et al. [12]). Hence, $|H| \leq O(C^2 \text{OPT})$.

Step 2: Constructing the Steiner Tree Problem : Recall that for each $p \in \mathcal{P}$, there is a cell $\text{cl}(p)$ such that $\sum_{v \in \text{gp}(p) \cap \text{cl}(p)} y_v^p \geq \Omega(1/C^2)$. Consider the collection $\Delta = \{\text{cl}(p) \mid p \in \mathcal{P}\}$ of all such cells (if there is a cell which contains the root r, we exclude it from Δ), from each cell $\text{cl} \in \Delta$, we pick an arbitrary point, called the *representative node* $v(\text{cl})$ of cl. From Equation 3 (i.e., $\sum_{v \in \text{gp}(p) \cap \text{cl}(p)} y_v^p \geq \Omega(1/C^2)$), we can see at least $\Omega(1/C^2)$ flow of commodity p that enters $\text{cl}(p)$. Now, we *reroute* such flow to the representative $v(\text{cl}(p))$, in order to create a Steiner tree LP. Consider the optimal fractional solution (x_{uv}, y_v) of Lp-flow. We would like to create another feasible fractional solution $(\widehat{x}_{uv}, \widehat{y}_v)$ for Lp-flow.

- (Flow Rerouting) For each node $u \in \text{gp}(p) \cap \text{cl}(p)$, let $\widetilde{x}^p_{uv(\text{cl}(p))} \leftarrow x^p_{uv(\text{cl}(p))} + y_u$. In other words, we route the flow excess at node u to node $v(\text{cl}(p))$. After such updates, for each $u \in \text{gp}(p) \cap \text{cl}(p), u \neq v(\text{cl}(p))$ we can see the flow excess is zero, or equivalently $\widetilde{y}_u^p = 0$. The flow excess at node $v(\text{cl}(p))$ is

$$\widetilde{y}^p_{v(\text{cl}(p))} = \sum_{v \in \text{gp}(p) \cap \text{cl}(p)} y_v^p \geq \Omega(1/C^2).$$

We repeat the above process for all $\text{cl} \in \Delta$.
- By uniformly increasing all variables, we obtain another feasible solution $(\widehat{x}_{uv}, \widehat{y}_v)$:

$$\widehat{x}_{uv}^p = \min\{C^2 \widetilde{x}_{uv}^p, 1\} \quad \text{and} \quad \widehat{y}_v^p = \min\{C^2 \widetilde{y}_v^p, 1\}.$$

Then, it is easy to see that $\widehat{y}^p_{v(\text{cl})} \geq 1$ for all $\text{cl} \in \Delta$. In equivalent words, at least 1 unit flow (thinking $\widehat{x}_{uv}^p$ as flow value on (u,v)) that enters $v(\text{cl}(p))$.

---

[3] A full version is available from the CS arXiv.

Now, consider the Steiner tree problem in $\mathcal{G}(\mathcal{C})$ in which the set of terminals is defined to be $\mathsf{Ter} = \{r\} \cup \{v(\mathsf{cl}) \mid \mathsf{cl} \in \Delta\}$. Let $\check{x}_e = \max_{p \in \mathcal{P}} \widehat{x}_{uv}^p + \max_{p \in \mathcal{P}} \widehat{x}_{vu}^p$ (Notice that Lp-flow is formulated on directed graphs and Steiner tree is formulated on undirected graphs. Here e is the undirected edge corresponding to directed edges uv and vu). It is easy to see that $\check{x}_e$ is a feasible solution for the following linear program relaxation for the Steiner tree problem (denoted as Lp-ST):

$$\text{minimize} \quad \sum_{e \in \mathcal{G}[\mathcal{C}]} x_e \tag{5}$$

$$\text{subject to} \quad \sum_{e \in \partial(S)} x_e \geq 1, \quad \text{for all } S \subset \mathcal{C} \text{ such that } r \in S \text{ and } \exists \mathsf{cl} \in \Delta, v(\mathsf{cl}) \notin S$$

$$x_e \in \{0, 1\}, \quad \forall e \in \mathcal{G}[\mathcal{C}].$$

Lemma 4. $\mathsf{Opt}(\mathsf{Lp\text{-}ST}) \leq O(C^2 \mathsf{Opt}(\mathsf{Lp\text{-}flow}))$.

It is well known that the integrality gap of the Steiner tree problem is a constant [34]. In particular, it is known that using the primal-dual method (based on Lp-ST) in [17] (see also [34, Chapter7.2]), we can obtain an integral solution $\bar{x}_e$ such that

$$\sum_{e \in \mathcal{G}[\mathcal{C}]} x_e \leq 2\mathsf{Opt}(\mathsf{Lp\text{-}ST}) \leq O(C^2 \mathsf{Opt}(\mathsf{Lp\text{-}flow})) \leq O(C^2 \mathsf{OPT}).$$

Let J be the set of vertices spanned by the integral Steiner tree $\{\bar{x}_e\}$. The above discussion shows that $|J| \leq O(C^2 \mathsf{OPT})$. Our final solution (the set of sensors we choose) is $\mathsf{Sol} = H \cup J$. The feasibility of Sol is proved in the following simple lemma.

Lemma 5. Sol *is a feasible solution.*

Proof. We only need to show that Sol induces a connected graph and covers all the target points. Obviously, H covers all target points, so does Sol. Since J is a Steiner tree, thus connected. Moreover, J connects all representatives $v(\mathsf{cl})$ for all $\mathsf{cl} \in \Delta$. H consists of only sensors in $\mathsf{cl} \in \Delta$. So every sensor in $v \in H$ (say $v \in \mathsf{cl}$) is connected to the representative $v(\mathsf{cl})$. So $H \cup J$ induces a connected subgraph. □

Lastly, we need to show the performance guarantee. This is easy since we have shown that both $|H| \leq O(C^2 \mathsf{OPT})$ and $|J| \leq O(C^2 \mathsf{OPT})$. So $|\mathsf{Sol}| = O(C^2 \mathsf{OPT}) = O(\mathsf{OPT})$ since C is assumed to be a constant.

4 Budgeted Connected Sensor Cover

Again we assume that $R_c = 1$ and $R_s = C$. Recall that our goal is to find a subset $\mathcal{S}' \subseteq \mathcal{S}$ of sensors with cardinality B which induces a connected subgraph and covers as many targets as possible. We first construct the communication graph $\mathcal{G}_c$ as in Section 3. Again, we only need to focus on a connected component of $\mathcal{G}_c$. Then we find a square Q in the Euclidean plane large enough such that all of the n sensors are inside Q. We partition Q into small square cells of equal size. Let the side length of each cell be $l = \frac{\sqrt{2}}{2}$. Denote the cell in the ith row and jth column of the partition as $\mathsf{cl}_{i,j}$. Let $V_{i,j} = \{v \in \mathcal{S} \mid v \in \mathsf{cl}_{i,j}\}$ be the collection of sensors in $\mathsf{cl}_{i,j}$. We then partition these cells into k^2 different cell groups $\mathsf{CG}_{a,b}$, where $k = \lceil 2C/l + 1 \rceil$. In particular, we let

$$\mathsf{CG}_{a,b} = \{\mathsf{cl}_{i,j} \mid i \equiv a(\bmod k), j \equiv b(\bmod k)\} \text{ for } a \in [k], b \in [k].$$

and $\mathcal{V}_{a,b} = \mathcal{S} \cap \mathsf{CG}_{a,b}$ be the collection of sensors in $\mathsf{CG}_{a,b}$.

With the above value k, we make a simple but useful observation as follows.

Observation 1. *There is no target covered by two different sensors contained in two different cells of* $\mathsf{CG}_{a,b}$.

Denote the optimal solution of Budgeted-CSC problem as OPT. In this section, we present an $O\left(\frac{1}{C^2}\right)$ factor approximation algorithm for the Budgeted-CSC problem.

4.1   The Algorithm

For $0 \leq a, b < k$, we repeat the following two steps, and output a tree T with $O(B)$ vertices (sensors) which covers the maximum number of targets. Then based on T, we find a subtree $\tilde{T}$ with exactly B vertices as our final output.

Step 1: Reassign profit : The profit $p(S)$ of a subset $S \subseteq \mathcal{S}$ is the number of targets covered by S. $p(S)$ is a submodular function. In this step, we design a new profit function (called *modified profit function*) $\hat{p} : \mathcal{S} \to \mathbb{Z}^+$ for the set of sensors. To some extent, $\hat{p}$ is a linearized version of p (module a constant approximation factor).

Now, we explain in details how $\hat{p}$ is defined. Fix a cell group $\mathsf{CG}_{a,b}$. [4] For the vertices in $\mathsf{CG}_{a,b}$, we use the greedy algorithm to reassign profits of the vertices in $\mathcal{V}_{a,b}$. Among all vertices in $\mathcal{V}_{a,b}$, we pick a vertex v_1 which can cover the most number of targets, and use this number as its modified profit $\hat{p}(v_1)$. Remove the chosen vertex and targets covered by it. We continue to pick the vertex v_2 in $\mathcal{V}_{a,b}$ which can cover the most number of uncovered targets. Set the modified profit $\hat{p}(v_2)$ to be the number of newly covered targets. Repeat the above steps until all the sensors in $\mathcal{V}_{a,b}$ have been picked out. For other vertices v which are not in $\mathcal{V}_{a,b}$, we simply set their modified profit $\hat{p}(v)$ as 0.

Let us first make some simple observations about p and $\hat{p}$. We use $\hat{p}(S)$ to denote $\sum_{v \in S} \hat{p}(v)$. First, it is not difficult to see that $\hat{p}(S) \leq p(S)$ for any subset $S \subseteq \mathcal{S}$. Second, we can see that it is equivalent to run the greedy algorithm for each cell in $\mathsf{CG}_{a,b}$ separately (due to Observation 1). Suppose $S_1 \subseteq \mathsf{cl}_{c,d}$,

---

[4] For each $\mathsf{CG}_{a,b}$, we define a modified profit function $\hat{p}_{a,b}$. For ease of notation, we omit the subscripts.

$S_2 \subseteq \mathsf{cl}_{c',d'}$ where $\mathsf{cl}_{c,d}$ and $\mathsf{cl}_{c',d'}$ are two different cells in $\mathsf{CG}_{a,b}$, then $p(S_1 \cup S_2) = p(S_1) + p(S_2)$ due to Observation 1.

Consider a cell $\mathsf{cl}_{c,d} \in \mathsf{CG}_{a,b}$. Let $D_{c,d} = \{v_1, v_2, ..., v_n\} \subseteq \mathsf{cl}_{c,d} \cap \mathcal{S}$, where the vertices are indexed by the order in which they were selected by the greedy algorithm. Let $D_{c,d}^i = \{v_1, v_2, ..., v_i\}$ be the first i vertices in $D_{c,d}$. By the following lemma, we can see that the modified profit function $\widehat{p}$ is a constant approximation to true profit function p over any vertex subset $V \subseteq \mathcal{V}_{a,b}$.

Lemma 6. *For a set of vertices V in the same cell $\mathsf{cl}_{c,d} \in \mathsf{CG}_{a,b}$, such that $|V| \leq i$, we have that $p(D_{c,d}^i) = \widehat{p}(D_{c,d}^i) \geq (1 - 1/e)p(V)$.*

Proof. By the greedy rule, we can see $p(D_{c,d}^i) = \widehat{p}(D_{c,d}^i)$. By Lemma 1, we know that $\widehat{p}(D_{c,d}^i) \geq (1 - 1/e) \max_{|V| \leq i} p(V)$. □

Step 2: Guess the optimal profit and calculate a tree T : Although the actual profit of OPT is unknown, we can guess the profit of OPT (by enumerating all possibilities). For each $0 \leq a, b < k$, we calculate in this step a tree T of size at most $4B$, using the QST algorithm (see Lemma 2). We can show that among these trees (for different a, b values), there must be one tree of profit no less than $\frac{1}{k^2} \left(1 - \frac{1}{e}\right)$ OPT.

After choosing the best tree T with the highest profit, we construct a subtree $\tilde{T}$ of size B based on T as our final solution of Budgeted-CSC.

We first show that there exists $0 \leq a, b < k$, such that based on the modified profit $\widehat{p}$ on $\mathsf{CG}_{a,b}$, there exists a tree with at most $2B$ vertices of total modified profit at least $\frac{1}{k^2} \left(1 - \frac{1}{e}\right)$ OPT. We use T_{OPT} to denote the set of vertices of the optimal solution.

Lemma 7. *There exists a tree T_0 in $\mathcal{G}_c$, $|T_0| \leq 2B$ such that $\widehat{p}(T_0) \geq \frac{1}{k^2} \left(1 - \frac{1}{e}\right)$ OPT.*

Then, by the Lemma 2 and Lemma 7, if we run the QST algorithm (with $\widehat{p}$ as the profit function), we can obtain the suitable tree T with at most $4B$ vertices of profit at least $\frac{1}{k^2} \left(1 - \frac{1}{e}\right) p(\mathsf{OPT})$. The pseudocode of the algorithm can be found in the full version of the paper.

Lemma 8. *Let T be the tree obtained by the above step, then $p(T) \geq \frac{1}{k^2} \left(1 - \frac{1}{e}\right)$ OPT*

Proof. By Lemma 7, we can obtain a tree T with at most $4B$. We also have $\widehat{p}(T) \geq \frac{1}{k^2} \left(1 - \frac{1}{e}\right)$ OPT. Since $p(S) \geq \widehat{p}(S)$ for any S, we have proved the lemma. □

Finally, we construct a subtree $\tilde{T}$ of B vertices based on tree T. This can be done using a simple dynamic program, in the same way as [23]. Denote the subtree with highest total profit as $\tilde{T}$. We can show the following lemma.[5]

---

[5] The proof is similar to that in [23] and can be find in the full version of the paper.

Lemma 9. $p(\tilde{T}) \geq \frac{1}{8}p(T)$.

Use the same dynamic programming algorithm in Khuller et al. [23], we can find $\tilde{T}$ from tree T. Combining Lemma 8 and Lemma 9, $p(\tilde{T}) \geq \frac{1}{8}\left(1 - \frac{1}{e}\right) \frac{1}{(2\sqrt{2}C+1)^2}\mathsf{OPT} = \frac{1}{12.66(8C^2+4\sqrt{2}C+1)}\mathsf{OPT} \geq \frac{1}{102C^2}\mathsf{OPT}$ (When C is large). Thus, we have obtained Theorem 2.

5 Conclusion and Future Work

There are several interesting future directions. The first obvious open question is that whether we can get constant approximations for MIN-CSC and Budgeted-CSC without Assumption 1 (it would be also interesting to obtain approximation ratios that have better dependency on C). Generalizing the problem further, an interesting future direction is the case where different sensors have different transmission ranges and sensing ranges. Whether the problems admit better approximation ratios than the (more general) graph theoretic counterparts is still wide open.

References

1. Ambühl, C., Erlebach, T., Mihalák, M., Nunkesser, M.: Constant-factor approximation for minimum-weight (connected) dominating sets in unit disk graphs. In: Díaz, J., Jansen, K., Rolim, J.D.P., Zwick, U. (eds.) APPROX 2006 and RANDOM 2006. LNCS, vol. 4110, pp. 3–14. Springer, Heidelberg (2006)
2. Brönnimann, H., Goodrich, M.T.: Almost optimal set covers in finite vc-dimension. DCG **14**(1), 463–479 (1995)
3. Calinescu, G., Chekuri, C., Pál, M., Vondrák, J.: Maximizing a monotone submodular function subject to a matroid constraint. SICOMP **40**(6), 1740–1766 (2011)
4. Calinescu, G., Zelikovsky, A.: The polymatroid steiner problems. JCO **9**(3), 281–294 (2005)
5. Chan, T.M., Grant, E., Könemann, J., Sharpe, M.: Weighted capacitated, priority, and geometric set cover via improved quasi-uniform sampling. In: SODA, pp. 1576–1585. SIAM (2012)
6. Chekuri, C., Even, G., Kortsarz, G.: A greedy approximation algorithm for the group steiner problem. Discrete Applied Mathematics **154**(1), 15–34 (2006)
7. Cheng, X., Huang, X., Li, D., Wu, W., Du, D.Z.: A polynomial-time approximation scheme for the minimum-connected dominating set in ad hoc wireless networks. Networks **42**(4), 202–208 (2003)
8. Clarkson, K.L., Varadarajan, K.: Improved approximation algorithms for geometric set cover. DCG **37**(1), 43–58 (2007)
9. Dai, D., Yu, C.: A 5+ ε-approximation algorithm for minimum weighted dominating set in unit disk graph. TCS **410**(8), 756–765 (2009)
10. Dinur, I., Steurer, D.: Analytical approach to parallel repetition. In: Proceedings of the 46th Annual ACM Symposium on Theory of Computing, pp. 624–633. ACM (2014)
11. Du, D.Z., Wan, P.J.: Connected Dominating Set: Theory and Applications, vol. 77. Springer Science & Business Media (2012)

12. Even, G., Rawitz, D., Shahar, S.M.: Hitting sets when the vc-dimension is small. IPL **95**(2), 358–362 (2005)
13. Fakcharoenphol, J., Rao, S., Talwar, K.: A tight bound on approximating arbitrary metrics by tree metrics. In: STOC, pp. 448–455. ACM (2003)
14. Feige, U.: A threshold of ln n for approximating set cover. JACM **45**(4), 634–652 (1998)
15. Garg, N.: Saving an epsilon: a 2-approximation for the k-mst problem in graphs. In: STOC, pp. 396–402. ACM (2005)
16. Garg, N., Konjevod, G., Ravi, R.: A polylogarithmic approximation algorithm for the group steiner tree problem. In: SODA, pp. 253–259. SIAM (1998)
17. Goemans, M.X., Williamson, D.P.: A general approximation technique for constrained forest problems. SICOMP **24**(2), 296–317 (1995)
18. Guha, S., Khuller, S.: Improved methods for approximating node weighted steiner trees and connected dominating sets. Information and computation **150**(1), 57–74 (1999)
19. Gupta, H., Zhou, Z., Das, S.R., Gu, Q.: Connected sensor cover: self-organization of sensor networks for efficient query execution. IEEE/ACM Transactions on Networking **14**(1), 55–67 (2006)
20. Hochbaum, D.S., Pathria, A.: Analysis of the greedy approach in problems of maximum k-coverage. Naval Research Logistics **45**(6), 615–627 (1998)
21. Huang, Y., Gao, X., Zhang, Z., Wu, W.: A better constant-factor approximation for weighted dominating set in unit disk graph. JCO **18**(2), 179–194 (2009)
22. Johnson, D.S., Minkoff, M., Phillips, S.: The prize collecting steiner tree problem: theory and practice. In: SODA, vol. 1, p. 4. Citeseer (2000)
23. Khuller, S., Purohit, M., Sarpatwar, K.K.: Analyzing the optimal neighborhood: algorithms for budgeted and partial connected dominating set problems. In: SODA, pp. 1702–1713. SIAM (2014)
24. Klein, P., Ravi, R.: A nearly best-possible approximation algorithm for node-weighted steiner trees. Journal of Algorithms **19**(1), 104–115 (1995)
25. Kuo, T.W., Lin, K.J., Tsai, M.J.: Maximizing submodular set function with connectivity constraint: theory and application to networks. In: INFOCOM, pp. 1977–1985. IEEE (2013)
26. Li, J., Jin, Y.: A PTAS for the weighted unit disk cover problem. In: Halldórsson, M.M., Iwama, K., Kobayashi, N., Speckmann, B. (eds.) ICALP 2015. LNCS, vol. 9134, pp. 898–909. Springer, Heidelberg (2015)
27. Lichtenstein, D.: Planar formulae and their uses. SICOMP **11**(2), 329–343 (1982)
28. Mustafa, N.H., Ray, S.: PTAS for geometric hitting set problems via local search. In: SCG, pp. 17–22. ACM (2009)
29. Nemhauser, G.L., Wolsey, L.A., Fisher, M.L.: An analysis of approximations for maximizing submodular set functions. Mathematical Programming **14**(1), 265–294 (1978)
30. Pyrga, E., Ray, S.: New existence proofs ε-nets. In: SCG, pp. 199–207. ACM (2008)
31. Varadarajan, K.: Weighted geometric set cover via quasi-uniform sampling. In: STOC, pp. 641–648. ACM (2010)
32. Vondrák, J., Chekuri, C., Zenklusen, R.: Submodular function maximization via the multilinear relaxation and contention resolution schemes. In: STOC, pp. 783–792. ACM (2011)
33. Wan, P.J., Alzoubi, K.M., Frieder, O.: Distributed construction of connected dominating set in wireless ad hoc networks. In: INFOCOM, vol. 3, pp. 1597–1604. IEEE (2002)

34. Williamson, D.P., Shmoys, D.B.: The design of approximation algorithms. Cambridge University Press (2011)
35. Willson, J., Ding, L., Wu, W., Wu, L., Lu, Z., Lee, W.: A better constant-approximation for coverage problem in wireless sensor networks (preprint)
36. Wu, L., Du, H., Wu, W., Li, D., Lv, J., Lee, W.: Approximations for minimum connected sensor covereq:cellsum. In: INFOCOM, pp. 1187–1194. IEEE (2013)
37. Zhang, W., Wu, W., Lee, W., Du, D.Z.: Complexity and approximation of the connected set-cover problem. Journal of Global Optimization 53(3), 563–572 (2012)
38. Zou, F., Wang, Y., Xu, X.H., Li, X., Du, H., Wan, P., Wu, W.: New approximations for minimum-weighted dominating sets and minimum-weighted connected dominating sets on unit disk graphs. TCS 412(3), 198–208 (2011)

Circuits Algorithms

Skew Circuits of Small Width

Nikhil Balaji[1]([✉]), Andreas Krebs[2], and Nutan Limaye[3]

[1] Chennai Mathematical Institute, Chennai, India
nikhil@cmi.ac.in
[2] University of Tübingen, Tübingen, Germany
mail@krebs-net.de
[3] Indian Institute of Technology, Bombay, India
nutan@cse.iitb.ac.in

Abstract. In this work, we study the power of bounded width branching programs by comparing them with bounded width skew circuits.

It is well known that branching programs of bounded width have the same power as skew circuit of bounded width. The naive approach converts a BP of width w to a skew circuit of width w^2. We improve this bound and show that BP of width $w \geq 5$ can be converted to a skew circuit of width 7. This also implies that skew circuits of bounded width are equal in power to skew circuits of width 7. For the other way, we prove that for any $w \geq 2$, a skew circuit of width w can be converted into an equivalent branching program of width w. We prove that width-2 skew circuits are not universal while width-3 skew circuits are universal and that any polynomial sized CNF or DNF is computable by width 3 skew circuits of polynomial size.

We prove that a width-3 skew circuit computing Parity requires exponential size. This gives an exponential separation between the power of width-3 skew circuits and width-4 skew circuits.

1 Introduction

The Boolean circuit complexity class NC^1 consists of Boolean functions computable by polynomial sized logarithmic depth circuits. Basic arithmetic operations like addition, multiplication and division are known to be in NC^1. All regular languages have uniform NC^1 families deciding them and there is a regular language which is NC^1-hard. Over the years, several useful characterizations of NC^1 have emerged: NC^1 contains exactly those regular languages that are characterized by having a monoid containing a non-solvable group. They are also equally expressive as Branching Programs of constant width. Our interest in NC^1 is motivated by the celebrated result of Barrington [Bar89], that Branching Programs of width 5 are sufficient to capture NC^1 in its entirety.

Branching programs have been pivotal to our understanding of computation with limited resources. They were first defined in [Lee59] and formally studied by Masek in his thesis [Mas76]. Borodin et al.[BDFP86] proved that AC^0 is contained in the class of functions computed by bounded width branching

© Springer International Publishing Switzerland 2015
D. Xu et al. (Eds.): COCOON 2015, LNCS 9198, pp. 199–210, 2015.
DOI: 10.1007/978-3-319-21398-9_16

programs and conjectured that Majority cannot be computed by them. In a surprising result, Barrington showed that in fact, width 5 branching programs can compute all of NC^1 and hence the Majority function.

After the strong lower bound results of [Raz87][Smo87] for AC^0, the question of proving lower bounds for NC^1 gained a lot of attention. However this has turned out to be a notorious open problem. The branching program characterization of NC^1 has provided an avenue to understand the power of classes that reside inside NC^1. Though proving lower bounds for width 5 branching programs is equivalent to proving lower bounds for NC^1, it is conceivable that proving lower bounds for width 4 branching programs is easier. In this regard, it is known [Bar85] that width 3 branching programs of a restricted type (permutation branching programs) require exponential size to compute the AND function. It is worthwhile to contrast this against the situation at width 5, where permutation branching programs are known to be as powerful as general branching programs of width 5 and hence NC^1 itself.

It is known that bounded width branching programs can be equivalently thought of as bounded width skew circuits (see for example [RR10]). Here, we take a closer look at this relationship. The folklore construction[1] converts a polynomial size branching program of width w into a polynomial size skew circuit of width w^2. We improve this construction and show that any bounded width branching program of width greater than or equal to 5 can be converted into an equivalent skew circuit of width 7. We also study the conversion of skew circuits into branching programs. Here, the known construction converts a skew circuit of width w into a branching program of width $w + 1$ [RR10]. We improve this construction and prove that a polynomial size skew circuit of width w can be converted into a polynomial size branching program of width w. These results prove that width 7 skew circuits of polynomial size characterize NC^1.

These structural results allow us to examine the set of languages in NC^1 by varying the width of skew circuits between 1 and 7. Like for permutation branching programs, some natural questions arise for bounded width skew circuits. We start by examining the power of width 2 skew circuits. We observe that they are not universal as they cannot compute parity of two bits.

We then study the power of width 3 skew circuits. Recall that a CNF (DNF) is an AND (OR) of ORs (ANDs) of variables, i.e. in a CNF the AND gate is (possibly) non-skew. We implement a CNF by a width 3 skew circuit. Formally, we prove that any k-CNF or any k-DNF of size s has width 3 skew circuits of length $O(sk)$. Given that any Boolean function on n variables has a CNF of exponential (in n) size, this also proves that width 3 skew circuits are universal.

We consider the problem of proving lower bound for width 3 skew circuits. A natural candidate is a function which has no polynomial sized CNF or DNF. It is known that Parity is one such function. We prove that Parity requires width 3 skew circuits of exponential size. We observe that Parity and Approximate Majority have respectively, linear and polynomial size width 4 skew circuits. This separates width 3 skew circuits from width 4 skew circuits.

---

[1] Replace each wire by an AND gate and each node by an OR gate.

2   Preliminaries

A directed acyclic graph $G = (V, E)$ is called layered if the vertex set of the graph can be partitioned, $V = V_1 \cup \ldots \cup V_\ell$ in such a way that for each edge $e = (u, v)$ there exists $1 \leq i < \ell$ such that $u \in V_i$ and $v \in V_{i+1}$. Given a layered graph G the *length* of the graph is the number of layers in it and the *width* of the graph is the maximum over $i \in [\ell]$, $|V_i|$.

Definition 1. *(Branching Programs) A Deterministic Branching Program (BP) is a layered directed acyclic graph G with the following properties:*

- *There is a designated source vertex s in the first layer (of in-degree 0) and a sink vertex t (of out-degree 0) in the last layer.*
- *The edges are labelled by an element of $X \cup \{0, 1\}$, where X is the set of input variables to the branching program.*

The branching program naturally computes a boolean function $f(X)$, where $f(X) = 1$ if and only if there is path from s to t in which each edge is labelled by a true literal or a constant 1 on input X. The length (width) of the BP is the length (respectively, width) of the underlying layered DAG.

We will denote the class of languages accepted by width-w BP by BP^w. Barrington [Bar85][Bar89] defined a restricted notion of branching programs called the Permutation Branching Program (PBP):

Definition 2. *(Permutation Branching Programs as a graph) A width-w PBP is a layered width w BP in which the following conditions hold:*

- *There are designated source vertices $s_1, s_2, \ldots, s_w$ in the first layer, say layer 1 (of in-degree 0) and sink vertices $t_1, t_2, \ldots, t_w$ (of out-degree 0) in the last layer, layer ℓ.*
- *Each layer has exactly w vertices.*
- *In each layer $1 \leq i < \ell$, all the edges are labelled by a unique variable, say x_{j_i}.*
- *In each layer $1 \leq i \leq \ell$ and $b \in \{0, 1\}$, the edges activated when $x_{j_i} = b$ forms a permutation/matching, say $\theta_{i,b}$.*

The permutation branching program naturally computes a boolean function $f(X)$, where $f(X) = 1$ if and only if there is path from s_1 to t_1, s_2 to t_2, and so on till s_w to t_w, where in each path each edge is labelled by a true literal or a constant 1 on input X. We will refer to the class of languages accepted by polynomial sized width-w PBP by PBP^w.

The above definition of PBP can be rephrased as follows:

Definition 3. *(Permutation Branching Programs as a set of instructions) A width-w length-ℓ PBP is a program given by a set of ℓ instructions in which for any $1 \leq i \leq \ell$, the ith instruction is a three tuple $\langle j_i, \theta^i, \sigma^i \rangle$, where j_i is an index from $\{1, 2, \ldots, |X|\}$, θ^i, σ^i are permutations of $\{1, 2, \ldots, w\}$. The output of the instruction is θ^i if $x_{j_i} = 1$ and it is σ^i if $x_{j_i} = 0$. The output of the program on input x is the product of the output of each instruction of the program on x.*

We say that a permutation branching program computes a function f if there exists a fixed permutation $\pi \neq id$ such that for every x such that $f(x) = 1$ the program outputs π and for every x such that $f(x) = 0$ the program outputs id^2.

It is easy to see that the above two definitions of PBP are equivalent.

Definition 4. *(Skew Circuits) An AND gate is called* skew *if all but one of its children are input variables. A Boolean circuit in which all the AND gates are skew is called a* skew circuit.

We assume that the skew circuits are layered. The width of the circuit is the maximum number of gates in any layer. The layer may have AND, OR or input gates. Each type of gate contributes towards the width. We assume that the fan-in of the AND gates is bounded by 2 and there are no NOT gates (negations appear only for the input variables)[3]. We denote the class of languages decided by width-w skew circuits by SK^w. The following lemma summarises some well known connections between BP^w, PBP^w, SK^w.

Lemma 1. *Let $w \in \mathbb{N}$. Then For any w, $\mathsf{PBP}^w \subseteq \mathsf{BP}^w$, for any $w \geq 5$, $\mathsf{BP}^w \subseteq \mathsf{PBP}^w$ [Bar89] and for any w, BP^w is contained in SK^{w^2} (see e.g. [RR10]).*

Definition 5. *(Approximate Majority) Approximate Majority* $\mathsf{ApproxMaj}_{a,n}$: $\{0,1\}^n \to \{0,1\}$ *is the promise problem defined as:* $\mathsf{maths}f\,ApproxMaj_{a,n}(x) := 0$ *if x has at most a no. of 1s and is 1 if x has at least $n - a$ no. of 1s.*

3   Branching Programs and Skew Circuits

Here we analyze the conversion from branching programs to skew circuits and vice versa. First we recall the following folklore lemma.

Lemma 2. *(Folklore) Let $f : \{0,1\}^n \to \{0,1\}$ be a Boolean function computed by a width-w length-ℓ branching program with at most k edges between any two consecutive layers and with the additional property that each layer reads at most one variable or its negation. Then there is a skew circuit of width $\max\{w + 2, k\}$ and size $O((k + w)\ell)$ computing f.*

3.1   Permutation Branching Programs to Skew Circuits

A permutation θ is called a *transposition* if either it is the identity permutation or there exists $i \neq j$ such that $\theta(i) = j$, $\theta(j) = i$ and for all $k \neq i \neq j$, $\theta(k) = k$. We call a transposition non-trivial if it is not the identity permutation, trivial otherwise.

---

[2] This is often called the strong acceptance condiction. Other notions of acceptance have been studied in the literature. See for example [Bro05].

[3] This assumption is not without loss of generality. However, we will see that when a branching program is converted into a skew circuits, exactly this type of skew circuits arise.

Definition 6. *(Transposition Branching Programs, TBP) A width-w length-ℓ TBP is a program given by a set of ℓ instructions in which for any $1 \leq i \leq \ell$, the ith instruction is a three tuple $\langle j_i, \theta^i, \sigma^i \rangle$, where j_i is an index from $\{1, 2, \ldots, |X|\}$, θ_i, σ_i are transpositions of $\{1, 2, \ldots, w\}$. The output of the instruction is θ_i if $x_{j_i} = 1$ and it is σ_i if $x_{j_i} = 0$. The output of the program on input x is the product of the output of each instruction of the program on x.*

Lemma 3. *Given a width-w PBP of length ℓ there is an equivalent width-w TBP of length $O(w\ell)$.*

Proof. It is known (see e.g. [Her06]) that any permutation of $\{1, 2, \ldots, w\}$ can be written as a product of W transpositions of $\{1, 2, \ldots, w\}$, where $W = O(w)$. Let P be a width-w PBP of length ℓ. Consider the ith instruction in the program, say $\langle j_i, \theta_i, \sigma_i \rangle$. We know that we can write θ_i as a product of W transpositions, i.e. $\theta_i = t_{i,1} \cdot t_{i,2} \ldots t_{i,W}$, where for $1 \leq j \leq W$ $t_{i,j}$ is a transposition. Similarly, we have $\sigma_i = s_{i,1} \cdot s_{i,2} \ldots s_{i,W}$, where $s_{i,j}$ is a transposition for $1 \leq j \leq W$.

To give a TBP equivalent to P, we replace every instruction $\langle j_i, \theta_i, \sigma_i \rangle$ in P by the following: $\langle j_i, t_{i,1}, id \rangle \cdot \langle j_i, t_{i,2}, id \rangle \ldots \langle j_i, t_{i,W}, id \rangle \cdot \langle j_i, id, s_{i,1} \rangle \cdot \langle j_i, id, s_{i,2} \rangle \ldots \langle j_i, id, s_{i,W} \rangle$. By a simple inductive argument we can prove that the the transposition branching program thus obtained is equivalent to P. As $W = O(w)$, the upper bound on the length of the resulting branching program follows. □

We defined TBPs as a set of instructions. Like in the case of PBPs, the definition of TBPs can be rephrased in terms of the underlying DAG. We observe the following about the DAG resulting from TBPs.

A width-w TBP is a layered width w PBP in which the following conditions hold: (1) There are designated source vertices $s_1, s_2, \ldots, s_w$ in the first layer, say layer 1 (of in-degree 0) and sink vertices $t_1, t_2, \ldots, t_w$ (of out-degree 0) in the last layer, layer ℓ, (2) Each layer has exactly w vertices, (3) In each layer all the edges are labelled by a unique variable, (4) In each layer $1 \leq i \leq \ell$, one of the following holds: (4a) either the edges corresponding to $x_{j_i} = 1$ form a non-trivial transposition and the edges coooresponding to $x_{j_i} = 0$ form the identity permutation or (4b) the edges corresponding to $x_{j_i} = 0$ form a non-trivial transposition and the edges coooresponding to $x_{j_i} = 1$ form the identity permutation.

Remark 1. *As a result of the above properties of the TBP the total number of distinct edges between any two layers in a width-w TBP is at most $w + 2$: there are w edges corresponding to the identity permutation, 2 edges corresponding to the transposition of two elements, and $w - 2$ edges corresponding to the identity maps for all but the two transposed elements. The $w - 2$ last edges overlap with the w edges corresponding to the identity permutation.*

Lemma 4. $\mathsf{PBP}^w \subseteq \mathsf{SK}^{w+2}$.

Proof. Given a PBP^w for L of size s, by Lemma 3 we know that L also has a width-w TBP. By Remark 1 the underlying DAG for the TBP has at most $w + 2$ edges between any two consecutive layers. Using Lemma 2 we get a skew circuit of width $w + 2$ for L. Note that the size of such a circuit is $O(ws)$.

Using Barrington's characterization of NC^1 and Lemma 4 we get the following: $NC^1 = BP^5 = PBP^5 \subseteq SK^7$.

3.2   Skew Circuits to Branching Programs

In this section we start from a skew circuit of bounded width and convert it into a branching program of bounded width. Formally, we prove the following:

Theorem 2. *If C is a skew circuit of width w and length ℓ then there is an equivalent branching program P of width w and size $O(w\ell)$.*

Proof. Recall that in a skew circuit C, AND gates have fan-in 2 and at least one child is an input variable whereas OR gates have arbitrary fan-in and arbitrary predecessors. Given a skew circuit C of width w and length ℓ_0 we will construct a branching program P of width w that will recognize the same language. Let $G_{i1}, \ldots, G_{iw}$ be the gates of C on layer i for $i = 1, \ldots, \ell_0$.

Let $X = \{\ell_{i_1}, \ell_{i_2}, \ldots, \ell_{i_L}\}$ ($|X| = L$) be the set of layers on which there is at least one input gate. Without loss of generality we assume that in each $j \in [L]$ the gate $G_{\ell_{i_j} w}$ is an input gate. (There may be other input gates as well.)

We will construct a branching program of length $s = L+2$ and width w. The nodes in the branching program in layer $\ell_{i_j} \in X$ will be called $N_{j0}, \ldots, N_{j(w-1)}$. The nodes N_{00} and $N_{(s-1)(w-1)}$ are respectively, initial and target nodes.

The nodes $N_{11}, \ldots, N_{(s-1)(w-1)}$ will, by our construction, compute the value of the nodes in a layer in X. More formally, for every input x, the gate $G_{\ell_j c}$ in layer ℓ_j of the circuit (and layer j in X) evaluates to 1 iff the node N_{jc} can be reached from the initial node. Since the gate G_{iw} in X is an input gate we will not add corresponding gate in the branching program. We have completely specified the vertex set of the branching program P.

We now describe the edge set of P. We add an edge from $N_{(j-1)0}$ to N_{j0} labeled by 1 for every $1 \leq j \leq s-1$. This ensures that all nodes N_{j0} are always reachable from the initial node.

Suppose that the layer ℓ_j and $\ell_j + 1$ are both in X, i.e. $\ell_{j+1} = \ell_j + 1$, then the edges between the nodes in the layer ℓ_j and ℓ_{j+1} in the branching program are easy to state. A node N_{j+1c} is connected to N_{jd} if there is an edge between the corresponding gates $G_{\ell_{j+1}c}$ and $G_{\ell_j d}$. Also the edge in the branching program is labeled by 1 if the gate $G_{\ell_j d}$ is an OR gate, and labeled by the variable x_i (or its negtion $\neg x_i$ if $G_{\ell_j d}$ is an AND gate querying x_i, resp. $\neg x_i$. If an OR gate in ℓ_j is connected to an input gate, we generate an edge to N_{j0} labeled by the literal queried by the input gate.

Now assume that the layer ℓ_j is in X and ℓ_{j+1} is the next layer in X and $\ell_{j+1} > \ell_j + 1$. Then in the skew circuit, no input gates occur strictly between the layers ℓ_j and ℓ_{j+1}. This implies that there are no AND gates in the layers $\ell_j + 2, \ldots, \ell_{j+1}$. Hence the functions computed by the gates in layer ℓ_{j+1} are ORs of some gates in layer $\ell_j + 1$. In layer gates in layer $\ell_j + 1$ are ORs of either ANDs of gates in layer $\ell_j + 1$ and an input variable or ORs of directly gates in layer $\ell_j + 1$. Therefore, we add the following edges in the branching program:

a node $N_{(j+1)c}$ is connected to N_{jd} if the OR function computed by $G_{\ell_{j+1}c}$ has $G_{\ell_j d}$ as one of the inputs. This edge in the branching program is labeled by 1 if this was a direct OR, it is labeled by the variable x_i (or its negtion $\neg x_i$) it it was an 'or' of an 'and' querying x_i resp. $\neg x_i$. e.

It is easy to verify by induction on the layers that N_{jc} is reachable from the inital gate if the corresponding gate evaluates true. Finally we add an edge from the node corresponding to the output gate to $N_{(s-1)(w-1)}$.

Putting together Lemma 4 and Theorem 2 we get the following corollary:

Corollary 1. $\mathsf{NC}^1 = \mathsf{BP}^5 = \mathsf{PBP}^5 = \mathsf{SK}^7$

4   Width ≤ 7 Skew Circuits

Here we study the structure of the languages in NC^1 by investigating properties of skew circuits of width 7 or less. By definition $\mathsf{SK}^i \subseteq \mathsf{SK}^{i+1}$ for $1 \leq i \leq 6$. We start by proving that width 2 circuits are not universal (We defer the proofs to the full version of the paper due to lack of space).

Lemma 5. *A width 2 skew circuit of any size cannot compute Parity of 2 bits.*

Recall that a k-*DNF of size s on n variables* is an OR of s terms, where each term is an AND of at most k literals from $\{x_1, x_2, \ldots, x_n, \neg x_1, \neg x_2, \ldots, \neg x_n\}$. Similarly, k-*CNF of size s on n variables* is an AND of s clauses, where each clause is an OR of at most k literals from $\{x_1, x_2, \ldots, x_n, \neg x_1, \neg x_2, \ldots, \neg x_n\}$.

Lemma 6. *Let f be a k-DNF of size s on n variables. Then f has a width-3 skew circuit of length $O(sk)$.*

We defer the proof of the above lemma to the full version of the paper due to lack of space. As any Boolean function on n variables has an n-DNF of size at most 2^n, we get the following corollary.

Corollary 2. *Let $f : \{0,1\}^n \to \{0,1\}$. Then f can be computed by width-3 skew circuit of length $O(n2^n)$, i.e. width-3 skew circuits are universal.*

Lemma 7. *Let f be a k-CNF of size s on n variables. Then f has a width-3 skew circuit of length $O(sk)$.*

Proof. Note that in a CNF, the top AND gate gets clauses as inputs. That is, the AND gate is not skew. However, it is still possible to get a skew circuit for CNFs. We prove this by induction on the number of clauses, i.e. s. The base case is $s = 1$. This is just an OR of literals, which is computable by width-2 skew circuit of length $O(k)$. Let $f_i(x_1, \ldots, x_n) = C_1 \wedge \ldots \wedge C_i$ be computable by width-3 skew circuit of length $O(ik)$. Now $f_{i+1} = f_i \wedge C_{i+1}$ (where $C_{i+1} = x_{j_1} \vee x_{j_2} \ldots \vee x_{j_k}$) is computed as follows: $f_{i+1} = (\ldots ((f_i \wedge x_{j_1}) \vee (f_i \wedge x_{j_2}) \ldots (f_i \wedge x_{j_k})) \ldots)$. Note that we need width 1 each for the f_i, the AND gate and the input variable. (Even though we require width 3 to compute f_i, after the computation, it just requires width 1 to carry around the value of the function to the next stage). (See Figure 1.) $\qquad\square$

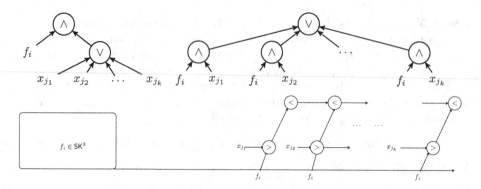

Fig. 1. Width 3 skew circuits for CNFs

By a result of Viola [Vio09], it is known that Approximate Majority is computable by P-uniform depth-3 circuits of polynomial size with alternating AND/OR layers with the output gate being an OR gate. This along with Lemma 7 above yields:

Corollary 3. ApproxMaj$_{a,n}$ *has skew circuits of width 4 and polynomial length.*

Razborov[Raz87] and Smolensky[Smo87] show that Parity does not have constant depth circuit of polynomial size. It also implies that Parity does not have polynomial size CNFs or DNFs . However, we now show that Parity has skew circuits of width 4 and polynomial size.

Lemma 8. *Parity on n variables has a skew circuit of width 4 and length $O(n)$.*

Proof. This is an easy observation which comes from the fact that Parity has a branching program of width 2 and length $O(n)$. This fact along with part 3 of Lemma 1 proves the result. ☐

5   Parity and SK³

In this section we prove that Parity does not have polynomial length width-3 skew circuits. As Parity has width 4 skew circuits of linear length (Lemma 8), this separates SK³ from SK⁴.

Theorem 3. $\mathsf{SK}^3 \subsetneq \mathsf{SK}^4$.

In order to prove this we first show that any width three skew circuit computing Parity can be converted into a normal form. We then show that any polynomial sized circuit of that normal form cannot compute Parity.

Lemma 9. *Let C be a Boolean width 3 skew circuit with size $s(n)$ computing* PARITY$_n$. *The circuit C can be converted into another circuit D such that D computes Parity of at least $n-2$ bits and has the following structure:*

1. *The top gate of D is an OR of at most $3s(n)^2$ disjoint skew circuits, say $C_1, C_2, \ldots, C_m$, where $m \le 3s(n)^2$.*
2. *The sum of sizes of all C_is is at most $O(s(n)^3)$*
3. *At most two of these circuits have width 3 and all the other have width at most 2.*

Lemma 10. *Let D be a circuit satisfying properties 1,2,3 from Lemma 9 and computing* PARITY_n. *Then there exists a constant c such that the size of D is at least $2^n/n^c$.*

Using Lemma 9 and Lemma 10 it is easy to see that the lower bound for Parity follows.

5.1   Proof of Lemma 9

Let C be a width three circuit computing Parity of n bits of size $s(n)$. The top gate of C cannot be AND as, by fixing the input wires of the AND gate, we can fix the output of the circuit, however, Parity of n bits cannot be fixed by fixing $< n$ bits. Therefore, we can assume that the top gate is an OR gate. We prove the following structural statement about the skew circuits (We defer the proof of the proposition to the full version of the paper due to lack of space):

Proposition 1. *Let C be any width three skew circuit computing Parity of n bits. Let k be the highest layer in C consisting of only AND gates, say g_k^1, g_k^2, g_k^3. We can convert this into another width 3 skew circuit D computing Parity of at least $n - 2$ bits such that no layer of D contains three AND gates.*

Therefore, we will now assume that we have a circuit D which computes parity of at least $n - 2$ bits in which no layer consists of only AND gates.

Let $G = (V, E)$ denote the DAG underlying the circuit D. Let $X \subseteq V$ denote the subset of gates which have a path to the top gate via only OR gates. Note that all the vertices in this set are themselves OR gates. We refer to the set of vertices X as ORSET.

Let $X_{in} \subseteq V$ denote the set of vertices in $V \setminus X$ which have an edge incident from it to some vertex in X. Similarly, let $X_{out} \subseteq V$ denote the set of vertices in $V \setminus X$ which have an edge incident to them from some vertex in X.

(a) Disconnect all edges incident to the set X_{out} from X. Let the new dangling wires be labelled with the constant 0 input.
(b) If after step (a) any AND gates receives constant input 0 then delete the gate and if any OR gate receives constant input 0 then delete this input of the OR gate.
(c) In the graph obtained after steps (a) and (b), consider $V \setminus X$. This disconnects the DAG G and gives rise to some connected components, say $X_1, X_2, \ldots, X_\ell$. For each $i \in [\ell]$ and edge (u, v) such that $u \in X_i$ and $v \in X$ let $C_{i,u}$ be the subgraph (DAG) of X_i with u as its sink.

Proposition 2. *The step (a) does not change the output function of the circuit.*

The proof of the above is deferred to the full version of the paper.Note that $V \setminus X$ is partitioned by X_is, $\therefore \ell \leq s(n)$. Also, $\cup_{i=1}^m X_i \subseteq V \setminus X$, hence $\sum_{i=1}^m |X_i| \leq s(n)$. The number of edges in G is $\leq 3s(n)$. Therefore, the total number of edges between any X_i and X is at most $3s(n)$ and the number of circuits $C_{u,i}$ is at most $s(n)^2$ and each such circuit is of size at most $s(n)$. This proves parts $1, 2$ of Lemma 9.

We will now prove part 3 of Lemma 9. The top gate of D is the same as the top gate of C, and it is an OR gate. Therefore, this gate is in the ORSET. Now as long as there are OR gates on the layers, we have at least one gate per layer in the ORSET. Finally, if there is a layer with no OR gates, then this layer must have at least one input variable (as D does not have any layer with three AND gates). The other two gates at this layer be g, h. Let X_g, X_h be two DAGs rooted at g and h, respectively, and C_g and C_h be the two corresponding circuits. These are possibly width three circuits. However, all other connected components of $V \setminus X$ are of width at most 2 due to step (b) above. This gives Part 3 of Lemma 9.

5.2   Proof of Lemma 10

Let D be the circuit given by Lemma 9. Let n_0 denote the number of unfixed variables. Let $C_1, C_2, \ldots, C_m$ be circuits given by Lemma 9. We know that at most two circuits among these have width 3. Let us assume without loss of generality that the two circuits are C_1, C_2. The output gates of these circuits are AND gates, say G_1 and G_2, respectively. These being skew circuits, all but one of the inputs of the AND gates are input gates. We will first prove the following proposition.

Proposition 3. *By fixing at most 2 variables both G_1 and G_2 can be set to 0.*

The proof of this proposition is deferred to the full version of the paper.

Let N denote the number of variables which were not set. Here, $N \geq n_0 - 2$. The new circuit, say D', is now an OR of $C_3, C_4, \ldots, C_m$ and by our assumption, it computes Parity of N variables. We will show that OR of polynomially many polynomial size width-2 skew circuits cannot compute Parity.

Let us fix some notation. Let $L_\oplus$ denote the 1 set of parity, i.e. $L_\oplus = \{x \in \{0,1\}^N \mid Parity(x) = 1\}$. We know that $|L_\oplus| = 2^{N-1}$. For any Boolean circuit C, let L_C denote $\{x \in \{0,1\}^N \mid C(x) = 1\}$. Note that as D' above is an OR of $C_3, C_4 \ldots, C_m$, we have $L_{D'} = \cup_{i=3}^m L_{C_i}$.

Definition 7. *We say that a Boolean circuit C α-approximates a function $f :$ $\{0,1\}^n \to \{0,1\}$ if the following conditions hold:*

– *$\forall x \in \{0,1\}^n$, if $f(x) = 0$ then $C(x) = 0$, i.e. C has no false positives.*
– *The ratio of $|\{x \mid C(x) = 1\}|$ to $|\{x \mid f(x) = 1\}|$ is at least α*

For the sake of contradiction we have assumed that D' computes parity of N bits. Assuming this and from the fact that $L_{D'} = \cup_{i=3}^m L_{C_i}$, we get that there exists an $i \in \{3, 4, \ldots, m\}$ such that C_i $1/m$-approximates parity of N bits. We will now prove that no such C_i exists, which will give us the contradiction. Formally, we prove the following:

Claim. Let D' and $C_3, C_4, \ldots, C_m$ be defined as above. There does not exists $i \in \{3, 4, \ldots, m\}$ such that C_i $1/m$-approximates Parity of N bits.

Proof. Suppose there exists a C_i which $1/m$-approximates parity of N bits. Recall that C_i is a width 2 skew circuit. Let the last layer be L and the first layer be 1. Let $\ell_{i_1}, \ell_{i_2}, \ldots, \ell_{i_t}$ be the layers in which there is one input gate, with ℓ_{i_1} being closest to layer 1 and ℓ_{i_t} being closest to layer L. (Note that, we can assume without loss of generality that layer 1 is the only layer which has two input gates.) Let the variables queried by these gates be $x_{i_1}, x_{i_2}, \ldots, x_{i_t}$, respectively. Let $h_{i_{t+1}}$ denote the output gate in layer L. Similarly, let h_{i_1} be the gate in layer ℓ_{i_1} (other than the input gate), h_{i_2} be the gate in layer ℓ_{i_2} (other than the input gate) and so on till h_{i_t} be the gate in layer ℓ_{i_t} (other than the input gate).

As there are no NOT gates in the circuit, $h_{i_{j+1}}$ is a monotone function of x_{i_j}, h_{i_j} for every $1 \leq j \leq t$. There is a unique value of x_{i_j}, say $b_{i_j} \in \{0, 1\}$, such that by setting $x_{i_j} = b_{i_j}$, $h_{i_{j+1}}$ becomes a non-trivial function of h_{i_j}. (This is because, there are at most 6 different monotone functions on two bits, two of which cannot occur in a minimal circuit. And the other four (AND, OR, NAND, NOR) have this property.)

Note that, the setting of $x_{i_t} = b_{i_t}$ will not fix the value of h_{i_t}. Suppose h_{i_t} gets fixed due to this setting. In that case, value of $h_{i_{t+1}}$ will also get fixed. Suppose the value of $h_{i_{t+1}}$ becomes 1, then for all settings of $x \neq x_{i_t}$, $h_{i_{t+1}}$ will continue to have value 1. But we have assumed that C_i has no false positives. Therefore, this is not possible. On the other hand, if the value of $h_{i_{t+1}}$ gets fixed to 0, then for all settings of variables $x \neq x_{i_t}$ the circuit will output 0. That is, for 2^{N-1} different inputs the circuit will output 0. However, we have assumed that the circuit outputs 1 for at least $2^{N-1}/m$ many inputs.

Assuming $x_{i_t} = b_{i_t}$ and $x_{i_t} \neq x_{i_{t-1}}$, we will repeat this argument for $x_{i_{t-1}}$. Let $x_{i_{t-1}} = b_{i_{t-1}}$ be the setting of $x_{i_{t-1}}$ which makes h_{i_t} a function of $h_{i_{t-1}}$. Suppose this setting of $x_{i_{t-1}}$ fixes h_{i_t} then that will inturn fix $h_{i_{t+1}}$. As before, to avoid false positives, the value of $h_{i_{t+1}}$ cannot be fixed to 1. And to ensure that the circuit evaluated to 1 on at least $2^{N-1}/m$ inputs, it cannot be fixed to 0.

In this way, we can repeat the argument for k distinct variables as long as $k < (N-1) - \lceil \log m \rceil$. Let k_0 be such that $k_0 = \omega(\log m)$ and $k_0 < (N-1) - \lceil \log m \rceil$. We fix k_0 distinct variables as above. But now note that any other setting of these k_0 variables fixes the value of $h_{i_{t+1}}$ to 0. Therefore, the circuit can be 1 on at most $O(2^{N-k_0})$ inputs. But this contradicts our assumption that $h_{i_{t+1}}$ evaluated to 1 on at least $2^N/m$ inputs from $L_\oplus$. $\qquad\qquad\square$

6 Discussion

The above study provides a wide range of interesting questions, answers to which may improve our understanding of functions in NC^1. Namely, the questions regarding lower bounds for width k skew circuits for $4 \leq k \leq 6$. Some of these questions could be more tractable than the daunting question of proving lower bounds for NC^1 circuits. We conjecture that Majority is a function with respect to which width 4 circuits will have exponential lower bound. Towards proving such a result, it may be interesting to obtain a normal form for width 4 skew circuits. It may also be possible that any function in NC^1 is computable by width k circuits for $4 \leq k \leq 6$. Such a result will tighten the connection between branching programs and skew circuits.

References

[Bar85]   Barrington, D.A.: Width-3 permutation branching programs. Technical Memo MIT/LCS/TM-293, Massachusetts Institute of Technology, Laboratory for Computer Science (1985)

[Bar89]   Barrington, D.A.: Bounded-width polynomial-size branching programs can recognize exactly those languages in NC^1. Journal of Computer and System Sciences **38**, 150–164 (1989)

[BDFP86] Borodin, A., Dolev, D., Fich, F.E., Paul, W.: Bounds for width two branching programs. SIAM Journal on Computing **15**(2), 549–560 (1986)

[Bro05]   Brodsky, A.: An impossibility gap between width-4 and width-5 permutation branching programs. Information Processing Letters **94**(4), 159–164 (2005)

[Her06]   Herstein, I.N.: Topics in algebra. John Wiley & Sons (2006)

[Lee59]   Lee, C.-Y.: Representation of switching circuits by binary-decision programs. Bell System Technical Journal **38**(4), 985–999 (1959)

[Mas76]   Masek, W.J.: A fast algorithm for the string editing problem and decision graph complexity. PhD thesis, Massachusetts Institute of Technology (1976)

[Raz87]   Razborov, A.A.: Lower bounds on the size of bounded depth circuits over a complete basis with logical addition. Mathematical Notes **41**(4), 333–338 (1987)

[RR10]   Rao, B.V.R.: A study of width bounded arithmetic circuits and the complexity of matroid isomorphism. [HBNI TH 17] (2010)

[Smo87]   Smolensky, R.: Algebraic methods in the theory of lower bounds for boolean circuit complexity. In: Proceedings of the nineteenth annual ACM symposium on Theory of computing, pp. 77–82. ACM (1987)

[Vio09]   Viola, E.: On approximate majority and probabilistic time. Computational Complexity **18**(3), 337–375 (2009)

Correlation Bounds and #SAT Algorithms for Small Linear-Size Circuits

Ruiwen Chen[1]($\boxtimes$) and Valentine Kabanets[2]

[1] School of Informatics, University of Edinburgh, Edinburgh, UK
rchen2@inf.ed.ac.uk
[2] School of Computing Science, Simon Fraser University, Burnaby, BC, Canada
kabanets@cs.sfu.ca

Abstract. We revisit the gate elimination method, generalize it to prove correlation bounds of boolean circuits with Parity, and also derive deterministic #SAT algorithms for small linear-size circuits. In particular, we prove that, for boolean circuits of size $3n - n^{0.51}$, the correlation with Parity is at most $2^{-n^{\Omega(1)}}$, and there is a #SAT algorithm running in time $2^{n-n^{\Omega(1)}}$; for circuit size $2.99n$, the correlation with Parity is at most $2^{-\Omega(n)}$, and there is a #SAT algorithm running in time $2^{n-\Omega(n)}$. Similar correlation bounds and algorithms are also proved for circuits of size almost $2.5n$ over the full binary basis B_2.

Keywords: Boolean circuit · Random restriction · Correlation bound · Satisfiability algorithm

1 Introduction

Connections between circuit lower bounds and efficient algorithms have been explicitly exploited in several recent breakthroughs. In particular, the "random restriction" technique, which was used to prove circuit lower bounds, was extended to get both satisfiability algorithms and average-case lower bounds for boolean formulas [6,16,17,22] and AC^0 circuits [2,13].

For de Morgan formulas, Santhanam [22] gave a #SAT algorithm running in time $2^{n-\Omega(n)}$ for formulas of linear size; the algorithm is based on a generalization of the "shrinkage under random restrictions" property, which was used to prove formula lower bounds [11,25]. Santhanam [22] observed that, one can define a random process of restrictions such that the formula size shrinks with high probability. This *concentrated shrinkage* implies not only #SAT algorithms but also correlation bounds. As shown in [22], a linear-size de Morgan formula has correlation at most $2^{-\Omega(n)}$ with Parity; the correlation of two n-input functions f and g is $|\mathbf{Pr}[f(x) = g(x)] - \mathbf{Pr}[f(x) \neq g(x)]|$, where x is chosen uniformly at random from $\{0,1\}^n$. Santhanam's algorithm was extended to $2^{n-n^{\Omega(1)}}$-time #SAT algorithms for de Morgan formulas of size $n^{2.49}$ in [6] and size $n^{2.63}$ in [7]. For formulas over the full binary basis B_2, Seto and Tamaki [24] extended [22]

© Springer International Publishing Switzerland 2015
D. Xu et al. (Eds.): COCOON 2015, LNCS 9198, pp. 211–222, 2015.
DOI: 10.1007/978-3-319-21398-9_17

Table 1. Worst-case and average-case lower bounds for computing Parity

	Worst-Case Lower Bounds	Average-Case Upper / Lower Bounds	
AC^0	$s = \exp(n^{\theta(\frac{1}{d-1})})$ [10,26]	$\epsilon = 2^{-\Omega(n/(\log s)^{d-1})}$ [12]	
De Morgan	$s = n^{2-\theta(1)}$ [25]	$\epsilon \geqslant 2^{-\Omega(n^2/s)}$	$\epsilon \leqslant 2^{-\Omega(n/\sqrt{s})}$ [1,21]
formulas			$\epsilon \leqslant 2^{-\Omega(n/c^2)}$ for $s = cn$ [22]
U_2-circuits	$s = 3n - \theta(1)$ [23]	$\epsilon \geqslant 2^{-\Omega(3n-s)}$	$\epsilon \leqslant 2^{-\Omega((3n-s)^2/n)}$ [This work]

to give a $2^{n-\Omega(n)}$-time #SAT algorithm for B_2-formulas of linear size, and also showed that such formulas cannot approximately compute affine extractors.

On the other hand, Komargodski, Raz, and Tal [16,17] also used the concentrated shrinkage property to generalize the worst-case formula lower bounds to the average case. They gave an explicit function (computable in polynomial time) such that de Morgan formulas of size $n^{2.99}$ can compute correctly on at most $1/2 + 2^{-n^{\Omega(1)}}$ fraction of inputs. Combining the techniques in [6,17], one can get a randomized $2^{n-n^{\Omega(1)}}$-time #SAT algorithm for de Morgan formulas of size $n^{2.99}$.

1.1 Our Results and Techniques

In this work, we get correlation bounds and #SAT algorithms for general boolean circuits. We consider circuits over the full binary basis B_2 and circuits over the basis $U_2 = B_2 \setminus \{\oplus, \equiv\}$.

We prove that, for U_2-circuits of size $3n - n^\epsilon$ for $\epsilon > 0.5$, the correlation with Parity is at most $2^{-n^{\Omega(1)}}$, and there is a #SAT algorithm running in time $2^{n-n^{\Omega(1)}}$; for U_2-circuits of size $3n - \epsilon n$ for $\epsilon > 0$, the correlation is at most $2^{-\Omega(n)}$, and there is a #SAT algorithm running in time $2^{n-\Omega(n)}$. For B_2-circuits, we give a similar #SAT algorithm for circuits of size almost $2.5n$, and show the average-case hardness of computing affine extractors using such circuits.

Our correlation bounds of U_2-circuits with Parity are almost optimal, up to constant factors in the exponents. In fact, one can construct a U_2-circuit of size $3n - l$ which computes Parity on at least $1/2 + 2^{-\Omega(l)}$ fraction of inputs. Table 1 summarizes the known worst-case and average-case lower bounds against Parity for several restricted circuit models. Note that, for the average-case bounds, we express the correlation ϵ as a function of the circuit size s.

However, there is still a gap between our average-case lower bounds and the worst-case lower bounds. The best known worst-case explicit lower bound is $5n - o(n)$ for U_2-circuits [15,18], and $3n - o(n)$ for B_2-circuits [3].

For #SAT algorithms, there is a known algorithm for B_2-circuits by Nurk [20] which runs in time $O(2^{0.4058s})$ for circuits of size s. The running time of our algorithm for B_2-circuits is almost the same as Nurk's [20]. We are not aware of any #SAT algorithm for U_2-circuits.

Our Techniques. We extend the gate elimination method which was previously used to prove worst-case circuit lower bounds [3,9,15,18,23,28]. We define a random process of restrictions such that the circuit size shrinks with high probability. This is similar to the concentrated shrinkage approach for boolean formulas [6,16,17,22,24]. We analyze this random process using the concentration bound given by a variant of Azuma's inequality as in [6]. This analysis is then used to get both correlation bounds and #SAT algorithms. The same approach works for both U_2-circuits and B_2-circuits, although we need different rules on defining restrictions.

As a byproduct of our algorithms, we show that small linear-size circuits have decision trees of non-trivial size. In particular, U_2-circuits of size s have equivalent decision trees of size $2^{n-\Omega((3n-s)^2/n)}$, and B_2-circuits of size s have parity decision trees of size $2^{n-\Omega((2.5n-s)^2/n)}$. Our correlation bounds follow directly from such non-trivial decision-tree representations.

Related Work. For U_2-circuits, the best known worst-case lower bound is $5n - o(n)$ by Iwama and Morizumi [15], improving upon a $4.5n - o(n)$ lower bound by Lachish and Raz [18], a $4n - c$ lower bound against symmetric functions by Zwick [28], and a $3n - c$ lower bound against Parity by Schnorr [23]. For B_2-circuits, the best known worst-case lower bound is $3n-o(n)$ by Blum [3]; Demenkov and Kulikov [9] gives an alternative proof of this lower bound against affine dispersers. Nurk [20] gave a satisfiability algorithm in time $O(2^{0.4058s})$ for B_2-circuits of size s. Nurk's algorithm [20] is also based on gate elimination and the running time is similar to ours, although we use a slightly different case analysis for gate elimination. We are not aware of any previous average-case lower bounds (correlation bounds) for general circuits.

2 Preliminaries

2.1 Circuits

Let B be a binary basis, i.e., a set of boolean functions on two variables. A *B-circuit* on n input variables is a directed acyclic graph with (1) nodes of in-degree 0 labeled by variables or constants, which we call *inputs*, and (2) nodes of in-degree 2 labeled by functions from B, which we call *gates*. There is a single node of out-degree 0, designated as the *output*. Without loss of generality, we assume, for each variable x_i, there is at most one input labeled by x_i. A circuit on n variables computes a boolean function $f\colon \{0,1\}^n \to \{0,1\}$. For two nodes u and v, we will write $u \to v$ if u feeds into v.

We consider two binary bases: the full basis B_2, which contains all boolean functions on two variables, and the basis $U_2 = B_2 \setminus \{\oplus, \equiv\}$. Specifically, the basis B_2 contains the following 16 functions $f(x,y)$: (1) six degenerate functions: 0, 1, x, $\neg x$, y, $\neg y$; (2) eight $\wedge$-type functions: $x \wedge y$, $x \vee y$, and the variations by negating one or both inputs; (3) two $\oplus$-type functions: $x \oplus y$, $x \equiv y$.

The *size* of a circuit C, denoted by $s(C)$, is the number of gates in C. The *circuit size* of a function $f\colon \{0,1\}^n \to \{0,1\}$ is the minimal size of a boolean circuit computing f. For convenience, we define $\mu(C) = s(C) + N(C)$, where

$N(C)$ is the number of inputs that C depends on. We let $\mu(C) = 0$ if C is constant, and $\mu(C) = 1$ if C is a literal.

A *restriction* ρ is a mapping from the input variables to $\{0, 1, *\}$. For a circuit C, the restricted circuit $C|_\rho$ is obtained by fixing $x_i = b$ for all x_i such that $\rho(x_i) = b \in \{0, 1\}$.

It is convenient to work with circuits without redundant nodes or wires. We will call a non-constant circuit (over U_2 or B_2) *simplified* if it does not have the following: (1) nodes labeled by constants, (2) gates labeled by degenerate functions, (3) non-output gates with out-degree 0, or (4) any input x and two gates u, v with three wires $x \to u$, $x \to v$, $u \to v$.

Lemma 1. *For any circuit C, there is a polynomial-time algorithm transforming C into an equivalent simplified circuit C' such that $s(C') \leqslant s(C)$ and $\mu(C') \leqslant \mu(C)$.*

Proof (Sketch). Cases (1)-(3) are trivial. For case (4), suppose w is the other node feeding into u. If C is over B_2, then v computes a binary function of x and w; if C is over U_2, then v computes an $\wedge$-type function of x and w (because a $\oplus$-type function requires at least 3 gates). In either case, we can connect w directly to v, remove the wire $u \to v$, and change the gate label of v. By checking through each input and gate, the transformation can be done in polynomial time. □

2.2   Correlation

Definition 1. *Let f and g be two boolean functions on n input variables. The correlation of f and g is defined as*

$$Corr(f, g) = |\mathbf{Pr}[f(x) = g(x)] - \mathbf{Pr}[f(x) \neq g(x)]| = |2\mathbf{Pr}[f(x) = g(x)] - 1|,$$

where x is chosen uniformly at random from $\{0, 1\}^n$.

The *correlation* of f with a circuit class $\mathcal{C}$ is the maximum of $Corr(f, C)$ for any $C \in \mathcal{C}$. Note that, a circuit C has correlation c with f if and only if C computes f or its negation correctly on a fraction $(1 + c)/2$ of all inputs. The correlation bound is also referred to as the *average-case lower bound* in the literature.

2.3   Decision Tree

A *decision tree* is a tree where (1) each internal node is labeled by a variable x, and has two outgoing edges labeled by $x = 0$ and $x = 1$, and (2) each leaf is labeled by a constant 0 or 1. A decision tree computes a boolean function by tracking the paths from the root to leaves. The *size* of a decision tree is the number of leaves of the tree.

A *parity decision tree* extends a decision tree such that each internal node is labeled by the parity of a subset of variables (including one single variable as a special case). We insist that, for each path from the root to a leaf, the parities appearing in the internal nodes are linearly independent.

2.4   Concentration Bounds

A sequence of random variables $X_0, X_1, \ldots, X_n$ is called a *supermartingale* with respect to a sequence of random variables $R_1, \ldots, R_n$ if $\mathbf{E}[X_i \mid R_{i-1}, \ldots, R_1] \leqslant X_{i-1}$, for $1 \leqslant i \leqslant n$. The following is a variant of Azuma's inequality which holds for supermartingales with one-side bounded differences.

Lemma 2 ([6]). *Let $\{X_i\}_{i=0}^{n}$ be a supermartingale with respect to $\{R_i\}_{i=1}^{n}$. Let $Y_i = X_i - X_{i-1}$. If, for every $1 \leqslant i \leqslant n$, the random variable Y_i (conditioned on $R_{i-1}, \ldots, R_1$) assumes two values with equal probability, and there exists $c_i \geqslant 0$ such that $Y_i \leqslant c_i$, then, for any $\lambda \geqslant 0$, we have $\mathbf{Pr}[X_n - X_0 \geqslant \lambda] \leqslant \exp\left(-\frac{\lambda^2}{2\sum_{i=1}^{n} c_i^2}\right).$*

3 U_2-circuits

All known lower bounds for U_2-circuits [15, 18, 23, 28] were proved using the gate elimination method. We will generalize this method by defining a random process of restrictions under which the circuit size reduces with high probability. This allows us to get a #SAT algorithm for U_2-circuits of size almost $3n$, and also prove a correlation bound against Parity.

3.1   Concentrated Shrinkage Under Restrictions

We call an $\wedge$-type function of two variables a *twig*. We now define a random process of restrictions where, at each step, we pick a variable or a twig and randomly assign it a value 0 or 1; we also simplify the circuit by eliminating unnecessary gates. The choice of variables or twigs at each step is determined by the following cases: (1) If the circuit is a literal, choose the variable in the literal. (2) If there is an input x with out-degree at least two, choose x. (3) Otherwise, there must be a gate u fed by two variables having out-degree 1; we choose u (which is a twig).

Let C be a simplified U_2-circuit on inputs $x_1, \ldots, x_n$. Let C' be the simplified circuit obtained after one step of restriction. Then we have the following lemma on the reduction of $\mu(C)$.

Lemma 3. *Suppose $\mu(C) \geqslant 4$. Let $\sigma = \mu(C) - \mu(C')$. Then we have $\sigma \geqslant 3$, and $\mathbf{E}[\sigma] \geqslant 4$.*

Proof. Consider the following cases (see also Figure 3.1):

(1) Suppose there is an input x_i feeding into two gates u and v. By Lemma 1, there is no edge between u and v. We randomly assign 0 or 1 to x_i, and consider the following sub-cases on the successors of u and v.

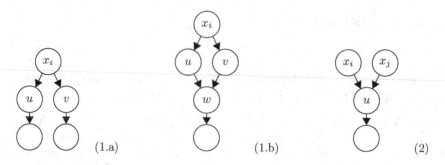

Fig. 1. Cases in Lemma 3

(a) If u and v feed into two different successors, we have the following possibilities. If under one assignment to x_i, none of u, v become constants, then we can eliminate x_i, u, v; and under the other assignment to x_i, since both of u, v will be constants, we can eliminate two more gates (successors of u, v); thus we have $\mathbf{Pr}[\sigma \geqslant 5] \geqslant 1/2$, and $\sigma \geqslant 3$. If under each assignment to x_i, only one of u, v becomes a constant, then we can eliminate x_i, u, v and one successor; thus $\sigma \geqslant 4$.

(b) If u and v feed into one single common successor w, we have similar situations as above. If under one assignment to x_i, both u and v become constants, then we can eliminate x_i, u, v, w and a successor of w; and under the other assignment to x_i, we can eliminate x_i, u, v. If under each assignment to x_i, only one of u, v becomes a constant, then we can eliminate x_i, u, v, w.

(2) If all inputs have out-degree 1, find a gate u fed by two inputs, say x_i and x_j. We randomly assign 0 and 1 to u; for each assignment, eliminate x_i, x_j, u and at least one successor of u. Then we have $\sigma \geqslant 4$.

In all cases, we have $\sigma \geqslant 3$, and $\mathbf{E}[\sigma] \geqslant 4$. $\square$

Next consider the reduction of $\mu(C)$ under a sequence of restrictions. Let $C_0 := C$, and, for $i = 1, \ldots, d$, let C_i be the circuit obtained after the i-th step. For convenience, we let $\mu_i := \mu(C_i)$. Let R_i be the random value assigned to the variable or twig at each step. We define a sequence of random variables $\{Z_i\}$ as follows:

$$Z_i = \begin{cases} \mu_i - (\mu_{i-1} - 4), & \mu_{i-1} \geqslant 4, \\ 0, & \mu_{i-1} < 4. \end{cases}$$

Note that $0 < \mu_{i-1} < 4$ holds only when C_{i-1} itself is a literal or a twig, which means C_i will be a constant.

Lemma 4. Let $X_0 = 0$ and $X_i = \sum_{j=1}^{i} Z_i$. Then we have $Z_i \leqslant 1$, and $\{X_i\}$ is a supermartingale with respect to $\{R_i\}$.

Proof. By Lemma 3, conditioning on $R_1, \ldots, R_{i-1}$, when $\mu_{i-1} \geqslant 4$, we have $\mu_i \leqslant \mu_{i-1} - 3$ and $\mathbf{E}[\mu_i] \leqslant \mu_{i-1} - 4$. Therefore, we get $Z_i \leqslant 1$, $\mathbf{E}[Z_i \mid R_{i-1}, \ldots, R_1] \leqslant 0$, and $\mathbf{E}[X_i \mid R_{i-1}, \ldots, R_1] \leqslant X_{i-1}$. Thus $\{X_i\}$ is a supermartingale with respect to $\{R_i\}$. $\square$

Lemma 5. *For* $\lambda \geqslant 0$, $\mathbf{Pr}[\mu_d \geqslant \max\{\mu_0 - 4d + \lambda, 1\}] \leqslant \exp(-\lambda^2/2d)$.

Proof. Conditioning on $R_1, \ldots, R_{i-1}$, the variable Z_i assumes two values with equal probability. By Lemma 4, we have $\{X_i\}$ is a supermartingale with respect to $\{R_i\}$, and $Z_i \leqslant c_i \equiv 1$. Applying the bound in Lemma 2, we have

$$\mathbf{Pr}\left[\sum_{i=1}^{d} Z_i \geqslant \lambda\right] \leqslant \exp\left(-\frac{\lambda^2}{2d}\right).$$

When $\mu_d > 0$, we have $\sum_{i=1}^{d} Z_i = \mu_d - \mu_0 + 4d$. Let E_1 be the event that $\mu_d > 0$; let E_2 be the event that $\sum_{i=1}^{d} Z_i \geqslant \lambda$. Then the final probability is $\mathbf{Pr}[E_1 \wedge E_2] \leqslant \mathbf{Pr}[E_2] \leqslant \exp(-\lambda^2/2d)$. $\square$

3.2   #SAT Algorithms

We now give a #SAT algorithm for circuits of size almost $3n$ based on the concentrated reduction of circuit size.

Theorem 1. *For U_2-circuits of size $s < 3n$, there is a deterministic #SAT algorithm running in time $2^{n - \Omega((3n-s)^2/n)}$.*

Proof. Let C be a circuit on n inputs $x_1, \ldots, x_n$ with size $s < 3n$. Let $\mu_0 := \mu(C) \leqslant s + n$. We use the following procedure to construct a generalized decision tree, where each internal node is labeled by a variable or a twig. We start with the root node and C.

- If C is a constant, label the current node by this constant and return.
- Use the cases in Lemma 3 to find either a variable or a twig; denote it by u. Label the current node by u.
- Build two outgoing edges labeled by $u = 0$ and $u = 1$. For each child node, simplify the circuit, and recurse.

We say a complete assignment to $x_1, \ldots, x_n$ is *consistent* with a path (from the root to a leaf) if it satisfies the restrictions along the path. Since each assignment $a \in \{0, 1\}^n$ is consistent with only one path, the paths give a disjoint partitioning of the boolean cube $\{0, 1\}^n$. To count the number of satisfying assignments for C, one can count for each path with leaf labeled by 1, and return the summation. Restrictions along each path is essentially a read-once 2-CNF, for which counting is easy. We next only need to bound the size of the tree.

We wish to bound the probability that a random path has length larger than $n - k$, for k to be chosen later. Let $\lambda = 4(n - k) - \mu_0 + 1$. Then by Lemma 5, at depth $n - k$, the restricted circuit becomes a constant with probability at least

$1 - \exp(-\lambda^2/2(n-k)) \geqslant 1 - 2^{-c\lambda^2/n}$ for a constant $c > 0$. The total number of paths with length larger than $n - k$ is at most $2^{n-k} \cdot 2^{-c\lambda^2/n} \cdot 2^k \leqslant 2^{n-c\lambda^2/n}$. Therefore, the size of the tree is at most $2^{n-k} + 2^{n-c\lambda^2/n}$. Choosing $k = (3n-s)/8$, both the tree size and the running time of the counting algorithm are bounded by $2^{n-\Omega((3n-s)^2/n)}$.   □

The following corollary is immediate.

Corollary 1. *(1) For U_2-circuits of size $3n - \epsilon n$ with $\epsilon > 0$, there is a deterministic #SAT algorithm running in time $2^{n-\Omega(n)}$. (2) For U_2-circuits of size $3n - n^\epsilon$ with $\epsilon > 0.5$, there is a deterministic #SAT algorithm running in time $2^{n-n^{\Omega(1)}}$.*

3.3   Correlation with Parity

Schnorr [23] proved a $3n - c$ lower bound for computing Parity using the following fact: a simplified U_2-circuit computing Parity cannot have any input variable with out-degree exactly 1. Indeed, if such an input x exists, one can fix all other variables such that the gate fed by x becomes a constant, but this makes the function independent of x, which is impossible for Parity.

We next generalize this lower bound to the average case by showing that a U_2-circuit of size $s < 3n$ cannot approximate well with Parity. The proof is by converting the generalized decision tree constructed in the proof of Theorem 1 into a normal decision tree without twigs, and argue that the tree size will not increase too much, as stated in the next lemma (the proof is left in the full version [5] of the paper).

Lemma 6. *Any function computed by a U_2-circuit of size $s < 3n$ has a decision tree of size $2^{n-\Omega((3n-s)^2/n)}$.*

The following lemma gives a simple relationship between the size of a decision tree and its correlation with Parity. It was previously used to derive correlation bounds for de Morgan formulas [22] and AC^0 circuits [13].

Lemma 7 ([22]). *A decision tree of size 2^{n-k} has correlation at most 2^{-k} with Parity.*

Theorem 2. *Let C be a U_2-circuit of size $s < 3n$. Then its correlation with Parity is at most $2^{-\Omega((3n-s)^2/n)}$. In particular, for $s = 3n - \epsilon n$ with $\epsilon > 0$, the correlation is at most $2^{-\Omega(n)}$; for $s = 3n - n^\epsilon$ with $\epsilon > 0.5$, the correlation is at most $2^{-n^{\Omega(1)}}$.*

Proof. The proof is immediate by Lemmas 6 and 7.   □

The above correlation bounds with Parity almost match with the upper bounds. To see this, we can construction an approximate circuit for Parity in the following way. Divide n inputs into l groups each of size n/l, use circuits of size $3(n/l - 1)$ to compute Parity exactly for each group, and then take the disjunction of the outputs from all groups. This circuit outputs 0 with probability 2^{-l}, but whenever it outputs 0, it agrees with Parity. Thus its correlation with Parity is at least 2^{-l}. The circuit size is $3(n/l - 1) \cdot l + l = 3n - 2l$.

4 B_2-circuits

In this section, we give #SAT algorithms and correlation bounds for B_2-circuits of size almost $2.5n$.

4.1   Concentrated Shrinkage and #SAT Algorithms

Given a simplified B_2-circuit C, we will construct a generalized parity decision tree, where each internal node is labeled by either a twig or a parity of a subset of variables. Starting from the root with the given circuit C, we use the following case analysis to identify labels and build branches recursively.

If the circuit becomes a constant, we label the current node by the constant; then this node is a leaf. If the circuit is a literal or a gate fed by two variables, then we choose the variable of the literal or the circuit itself as the label, and build two branches. Otherwise, consider a topological order on the gates of the circuit, and let u be the first gate which is either $\oplus$-type of out-degree at least 2 or $\wedge$-type. Consider the following cases (see also Figure 4.1):

(1) If u is a $\oplus$-type gate of out-degree at least 2, then it computes $\oplus_{i \in I} x_i$ (or its negation) for some subset $I \subseteq [n]$. We choose $\oplus_{i \in I} x_i$ as the label, and build two branches; for the branch $\oplus_{i \in I} x_i = b \in \{0, 1\}$, we replace u by a constant, and substitute an arbitrary variable x_j for $j \in I$ by a sub-circuit $\oplus_{i \in I \setminus \{j\}} x_i \oplus b$. In both branches, we can eliminate one variable x_j, and at least 3 gates (u and its two successors).

(2) If u is an $\wedge$-type gate fed by some $\oplus$-type gate v, suppose w is the other node feeding into u.
   - If w has out-degree 1, then we choose the parity function computed at v as the label, and build two branches similar to Case (1). In one branch, we can eliminate some input x_j and two gates v, u; in the other branch, we can eliminate two more nodes: w and a successor of u.
   - If w has out-degree at least 2, then it must be a variable. We choose w as the label, and build two branches. In one branch, we can eliminate w and its two successors; in the other branch, we can eliminate two more gates: v and a successor of u.

(3) If u is an $\wedge$-type gate fed by two inputs x_i and x_j where at least one of them, say x_i, has out-degree at least 2, then we choose x_i as the label and build two branches. In one branch, we can eliminate x_i and its two successors; in the other branch, we can eliminate one more gate: a successor of u.

(4) If u is an $\wedge$-type gate fed by two inputs each of out-degree 1, then choose the twig computed at u as the label. In both branches, we can eliminate x_i, x_j, u and a successor of u.

Consider a random path from the root of the decision tree to its leaves. Let $C_0 := C$, and let C_i be the restricted circuit obtained at depth i. Let $\mu_i := \mu(C_i)$. The next lemma follows directly from the above case analysis.

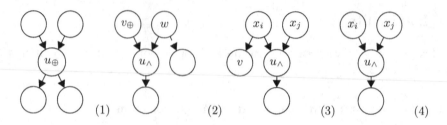

Fig. 2. Cases for eliminating gates in B_2-circuits

Lemma 8. *If $\mu_i > 4$, then $\mu_i - \mu_{i+1} \geqslant 3$, and $\mathbf{E}[\mu_i - \mu_{i+1}] \geqslant 3.5$. If $\mu_i \leqslant 4$, then $\mu_{i+1} = 0$.*

Then we have the following concentrated shrinkage.

Lemma 9. *For $\lambda \geqslant 0$, $\mathbf{Pr}\left[\mu_d \geqslant \max\{\mu_0 - 3.5d + \lambda, 1\}\right] \leqslant \exp(-\lambda^2/2d)$.*

Theorem 3. *For B_2-circuits of size $s < 2.5n$, there is a deterministic #SAT algorithm running in time $2^{n-\Omega((2.5n-s)^2/n)}$. In particular, for $s = 2.5n - \epsilon n$ with $\epsilon > 0$, the algorithm runs in time $2^{n-\Omega(n)}$; for $s = 2.5n - n^\epsilon$ with $\epsilon > 0.5$, the algorithm runs in time $2^{n-n^{\Omega(1)}}$.*

We omit the proofs of Lemma 9 and Theorem 3 since they are similar to the proofs of Lemma 5 and Theorem 1.

4.2   Correlation Bounds

Demenkov and Kulikov [9] proved that affine dispersers for sources of dimension d requires B_2-circuits of size $3n - \Omega(d)$. We next extend this result to the average case by showing that affine extractors have small correlations with B_2-circuits of size less than $2.5n$.

Definition 2. *Let F_2 be the finite field with elements $\{0,1\}$. A function $\mathsf{AE} \colon F_2^n \to F_2$ is a (k,ϵ)-affine extractor if for any uniform distribution X over some k-dimensional affine subspace of F_2^n, $|\mathbf{Pr}[\mathsf{AE}(X) = 1] - 1/2| \leqslant \epsilon$.*

We will need the following constructions of affine extractors.

Theorem 4 ([4,19,27]). *(1) For any $\delta > 0$ there exists a polynomial-time computable (k,ϵ)-affine extractor $\mathsf{AE}_1 \colon \{0,1\}^n \to \{0,1\}$ with $k = \delta n$ and $\epsilon = 2^{-\Omega(n)}$. (2) There exists a constant $c > 0$ and a polynomial-time computable (k,ϵ)-affine extractor $\mathsf{AE}_2 \colon \{0,1\}^n \to \{0,1\}$ with $k = cn/\sqrt{\log\log n}$ and $\epsilon = 2^{-n^{\Omega(1)}}$.*

We will prove our correlation bounds using the following representation of B_2-circuits by parity decision trees.

Lemma 10. *Any function computed by a B_2-circuit of size $s < 2.5n$ is computable by a parity decision tree of size $2^{n-\Omega((2.5n-s)^2/n)}$.*

The proof, which we omit here, is almost the same as the proof of Lemma 6. That is, using the algorithm in Theorem 3, one can construct a generalized parity decision tree which may have twigs, and then expand the twigs and argue that the tree size does not increase much. Note that, when we restrict a twig, the two variables in the twig are completely eliminated; when we restrict a parity, since one variable is substituted, all parity restrictions are linearly independent.

The following lemma gives the correlation of (relatively small) parity decision trees with affine extractors given in Theorem 4. It was implicit in [24] and was also given in [8].

Lemma 11 ([8,24]). *(1) For any $\delta > 0$, a parity decision tree of size 2^{n-k} for $k = \delta n$ has correlation at most $2^{-\Omega(n)}$ with AE_1. (2) There is a constant $c > 0$ such that a parity decision tree of size 2^{n-k} for $k = cn/\sqrt{\log \log n}$ has correlation at most $2^{-n^{\Omega(1)}}$ with AE_2.*

The next theorem follows by Lemma 10 and Lemma 11.

Theorem 5. *(1) For any $\delta > 0$ and B_2-circuit of size $2.5n - \delta n$, its correlation with AE_1 is at most $2^{-\Omega(n)}$. (2) There exists a constant $c > 0$ such that, for any B_2-circuit of size $2.5n - cn/\sqrt[4]{\log \log n}$, its correlation with AE_2 is at most $2^{-n^{\Omega(1)}}$.*

5   Open Questions

It is open whether our correlation bounds (for the size almost $3n$ for U_2-circuits, and almost $2.5n$ for B_2-circuits) can be improved to match with the best known worst-case lower bounds (for the size almost $5n$ for U_2-circuits, and almost $3n$ for B_2-circuits). Pseudorandom generators for boolean formulas were constructed in [14] based on concentrated shrinkage and decomposition of the formula tree. It would be interesting to get pseudorandom generators for general boolean circuits.

References

1. Beals, R., Buhrman, H., Cleve, R., Mosca, M., de Wolf, R.: Quantum lower bounds by polynomials. JACM **48**(4), 778–797 (2001)
2. Beame, P., Impagliazzo, R., Srinivasan, S.: Approximating ac^0 by small height decision trees and a deterministic algorithm for #ac^0 sat. In: Proceedings of the 2012 IEEE Conference on Computational Complexity, CCC 2012 (2012)
3. Blum, N.: A Boolean function requiring $3n$ network size. Theoretical Computer Science **28**, 337–345 (1984)
4. Bourgain, J.: On the construction of affine-source extractors. Geometric and Functional Analysis **17**(1), 33–57 (2007)
5. Chen, R., Kabanets, V.: Correlation bounds and #sat algorithms for small linear-size circuits. ECCC **21**, 184 (2014)

6. Chen, R., Kabanets, V., Kolokolova, A., Shaltiel, R., Zuckerman, D.: Mining circuit lower bound proofs for meta-algorithms. In: CCC 2014 (2014)
7. Chen, R., Kabanets, V., Saurabh, N.: An improved deterministic #SAT algorithm for small de Morgan formulas. In: Csuhaj-Varjú, E., Dietzfelbinger, M., Ésik, Z. (eds.) MFCS 2014, Part II. LNCS, vol. 8635, pp. 165–176. Springer, Heidelberg (2014)
8. Cohen, G., Shinkar, I.: The complexity of DNF of parities. ECCC **21**, 99 (2014)
9. Demenkov, E., Kulikov, A.S.: An elementary proof of a $3n-o(n)$ lower bound on the circuit complexity of affine dispersers. In: Murlak, F., Sankowski, P. (eds.) MFCS 2011. LNCS, vol. 6907, pp. 256–265. Springer, Heidelberg (2011)
10. Håstad, J.: Almost optimal lower bounds for small depth circuits. In: STOC 1986, pp. 6–20 (1986)
11. Håstad, J.: The shrinkage exponent of de Morgan formulae is 2. SIAM Journal on Computing **27**, 48–64 (1998)
12. Håstad, J.: On the correlation of parity and small-depth circuits. ECCC **19**, 137 (2012)
13. Impagliazzo, R., Matthews, W., Paturi, R.: A satisfiability algorithm for AC^0. In: SODA 2012, pp. 961–972 (2012)
14. Impagliazzo, R., Meka, R., Zuckerman, D.: Pseudorandomness from shrinkage. In: FOCS 2012, pp. 111–119 (2012)
15. Iwama, K., Morizumi, H.: An explicit lower bound of $5n-o(n)$ for boolean circuits. In: MFCS 2002, pp. 353–364 (2002)
16. Komargodski, I., Raz, R.: Average-case lower bounds for formula size. In: STOC 2013, pp. 171–180 (2013)
17. Komargodski, I., Raz, R., Tal, A.: Improved average-case lower bounds for demorgan formula size. In: FOCS 2013, pp. 588–597 (2013)
18. Lachish, O., Raz, R.: Explicit lower bound of $4.5n-o(n)$ for boolena circuits. In: STOC 2001, pp. 399–408. ACM, New York (2001)
19. Li, X.: A new approach to affine extractors and dispersers. In: CCC 2011, pp. 137–147 (2011)
20. Nurk, S.: An $o(2^{0.4058m})$ upper bound for circuit sat. PDMI Preprint (2009)
21. Reichardt, B.: Reflections for quantum query algorithms. In: SODA 2011, pp. 560–569 (2011)
22. Santhanam, R.: Fighting perebor: new and improved algorithms for formula and qbf satisfiability. In: FOCS 2010, pp. 183–192 (2010)
23. Schnorr, C.: Zwei lineare untere schranken für die komplexität boolescher funktionen. Computing **13**(2), 155–171 (1974)
24. Seto, K., Tamaki, S.: A satisfiability algorithm and average-case hardness for formulas over the full binary basis. In: CCC 2012, pp. 107–116 (2012)
25. Subbotovskaya, B.A.: Realizations of linear functions by formulas using and or, not. Soviet Math. Doklady **2**, 110–112 (1961)
26. Yao, A.C.: Separating the polynomial-time hierarchy by oracles. In: FOCS 1985, pp. 1–10 (1985)
27. Yehudayoff, A.: Affine extractors over prime fields. Combinatorica **31**(2), 245–256 (2011)
28. Zwick, U.: A 4n lower bound on the combinational complexity of certain symmetric boolean functions over the basis of unate dyadic boolean functions. SIAM J. Comput. **20**(3), 499–505 (1991)

Commuting Quantum Circuits
with Few Outputs are Unlikely
to be Classically Simulatable

Yasuhiro Takahashi[1]([✉]), Seiichiro Tani[1], Takeshi Yamazaki[2],
and Kazuyuki Tanaka[2]

[1] NTT Communication Science Laboratories, NTT Corporation,
Atsugi 243-0198, Japan
{takahashi.yasuhiro,tani.seiichiro}@lab.ntt.co.jp
[2] Mathematical Institute, Tohoku University, Sendai 980-8578, Japan
{yamazaki,tanaka}@math.tohoku.ac.jp

Abstract. We study the classical simulatability of commuting quantum circuits with n input qubits and $O(\log n)$ output qubits, where a quantum circuit is classically simulatable if its output probability distribution can be sampled up to an exponentially small additive error in classical polynomial time. Our main result is that there exists a commuting quantum circuit that is not classically simulatable unless the polynomial hierarchy collapses to the third level. This is the first formal evidence that a commuting quantum circuit is not classically simulatable even when the number of output qubits is exponentially small. Then, we consider a generalized version of the circuit and clarify the condition under which it is classically simulatable. We apply these results to examining the ability of IQP circuits to implement fundamental operations, and to examining the classical simulatability of slightly extended Clifford circuits.

1 Introduction and Summary of Results

One of the most important challenges in quantum information processing is to understand the difference between quantum and classical computation. An approach to meeting this challenge is to study the classical simulatability of quantum computation. Previous studies have shown that restricted models of quantum computation, such as commuting quantum circuits, contribute to this purpose [1,2,4,6,7,9,10]. Because of the simplicity of such restricted models, they also contribute to identifying the source of the computational power of quantum computers. It is thus of interest to study their classical simulatability.

We study the classical simulatability of commuting quantum circuits with n input qubits and $O(\text{poly}(n))$ ancillary qubits initialized to $|0\rangle$, where a commuting quantum circuit is a quantum circuit consisting of pairwise commuting gates, each of which acts on a constant number of qubits. When every gate in a commuting quantum circuit acts on at most c qubits, the circuit is said to be c-local. A commuting quantum circuit is a restricted model of quantum computation in the

© Springer International Publishing Switzerland 2015
D. Xu et al. (Eds.): COCOON 2015, LNCS 9198, pp. 223–234, 2015.
DOI: 10.1007/978-3-319-21398-9_18

sense that all the gates in the circuit can be applied in an arbitrary order. Moreover, there exists a basis in which all the gates are diagonal and thus, under some conditions, they can be implemented simultaneously [5]. In spite of these severe restrictions, as mentioned below, there are evidences that commuting quantum circuits are not classically simulatable in various settings [2,7]. This remarkable feature makes such circuits particularly interesting for study.

For considering the classical simulatability, we adopt strong and weak simulations. The strong simulation of a quantum circuit is to compute its output probability in classical polynomial time and the weak one is to sample its output probability distribution likewise. Any strongly simulatable quantum circuit is easily shown to be weakly simulatable. Our main focus is on the hardness of classically simulating quantum circuits and thus we mainly consider weak simulatability, which yields a stronger result. Previous hardness results on the weak simulatability are usually obtained with respect to multiplicative errors [2,6,10]. They can usually be turned into hardness results with respect to exponentially small additive errors, although in general it is difficult to exactly determine the relative strength of these error settings. In this paper, we deal with exponentially small additive errors. We note that our hardness results in this paper can be turned into hardness results with respect to multiplicative errors.

Bremner et al. [2] showed that there exists a 2-local IQP circuit with $O(\text{poly}(n))$ output qubits such that it is not weakly simulatable (under a plausible assumption), where an IQP circuit is a commuting quantum circuit such that each commuting gate is diagonal in the X-basis $\{(|0\rangle \pm |1\rangle)/\sqrt{2}\}$. This means that, when the number of output qubits is large, even a simple commuting quantum circuit is powerful. On the other hand, Ni et al. [7] showed that any 2-local commuting quantum circuit with $O(\log n)$ output qubits is strongly simulatable, whereas there exists a 3-local commuting quantum circuit with only one output qubit such that it is not strongly simulatable (under a plausible assumption). Thus, when the number of output qubits is $O(\log n)$, the classical simulatability of commuting quantum circuits depends on the number of qubits affected by each gate. A natural question is whether there exists a commuting quantum circuit with $O(\log n)$ output qubits such that it is not weakly simulatable.

We provide the first formal evidence for answering the question affirmatively:

Theorem 1. *There exists a 5-local commuting quantum circuit with $O(\log n)$ output qubits such that it is not weakly simulatable unless the polynomial hierarchy* PH *collapses to the third level, i.e., unless* PH $= \Delta_3^p$.

It is believed that PH does not collapse to any level. Thus, the circuit in Theorem 1 is a desired evidence. To prove Theorem 1, we first show the existence of a depth-3 quantum circuit A_n with $O(\text{poly}(n))$ output qubits such that it is not weakly simulatable unless PH $= \Delta_3^p$. Our idea for constructing the circuit in Theorem 1 is to decrease the number of the output qubits by combining A_n with the OR reduction quantum circuit [5], which reduces the computation of the OR function on k bits to that on $O(\log k)$ bits. The resulting circuit has $O(\log n)$ output qubits and is not weakly simulatable unless PH $= \Delta_3^p$, but it

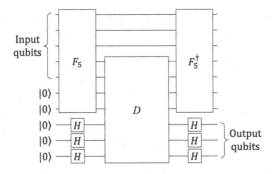

Fig. 1. Circuit $(F_n^\dagger \otimes H^{\otimes l})D(F_n \otimes H^{\otimes l})$, where $n = 5$, $s = 2$, $t = 4$, and $l = 3$

is not a commuting quantum circuit. An important observation is that the OR reduction circuit can be transformed into a 2-local commuting quantum circuit. We regard a quantum circuit consisting of A_n, g, and $A_n^\dagger$ as a single gate $A_n^\dagger g A_n$ for any gate g that is either a ΛX gate or a commuting gate in the commuting OR reduction circuit, where $A_n^\dagger$ is the circuit obtained from A_n by reversing the order of the gates and replacing each gate with its inverse. A rigorous analysis of a quantum circuit consisting of the gates $A_n^\dagger g A_n$ implies Theorem 1.

Then, we study the weak simulatability of a generalized version of the circuit in Theorem 1. We assume that we are given two quantum circuits F_n and D: F_n has n input qubits, $s = O(\mathrm{poly}(n))$ ancillary qubits, and t output qubits, and D is a commuting quantum circuit on $t + l$ qubits such that each commuting gate is diagonal in the Z-basis $\{|0\rangle, |1\rangle\}$, where $l = O(\log n)$. We consider the circuit of the form depicted in Fig. 1, which we denote as $(F_n^\dagger \otimes H^{\otimes l})D(F_n \otimes H^{\otimes l})$, although its precise definition is provided in Section 3.2. The input qubits and output qubits of the circuit are the input qubits of F_n and the l qubits on which H gates are applied, respectively. In particular, when F_n is A_n and D consists only of controlled phase-shift gates, a commuting version of the whole circuit is the circuit in Theorem 1. We show the following relationship:

Theorem 2. *If F_n is weakly simulatable, then $(F_n^\dagger \otimes H^{\otimes l})D(F_n \otimes H^{\otimes l})$ with $l = O(\log n)$ output qubits is also weakly simulatable.*

This is a generalization of the previous result that any IQP circuit with $O(\log n)$ output qubits is weakly simulatable [2], which corresponds to the case when F_n is a layer of H gates. We show Theorem 2 by generalizing the proof of the previous result. Theorem 2 implies a suggestion on how to improve Theorem 1 in terms of locality as follows. Choosing a depth-3 quantum circuit as F_n yields the 5-local commuting quantum circuit in Theorem 1 and a possible way to construct a 3- or 4-local one that is not weakly simulatable would be to somehow choose a depth-2 quantum circuit as F_n. By Theorem 2, such a construction is impossible. This is because, since any depth-2 quantum circuit is weakly simulatable [10], choosing a depth-2 quantum circuit as F_n yields only a weakly simulatable quantum circuit.

We consider two applications of our results. The first one is to examine the ability of IQP circuits to implement a fundamental operation, the OR reduction operation defined as follows: a quantum operation on $n + m$ qubits is called an OR reduction operation if it maps $|x\rangle|0^m\rangle$ to $|x\rangle|\eta\rangle$ for any $x \in \{0,1\}^n$, where $m = O(\log n)$, $|\eta\rangle = |0^m\rangle$ if $x = 0^n$, and $\langle 0^m|\eta\rangle = 0$ otherwise [5]. This operation is shown to be quite useful for significantly reducing the depth of quantum circuits with a certain gate set [5,8]. A 2-local commuting quantum circuit can implement an OR reduction operation as will be shown in the proof of Theorem 1, but we provide an evidence that IQP circuits cannot:

Theorem 3. *For any OR reduction operation, there does not exist an IQP circuit for implementing it unless* $\mathsf{PH} = \Delta_3^p$.

Theorem 3 provides the reason for the difference, which is shown by Theorem 1, between IQP circuits and 5-local commuting quantum circuits in terms of the weak simulatability. It also shows the limitation of the computational power of IQP circuits. The proof of Theorem 3 is a direct application of Theorems 1 and 2.

The second application is to examine the weak simulatability of slightly extended Clifford circuits. For comparison, let us consider Clifford circuits with n input qubits, $O(\text{poly}(n))$ ancillary qubits in a product state, and $O(\log n)$ output qubits. A simple extension of the proof in Refs. [3,6] implies that any Clifford circuit in this setting is strongly simulatable. We show the following theorem:

Theorem 4. *There exists a Clifford circuit augmented by a depth-1 non-Clifford layer with $O(\text{poly}(n))$ ancillary qubits in a particular product state and with $O(\log n)$ output qubits such that it is not weakly simulatable unless* $\mathsf{PH} = \Delta_3^p$.

Just like Theorems 1 and 2, Theorem 4 contributes to understanding a subtle difference between quantum and classical computation. The proof of Theorem 4 is a simple modification of the proof of Theorem 1.

2   Preliminaries

2.1   Quantum Circuits

The elementary gates in this paper are a Hadamard gate H, a phase-shift gate $R(\theta)$ with angle $\theta = \pm 2\pi/2^k$ for any $k \in \mathbb{N}$, and a controlled-Z gate ΛZ, where

$$H = \frac{1}{\sqrt{2}}\begin{pmatrix} 1 & 1 \\ 1 & -1 \end{pmatrix}, \ R(\theta) = \begin{pmatrix} 1 & 0 \\ 0 & e^{i\theta} \end{pmatrix}, \ \Lambda Z = \begin{pmatrix} 1 & 0 & 0 & 0 \\ 0 & 1 & 0 & 0 \\ 0 & 0 & 1 & 0 \\ 0 & 0 & 0 & -1 \end{pmatrix}.$$

We denote $R(\pi)$, $R(\pi/2)$, and $HR(\pi)H$ as Z, P, and X, respectively, where Z and X (with $Y = iXZ$ and identity I) are called Pauli gates. We also denote $H\Lambda ZH$ as ΛX, which is a CNOT gate, where H acts on the target qubit. A quantum circuit consists of the elementary gates. A Clifford circuit is a quantum circuit consisting only of H, P, and ΛZ. A commuting quantum circuit is a quantum circuit consisting of pairwise commuting gates, where we do not require that each commuting gate be one of the elementary gates. In other words, when we think of a quantum circuit as a commuting quantum circuit, we are allowed to regard a group of elementary gates in the circuit as a single gate and we require that such gates be pairwise commuting. An IQP circuit is a commuting quantum circuit such that each commuting gate is diagonal in the X-basis $\{(|0\rangle \pm |1\rangle)/\sqrt{2}\}$ [2]. Any IQP circuit on s qubits can be represented as follows: the first part consists of H gates on s qubits, the middle part a commuting quantum circuit D, and the last part H gates on s qubits, where each commuting gate in D is diagonal in the Z-basis $\{|0\rangle, |1\rangle\}$.

The complexity measures of a quantum circuit are its size and depth. The size of a quantum circuit is the number of elementary gates in the circuit. To define the depth, we consider the circuit as a set of layers $1, \ldots, d$ consisting of one-qubit or two-qubit gates, where gates in the same layer act on pairwise disjoint sets of qubits and any gate in layer j is applied before any gate in layer $j + 1$. The depth of the circuit is the smallest possible value of d [4]. It might be natural to require that each gate in a layer be one of the elementary gates, but for simplicity, when we count the depth, we allow any one-qubit or two-qubit gates that can be obtained as a sequence of elementary gates in the circuit. This does not essentially affect our results, since, regardless of whether we adopt the requirement or not, the depth of the circuit we are interested in is a constant. A quantum circuit can use ancillary qubits initialized to $|0\rangle$.

We deal with a uniform family of polynomial-size quantum circuits $\{C_n\}_{n \geq 1}$, where each C_n has n input qubits and $O(\text{poly}(n))$ ancillary qubits, and angles θ of phase-shift gates in C_n are restricted to $\pm 2\pi/2^k$ with $k = O(\text{poly}(n))$. Some of the input and ancillary qubits are called output qubits. At the end of the computation, Z-measurements, i.e., measurements in the Z-basis, are performed on the output qubits. The uniformity means that there exists a polynomial-time deterministic classical algorithm for computing the function $1^n \mapsto \overline{C_n}$, where $\overline{C_n}$ is the classical description of C_n. A symbol denoting a quantum circuit, such as C_n, also denotes its matrix representation in some fixed basis. Any quantum circuit in this paper is understood to be an element of a uniform family of quantum circuits and thus, for simplicity, we deal with a quantum circuit C_n in place of a family $\{C_n\}_{n \geq 1}$. We require that each commuting gate in a commuting quantum circuit act on a constant number of qubits. When every commuting gate acts on at most c qubits, the circuit is said to be c-local [7].

2.2   Classical Simulatability and Complexity Classes

We deal with polynomial-size classical circuits and randomized classical circuits [2] to model polynomial-time deterministic classical algorithms and their probabilistic versions, respectively. Let C_n be a quantum circuit with n input qubits, $O(\text{poly}(n))$ ancillary qubits, and m output qubits. For any $x \in \{0,1\}^n$, there exists an output probability distribution $\{(y, \Pr[C_n(x) = y])\}_{y \in \{0,1\}^m}$, where $\Pr[C_n(x) = y]$ is the probability of obtaining $y \in \{0,1\}^m$ by

Z-measurements on the output qubits of C_n with the input state $|x\rangle$. The classical simulatability is defined as follows [2,6,7,9,10]:

Definition 1. – C_n *is strongly simulatable if* $\Pr[C_n(x) = y]$ *and its marginal output probability can be computed up to an exponentially small additive error in classical* $O(\mathrm{poly}(n))$ *time. More precisely, for any* $0 < m' \leq m$, m' *output qubits chosen from the* m *output qubits of* C_n, *and polynomial* p, *there exists a polynomial-size classical circuit* D_n *such that, for any* $x \in \{0,1\}^n$ *and* $y' \in \{0,1\}^{m'}$, $|D_n(x, y') - \Pr[C_n(x) = y']| \leq 1/2^{p(n)}$.

– C_n *is weakly simulatable if* $\{(y, \Pr[C_n(x) = y])\}_{y \in \{0,1\}^m}$ *can be sampled up to an exponentially small additive error in classical* $O(\mathrm{poly}(n))$ *time. More precisely, for any polynomial* p, *there exists a polynomial-size randomized classical circuit* R_n *such that, for any* $x \in \{0,1\}^n$ *and* $y \in \{0,1\}^m$, $|\Pr[R_n(x) = y] - \Pr[C_n(x) = y]| \leq 1/2^{p(n)}$.

Any strongly simulatable quantum circuit is weakly simulatable [2,10].

The following two complexity classes are important for our discussion [2]:

Definition 2. *Let* L *be a language, i.e.,* $L \subseteq \{0,1\}^*$.

– $L \in \mathsf{PostBQP}$ *if there exists a polynomial-size quantum circuit* C_n *with* n *input qubits,* $O(\mathrm{poly}(n))$ *ancillary qubits, one output qubit, and one particular qubit (other than the output qubit) called the postselection qubit such that, for any* $x \in \{0,1\}^n$, $\Pr[\mathrm{post}_n(x) = 0] > 0$, $\Pr[C_n(x) = 1 | \mathrm{post}_n(x) = 0] \geq 2/3$ *if* $x \in L$, *and* $\Pr[C_n(x) = 1 | \mathrm{post}_n(x) = 0] \leq 1/3$ *if* $x \notin L$, *where the event "*$\mathrm{post}_n(x) = 0$*" means that the classical outcome of the Z-measurement on the postselection qubit is 0.*

– $L \in \mathsf{PostBPP}$ *if there exists a polynomial-size randomized classical circuit* R_n *with* n *input bits that, for any* $x \in \{0,1\}^n$, *outputs* $R_n(x), \mathrm{post}_n(x) \in \{0,1\}$ *such that* $\Pr[\mathrm{post}_n(x) = 0] > 0$, $\Pr[R_n(x) = 1 | \mathrm{post}_n(x) = 0] \geq 2/3$ *if* $x \in L$, *and* $\Pr[R_n(x) = 1 | \mathrm{post}_n(x) = 0] \leq 1/3$ *if* $x \notin L$.

We use the notation $\mathrm{post}_n(x) = 0$ both in the quantum and classical settings, but the meaning will be clear from the context.

Another important class is the polynomial hierarchy $\mathsf{PH} = \bigcup_{j \geq 1} \Delta_j^p$. Here, $\Delta_1^p = \mathsf{P}$ and $\Delta_{j+1}^p = \mathsf{P}^{\mathsf{N}\Delta_j^p}$ for any $j \geq 1$, where P is the class of languages decided by polynomial-size classical circuits and $\mathsf{N}\Delta_j^p$ is the non-deterministic class associated to Δ_j^p [2]. It is believed that $\mathsf{PH} \neq \Delta_j^p$ for any $j \geq 1$. As shown in Ref. [2], if $\mathsf{PostBQP} \subseteq \mathsf{PostBPP}$, then $\mathsf{PH} = \Delta_3^p$. It can be shown that, in our setting of elementary gates and quantum circuits, this relationship also holds when the condition $\Pr[\mathrm{post}_n(x) = 0] > 0$ in the definition of $\mathsf{PostBQP}$ is replaced with the condition that, for some polynomial q (depending only on C_n), $\Pr[\mathrm{post}_n(x) = 0] \geq 1/2^{q(n)}$. In the following, we adopt the latter condition.

3   Commuting Quantum Circuits

3.1   Hardness of the Weak Simulation

A key component of the circuit in Theorem 1 is a depth-3 quantum circuit with $O(\mathrm{poly}(n))$ output qubits such that it is not weakly simulatable unless

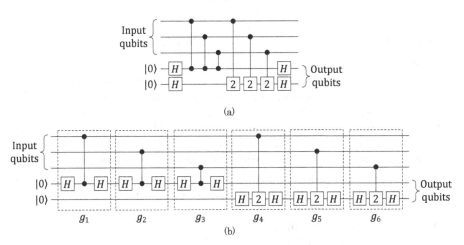

Fig. 2. (a): The non-commuting OR reduction circuit, where the gate represented by two black circles connected by a vertical line is a ΛZ gate, i.e., a controlled-$R(2\pi/2^1)$ gate, and the gate "2" is an $R(2\pi/2^2)$ gate. (b): The commuting OR reduction circuit.

PH $= \Delta_3^p$. Such a circuit exists when weak simulatability is defined with respect to a multiplicative error [2]. We analyze its weak simulatability with an exponentially small additive error (Definition 1) and show the following lemma:

Lemma 1. *There exists a depth-3 polynomial-size quantum circuit with $O(\mathrm{poly}(n))$ output qubits such that it is not weakly simulatable unless PH $= \Delta_3^p$.*

The proof is omitted in this paper. We decrease the number of the output qubits of the circuit in Lemma 1 using the OR reduction circuit [5]. The OR reduction circuit with b input qubits has $m = \lceil \log(b+1) \rceil$ ancillary qubits, which are also output qubits. For any input state $|x\rangle|0^m\rangle$ with $x \in \{0,1\}^b$, the circuit outputs $|x\rangle|\eta\rangle$, where $|\eta\rangle = |0^m\rangle$ if $x = 0^b$ and $\langle 0^m|\eta\rangle = 0$ otherwise. The first part consists of H gates on the ancillary qubits. The middle part consists of b controlled-$R(2\pi/2^k)$ gates over all $1 \le k \le m$, where each gate uses an input qubit as the control qubit and an ancillary qubit as the target qubit. The last part is the same as the first one. We call the circuit the *non-commuting* OR reduction circuit. It is depicted in Fig. 2(a), where $b = 3$.

An important observation is that the non-commuting OR reduction circuit can be transformed into a 2-local commuting quantum circuit. This is shown by considering a quantum circuit consisting of gates g_j on two qubits, where g_j is a controlled-$R(2\pi/2^k)$ gate sandwiched between H gates on the target qubit. Since $H^2 = I$ and controlled-$R(2\pi/2^k)$ gates are pairwise commuting gates on two qubits, the operation implemented by the circuit is the same as that implemented by the non-commuting OR reduction circuit and the gates g_j are pairwise commuting gates on two qubits. We call the circuit the *commuting* OR reduction circuit. It is depicted in Fig. 2(b), where $b = 3$. Combining this circuit with the circuit in Lemma 1 implies the following lemma:

Lemma 2. *There exists a commuting quantum circuit with $O(\log n)$ output qubits such that it is not weakly simulatable unless* PH $= \Delta_3^p$.

Proof. We assume that PH $\neq \Delta_3^p$. Lemma 1 and its proof imply that there exists a depth-3 polynomial-size quantum circuit A_n with n input qubits, $a + b$ ancillary qubits, and $b + 2$ output qubits such that it is not weakly simulatable, where $a = O(\mathrm{poly}(n))$ and $b = O(\mathrm{poly}(n))$. Moreover, the proof that A_n is not weakly simulatable depends only on $\Pr[A_n(x) = 0^{b+1}1]$ and $\Pr[A_n(x) = 0^{b+1}0]$ for any $x \in \{0,1\}^n$. We decrease the number of the first $b + 1$ output qubits, which are the postselection qubits, using the commuting OR reduction circuit. To do so, we construct a quantum circuit E_n with n input qubits, $a + b + m + 1$ ancillary qubits, and $m + 1$ output qubits as follows, where $m = \lceil \log(b+2) \rceil$. As an example, E_n is depicted in Fig. 3(a), where $n = 5$, $a = 0$, and $b = 2$.

1. Apply A_n on n input qubits and $a + b$ ancillary qubits, where the input qubits of E_n are those of A_n.
2. Apply a ΛX gate on the last output qubit of A_n and on an ancillary qubit (other than the ancillary qubits in Step 1), where the output qubit is the control qubit.
3. Apply a commuting OR reduction circuit on the $b + 1$ postselection qubits and m ancillary qubits (other than the ancillary qubits in Steps 1 and 2), where the $b + 1$ qubits are the input qubits of the OR reduction circuit.
4. Apply $A_n^\dagger$ as in Step 1, where $A_n^\dagger$ is the circuit obtained from A_n by reversing the order of the gates and replacing each gate with its inverse.

The output qubits of E_n are the $m + 1$ ancillary qubits used in Steps 2 and 3. Step 4 does not affect the output probability distribution of E_n, but it allows us to construct the commuting quantum circuit described below. By the construction of E_n, for any $x \in \{0,1\}^n$, $\Pr[A_n(x) = 0^{b+1}1] = \Pr[E_n(x) = 0^m1]$ and $\Pr[A_n(x) = 0^{b+1}0] = \Pr[E_n(x) = 0^m0]$. This implies that E_n is not weakly simulatable. The proof is the same as that of Lemma 1 except that the number of output qubits we need to consider is only $m + 1 = O(\log n)$.

We regard a quantum circuit consisting of A_n, g, and $A_n^\dagger$ (in this order) as a single gate $A_n^\dagger g A_n$ for any gate g that is either a ΛX gate in Step 2 of E_n or g_j in the commuting OR reduction circuit. We consider a quantum circuit consisting of the gates $A_n^\dagger g A_n$. The input qubits and output qubits of E_n are naturally considered as the input qubits and output qubits of the new circuit, respectively. The new circuit based on E_n in Fig. 3(a) is depicted in Fig. 3(b). Since these gates g in E_n are pairwise commuting, so are the gates $A_n^\dagger g A_n$. Moreover, since the depth of A_n is three and g acts on two qubits, $A_n^\dagger g A_n$ acts on a constant number of qubits. By the construction of the new circuit, its output probability distribution is the same as that of E_n. Thus, the new circuit is not weakly simulatable.   □

We analyze $A_n^\dagger g A_n$ in the above proof, which implies the following lemma:

Lemma 3. *For any gate $A_n^\dagger g A_n$ in the proof of Lemma 2, there exists a quantum circuit on at most five qubits that implements the gate.*

The proof is omitted in this paper. The above lemmas imply Theorem 1.

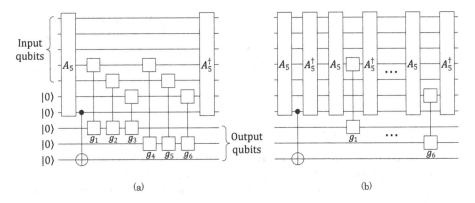

Fig. 3. (a): Circuit E_n, where $n = 5$, $a = 0$, and $b = 2$. The gate represented by a black circle and $\oplus$ connected by a vertical line is a ΛX gate. The gates g_j are the ones in Fig. 2(b). (b): The commuting quantum circuit based on E_n in (a).

3.2   Weak Simulatability of a Generalized Version

As in Fig. 2(a), a non-commuting OR reduction circuit with $b + 1$ input qubits can be represented as three parts: the first part consists of H gates on m qubits, the middle part a quantum circuit D', and the last part H gates on m qubits, where $m = \lceil \log(b+2) \rceil$ and D' consists only of controlled-$R(2\pi/2^k)$ gates. Since ΛX (in Step 2 of E_n in the proof of Lemma 2) is $H\Lambda Z H$, the circuit in Theorem 1 can be represented similarly: the first part consists of A_n and H gates on $m + 1$ qubits, the middle part D'', and the last part $A_n^\dagger$ and H gates on $m + 1$ qubits, where D'' consists of D' and ΛZ, and A_n is the depth-3 quantum circuit (with $a + b$ ancillary qubits and $b + 2$ output qubits) described above. The output qubits of the whole circuit are the $m + 1$ qubits on which H gates are applied.

We consider a generalized version of the circuit in Theorem 1. We assume that we are given two quantum circuits F_n and D: F_n has n input qubits, $s = O(\mathrm{poly}(n))$ ancillary qubits, and t output qubits, and D is a commuting quantum circuit on $t + l$ qubits such that each commuting gate is diagonal in the Z-basis, where $l = O(\log n)$. We construct a quantum circuit with n input qubits, $s + l$ ancillary qubits, and l output qubits as follows, where we denote this circuit as $(F_n^\dagger \otimes H^{\otimes l}) D (F_n \otimes H^{\otimes l})$. As an example, the circuit is depicted in Fig. 1, where $n = 5$, $s = 2$, $t = 4$, and $l = 3$.

1. Apply F_n on n input qubits and s ancillary qubits, where the input qubits of the whole circuit are those of F_n.
2. Apply H gates on l ancillary qubits (other than the ancillary qubits in Step 1).
3. Apply D on $t + l$ qubits, which are the output qubits of F_n and the ancillary qubits in Step 2.
4. Apply H gates as in Step 2 and then $F_n^\dagger$ as in Step 1.

The output qubits of the whole circuit are the l qubits on which H gates are applied. The circuit in Theorem 1 corresponds to the case when $F_n = A_n$,

$D = D''$, $s = a + b$, $t = b + 2$, and $l = m + 1$. When F_n is a layer of H gates with arbitrary s and t, the circuit $(F_n^\dagger \otimes H^{\otimes l})D(F_n \otimes H^{\otimes l})$, which becomes an IQP circuit, is weakly simulatable [2]. A simple generalization of the proof of the previous result implies Theorem 2. The detailed proof is omitted in this paper.

4   Applications

4.1   IQP Circuits and 2-local Commuting Quantum Circuits

To exhibit the difference between IQP circuits and 2-local commuting quantum circuits, we consider OR reduction operations (defined in Section 1). For example, the operation implemented by the commuting OR reduction circuit in Section 3 is an OR reduction operation. This immediately shows that there exists a 2-local commuting quantum circuit for implementing an OR reduction operation. In contrast, Theorem 3 shows that IQP circuits cannot implement any OR reduction operation unless $\mathsf{PH} = \Delta_3^p$. We now prove the theorem:

Proof. We assume that $\mathsf{PH} \neq \Delta_3^p$ and there exists an IQP circuit (on s qubits) for implementing an OR reduction operation. We use this IQP circuit in place of the commuting OR reduction circuit in the proof of Lemma 2. The resulting circuit B_n is the same as the one depicted in Fig. 3(a) except that the middle part (excluding a ΛX gate) becomes the IQP circuit. The proof of the lemma implies that B_n is not weakly simulatable.

We simplify B_n. Recall that, by the definition, the IQP circuit on s qubits can be represented as follows: the first part consists of H gates on the s qubits, the middle part a commuting quantum circuit D, and the last part H gates on the s qubits, where each commuting gate in D is diagonal in the Z-basis. The third layer of A_n included in B_n can be decomposed into two sublayers: the first sublayer consists of ΛZ gates and the second sublayer H gates. All the H gates are cancelled out by the corresponding H gates in the first part of the IQP circuit. Similarly, all the H gates in the first layer of $A_n^\dagger$ are cancelled out by the corresponding H gates in the last part of the IQP circuit. When an H gate in the IQP circuit is applied on a qubit on which no gate is applied in the third layer of A_n, it can be regarded as a gate in the second layer of A_n or $A_n^\dagger$. Thus, by adding the ΛZ gates in the third layer of A_n and in the first layer of $A_n^\dagger$ to D, we can regard A_n as a depth-2 quantum circuit and thus B_n is a quantum circuit of the form in Theorem 2 with $F_n = A_n$ of depth 2. Since any depth-2 quantum circuit is weakly simulatable [10], B_n is weakly simulatable. This contradicts the above observation. Thus, if $\mathsf{PH} \neq \Delta_3^p$, there does not exist an IQP circuit for implementing an OR reduction operation.   □

4.2   Clifford Circuits

We consider Clifford circuits with n input qubits, $O(\mathrm{poly}(n))$ ancillary qubits, and $O(\log n)$ output qubits, where the ancillary qubits are allowed to be in a general product state. Such a Clifford circuit with only one output qubit is strongly simulatable [3,6]. We can simply extend this property as follows:

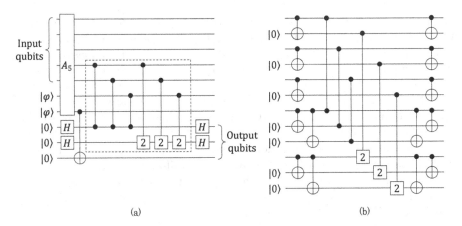

Fig. 4. (a): Circuit E'_n, where $n = 5$, $a = 0$, and $b = 2$. The dashed box represents the middle part of the non-commuting OR reduction circuit. (b): The circuit obtained from the middle part in (a). The qubits in state $|0\rangle$ are new ancillary qubits.

Lemma 4. *Any Clifford circuit with $O(\text{poly}(n))$ ancillary qubits in a general product state and with $O(\log n)$ output qubits is strongly simulatable.*

The proof is omitted in this paper.

Just like Lemma 1, we can show the following lemma:

Lemma 5. *There exists a Clifford circuit with $O(\text{poly}(n))$ ancillary qubits in a particular product state and with $O(\text{poly}(n))$ output qubits such that it is not weakly simulatable unless $\mathsf{PH} = \Delta_3^p$.*

The proof is also omitted. The proof of Lemma 2 leads to the following lemma:

Lemma 6. *There exists a Clifford circuit combined with an OR reduction circuit with $O(\text{poly}(n))$ ancillary qubits in a particular product state and with $O(\log n)$ output qubits such that it is not weakly simulatable unless $\mathsf{PH} = \Delta_3^p$.*

Proof. We assume that $\mathsf{PH} \neq \Delta_3^p$. Lemma 5 and its proof imply that there exists a Clifford circuit A_n with n input qubits, $a = O(\text{poly}(n))$ ancillary qubits initialized to $|0\rangle$, $b = O(\text{poly}(n))$ ancillary qubits initialized to $|\varphi\rangle = (|0\rangle + e^{i\pi/4}|1\rangle)/\sqrt{2}$, and $b + 2$ output qubits such that it is not weakly simulatable. Moreover, the proof that A_n is not weakly simulatable depends only on $\Pr[A_n(x) = 0^{b+1}1]$ and $\Pr[A_n(x) = 0^{b+1}0]$ for any $x \in \{0,1\}^n$. Just like E_n in the proof of Lemma 2, we construct a quantum circuit E'_n with n input qubits and $a + b + m + 1$ ancillary qubits as follows, where $m = \lceil \log(b + 2) \rceil$. As an example, E'_n is depicted in Fig. 4(a), where $n = 5$, $a = 0$, and $b = 2$.

1. Apply A_n on n input qubits, a ancillary qubits initialized to $|0\rangle$, and b ancillary qubits initialized to $|\varphi\rangle$, where the input qubits of E'_n are those of A_n.
2. Apply a ΛX gate on the last output qubit of A_n and on an ancillary qubit.
3. Apply a *non-commuting* OR reduction circuit on the $b + 1$ postselection qubits and m ancillary qubits.

The output qubits of E'_n are the $m + 1$ ancillary qubits used in Steps 2 and 3. We can show that E'_n is not weakly simulatable as in the proof of Lemma 2. □

We replace the non-commuting OR reduction circuit with a constant-depth OR reduction circuit with unbounded fan-out gates [5], where an unbounded fan-out gate can be considered as a sequence of ΛX gates with the same control qubit. Decomposing the unbounded fan-out gates into ΛX gates in the constant-depth OR reduction circuit yields a Clifford circuit augmented by a depth-1 non-Clifford layer, which consists only of controlled phase-shift gates. This procedure transforms the middle part of the non-commuting OR reduction circuit in Step 3, which is the only part in E'_n that includes non-Clifford gates, into a quantum circuit that has ΛX gates and a depth-1 layer consisting of all the gates in the middle part. The circuit obtained in this way from the middle part in Fig. 4(a) is depicted in Fig. 4(b). This transformation with Lemma 6 implies Theorem 4.

Acknowledgments. We thank the anonymous referees for valuable comments. S.T. is deeply grateful to the ELC project (Grant-in-Aid for Scientific Research on Innovative Areas MEXT Japan) for encouraging the research presented in this paper.

References

1. Aaronson, S., Arkhipov, A.: The computational complexity of linear optics. In: Proceedings of the 43rd ACM Symposium on Theory of Computing (STOC), pp. 333–342 (2011)
2. Bremner, M.J., Jozsa, R., Shepherd, D.J.: Classical simulation of commuting quantum computations implies collapse of the polynomial hierarchy. Proceedings of the Royal Society A **467**, 459–472 (2011)
3. Clark, S., Jozsa, R., Linden, N.: Generalized Clifford groups and simulation of associated quantum circuits. Quantum Information and Computation **8**(1&2), 106–126 (2008)
4. Fenner, S.A., Green, F., Homer, S., Zhang, Y.: Bounds on the power of constant-depth quantum circuits. In: Liśkiewicz, M., Reischuk, R. (eds.) FCT 2005. LNCS, vol. 3623, pp. 44–55. Springer, Heidelberg (2005)
5. Høyer, P., Špalek, R.: Quantum fan-out is powerful. Theory of Computing **1**(5), 81–103 (2005)
6. Jozsa, R., van den Nest, M.: Classical simulation complexity of extended Clifford circuits. Quantum Information and Computation **14**(7&8), 633–648 (2014)
7. Ni, X., van den Nest, M.: Commuting quantum circuits: efficient classical simulations versus hardness results. Quantum Information and Computation **13**(1&2), 54–72 (2013)
8. Takahashi, Y., Tani, S.: Collapse of the hierarchy of constant-depth exact quantum circuits. In: Proceedings of the 28th IEEE Conference on Computational Complexity (CCC), pp. 168–178 (2013)
9. Takahashi, Y., Yamazaki, T., Tanaka, K.: Hardness of classically simulating quantum circuits with unbounded Toffoli and fan-out gates. Quantum Information and Computation **14**(13&14), 1149–1164 (2014)
10. Terhal, B.M., DiVincenzo, D.P.: Adaptive quantum computation, constant-depth quantum circuits and Arthur-Merlin games. Quantum Information and Computation **4**(2), 134–145 (2004)

Evaluating Matrix Circuits

Daniel König and Markus Lohrey[✉]

Universität Siegen, Siegen, Germany
lohrey@eti.uni-siegen.de

Abstract. The circuit evaluation problem (also known as the compressed word problem) for finitely generated linear groups is studied. The best upper bound for this problem is coRP, which is shown by a reduction to polynomial identity testing (PIT). Conversely, the compressed word problem for the linear group $SL_3(\mathbb{Z})$ is equivalent to PIT. In the paper, it is shown that the compressed word problem for every finitely generated nilpotent group is in $DET \subseteq NC^2$. Within the larger class of polycyclic groups we find examples where the compressed word problem is at least as hard as PIT for skew arithmetical circuits.

1 Introduction

The study of circuit evaluation problems has a long tradition in theoretical computer science and is tightly connected to many aspects in computational complexity theory. One of the most important circuit evaluation problems is *polynomial identity testing (PIT)*: The input is an arithmetic circuit, whose internal gates are labelled with either addition or multiplication and its input gates are labelled with variables $(x_1, x_2, \ldots)$ or constants $(-1, 0, 1)$, and it is asked whether the output gate evaluates to the zero polynomial (in this paper, we always work in the polynomial ring over the coefficient ring $\mathbb{Z}$ or $\mathbb{Z}_p$ for a prime p). Based on the Schwartz-Zippel-DeMillo-Lipton Lemma, Ibarra and Moran [10] proved that PIT over $\mathbb{Z}$ or $\mathbb{Z}_p$ belongs to the class coRP (co-randomized polynomial time). Whether PIT $\in$ P is an important problem. In [11] it was shown that if there is a language in $DTIME(2^{\mathcal{O}(n)})$ with circuit complexity $2^{\Omega(n)}$, then P = BPP (and hence P = RP = coRP). On the other hand, Kabanets and Impagliazzo [12] proved that if PIT belongs to P, then (i) there is a language in NEXPTIME that does not have polynomial size circuits, or (ii) the permanent is not computable by polynomial size arithmetic circuits. Both conclusions are major open problem in complexity theory. Hence, although it is quite plausible that PIT $\in$ P, it is difficult to prove.

Circuit evaluation problems can be also studied for other structures than polynomial rings, in particular non-commutative structures. For finite monoids, the circuit evaluation was studied in [6], where it was shown that for every non-solvable finite monoid the circuit evaluation problem is P-complete, whereas for every solvable finite monoid, the circuit evaluation problem belongs to the parallel complexity class $DET \subseteq NC^2$. Starting with [17] the circuit evaluation problem has been also studied for infinite finitely generated (f.g) monoids, in

© Springer International Publishing Switzerland 2015
D. Xu et al. (Eds.): COCOON 2015, LNCS 9198, pp. 235–248, 2015.
DOI: 10.1007/978-3-319-21398-9_19

particular infinite f.g. groups. In this context, the input gates of the circuit are labelled with generators of the monoid, the internal gates compute the product of the two input gates, and it is asked whether the circuit evaluates to the identity element. In [17] and subsequent work, the circuit evaluation problem is also called the *compressed word problem (CWP)*. This is due to the fact that if one forgets the underlying monoid structure of a multiplicative circuit, the circuit simply evaluates to a word over the monoid generators labelling the input gates. This word can be of length exponential in the number of circuit gates, and the circuit can be seen as a compressed representation of this word. In this context, circuits are also known as *straight-line programs* and are intensively studied in the area of string compression [18].

Concerning the CWP, polynomial time algorithms have been developed for many important classes of groups, e.g., finite groups, f.g. nilpotent groups, f.g. free groups, graph groups (also known as right-angled Artin groups), and virtually special groups. The latter contain all Coxeter groups, one-relator groups with torsion, fully residually free groups, and fundamental groups of hyperbolic 3-manifolds; see [19]. For the important class of f.g. linear groups, i.e., f.g. groups of matrices over a field, the CWP reduces to PIT (over $\mathbb{Z}$ or $\mathbb{Z}_p$, depending on the characteristic of the field) and hence belongs to coRP [19]. Vice versa, in [19] it was shown that PIT over $\mathbb{Z}$ reduces to the CWP for the linear group $SL_3(\mathbb{Z})$. This result indicates that derandomizing the CWP for a f.g. linear group will be in general very difficult.

In this paper, we further investigate the tight correspondence between commutative circuits over rings and non-commutative circuits over linear groups. In Sec. 6 we study the complexity of the CWP for f.g. nilpotent groups. It is known to be in P [19]. Here, we show that for every f.g. nilpotent group the CWP belongs to the parallel complexity class $DET \subseteq NC^2$, which is the class of all problems that are NC^1-reducible to the computation of the determinant of an integer matrix, see [8]. To the knowledge of the authors, f.g. nilpotent groups are the only examples of infinite groups for which the CWP belongs to NC. Even for free groups, the CWP is P-complete [17]. The main step of our proof for f.g. nilpotent groups is to show that for a torsion-free f.g. nilpotent group G the CWP belongs to the logspace counting class $C_=L$ (and is in fact $C_=L$-complete if $G \neq 1$). To show this, we use the fact that a f.g. torsion-free nilpotent group embeds into the group $UT_d(\mathbb{Z})$ of d-dimensional unitriangular matrices over $\mathbb{Z}$ for some d. Then, we reduce the CWP for $UT_d(\mathbb{Z})$ to the $C_=L$-complete problem, whether two additive circuits over the naturals evaluate to the same number. Let us mention that there are several $C_=L$-complete problems related to linear algebra [1].

We also study the CWP for the matrix group $UT_d(\mathbb{Z})$ for the case that the dimension d is not fixed, i.e., part of the input (Sec. 7). In this case, the CWP turns out to be complete for the counting class $C_=LogCFL$, which is the LogCFL-analogue of $C_=L$.

Finally, in Sec. 8 we move from nilpotent groups to polycyclic groups. These are solvable groups, where every subgroup is finitely generated. By results of

Maltsev, Auslander, and Swan these are exactly the solvable subgroups of $\mathsf{GL}_d(\mathbb{Z})$ for some d. We prove that polynomial identity testing for skew arithmetical circuits reduces to the CWP for a specific 2-generator polycyclic group of Hirsch length 3. A skew arithmetical circuit is an arithmetic circuit (as defined in the first paragraph of the introduction) such that for every multiplication gate, one of its input gates is an input gate of the circuit, i.e., a variable or a constant. These circuits exactly correspond to algebraic branching programs. Even for skew arithmetical circuits, no polynomial time algorithm is currently known (although the problem belongs to coRNC).

Full proofs can be found in the long version [14].

2   Arithmetical Circuits

We use the standard notion of (division-free) arithmetical circuits. Let us fix a set $X = \{x_1, x_2, \ldots\}$ of variables. An *arithmetical circuit* is a triple $C = (V, S, \mathsf{rhs})$, where (i) V is a finite set of *gates*, (ii) $S \in V$ is the *output gate*, and (iii) for every gate A, $\mathsf{rhs}(A)$ (the *right-hand side of A*) is either a variable from X, one of the constants -1, 0, 1, or an expression of the form $B + C$ (then A is an addition gate) or $B \cdot C$ (then A is a multiplication gate), where B and C are gates. Moreover, there must exist a linear order $<$ on V such that $B < A$ whenever B occurs in $\mathsf{rhs}(A)$. A gate A with $\mathsf{rhs}(A) \in X \cup \{0, -1, 1\}$ is an *input gate*. Over a fixed ring $(R, +, \cdot)$ (which will be $(\mathbb{Z}, +, \cdot)$ in most cases) we can evaluate every gate $A \in V$ to a polynomial $\mathsf{val}_C(A)$ with coefficients from R and variables from X (val stands for "value"). Moreover let $\mathsf{val}(C) = \mathsf{val}_C(S)$ be the polynomial to which C evaluates. Two arithmetical circuits C_1 and C_2 are equivalent over the ring $(R, +, \cdot)$ if $\mathsf{val}(C_1) = \mathsf{val}(C_2)$.

Fix an arithmetical circuit $C = (V, S, \mathsf{rhs})$. We can view C as a directed acyclic graph (dag), where every node is labelled with a variable or a constant or an operator $+$, $\cdot$. If $\mathsf{rhs}(A) = B \circ C$ (for $\circ$ one of the operators), then there is an edge from B to A and C to A. The *depth* $\mathsf{depth}(A)$ (resp., *multiplication depth* $\mathsf{mdepth}(A)$) of the gate A is the maximal number of gates (resp., multiplication gates) along a path from an input gate to A. So, input gates have depth one and multiplication depth zero. The *depth* (resp., *multiplication depth*) of C is $\mathsf{depth}(C) = \mathsf{depth}(S)$ (resp., $\mathsf{mdepth}(C) = \mathsf{mdepth}(S)$). The *formal degree* $\mathsf{deg}(A)$ of a gate A is 1 if A is an input gate, $\max\{\mathsf{deg}(B), \mathsf{deg}(C)\}$ if $\mathsf{rhs}(A) = B+C$, and $\mathsf{deg}(B) + \mathsf{deg}(C)$ if $\mathsf{rhs}(A) = B \cdot C$. The formal degree of C is $\mathsf{deg}(C) = \mathsf{deg}(S)$. A *positive circuit* is an arithmetical circuit without input gates labelled by the constant -1. An *addition circuit* is a positive circuit without multiplication gates. A *variable-free circuit* is a circuit without variables. It evaluates to an element of the underlying ring. A *skew circuit* is an arithmetical circuit such that for every multiplication gate A with $\mathsf{rhs}(A) = B \cdot C$, one of the gates B, C is an input gate.

In the rest of the paper we will also allow more complicated expressions in right-hand sides for gates. For instance, we may have a gate with $\mathsf{rhs}(A) = (B + C) \cdot (D + E)$. When writing down such a right-hand side, we implicitly assume that there are additional gates in the circuit, with (in our example) right hand sides $B + C$ and $D + E$, respectively. The proof of the following lemma uses standard ideas.

Lemma 1. *Given an arithmetical circuit C one can compute in logspace positive circuits C_1, C_2 such that $\text{val}(C) = \text{val}(C_1) - \text{val}(C_2)$ for every ring. Moreover, for $i \in \{1, 2\}$ we have $\deg(C_i) \leq \deg(C)$, $\text{depth}(C_i) \leq 2 \cdot \text{depth}(C)$, and $\text{mdepth}(C_i) \leq \text{mdepth}(C)$.*

Polynomial identity testing (PIT) for a ring R is the following computational problem: Given an arithmetical circuit C (with variables $x_1, \ldots, x_n$), does $\text{val}(C) = 0$ hold, i.e., does C evaluate to the zero-polynomial in $R[x_1, \ldots, x_n]$? It is an outstanding open problem in complexity theory, whether PIT for $\mathbb{Z}$ can be solved in polynomial time.

3   Complexity Classes

The counting class #L consists of all functions $f : \Sigma^* \to \mathbb{N}$ for which there is a logspace bounded nondeterministic Turing machine M such that for every $w \in \Sigma^*$, $f(w)$ is the number of accepting computation paths of M on input x. The class $C_=L$ contains all languages A for which there are two functions $f_1, f_2 \in \#L$ such that for every $w \in \Sigma^*$, $w \in A$ if and only if $f_1(w) = f_2(w)$. The class $C_=L$ is closed under logspace many-one reductions. The canonical $C_=L$-complete problem is the following: The input consists of two dags G_1 and G_2 and vertices s_1, t_1 (in G_1) and s_2, t_2 (in G_2), and it is asked whether the number of paths from s_1 to t_1 in G_1 is equal to the number of paths from s_2 to t_2 in G_2. A reformulation of this problem is: Given two variable-free addition circuits C_1 and C_2, does $\text{val}(C_1) = \text{val}(C_2)$ hold?

We use standard definitions concerning circuit complexity, see e.g. [26]. In particular we will consider the class TC^0 of all problems that can be solved by a polynomial size circuit family of constant depth that uses NOT-gates and unbounded fan-in AND-gates, OR-gates, and majority-gates. For DLOGTIME-uniform TC^0 it is required in addition that for binary coded gate numbers u and v, one can (i) compute the type of gate u in time $O(|u|)$ and (ii) check in time $O(|u| + |v|)$ whether u is an input gate for v. Note that the circuit for inputs of length n has at most $p(n)$ gates for a polynomial $p(n)$. Hence, the binary codings u and v have length $O(\log n)$, i.e., the above computations can be done in $\mathsf{DTIME}(\log n)$. This is the reason for using the term "DLOGTIME-uniform". If majority gates are not allowed, we obtain the class (DLOGTIME-uniform) AC^0. The class (DLOGTIME-uniform) NC^1 is defined by (DLOGTIME-uniform) polynomial size circuit families of logarithmic depth that use NOT-gates and fan-in-2 AND-gates and OR-gates. A language A is AC^0-reducible to languages $B_1, \ldots, B_k$ if A can be solved with a DLOGTIME-uniform polynomial size circuit family of constant depth that uses NOT-gates and unbounded fan-in AND-gates, OR-gates, and B_i-gates ($1 \leq i \leq k$). Here, a B_i-gate (it is also called an oracle gate) receives an ordered tuple of inputs $x_1, x_2, \ldots, x_n$ and outputs 1 if and only if $x_1 x_2 \cdots x_n \in B_i$. Sometimes, also the term "uniform constant depth reducibility" is used for this type of reductions. In the same way, the weaker NC^1-reducibility can be defined. Here, one counts the depth of a B_i-gate with inputs $x_1, x_2, \ldots, x_n$ as $\log n$. The class DET contains all problems that are NC^1-reducible to the computation of the determinant of an integer matrix, see [8]. It is known that $\mathsf{C}_=\mathsf{L} \subseteq \mathsf{DET} \subseteq \mathsf{NC}^2$, see e.g. [4, Sec.4].

An *NAuxPDA* is a nondeterministic Turing machine with an additional pushdown store. The class $\mathsf{LogCFL} \subseteq \mathsf{NC}^2$ is the class of all languages that can be accepted by a polynomial time bounded NAuxPDA whose work tape is logarithmically bounded (but the pushdown store is unbounded). If we assign to the input the number of accepting computation paths of such an NAuxPDA, we obtain the counting class $\#\mathsf{LogCFL}$. In [25] it is shown that a function $f : \{0,1\}^* \to \mathbb{N}$ belongs to $\#\mathsf{LogCFL}$ if and only if there exists a logspace-uniform family $(\mathcal{C}_n)_{n \geq 1}$ of positive arithmetic circuits such that $\mathcal{C}_n$ computes the mapping f restricted to $\{0,1\}^n$ and there is a polynomial $p(n)$ such that the formal degree of $\mathcal{C}_n$ is bounded by $p(n)$. The class $\mathsf{C}_=\mathsf{LogCFL}$ contains all languages A for which there are two functions $f_1, f_2 \in \#\mathsf{LogCFL}$ such that for every $w \in \Sigma^*$, $w \in A$ if and only if $f_1(w) = f_2(w)$. We need the following lemma, whose proof is based on folklore ideas:

Lemma 2. *There is an NAuxPDA $\mathcal{P}$ that gets as input a positive variable-free arithmetic circuit $\mathcal{C}$ and such that the number of accepting computations of $\mathcal{P}$ on input $\mathcal{C}$ is $\mathsf{val}(\mathcal{C})$. Moreover, the running time is bounded polynomially in $\mathsf{depth}(\mathcal{C}) \cdot \mathsf{deg}(\mathcal{C})$.*

4   Matrices and Groups

In this paper we are concerned with certain subclasses of *linear groups*. A group is linear if it is isomorphic to a subgroup of $\mathsf{GL}_d(F)$ (the group of all invertible $(d \times d)$-matrices over the field F) for some field F.

A (n-step) solvable group G is a group G, which has a a subnormal series $G = G_n \rhd G_{n-1} \rhd G_{n-2} \rhd \cdots \rhd G_1 \rhd G_0 = 1$ (i.e., G_i is a normal subgroup of G_{i+1} for all $0 \leq i \leq n-1$) such that every quotient G_{i+1}/G_i is abelian ($0 \leq i \leq n-1$). If every quotient G_{i+1}/G_i is cyclic, then G is called *polycyclic*. The number of $0 \leq i \leq n-1$ such that $G_{i+1}/G_i \cong \mathbb{Z}$ is called the *Hirsch length* of G; it does not depend on the chosen subnormal series. If $G_{i+1}/G_i \cong \mathbb{Z}$ for all $0 \leq i \leq n-1$ then G is called *strongly polycyclic*. A group is polycyclic if and only if it is solvable and every subgroup is finitely generated. Polycyclic groups are linear. Auslander and Swan [5,24] proved that the polycyclic groups are exactly the solvable groups of integer matrices.

For a group G its *lower central series* is the series $G = G_1 \rhd G_2 \rhd G_3 \rhd \cdots$ of subgroups, where $G_{i+1} = [G_i, G]$, which is the subgroup generated by all commutators $[g, h]$ with $g \in G_i$ and $h \in G$. Indeed, G_{i+1} is a normal subgroup of G_i. The group G is *nilpotent*, if its lower central series terminates after finitely many steps in the trivial group 1. Every f.g. nilpotent group is polycyclic.

Let G be a f.g. group and let G be finitely generated as a group by Σ. Then, as a monoid G is finitely generated by $\Sigma \cup \Sigma^{-1}$ (where $\Sigma^{-1} = \{a^{-1} \mid a \in \Sigma\}$ is a disjoint copy of Σ and a^{-1} stands for the inverse of the generator $a \in \Sigma$). Recall that the *word problem* for G is the following computational problem: Given a string $w \in (\Sigma \cup \Sigma^{-1})^*$, does w evaluate to the identity of G. Kharlampovich proved that there exist finitely presented 3-step solvable groups with an undecidable word problem. On the other hand, for every f.g. linear group the word problem can be solved in deterministic logarithmic space by results of Lipton and Zalcstein [16] and Simon [23]. This applies in particular to polycyclic groups. Robinson proved in his thesis that the word problem for a polycyclic group belongs to TC^0 [21], but his circuits are not uniform. Waack considered in [27] arbitrary f.g. solvable linear groups (which include the polycyclic groups) and proved that their word problems belong to logspace-uniform NC^1. In [14] we combine Waack's technique with the famous division breakthrough result by Hesse, Allender, and Barrington [9] to show that for every f.g. solvable linear group the word problem belongs to DLOGTIME-uniform TC^0.

5   Straight-Line Programs and the Compressed Word Problem

A straight-line program (briefly, SLP) is basically a multiplicative circuit over a monoid. We define an SLP over the finite alphabet Σ as a triple $\mathcal{G} = (V, S, \mathsf{rhs})$, where V is a finite set of variables (or gates), $S \in V$ is the start variable (or output gate), and rhs maps every variable to a right-hand side $\mathsf{rhs}(A)$, which is either a symbol $a \in \Sigma$, or of the form BC, where $B, C \in V$. As for arithmetical circuits we require that there is a linear order $<$ on V such that $B < A$, whenever B occurs in $\mathsf{rhs}(A)$. The terminology "(start) variable" (instead of "(output) gate") comes from the fact that an SLP is quite often defined as a context-free grammar that produces a single string over Σ. This string is defined in the obvious way by iteratively replacing variables by the corresponding right-hand sides, starting with the start variable. We denote this string with $\mathsf{val}(\mathcal{G})$. The unique string over Σ, derived from the variable $A \in V$, is denoted with $\mathsf{val}_{\mathcal{G}}(A)$. We will also allow more general right-hand sides from $(V \cup \Sigma)^*$, but by introducing new variables we can always obtain an equivalent SLP in the above form.

If we have a monoid M, which is finitely generated by the set Σ, then there exists a canonical monoid homomorphism $h : \Sigma^* \to M$. Then, an SLP $\mathcal{G}$ over the alphabet Σ can be evaluated over the monoid M, which yields the monoid element $h(\mathsf{val}(\mathcal{G}))$. In this paper, we are only interested in the case that the monoid M is a f.g. group G. Let G be finitely generated as a group by Σ. An SLP over the alphabet $\Sigma \cup \Sigma^{-1}$ is also called an SLP over the group G. In this case, we will quite often identify the string $\mathsf{val}(\mathcal{G}) \in (\Sigma \cup \Sigma^{-1})^*$ with the group element $g \in G$ to which it evaluates. We will briefly write "$\mathsf{val}(\mathcal{G}) = g$ in G" in this situation.

The main computational problem we are interested in is the *compressed word problem* for a f.g. group G (with a finite generating set Σ), briefly $\mathsf{CWP}(G)$. The input for this problem is an SLP $\mathcal{G}$ over the alphabet $\Sigma \cup \Sigma^{-1}$, and it is asked whether $\mathsf{val}(\mathcal{G}) = 1$ in G (where of course 1 denotes the group identity). The term "compressed word problem" comes from the fact that this problem can be seen as a succinct version of the classical word problem for G, where the input is an explicitly given string $w \in (\Sigma \cup \Sigma^{-1})^*$ instead of an SLP-compressed string.

The compressed word problem is related to the classical word problem. For instance, the classical word problem for a f.g. subgroup of the automorphism group of a group G can be reduced to the compressed word problem for G, and similar results are known for certain group extensions, see [19] for more details. There are several important classes of groups, for which the compressed word problem can be solved in polynomial time, and for finitely generated linear groups the compressed word problem belongs to co-randomized polynomial time, see the introduction. In [6] the parallel complexity of the compressed word problem (there, called the circuit evaluation problem) for finite groups was studied, and the following result was shown:

Theorem 1 ([6]). *Let G be a finite group. If G is solvable, then $\mathsf{CWP}(G)$ belongs to the class $\mathsf{DET} \subseteq \mathsf{NC}^2$. If G is not solvable, then $\mathsf{CWP}(G)$ is P-complete.*

6   CWP for Finitely Generated Nilpotent Groups

In [19] it was shown that the compressed word problem for a finitely generated nilpotent group can be solved in polynomial time. The main result of this section is:

Theorem 2. *Let $G \neq 1$ be a f.g. torsion-free nilpotent group. Then $\mathsf{CWP}(G)$ is complete for the class $\mathsf{C_{=}L}$.*

For the lower bound let G be a non-trivial f.g. torsion-free nilpotent group. Since $G \neq 1$, G contains $\mathbb{Z}$. Hence, it suffices to prove the following:

Lemma 3. $\mathsf{CWP}(\mathbb{Z})$ *is hard for* $\mathsf{C_{=}L}$.

Proof. An SLP $\mathcal{G}$ over the generator 1 of $\mathbb{Z}$ and its inverse -1 is nothing else than a variable-free arithmetical circuit $\mathcal{C}$ without multiplication gates. Using Lemma 1 we can construct in logspace two addition circuits $\mathcal{C}_1$ and $\mathcal{C}_2$ such that $\mathsf{val}(\mathcal{C}) = 0$ if and only if $\mathsf{val}(\mathcal{C}_1) = \mathsf{val}(\mathcal{C}_2)$. Checking the latter is complete for $\mathsf{C_{=}L}$ as remarked in Sec. 3. ☐

For the upper bound in Thm. 2, we use the fact that every torsion-free f.g. nilpotent group can be represented by unitriangluar integer matrices. Let A be a $(d \times d)$-matrix over $\mathbb{Z}$. With $A[i, j]$ we denote the entry of A in row i and column j. The matrix A is *triangular* if $A[i, j] = 0$ whenever $i > j$, i.e., all entries below the main diagonal are 0. A *unitriangular matrix* is a triangular matrix A such that $A[i, i] = 1$ for all $1 \leq i \leq d$, i.e., all entries on the main diagonal are 1. We denote the set of unitriangular $(d \times d)$-matrices over $\mathbb{Z}$ with $\mathsf{UT}_d(\mathbb{Z})$. This is a group with respect to matrix multiplication. Let $1 \leq i < j \leq d$. With $T_{i,j}$ we denote the matrix from $\mathsf{UT}_d(\mathbb{Z})$ such that $T_{i,j}[i, j] = 1$ and $T_{i,j}[k, l] = 0$ for all k, l with $1 \leq k < l \leq d$ and $(k, l) \neq (i, j)$. The notation $T_{i,j}$ does not specify the dimension d of the matrix, but the dimension will be always clear from the context. The group $\mathsf{UT}_d(\mathbb{Z})$ is generated by the finite set $\Gamma_d = \{T_{i,i+1} \mid 1 \leq i < d\}$, see e.g. [7]. For every torsion-free f.g. nilpotent group G there exists some $d \geq 1$ such that $G \leq \mathsf{UT}_d(\mathbb{Z})$ [13, Thm.17.2.5]. Hence, the upper bound in Thm. 2 follows from:

Lemma 4. *For every $d \geq 1$, $\mathsf{CWP}(\mathsf{UT}_d(\mathbb{Z}))$ belongs to $\mathsf{C}_=\mathsf{L}$.*

For the rest of this section let us fix a number $d \geq 1$ and consider the unitriangluar matrix group $\mathsf{UT}_d(\mathbb{Z})$. Consider an SLP $\mathcal{G} = (V, S, \mathsf{rhs})$ over the alphabet $\Gamma_d \cup \Gamma_d^{-1}$, where Γ_d is the finite generating set of $\mathsf{UT}_d(\mathbb{Z})$ from Sec. 4. Note that for every variable $A \in V$, $\mathsf{val}_\mathcal{G}(A)$ is a word over the alphabet $\Gamma_d \cup \Gamma_d^{-1}$. We identify in the following this word with the matrix to which it evaluates. Thus, $\mathsf{val}_\mathcal{G}(A) \in \mathsf{UT}_d(\mathbb{Z})$.

Assume we have given an arithmetical circuit $\mathcal{C}$. A partition $\biguplus_{i=1}^m V_i$ of the set of all multiplication gates of $\mathcal{C}$ is called *structure-preserving* if for all multiplication gates u, v of $\mathcal{C}$ the following holds: If there is a non-empty path from u to v in (the dag corresponding to) $\mathcal{C}$ then there exist $1 \leq i < j \leq d$ such that $u \in V_i$ and $v \in V_j$. In a first step, we transform our SLP $\mathcal{G}$ in logspace into a variable-free arithmetical circuit $\mathcal{C}$ of multiplication depth at most d such that $\mathcal{G}$ evaluates to the identity matrix if and only if $\mathcal{C}$ evaluates to 0. Moreover, we also compute a structure-preserving partition of the multiplication gates of $\mathcal{C}$. This partition will be needed for the further computations. The degree bound in the following lemma will be needed in Sec. 7.

Lemma 5. *From the SLP $\mathcal{G} = (V, S, \mathsf{rhs})$ we can compute in logspace a variable-free arithmetical circuit $\mathcal{C}$ with $\mathsf{mdepth}(\mathcal{C}) \leq d$ and $\deg(\mathcal{C}) \leq 2(d-1)$, such that $\mathsf{val}(\mathcal{G}) = \mathsf{Id}_d$ if and only if $\mathsf{val}(\mathcal{C}) = 0$. In addition we can compute in logspace a structure-preserving partition $\biguplus_{i=1}^d V_i$ of the set of all multiplication gates of $\mathcal{C}$.*

Proof. The set of gates of $\mathcal{C}$ is $W = \{A_{i,j} \mid A \in V, 1 \leq i < j \leq d\} \uplus \{T\}$, where T is the output gate. The idea is simple: Gate $A_{i,j}$ will evaluate to the matrix entry $\mathsf{val}_\mathcal{G}(A)[i, j]$. To achieve this, we define the right-hand side mapping of the circuit $\mathcal{G}$ (which we denote again with rhs) as follows: If $\mathsf{rhs}(A) = M \in \Gamma_d \cup \Gamma_d^{-1}$, then $\mathsf{rhs}(A_{i,j}) = M[i, j] \in \{-1, 0, 1\}$, and if $\mathsf{rhs}(A) = BC$, then $\mathsf{rhs}(A_{i,j}) = B_{i,j} + C_{i,j} + \sum_{i<k<j} B_{i,k} \cdot C_{k,j}$ (which is the rule for matrix multiplication taking into account that all matrices are unitriangular). Finally, we set $\mathsf{rhs}(T) = \sum_{1 \leq i < j \leq d} S_{i,j}^2$. Then, $\mathsf{val}(\mathcal{C}) = 0$ iff $\mathsf{val}_\mathcal{G}(S)[i, j] = 0$ for all $1 \leq i < j \leq d$ iff $\mathsf{val}(\mathcal{G})$ is the identity matrix.

Concerning the multiplication depth, note that the multiplication depth of the gate $A_{i,j}$ is bounded by $j - i$: The only multiplications in $\mathsf{rhs}(A_{i,j})$ are of the form $B_{i,k}C_{k,j}$ (and these multiplications are not nested). Hence, by induction, the multiplication depth of $A_{i,j}$ is bounded by $1 + \max\{k - i, j - k \mid i < k < j\} =$

$j - i$. It follows that every gate $S_{i,j}$ has multiplication depth at most $d-1$, which implies that the output gate T has multiplication depth at most d. Similarly, it can be shown by induction that $\deg(A_{i,j}) \leq j - i$. Hence, $\deg(A_{i,j}) \leq d - 1$ for all $1 \leq i < j \leq d$, which implies that the formal degree of the circuit is bounded by $2(d-1)$.

The structure-preserving partition $\biguplus_{i=1}^{d} V_i$ of the set of all multiplication gates of $\mathcal{C}$ can be defined as follows: All gates corresponding to multiplications $B_{i,k} \cdot C_{k,j}$ in $\mathsf{rhs}(A_{i,j})$ are put into the set V_{j-i}. Finally, all gates corresponding to multiplications $S_{i,j}^2$ in $\mathsf{rhs}(T)$ are put into V_d. It is obvious that this partition is structure-preserving.   □

In a second step we apply Lemma 1 and construct from the above circuit $\mathcal{C}$ two variable-free positive circuits $\mathcal{C}_1$ and $\mathcal{C}_2$, both having multiplication depth at most d such that $\mathsf{val}(\mathcal{C}) = \mathsf{val}(\mathcal{C}_1) - \mathsf{val}(\mathcal{C}_2)$. Hence, our input SLP $\mathcal{G}$ evaluates to the indentity matrix if and only if $\mathsf{val}(\mathcal{C}_1) = \mathsf{val}(\mathcal{C}_2)$. Moreover, using the construction from Lemma 1 it is straightforward to compute in logspace a structure-preserving partition $\biguplus_{i=1}^{d} V_{k,i}$ of the the set of all multiplication gates of $\mathcal{C}_k$ ($k \in \{1,2\}$).

The following lemma concludes the proof that $\mathsf{CWP}(\mathsf{UT}_d(\mathbb{Z}))$ belongs to $\mathsf{C}_=\mathsf{L}$. For the proof one eliminates in a single phase all multiplication gates in a layer. This can be achieved by a logspace reduction, and since the total number of layers is constant, the whole elimination procedure works in logspace.

Lemma 6. *Let d be constant. From a given variable-free positive circuit $\mathcal{C}$ of multiplication depth d together with a structure-preserving partition $\biguplus_{i=1}^{d} V_i$ of the set of all multiplication gates of $\mathcal{C}$, we can compute in logarithmic space a variable-free addition circuit $\mathcal{D}$ such that $\mathsf{val}(\mathcal{C}) = \mathsf{val}(\mathcal{D})$.*

So far, we have restricted to *torsion-free* f.g. nilpotent groups. For general f.g. nilpotent groups, we use the fact that every f.g. nilpotent group contains a torsion-free normal f.g. nilpotent subgroup of finite index [13, Thm. 17.2.2], in order to show that the compressed word problem for every f.g. nilpotent group belongs to the complexity class DET: To do this we need the following result. For the proof one can adopt the proof of [19, Thm.4.4], where the statement is shown for polynomial time many-one reducibility instead of AC^0-reducibility.

Theorem 3. *Let G be a finitely generated group. For every normal subgroup H of G with a finite index, $\mathsf{CWP}(G)$ is AC^0-reducible to $\mathsf{CWP}(H)$ and $\mathsf{CWP}(G/H)$.*

We can now show:

Theorem 4. *For every f.g. nilpotent group, the compressed word problem is in* DET.

Proof. Let G be a f.g. nilpotent group. If G is finite, then the result follows from Thm. 1 (every nilpotent group is solvable). If G is infinite, then G has a f.g. torsion-free normal subgroup H of finite index [13, Thm. 17.2.2]. Subgroups and quotients of nilpotent groups are nilpotent too [22, Chapter5], hence H and G/H are nilpotent; moreover H is finitely generated. By Thm. 2, $\mathsf{CWP}(H)$ belongs to

$C_=L \subseteq DET$. Moreover, by Thm. 1, $CWP(G/H) \in DET$ as well. Finally, Thm. 3 implies $CWP(G) \in DET$.   □

Actually, Thm. 4 can be slightly extended to groups that are (f.g. nilpotent)-by-(finite solvable) (i.e., groups that have a normal subgroup, which is f.g. nilpotent, and where the quotient is finite solvable. This follows from Thm. 3 and the fact that the compressed word problem for a finite solvable group belongs to DET (Thm. 1).

7 The Uniform CWP for Unitriangular Groups

For Lemma 4 it is crucial that the dimension d is a constant. In this section, we consider a uniform variant of the compressed word problem for $UT_d(\mathbb{Z})$. We denote this problem with $CWP(UT_*(\mathbb{Z}))$. The input consists of a unary encoded number d and an SLP, whose terminal symbols are generators of $UT_d(\mathbb{Z})$ or their inverses. Alternatively, we can assume that the terminal symbols are arbitrary matrices from $UT_d(\mathbb{Z})$ with binary encoded entries (given such a matrix M, it is easy to construct an SLP over the generator matrices that produces M). The question is whether the SLP evaluates to the identity matrix. We show that this problem is complete for the complexity class $C_=LogCFL$.

Theorem 5. *The problem* $CWP(UT_*(\mathbb{Z}))$ *is complete for* $C_=LogCFL$.

Proof. We start with the upper bound. Consider an SLP $\mathcal{G}$, whose terminal symbols are generators of $UT_d(\mathbb{Z})$ or their inverses. The dimension d is clearly bounded by the input size. Consider the variable-free arithmetic circuit $\mathcal{C}$ constructed from $\mathcal{G}$ in Lemma 5 and let $\mathcal{C}_1$ and $\mathcal{C}_2$ be the two variable-free positive arithmetic circuits obtained from $\mathcal{C}$ using Lemma 1. Then $\mathcal{G}$ evaluates to the identity matrix if and only if $val(\mathcal{C}_1) = val(\mathcal{C}_2)$. Moreover, the formal degrees $\deg(\mathcal{C}_1)$ and $\deg(\mathcal{C}_2)$ are bounded by $2(d-1)$, i.e., polynomially bounded in the input length. Finally, we compose a logspace machine that computes from the input SLP $\mathcal{G}$ the circuit $\mathcal{C}_i$ with the NAuxPDA from Lemma 2 to get an NAux-PDA $\mathcal{P}_i$ such that the number of accepting computation paths of $\mathcal{P}_i$ on input $\mathcal{G}$ is exactly $val(\mathcal{C}_i)$. Moreover, the running time of $\mathcal{P}_i$ on input $\mathcal{G}$ is bounded polynomially in $(2d-1) \cdot depth(\mathcal{C}_i) \in O(d \cdot |\mathcal{G}|)$.

Let us now show that $CWP(UT_*(\mathbb{Z}))$ is hard for $C_=LogCFL$. Let $(\mathcal{C}_{1,n})_{n \geq 0}$ and $(\mathcal{C}_{2,n})_{n \geq 0}$ be two logspace-uniform families of positive arithmetical circuits of polynomially bounded size and formal degree. Let $w = a_1 a_2 \cdots a_n \in \{0,1\}^n$ be an input for the circuits $\mathcal{C}_{1,n}$ and $\mathcal{C}_{2,n}$. Let $\mathcal{C}_i$ be the variable-free positive arithmetical circuit obtained from $\mathcal{C}_{i,n}$ by replacing every x_j-labelled input gate by $a_j \in \{0,1\}$. By [3, Lemma 3.2] we can assume that every gate of $\mathcal{C}_i$ is labelled by its formal degree. By adding if necessary additional multiplication gates, where one input is set to 1, we can assume that $\mathcal{C}_1$ and $\mathcal{C}_2$ have the same formal degree $d \leq p(n)$ for a polynomial p. Analogously, we can assume that if A is an addition gate in $\mathcal{C}_1$ or $\mathcal{C}_2$ with right-hand side $B + C$, then $\deg(B) = \deg(C) = \deg(A)$. All these preprocessing steps can be carried out in logarithmic space.

We will construct in logarithmic space an SLP $\mathcal{G}$ over the alphabet $\Gamma_{d+1} \cup \Gamma_{d+1}^{-1}$, where Γ_{d+1} is our canonical generating set for the matrix group $\mathsf{UT}_{d+1}(\mathbb{Z})$, such that $\mathcal{G}$ evaluates to the identity matrix if and only if $\mathsf{val}(\mathcal{C}_1) = \mathsf{val}(\mathcal{C}_2)$. Let v_i be the output value of $\mathcal{C}_i$. We first construct in logspace an SLP $\mathcal{G}_1$ that evaluates to the matrix $T_{1,d}^{v_1}$. In the same way we can construct in logspace a second SLP $\mathcal{G}_2$ that evaluates to $T_{1,d}^{-v_2}$. Then, by concatenating the two SLPs $\mathcal{G}_1$ and $\mathcal{G}_2$ we obtain the desired SLP.

The variables of $\mathcal{G}_1$ are $A_{i,j}^b$, where A is a gate of $\mathcal{C}_1$, $b \in \{-1, 1\}$, and $1 \leq i < j \leq d$ such that $j - i$ is the formal degree of A. The SLP $\mathcal{G}_1$ will be constructed in such a way that $\mathsf{val}_{\mathcal{G}_1}(A_{i,j}^b) = T_{i,j}^{b \cdot v}$, where $v = \mathsf{val}_{\mathcal{C}_1}(A)$. If $\mathsf{rhs}_{\mathcal{C}_1}(A) = 0$, then we set $\mathsf{rhs}_{\mathcal{G}_1}(A_{i,j}^b) = \mathsf{Id}$ and if $\mathsf{rhs}_{\mathcal{C}_1}(A) = 1$, then we set $\mathsf{rhs}_{\mathcal{G}_1}(A_{i,j}^b) = T_{i,j}^b$. If $\mathsf{rhs}_{\mathcal{C}_1}(A) = B + C$, then we set $\mathsf{rhs}_{\mathcal{G}_1}(A_{i,j}^b) = B_{i,j}^b C_{i,j}^b$. Correctness follows immediately by induction. Note that $\deg(B) = \deg(C) = \deg(A) = j - i$, which implies that the gates $B_{i,j}^b$ and $C_{i,j}^b$ exist. Finally, if $\mathsf{rhs}_{\mathcal{C}_1}(A) = B \cdot C$, then we set $\mathsf{rhs}_{\mathcal{G}_1}(A_{i,j}^1) = B_{i,k}^{-1} C_{k,j}^{-1} B_{i,k}^1 C_{k,j}^1$ and $\mathsf{rhs}_{\mathcal{G}_1}(A_{i,j}^{-1}) = C_{k,j}^{-1} B_{i,k}^{-1} C_{k,j}^1 B_{i,k}^1$, where k is such that $\deg(B) = k - i$ and $\deg(B) = j - k$. Such a k exists since $j - i = \deg(A) = \deg(B) + \deg(C)$. Correctness follows by induction and the simple fact that $T_{i,j}^{-a}, T_{j,k}^{-b} T_{i,j}^a, T_{j,k}^b = T_{i,k}^{ab}$ for all $a, b \in \mathbb{Z}$ and $1 \leq i < j < k \leq d$; see [20]. $\qquad\square$

8 CWP for Polycyclic Groups

In this section we look at the compressed word problem for polycyclic groups. Since every polycyclic group is f.g. linear, the compressed word problem for a polycyclic group can be reduced to PIT. Here, we show a lower bound: There is a polycyclic group G such that PIT for skew arithmetical circuits can be reduced to $\mathsf{CWP}(G)$. In this context, it is interesting to note that PIT for arbitrary circuits can be reduced to the compressed word problem to the linear (but not polycyclic) group $\mathsf{SL}_3(\mathbb{Z})$ [19, Thm.4.16].

Let us start with a specific example of a polycyclic group. Consider the two matrices

$$g_a = \begin{pmatrix} a & 0 \\ 0 & 1 \end{pmatrix} \text{ and } h = \begin{pmatrix} 1 & 1 \\ 0 & 1 \end{pmatrix}, \tag{1}$$

where $a \in \mathbb{R}$, $a \geq 2$. Let $G_a = \langle g_a, h \rangle \leq \mathsf{GL}_2(\mathbb{R})$. Let us remark that, for instance, the group G_2 is not polycyclic, see e.g. [28, p. 56]. On the other hand, we have:

Proposition 1. *The group $G = G_{1+\sqrt{2}}$ is polycyclic.*

The main result of this section is:

Theorem 6. *Let $a \geq 2$. Polynomial identity testing for skew arithmetical circuits is logspace-reducible to the compressed word problem for the group G_a.*

In particular, there exist polycyclic groups for which the compressed word problem is at least as hard as polynomial identity testing for skew circuits. Recall that it is not known, whether there exists a polynomial time algorithm for polynomial identity testing restricted to skew arithmetical circuits.

For the proof of Thm. 6, we use the following result from [2] (see the proof of [2, Prop. 2.2], where the result is shown for $a = 2$, but the proof works for any $a \geq 2$):

Lemma 7. *Let C be an arithmetical circuit of size n with variables $x_1, \ldots, x_m$ and let $p(x_1, \ldots, x_m) = \mathsf{val}(C)$. Let $a \geq 2$ be a real number. Then $p(x_1, \ldots, x_n)$ is the zero-polynomial if and only if $p(\alpha_1, \ldots, \alpha_n) = 0$, where $\alpha_i = a^{2^{i \cdot n^2}}$ for $1 \leq i \leq m$.*

Proof of Thereom 6. Let us fix a skew arithmetical circuit C of size n with m variables $x_1, \ldots, x_m$. We will define an SLP $\mathcal{G}$ over the alphabet $\{g_a, g_a^{-1}, h, h^{-1}\}$ such that $\mathsf{val}(\mathcal{G}) = \mathsf{Id}$ in G_a if and only if $\mathsf{val}(C) = 0$. First of all, using iterated squaring, we can construct an SLP $\mathcal{H}$ with variables $A_1, A_1^{-1} \ldots, A_m, A_m^{-1}$ (and some other auxiliary variables) such that

$$\mathsf{val}_{\mathcal{H}}(A_i) = g_a^{2^{i \cdot n^2}} = \begin{pmatrix} \alpha_i & 0 \\ 0 & 1 \end{pmatrix} \quad \text{and} \quad \mathsf{val}_{\mathcal{H}}(A_i^{-1}) = g_a^{-2^{i \cdot n^2}} = \begin{pmatrix} \alpha_i^{-1} & 0 \\ 0 & 1 \end{pmatrix}.$$

We now construct the SLP $\mathcal{G}$ as follows: The set of variables of $\mathcal{G}$ consists of the gates of C and the variables of $\mathcal{H}$. We copy the right-hand sides from $\mathcal{H}$ and define the right-hand side for a gate A of C as follows: (i) $\mathsf{rhs}_{\mathcal{G}}(A) = h^n$ if $\mathsf{rhs}_C(A) = n \in \{0, -1, 1\}$, (ii) $\mathsf{rhs}_{\mathcal{G}}(A) = BC$ if $\mathsf{rhs}_C(A) = B + C$, and (iii) $\mathsf{rhs}_{\mathcal{G}}(A) = A_i B A_i^{-1}$ if $\mathsf{rhs}_C(A) = x_i \cdot B$.

A straightforward induction shows that for every gate A of C we have the following, where we denote for better readability the polynomial $\mathsf{val}_C(A)$ to which gate A evaluates with p_A:

$$\mathsf{val}_{\mathcal{G}}(A) = \begin{pmatrix} 1 & p_A(\alpha_1, \ldots, \alpha_n) \\ 0 & 1 \end{pmatrix}$$

We finally take the output gate S of the skew circuit C as the start variable of $\mathcal{G}$. Then, $\mathsf{val}(\mathcal{G})$ yields the identity matrix in the group G_a if and only if $p_S(\alpha_1, \ldots, \alpha_n) = 0$. By Lemma 7 this is equivalent to $\mathsf{val}(C) = p_S(x_1, \ldots, x_n) = 0$. □

Actually, we can carry out the above reduction for a class of arithmetical circuits that is slightly larger than the class of skew arithmetical circuits. Let us define a *powerful skew circuit* as an arithmetical circuit, where for every multiplication gate A, $\mathsf{rhs}(A)$ is of the form $x_i^e \cdot B$, where $e \geq 0$ is a binary coded number. Such a circuit can be converted into an ordinary arithmetical circuit, which, however is no longer skew. To extend the reduction from the proof of Thereom 6 to powerful skew circuits, we set for a gate A with $\mathsf{rhs}_C(A) = x_i^e \cdot B$: $\mathsf{rhs}_{\mathcal{G}}(A) = A_i^e B A_i^{-e}$. The powers A_i^e and A_i^{-e} can be defined using additional multiplication gates. In our recent paper [15], we introduced powerful skew circuits, and proved that for this class, PIT belongs to coRNC. We applied this result to the compressed word problem for wreath products.

Let us look again at the group $G = G_{1+\sqrt{2}}$ from Prop. 1. A closer inspection (see [14]) shows that $[G, G] \cong \mathbb{Z} \times \mathbb{Z}$ and $G/[G, G] \cong \mathbb{Z} \times \mathbb{Z}_2$. Hence, G has a subnormal series of the form $G \rhd H \rhd \mathbb{Z} \times \mathbb{Z} \rhd \mathbb{Z} \rhd 1$, where H has index 2 in G and $H/(\mathbb{Z} \times \mathbb{Z}) \cong \mathbb{Z}$. The group H is strongly polycyclic and has Hirsch length 3. By Thm. 3 we obtain:

Corollary 1. *There is a strongly polycyclic group H of Hirsch length 3 such that polynomial identity testing for skew circuits is polynomial time reducible to* CWP(H).

References

1. Allender, E., Beals, R., Ogihara, M.: The complexity of matrix rank and feasible systems of linear equations. Comput. Complex. **8**(2), 99–126 (1999)
2. Allender, E., Bürgisser, P., Kjeldgaard-Pedersen, J., Miltersen, P.B.: On the complexity of numerical analysis. SIAM J. Comput. **38**(5), 1987–2006 (2009)
3. Allender, E., Jiao, J., Mahajan, M., Vinay, V.: Non-commutative arithmetic circuits: Depth reduction and size lower bounds. Theor. Comput. Sci. **209**(1–2), 47–86 (1998)
4. Àlvarez, C., Jenner, B.: A very hard log-space counting class. Theor. Comput. Sci. **107**(1), 3–30 (1993)
5. Auslander, L.: On a problem of Philip Hall. Annals of Mathematics **86**(2), 112–116 (1967)
6. Beaudry, M., McKenzie, P., Péladeau, P., Thérien, D.: Finite monoids: From word to circuit evaluation. SIAM J. Comput. **26**(1), 138–152 (1997)
7. Biss, D.K., Dasgupta, S.: A presentation for the unipotent group over rings with identity. Journal of Algebra **237**(2), 691–707 (2001)
8. Cook, S.A.: A taxonomy of problems with fast parallel algorithms. Inform. Control **64**, 2–22 (1985)
9. Hesse, W., Allender, E., Barrington, D.A.M.: Uniform constant-depth threshold circuits for division and iterated multiplication. J. Comput. System Sci. **65**, 695–716 (2002)
10. Ibarra, O.H., Moran, S.: Probabilistic algorithms for deciding equivalence of straight-line programs. J. Assoc. Comput. Mach. **30**(1), 217–228 (1983)
11. Impagliazzo, R., Wigderson, A.: P = BPP if E requires exponential circuits: derandomizing the XOR lemma. In: Proc. STOC 1997, pp. 220–229. ACM Press (1997)
12. Kabanets, V., Impagliazzo, R.: Derandomizing polynomial identity tests means proving circuit lower bounds. Comput. Complex. **13**(1–2), 1–46 (2004)
13. Kargapolov, M.I., Merzljakov, J.I.: Fundamentals of the Theory of Groups. Springer (1979)
14. König, D., Lohrey, M.: Evaluating matrix circuits (2015). arXiv.org
15. König, D., Lohrey, M.: Parallel identity testing for algebraic branching programs with big powers and applications (2015). arXiv.org
16. Lipton, R.J., Zalcstein, Y.: Word problems solvable in logspace. J. Assoc. Comput. Mach. **24**(3), 522–526 (1977)
17. Lohrey, M.: Word problems and membership problems on compressed words. SIAM J. Comput. **35**(5), 1210–1240 (2006)

18. Lohrey, M.: Algorithmics on SLP-compressed strings: A survey. Groups Complexity Cryptology **4**(2), 241–299 (2012)
19. Lohrey, M.: The Compressed Word Problem for Groups. SpringerBriefs in Mathematics. Springer (2014)
20. Lohrey, M.: Rational subsets of unitriangular groups. International Journal of Algebra and Computation (2015). doi:10.1142/S0218196715400068
21. Robinson, D.: Parallel Algorithms for Group Word Problems. Ph.D. thesis, UCSD (1993)
22. Rotman, J.J.: An Introduction to the Theory of Groups, 4th edn. Springer (1995)
23. Simon, H.-U.: Word problems for groups and contextfree recognition. In: Proceedings of Fundamentals of Computation Theory, FCT 1979, pp. 417–422. Akademie-Verlag (1979)
24. Swan, R.: Representations of polycyclic groups. Proc. Am. Math. Soc. **18**, 573–574 (1967)
25. Vinay, V.: Counting auxiliary pushdown automata and semi-unbounded arithmetic circuits. In: Proc. Structure in Complexity Theory Conference, pp. 270–284. IEEE Computer Society (1991)
26. Vollmer, H.: Introduction to Circuit Complexity. Springer (1999)
27. Waack, S.: On the parallel complexity of linear groups. ITA **25**(4), 265–281 (1991)
28. Wehrfritz, B.A.F.: Infinite Linear Groups. Springer (1977)

Computing and Graph

Approximation and Nonapproximability for the One-Sided Scaffold Filling Problem

Haitao Jiang[✉], Jingjing Ma, Junfeng Luan, and Daming Zhu

School of Computer Science and Technology, Shandong University,
Jinan 250101, People's Republic of China
{htjiang,jfluan,dmzhu}@sdu.edu.cn, mjjdha@163.com

Abstract. Scaffold filling is an interesting combinatorial optimization problem from genome sequencing. The one-sided scaffold filling problem can be stated as: given an incomplete scaffold with some genes missing and a reference scaffold, the purpose is to insert the missing genes back into the incomplete scaffold(called "filling the scaffold"), such that the number of common adjacencies between the filled scaffold and the reference scaffold is maximized. This problem is NP-hard for genome with duplicated genes, and can be approximated within 1.25 by a very complicated combinatorial method. In this paper, we firstly improve the approximation factor to 6/5 by not-oblivious local search; then we show that this problem is MAX-SNP-complete.

1 Introduction

Motivation

Scaffold filling is a necessary step in the Next Generation Sequencing (NGS), which has greatly improved the speed of genome sequencing. Though the speed and quantity grows hugely by NGS, the accuracy is still not high enough. The bottleneck is that these sequences are often only a part of the complete genome, but concatenating them together to a whole genome, in general, still an intractable problem in the sense of computer algorithm. Currently, most sequencing results for genomes usually are in the form of scaffolds or contigs. Sometimes, applying these incomplete genomes for genomic analysis will introduce unnecessary errors. So it is natural to fill the missing gene fragments into the incomplete genome in a combinatorial way, and to obtain an 'augmented' genome which is closer to some reference genome.

Known Results

In reference [13], Muñoz *et al.* proposed the one-sided permutation scaffold filling problem, and devised an exact algorithm to minimize the genome rearrangement distance, i.e., the Double Cut and Join distance. Subsequently, Jiang *et al.* considered the permutation scaffold filling problem without any reference genome, they show that it is polynomially solvable under the breakpoint distance.

For genomes containing some duplicated genes, their scaffold filling becomes more difficult. Firstly, to measure the similarity between genomes with gene

© Springer International Publishing Switzerland 2015
D. Xu et al. (Eds.): COCOON 2015, LNCS 9198, pp. 251–263, 2015.
DOI: 10.1007/978-3-319-21398-9_20

repetition, there are three general criterion: the exemplar genomic distance [14], the minimum common string partition (MCSP) distance [3] and the maximum number of common string adjacencies [1,11,12]. Unfortunately, unless P=NP, the former two criterion are hard to be computed and even hard to be approximated [2–4,7–9]. Therefore, the measure of the maximum number of common adjacenciesis is meaningful in itself.

Though the breakpoint distance on genomes with duplicated genes makes no sense, its complementary distance, the number of common adjacencies, formly defined by Jiang and Zhu [11,12], shows very good computable properties. Under the measure of common adjacencies, there are a series of related works. For the one-sided scaffold filling problem, Jiang et al. proved that its decision problem is NP-complete, and designed a 1.33-approximation algorithm with a greedy strategy [11,12]. Recently, Liu et al. improve the approximation factor to 1.25 by a combinatorial method, their key idea is a step to compute a maximum matching of a constructed bipartite graph, as well as a step of local improvement [15]. For the two-sided scaffold filling problem, as a generalization of the one-sided problem, it is also NP-hard. In [16], Liu et al. propose the first non-trivial polynomial time approximation algorithm with a ratio 3/2.

Our Contributions

1. An 6/5-approximation algorithm for the one-sided scaffold filling problem by a not-oblivious local search method. The "not-oblivious" local search was systematic described in [5]. The main idea is not improving the objective function directly , but improving another function that contains the same parameters to objective function but different indexes. With respect to local search, we just use 1-substitution;
2. We prove that the scaffold filling problem is MAX-SNP-complete by an L-reduction from the Maximum 3-Dimensional Matching problem [17].

2 Preliminaries

Firstly, we review some necessary definitions, which are also defined in [12]. Throughout this paper, we only consider unsigned genes and genomes. Given a gene family Σ, a string P is called *permutation* if each element in Σ appears exactly once in P. We use $c(P)$ to denote the set of elements in permutation P. A string S is called *sequence* if some genes appear more than once in S, and $c(S)$ denotes genes of S, which is a multi-set of elements in Σ. For example, $\Sigma = \{a, b, c, d\}$, $S = acdabcd$, $c(S) = \{a, a, b, c, c, d, d\}$. A *scaffold* is an *sequence*, typically obtained by some sequencing and assembling process. A substring with m genes is called an *m-substring*, and a 2-substring is also called a *pair*, as the genes are unsigned, the relative order of the two genes of a pair does not matter, i.e., the pair xy is equal to the pair yx. Given a scaffold $S=s_1s_2s_3\cdots s_n$, let $PAIR_S = \{s_1s_2, s_2s_3, \ldots, s_{n-1}s_n\}$ be the set of pairs in S.

Given two scaffolds $S=s_1s_2\cdots s_n$ and $T=t_1t_2\cdots t_m$, if $s_is_{i+1} = t_jt_{j+1}$ (or $s_is_{i+1} = t_{j+1}t_j$), where $s_is_{i+1} \in PAIR_S$ and $t_jt_{j+1} \in PAIR_T$, we say that

$s_i s_{i+1}$ and $t_j t_{j+1}$ (or $t_{j+1} t_j$) are matched to each other. Once $s_i s_{i+1}$ is matched to a pair $t_j t_{j+1}$ in $PAIR_T$, it will never be matched to another pair $t_k t_{k+1}$ ($k \neq j$) in $PAIR_T$. Also, we claim that each pair can be matched at most one time, and any pair should be matched if possible. We denote the match between $PAIR_S$ and $PAIR_T$ as $R = \{(s,t) \mid s \text{ is matched to } t, \text{ and } s \in PAIR_S, t \in PAIR_T\}$. If the cardinality of R is maximized, we call it maximum match. We will base it to identify a pair in S and T as a common adjacency or a breakpoint.

Definition 1. *Given two scaffolds $S = s_1 s_2 \cdots s_n$ and $T = t_1 t_2 \cdots t_m$, let R be a maximum match between $PAIR_S$ and $PAIR_T$. A matched pair in S or T with respect to R is called an **adjacency**, and an unmatched pair with respect to R is called a **breakpoint** in S and T respectively.*

It follows from the definition that scaffolds S and T contain the same adjaciencies but distinct breakpoints. The maximum matched pairs in T (or equally, in S) is called the *adjacency set* between S and T, denoted as $a(S,T)$. By $b_S(S,T)$ and $b_T(S,T)$ we denote the set of breakpoints in S and T respectively. We will illuminate the above involved definitions by Fig.1.

$$scaffold \ \ T = \langle a \ c \ d \ a \ b \ c \ d \rangle$$
$$scaffold \ \ S = \langle a \ b \ d \ c \ a \ c \ d \rangle$$
$$PAIR_T = \{ac, cd, da, ab, bc, cd\}$$
$$PAIR_S = \{ab, bd, dc, ca, ac, cd\}$$
$$matched \ \ pairs \ : \ (ac \leftrightarrow ca), (cd \leftrightarrow dc), (ab \leftrightarrow ab), (cd \leftrightarrow cd)$$
$$a(S,T) = \{ac, cd, ab, cd\}$$
$$b_T(S,T) = \{da, bc\}$$
$$b_S(S,T) = \{bd, ac\}$$

Fig. 1. An example for adjacency, breakpoint and related definitions

Given two scaffolds $S = s_1 s_2 \cdots s_n$ and $T = t_1 t_2 \cdots t_m$, as we can see, each gene except the four ending ones is involved in two adjaciencies or two breakpoints or one adjacency and one breakpoint. To get rid of this unbalance, we add "$*$" to both the ends of S and T. Then, in the following part of this paper, we assume that $S = s_0 s_1 \cdots s_n s_{n+1}$ and $T = t_0 t_1 \cdots t_m t_{m+1}$, where $s_0 = s_{n+1} = t_0 = t_{m+1} = *$.

For a sequence S and a multi-set of elements X, let $S + X$ be the set of all possible resulting sequences after filling all the elements in X into S. Now, we put forward the problems we study in this paper formally.

Definition 2. *Two-sided Scaffold Filling Problem(TSSF-MNA problem).*
Input: *two scaffolds S and T over a gene set Σ and two multi-sets of elements X and Y, where $X = c(T) - c(S)$ and $Y = c(S) - c(T)$.*
Question: *Find $S' \in S + X$ and $T' \in T + Y$ such that $|a(S',T')|$ is maximized.*

The one-sided SF-MNA problem is a special instance of the TSSF-MNA problem where Y is empty, and is the problem we studied in this paper. The definition of it is described below.

Definition 3. *One-sided Scaffold Filling Problem(OSSF-MNA problem).*
Input: *a complete scaffold T and an incomplete scaffold S over a gene set Σ, a multi-set $X = c(T) - c(S) \neq \varnothing$, while $c(S) - c(T) = \varnothing$.*
Question: *Find $S' \in S + X$ such that $|a(S', T)|$ is maximized.*

Note that while the TSSF-MNA problem is more general and more difficult, the OSSF-MNA is more practical as a lot of genome analysis are based on some reference genome [13].

3   An Approximation Algorithm by Local Search

In this section, we show a local search algorithm for the one-sided scaffold filling problem, before that, we recall a key algorithm presented in [15] by Liu et al., which is actually a method of filling a scaffold, guaranteeing that on average each insertion of a single gene would generate at least one common adjacency. In our algorithm, after the local search step, we call that algorithm *Insert-Whole-String* in [15].

Theorem 1. *There exists a polynomial time algorithm to fill a scaffold, guaranteeing that the number of adjacencies increased is not smaller than the number of genes inserted.*

3.1   An Equivalent Goal

The goal of solving this problem is to insert the genes of X into scaffold S and obtain as many adjacencies as possible. No matter in what order the genes are inserted, they appears in groups in the final $S' \in S + X$, so we can consider that S' is obtained by inserting strings (composed of genes of X) into S.

Obviously, inserting a string of length one (i.e., a single gene) will generate at most two more adjacencies, and inserting a string of length m will generate at most $m+1$ more adjacencies. Therefore, there would be two types of inserted strings.

- good string: a string of k missing genes $x_1, x_2, \ldots, x_k$ are inserted in between $s_i s_{i+1}$ in scaffold S to obtain $k+1$ adjacencies (i.e., $s_i x_1, x_1 x_2, \ldots, x_{k-1} x_k, x_k s_{i+1}$), where $s_i s_{i+1}$ is a breakpoint.
  In this case, $x_1, x_2, \ldots, x_k$ is called a k-good string, $s_i s_{i+1}$ is called a *dock*, and we also say that $s_i s_{i+1}$ *docks* the corresponding k-good string $x_1 \ldots x_k$.
- bad string: a sequence of l missing genes $y_1, y_2, \ldots, y_l$ are inserted in between $s_j s_{j+1}$ in scaffold S to obtain l adjacencies (i.e., $s_j y_1$ or $y_l s_{j+1}, y_1 y_2, \ldots, y_{l-1} y_l$), where $s_j s_{j+1}$ is a breakpoint;
  or a sequence of l missing genes $y_1, y_2, \ldots, y_l$ are inserted in between $s_j s_{j+1}$ in scaffold S to obtain $l+1$ adjacencies (i.e., $s_j y_1, y_1 y_2, \ldots, y_{l-1} y_l, y_l s_{j+1}$), where $s_j s_{j+1}$ is an adjacency.

$$scaffold \ T = \langle ...a \ x \ b \ ...a \ y \ z \ b...c \ y \ z \ d...c \ y \ d \ \rangle$$

$$scaffold \ S = \langle a \ b...c \ d...c \ z \ d...a....b...y...y...\rangle$$

$$missing \ genes \ X = \{x, y, z\}$$

$$good \ strings \quad : \{x \ : \ docked \ by \ ab, \ yz : \ docked \ by \ ab, y : \ docked \ by \ cz\}$$

$$some \ bad \ strings \quad : \{x : inserted \ right \ after \ the \ second \ a, y : inserted \ inbetween \ zd\}$$

Fig. 2. An example for good strings and bad strings

The following figure shows some examples of good strings and bad strings.

From Theorem 1 and the definition of good strings, we can describe the optimal solution in the following formula. Let $N_0 = |a(S,T)|$ be the number of adjacencies between S and T, $|OPT|$ be the number of adjacencies in some optimal solution, and b_i be the number of i-good strings in this optimal solution. Assume the length of good strings is at most p. Then we have,

$$|OPT| = N_0 + |X| + \sum_{i=0}^{p} b_i$$

Since Theorem 1 guarantees at least $N_0 + |X|$ adjacencies are increased, to obtain a better solution, we focus on searching enough number of good strings. The following Lemma shows the bound for the optimal number of adjacencies, and how many good strings we should obtain to reach a approximation factor of 6/5.

Lemma 1. *Let $|APP|$ be the number of adjacencies obtained by an algorithm and b'_i be the number of i-good strings in this algorithm, then the approximation factor is 6/5, provided that $4b_1 + 3b_3 + 2b_3 + b_4 \leq 6b'_1 + 6b'_2 + 6b'_3 + 6b'_4$.*

Proof. Since $|OPT| = N_0 + |X| + \sum_{i=0}^{p} b_i$ and $|APP| = N_0 + |X| + \sum_{i=0}^{q} b'_i$, where q is the greatest length of good strings by the algorithm. Then we have,

$$|OPT| = N_0 + |X| + \sum_{i=1}^{q} b_i$$

$$= N_0 + |X| + b_1 + b_2 + b_3 + b_4 + \sum_{i=5}^{q} b_i$$

$$\leq N_0 + |X| + b_1 + b_2 + b_3 + b_4 + (|X| - b_1 - 2b_2 - 3b_3 - 4b_4)/5$$

$$\leq N_0 + \frac{6}{5}(|X| + \frac{4}{6}b_1 + \frac{3}{6}b_2 + \frac{2}{6}b_3 + \frac{1}{6}b_4)$$

$$\leq N_0 + \frac{6}{5}(|X| + b'_1 + b'_2 + b'_3 + b'_4)$$

$$\leq \frac{6}{5}|APP|$$

□

From Lemma 1, we will obtain the approximation factor of 6/5, as long as the number of good strings in our algorithm is at least $4/6b_1 + 3/6b_2 + 2/6b_3 + 1/6b_4$, i.e., $4b_1 + 3b_3 + 2b_3 + b_4 \leq 6b'_1 + 6b'_2 + 6b'_3 + 6b'_4$.

3.2   Description of the Local Search Algorithm

From Lemma 1, the key purpose of our algorithm is to obtain enough number of good strings of length 1, 2, 3, and 4. Let b'_i be the number of i-good strings computed by our algorithm. There are three stages in our algorithm: the first is a greedy stage, which provides an initial solution for the local search step; then the algorithm improves the solution by a local search method, during which, the algorithm iteratively optimizes the function $D = b'_1 + b'_2/2 + b'_3/6$ by 1-substitution; finally, the algorithm inserts the remaining genes, guaranteeing that each insertion of a gene will generate at least one more adjacency. we show the algorithm as follows.

Algorithm 1. Scaffold Filling by Local Search

1: **for** (i from 1 to 4) **do**
2:     Identify the breakpoints and adjacencies in T and S, $S_{i-good} = \phi$
3:     **while** ($X \neq \phi$, and there is an i-string $x_1 \cdots x_i$ composed of i genes in X is docked at $s_w s_{w+1}$, where $s_w s_{w+1} \in b_T(S, T)$) **do**
4:         { insert the i-string $x_1, \cdots, x_i$ between s_w and s_{w+1} in S, and obtain S'.
5:         $S = S'$, $S_{i-good} = S_{i-good} \cup \{x_1, \cdots, x_i\}$
6:         $b_T(S, T) = b_T(S, T) - \{s_w s_{w+1}\}$, $X = X - \{x_1, \cdots, x_i\}$ }
7:     **end while**
8: **end for**
9: Set b'_i be the number of i-good strings currently.
10: b'_i be the number of i-good strings currently.
11: $S_{good} = S_{1-good} \cup S_{2-good} \cup S_{3-good}$
12: **for** (i from 1 to 4) **do**
13:     $b'_i = |S_{i-good}|$
14: **end for**
15: Set $D = b'_1 + b'_2/2 + b'_3/6$.
16: **while** there exists a good string α, whose deletion will bring at least one other good string β, and increase D **do**
17:     Delete α from S_{good}; add the new good strings α brings to S_{good}.
18: **end while**
19: Insert the remaining genes according to the Insert-Whole-String() method in [15].

---

4   Proof of the Approximation Factor

In our algorithm, we try to find good strings as many as possible. But a good string (say I_s) found by our algorithm may make other good strings in the optimal solution infeasible, we say I_s *destroys* them. The following lemma shows that the number of good strings that could be destroyed by a given good string is bounded.

Lemma 2. *An i-good string can destroy at most $i + 1$ good strings in some optimal solution.*

Proof. Assume that an i-good string I_s is inserted in between some breakpoint $s_j s_{j+1}$ in S. Then each of its genes, if was not occupied by I_s, can be part of a distinct good string respectively in some optimal solution. Also, there may exist another good string that could be inserted in between the breakpoint $s_j s_{j+1}$ in the optimal solution. Totally, at most $i+1$ good strings in the optimal solution could be destroyed by I_s. □

Corollary 1. *An i-good string in the optimal solution can be destroyed at most $i + 1$ times by good strings from other solutions.*

To prove the approximation factor, we only need to compare the solution of our algorithm with a fixed optimal solution. If a good string appears both in our solution and the optimal solution, it will not affect our analysis, so we only consider the distinct good strings between our solution and the optimal solution.

Let S_i' be the set of i-good strings found by our algorithm, and $|S_i'| = b_i'$, $i \in \{1, 2, 3, 4\}$. Let S_i be the set of i-good strings in the optimal solution, and $|S_i| = b_i$, $i \in \{1, 2, 3, 4\}$. To analyze the relationship between the number of good strings we found and good strings in the optimal solution, we construct an imaginary bipartite graph $G = (L, R, E)$. $L = \bigcup_{i=1}^{4} S_i'$, $R = \bigcup_{i=1}^{4} S_i$, and if a good string $\alpha \in L$ destroys a good string $\beta \in R$, there would be an edge between them in G. (See Fig-3.)

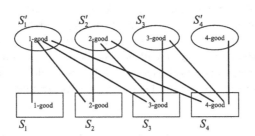

Fig. 3. Sketch of the imaginary graph

For a vertex $\alpha \in L \cup R$, let $d(\alpha)$ be the degree of α in G. We found that the graph G has the following properties.

Lemma 3. *In R, there is no isolated vertex.*

Proof. Since the *Scaffold Filling by Local Search* algorithm searches for good strings greedily first, any good string of length 1,2,3 and 4 in the optimal solution would other be chosen or be destroyed. If chose, it will not appear in R; If destroyed, it will appear in R, connecting to some good strings in L. The local search stage will not left a good string unchosen. □

Lemma 4. *For any vertex $c'_j \in S'_j \subseteq L$ ($j=1,2,3$), no vertex $c_i \in S_i$ with $i < j$ and $d(c_i) = 1$ can connect to c'_j, and at most one vertex $c_j \in S_j$ with $d(c_j) = 1$ can connect to c'_j.*

Proof. The local search stage aims at maximizing $D = b'_1 + b'_2/2 + b'_3/6$ by 1-substitution, so whenever there is a good string of length less than j whose degree is one, or more than one good strings of length j whose degree are one connecting to c'_j, a 1-substitution would be applied. □

Lemma 5. *For any vertex $c'_j \in S'_j \subseteq L$ ($j = 1, 2, 3$), at most $j + 1$ vertices in R, the degree of which is one and the length of which are greater than j, can connect to c'_j.*

Proof. From Lemma 4, at most one j-good strings with one degree in R can connect to a j-good string in L. The local search stage aims at maximizing $D = b'_1 + b'_2/2 + b'_3/6$ by 1-substitution, so if there is a j-good string in R, it would be impossible that a good string of length less than j whose degree is one in R connect to it. From Lemma 2, c'_j can destroy at most $j + 1$ good strings of the optimal solution. So the lemma follows. □

Lemma 6. *For any vertex $c_i \in S_i \subseteq R(i = 1, 2, 3, 4)$, c_i is connecting to at least one vertex $c'_j \in S'_j \subseteq L(j = 1, 2, 3, 4)$ with $j \leq i$.*

Proof. From the construction of the imaginary graph, we know c_i is connecting to at least one vertex $c'_j \in S'_j$. Assume to the contrary that there exists a vertex $c_i \in S_i$, which connects to some vertices ($\in S'_j$) with $j \leq i$. Then c_i is destroyed by some good strings of length greater than i. Since the algorithm *Scaffold Filling by Local Search* search for good strings greedily first, then c_i would other be chosen or be destroyed by at least one chosen good strings of length i or less. Once it was chosen, since it would destroy at most $i + 1$ good strings, from the function $D = b'_1 + b'_2/2 + b'_3/6$, it will not be replaced by some good strings of length greater than i, so the exclusive possibility to replace it is replacing it by a good strings of length i or less and some other good strings. So no matter how many times it was chosen and replaced, it will keep connecting to at least one good string of length i or less. □

Let X_i be the number of i-good strings in the optimal solution, and Y_i be the number of other i-good strings, i.e., $b_i = X_i + Y_i$. Let K_{ij} be the number of j-good stings whose degree are one in the optimal solution connecting to i-good stings in our solution. from Lemma 6, we know that $1 \leq i \leq j \leq 4$. Let H_{ij} be the number of edges between i-good stings in our solution and j-good stings whose degree are not one in the optimal solution, where $j > i$.

Lemma 7. $4b_1 + 3b_2 + 2b_3 + b_4 \leq 6b'_1 + 6b'_2 + 5b'_3 + 5b'_4$.

Proof. From Lemma 4,5, we have,

$$X_1 \leq b'_1 - (K_{12} + K_{13} + K_{14})/2 \tag{1}$$

Since other 1-good strings has degree at least 2, and from Lemma 6, all the degrees are contributed by the 1-good string in our solution, and each 1-good sting can contribute at most 2 degree. Thus we have,

$$2Y_1 + X_1 \leq 2b_1' - (K_{12} + K_{13} + K_{14} + H_{12} + H_{13} + H_{14}) \tag{2}$$

Similar argument holds for X_2, Y_2, X_3, Y_3, then we have,

$$X_2 - K_{12} \leq b_2' - (K_{23} + K_{24})/3 \tag{3}$$

$$2Y_2 + X_2 - K_{12} - H_{12} \leq 3b_2' - (K_{23} + K_{24} + H_{23} + H_{24}) \tag{4}$$

$$X_3 - K_{13} - K_{23} \leq b_3' - K_{34}/4 \tag{5}$$

$$2Y_3 + X_3 - K_{13} - K_{23} - H_{13} - H_{23} \leq 4b_3' - (K_{34} + H_{34}) \tag{6}$$

Use $\frac{(1)+(2)}{2}$, $\frac{(3)+(4)}{2}$, $\frac{(5)+(6)}{2}$, we have,

$$b_1 \leq 3b_1'/2 - 3(K_{12} + K_{13} + K_{14})/4 - (H_{12} + H_{13} + H_{14})/2 \tag{7}$$

$$b_2 \leq 2b_2' + K_{12} - 2(K_{23} + K_{24})/3 - (H_{23} + H_{24})/2 + H_{12}/2 \tag{8}$$

$$b_3 \leq 5b_2'/2 + K_{13} + K_{23} - 5K_{34}/8 - H_{34}/2 + (H_{13} + H_{23})/2 \tag{9}$$

As for the 4-good strings, since each 4-good string in our solution can connect to at most 5 4-good strings of the optimal solution, thus we have

$$b_4 \leq 5b_4' + (K_{14} + K_{24} + K_{34}) + (H_{14} + H_{24} + H_{34})/2 \tag{10}$$

Use $4 \times (7) + 3 \times (8) + 2 \times (9) + (10)$, we will have,

$$4b_1 + 3b_2 + 2b_3 + b_4 \leq 6b_1' + 6b_2' + 5b_3' + 5b_4' \tag{11}$$

Directly form Lemma 1 and Lemma 7, we have,

Theorem 2. *The algorithm* Scaffold Filling by Local Search *guarantees an approximation factor of 6/5.*

5   The Scaffold Filling Problem is MAX-SNP-complete

In reference [12], Jiang et al., has proved that the Scaffold Filling Problem is NP-hard. In this section, we show that this problem is MAX-SNP-complete, which makes a polynomial time approximation schemes nearly impossible. To complete this proof, we need to construct an L-reduction from a known MAX-SNP-complete problem (the Maximum 3-Dimensional Matching problem in this paper), to the One-sided Scaffold Filling Problem. Though the reduction method here is similar to that of [12], we conduct a stricter analysis, and produce a stronger negative result.

Definition 4. *Maximum 3-Dimensional Matching* [17].
Instance:*Set $T \subseteq X \times Y \times Z$, where X, Y, and Z are disjoint.*
Solution*: A matching for T, i.e., a subset $M \subseteq T$ such that no elements in M agree in any coordinate.*
Measure*: Cardinality of the matching, i.e., $|M|$.*

The variation in which the number of occurrences of any element in X, Y or Z is bounded by 3, is called the 3-bounded Maximum 3-Dimensional Matching problem, abbreviated as $3DM - 3$. The 3-bounded Maximum 3-Dimensional Matching problem is Shown to be MAX SNP-complete in [18].

We consider a more special variation of the 3-bounded Maximum 3-Dimensional Matching problem, in which each pair of triples in T can share at most one common element, we call this problem 3-bounded 1-common Maximum 3-Dimensional Matching, abbreviated as $(3DM - 3)^1$.

Theorem 3. $(3DM - 3)^1$ *is MAX SNP-complete.*

Proof. It suffices to show that the 3-bounded Maximum 3-Dimensional Matching problem can be L-reduced to $(3DM - 3)^1$. Construct an instance of $(3DM - 3)^1$ from an instance of 3-bounded Maximum 3-Dimensional Matching as follows. Given an instance I of 3-bounded Maximum 3-Dimensional Matching with $|X| = |Y| = |Z| = q$ and m triples ($3q \geq m$), for each triple $T_u = (x_i, y_j, z_k)$, construct the following 5 triples: $T'_u = \{(x_i, y_{u,1}, z_{u,2}), (x_{u,1}, y_{u,2}, z_k), (x_{u,2}, y_j, z_{u,1}), (x_{u,1}, y_{u,1}, z_{u,1}), (x_{u,2}, y_{u,2}, z_{u,2})\}$. Then, in the $(3DM - 3)^1$ instance I', we have $|X'| = |Y'| = |Z'| = q + 2m$ and $5m$ triples. It remains to show that this reduction satisfies the two inequalities of an L-reduction.

Firstly, if T_u was chosen as a triple of $OPT(I)$, we can choose the former three triples of T'_u into $OPT(I')$, note that this is the unique way to choose three disjoint triples from T'_u; and if not, we can choose the latter two triples of T'_u into $OPT(I')$. So, $|OPT(I')| = |OPT(I)| + 2m$. Since each triple of I can share common elements with at most 6 other triples, then $OPT(I) \geq m/7$. Thus, $|OPT(I')| \leq |OPT(I)| + 14|OPT(I)| = 15|OPT(I)|$.

For a solution $c(I')$ of I', we choose T_u into a corresponding solution $c(I)$ of I if and only if $c(I')$ contains three triples of T'_u. If $|c(I')| \leq 2m$, we have $|OPT(I)| - |c(I)| \leq |OPT(I)| - (|OPT(I')| - 2m) \leq |OPT(I')| - |c(I')|$. On the other hand, if $|c(I')| > 2m$, since there are only m groups of T'_us, there should be at least $|c(I')| - 2m$ groups of T'_us, from which we choose 3 triples, so $|c(I)| \geq |c(I')| - 2m$. Therefore, $|OPT(I)| - |c(I)| \leq (|OPT(I')| - 2m) - (|c(I')| - 2m) = |OPT(I')| - |c(I')|$. This completes the proof. $\square$

Now, we L-reduce $(3DM - 3)^1$ to the One-sided Scaffold Filling problem.

Theorem 4. *The One-sided Scaffold Filling problem is MAX SNP-complete.*

Proof. Suppose that we are given a $(3DM - 3)^1$ instance I with $|X| = |Y| = |Z| = q$ and m triples ($3q \geq m$). Assume that each element appears at least once in the triples,i.e., $q \leq m$. We construct an instance J of the One-sided Scaffold Filling problem as follows. Let the elements to be inserted are $X \cup Y \cup Z$. For each triple $T_u = (x_i, y_j, z_k)$, we construct two substrings $T'_u = g_u f_u x_i y_j z_k f'_u g'_u$ and $M'_u = g'_u f_u f'_u g_u$. For an element of X, say x_i, appearing $l \leq 3$ times in the triples, we denote each appearance by $x_i^1, x_i^2, \ldots, x_i^l$ respectively. Similar denotations are used for the elements of Y and Z. There are also $3m - 3q + 4$ new elements as separators. We denote the concatenation of strings by Π.

$$S = \Pi_{i=1}^q (x_i^1 r_x^1 x_i^2 r_x^2 y_i^1 r_y^1 y_i^2 r_y^2 z_i^1 r_z^1 z_i^2 r_z^2) R_1 R_2 \Pi_{i=1}^{\lceil m/2 \rceil} (M'_{2i-1}) R_3 R_4 \Pi_{i=1}^{\lfloor m/2 \rfloor} (M'_{2i})$$

$$T = \Pi_{i=1}^q (r_x^1 r_x^2 r_y^1 r_y^2 r_z^1 r_z^2) R_3 R_2 R_4 R_1 \Pi_{i=1}^m (T'_i)$$

Note that $x_i^1 = x_i^2 = x_i$, and if x_i appears only once, then $x_i^1, r_x^1, x_i^2, r_x^2$ are all empty strings; if x_i appears twice, then then x_i^2, r_x^2 are all empty strings. We can observe that there is no common adjacency between S and T, and S contains $10m - 6q + 4$ elements while T has $10m - 3q + 4$ elements. Next, we show that this reduction fulfills the two inequalities of an L-reduction.

Firstly, if T_u was chosen as a triple of $OPT(I)$, we can insert x_i, y_j, z_k into M'_u, which brings 4 common adjacencies. For these elements not covered by $OPT(I)$, Theorem 1 ensures that each insertion of an element will bring at least one common adjacencies. Since the common adjacencies of $OPT(J)$ must be the following four forms: $f'_u x_i, x_i y_j, y_j z_k, z_k f'_u$, since each pair of triples shares at most one element, then $x_i y_j, y_j z_k$ can be obtained at most once, the unique way to obtain such four common adjacencies, is to insert the three elements of T_u into M'_u.

Then, we prove that there is no good strings of length greater than 3 in $OPT(J)$. Observe that the unique to obtain one more common adjacency to to insert some elements into a M'_u. Assume to the contrary that $s_1 s_2 \cdots s_r$ ($r \geq 4$) be a r-good string by inserting it into M'_v, then $s_1 \in X \cap T_v, s_r \in Z \cap T_v$, which means that $s_2 \in Y \cap T_v$ and $s_{r-1} \in Y \cap T_v$, a contradiction. Therefore, $|OPT(J)| = |OPT(I)| + 3q \leq |OPT(I)| + 3m$. Since each triple of I can share common elements with at most 6 other triples, then $OPT(I) \geq m/7$. Thus, $|OPT(J)| \leq |OPT(I)| + 21|OPT(I)| = 22|OPT(I)|$.

For a solution $c(J)$ of J, we choose T_u into a corresponding solution $c(I)$ of I if and only if $c(J)$ contains four common adjacencies $f'_u x_i, x_i y_j, y_j z_k, z_k f'_u$. If $|c(J)| \leq 3q$, we have $|OPT(I)| - |c(I)| \leq |OPT(I)| = |OPT(J)| - 3q \leq |OPT(J)| - |c(J)|$. On the other hand, if $|c(J)| > 3q$, there should be at least $|c(J)| - 3q$ M'_us, from which we obtain 4 common adjacencies by insertions of the three elements of a triple, so $|c(I)| \geq |c(J)| - 3q$. Therefore, $|OPT(I)| - |c(I)| \leq (|OPT(J)| - 3q) - (|c(J)| - 3q) = |OPT(J)| - |c(J)|$. This completes the proof. $\square$

6   Conclusion

In this paper, we investigate the One-sided Scaffold Filling problem, we improve the approximation factor from 1.25 to 1.2 by a not-oblivious local search method.

Our algorithm is quite implementable since we only use 1-substitution. A meaningful future work is to analyze the performance by c-substitution for any constant c, We believe that, with c increasing, the approximation ratio could be improved slightly. but this would stop for some c.

We define a term called "good string", which help us describe the optimal solution very well, upon this, by intuition, it seems that this problem could admit a $PTAS$. But our MAX-SNP-complete proof prevents it.

Acknowledgments. This research is partially supported NSF of China under grant 61202014 and 61472222, by NSF of Shandong Provence (China) under grant ZR2012FQ008 and ZR2012FZ002, by China Postdoctoral Science Foundation funded project under grant 2011M501133 and 2012T50614.

References

1. Angibaud, S., Fertin, G., Rusu, I., Thevenin, A., Vialette, S.: On the approximability of comparing genomes with duplicates. J. Graph Algorithms and Applications **13**(1), 19–53 (2009)
2. Blin, G., Fertin, G., Sikora, F., Vialette, S.: The EXEMPLAR BREAKPOINT DISTANCE for non-trivial genomes cannot be approximated. In: Das, S., Uehara, R. (eds.) WALCOM 2009. LNCS, vol. 5431, pp. 357–368. Springer, Heidelberg (2009)
3. Cormode, G., Muthukrishnan, S.: The string edit distance matching problem with moves. In: Proc. 13th ACM-SIAM Symp. on Discrete Algorithms (SODA 2002), pp. 667–676 (2002)
4. Chen, Z., Fowler, R., Fu, B., Zhu, B.: On the inapproximability of the exemplar conserved interval distance problem of genomes. J. Combinatorial Optimization **15**(2), 201–221 (2008)
5. Khanna, S., Motwani, R., Madhu, S., Umesh, V.: On syntactic versus computational views of approximability. SIAM Journal on Computing **28**(1), 164–191 (1998)
6. Chen, Z., Fu, B., Xu, J., Yang, B., Zhao, Z., Zhu, B.: Non-breaking similarity of genomes with gene repetitions. In: Ma, B., Zhang, K. (eds.) CPM 2007. LNCS, vol. 4580, pp. 119–130. Springer, Heidelberg (2007)
7. Chen, Z., Fu, B., Zhu, B.: The approximability of the exemplar breakpoint distance problem. In: Cheng, S.-W., Poon, C.K. (eds.) AAIM 2006. LNCS, vol. 4041, pp. 291–302. Springer, Heidelberg (2006)
8. Goldstein, A., Kolman, P., Zheng, J.: Minimum common string partitioning problem: hardness and approximations. In: Fleischer, R., Trippen, G. (eds.) ISAAC 2004. LNCS, vol. 3341, pp. 484–495. Springer, Heidelberg (2004). also in: The Electronic Journal of Combinatorics **12** (2005), paper R50
9. Jiang, M.: The zero exemplar distance problem. In: Tannier, E. (ed.) RECOMB-CG 2010. LNCS, vol. 6398, pp. 74–82. Springer, Heidelberg (2010)
10. Jiang, H., Zheng, C., Sankoff, D., Zhu, B.: Scaffold filling under the breakpoint distance. In: Tannier, E. (ed.) RECOMB-CG 2010. LNCS, vol. 6398, pp. 83–92. Springer, Heidelberg (2010)
11. Jiang, H., Zhong, F., Zhu, B.: Filling scaffolds with gene repetitions: maximizing the number of adjacencies. In: Giancarlo, R., Manzini, G. (eds.) CPM 2011. LNCS, vol. 6661, pp. 55–64. Springer, Heidelberg (2011)

12. Jiang, H., Zheng, C., Sankoff, D., Zhu, B.: Scaffold filling under the breakpoint and related distances. IEEE/ACM Trans. Bioinformatics and Comput. Biology 9(4), 1220–1229 (2012)
13. Muñoz, A., Zheng, C., Zhu, Q., Albert, V., Rounsley, S., Sankoff, D.: Scaffold filling, contig fusion and gene order comparison. BMC Bioinformatics 11, 304 (2010)
14. Sankoff, D.: Genome rearrangement with gene families. Bioinformatics 15(11), 909–917 (1999)
15. Liu, N., Jiang, H., Zhu, D., Zhu, B.: An Improved Approximation Algorithm for Scaffold Filling to Maximize the Common Adjacencies. IEEE/ACM Trans. Comput. Biology Bioinform. 10(4), 905–913 (2013)
16. Liu, N., Zhu, D.: The algorithm for the two-sided scaffold filling problem. In: Chan, T.-H.H., Lau, L.C., Trevisan, L. (eds.) TAMC 2013. LNCS, vol. 7876, pp. 236–247. Springer, Heidelberg (2013)
17. Garey, M.R., Johnson, D.S.: Computers and Intractability: A Guide to the Theory of NP-Completeness. W.H. Freeman (1979)
18. Kann, V.: Maximum bounded 3-dimensional matching is MAX SNP-complete. Inform. Process. Lett. 37, 27–35 (1991)

Packing Cubes into a Cube in (*D*>3)-Dimensions

Yiping Lu[1(⊠)], Danny Z. Chen[2], and Jianzhong Cha[1]

[1] School of Mechanical, Electronic and Control Engineering,
Beijing Jiaotong University, Beijing 100044, China
{yplu,jzcha}@bjtu.edu.cn
[2] Department of Computer Science and Engineering,
University of Notre Dame, Notre Dame, IN 46556, USA
dchen@cse.nd.edu

Abstract. The problem of determining whether a set of cubes can be orthogonally packed into a cube has been studied in 2-diminson, 3-dimension, and (*d*>3)-dimensions. Open questions were asked on whether this problem is NP-complete in three articles in 1989, 2005, and 2009, respectively. In 1990, the problem of packing squares into a square was shown to be NP-complete by Leung et al. Recently, the problem of packing cubes into a cube in 3-D was shown to be NP-complete by Lu et al. In this paper, we show that the problem in (*d*>3)-dimensions is NP-complete in the strong sense, thus settling the related open question posed by previous researchers.

Keywords: Packing problems · Cube packing · NP-completeness

1 Introduction

2-D square packing and 3-D and (d>3)-D cube packing have been intensely studied (e.g., see [1, 2, 3, 4, 5, 7, 8, 9, 10, 11]). All known solutions to the cube packing problem are mainly based on approximation or heuristic methods. Determining the complexity of the cube packing problem is needed for developing effective algorithms for it. The common belief is that for all (d ≥2)-D, the cube packing problem is NP-complete. In fact, NP-complete proofs have been given for the problem in 2-D and 3-D. However, it should be pointed out that, unlike many other geometric problems, the known NP-complete proofs of the cube packing problem in 2-D and 3-D do not immediately translate into a satisfactory NP-complete proof for the version in (d>3)-D. Thus, proving the NP-completeness of the problem in (d>3)-D is still needed.

The NP-completeness of the problem of 'packing multiple rectangles into a rectangle' follows trivially because one of its special cases can be degenerated to the well-known knapsack problem [6].

In 1989, Li and Cheng [10] showed that both the problem of 'packing multiple squares into a rectangle' and the problem of 'packing multiple rectangles into a square' are NP-complete, but posted a question on the NP-completeness of the problem of 'packing squares into a square'.

Li and Cheng's question [10] was answered by Leung et al. [9] in 1990, who proved that the problem of 'packing squares into a square' is indeed NP-complete;

© Springer International Publishing Switzerland 2015
D. Xu et al. (Eds.): COCOON 2015, LNCS 9198, pp. 264–276, 2015.
DOI: 10.1007/978-3-319-21398-9_21

but, the NP-completeness of the problem version in (d>2)-D (i.e., the problem of packing multiple cubes into a cube in (d>2)-dimension) was still not settled.

In 2005 and 2009, Epstein and van Stee [5] and Harren [7], respectively, posed the determination of NP-completeness of this cube packing problem in (d>2)-D as an open problem.

The 3-D version of the problem was proved to be NP-complete recently [11].

In this paper, we show that the (d>3)-D version of the problem is strongly NP-complete, thus completely settling the open problem posed by Epstein and van Stee [5] and Harren [7]. Our proof is based on a unified construction scheme for the problem in all (d≥2)-dimensions, which actually simplifies the proofs for the 2-D and 3-D cases in [9, 11].

1.1   The Problem of Packing Multiple Cubes into a Cube

The problem of packing multiple cubes into a cube in (d>1)-D is that of orthogonally packing a set of small cubic items into a big cubic volume without any interior overlapping among the small cubic items. The problem is formally defined as follows.

THE PROBLEM OF PACKING CUBES INTO A CUBE

INSTANCE: A big cube C and a set of small cubes, $L = \{c_1, c_2, ..., c_n\}$, whose sizes are all positive integers.

QUESTION: Can all cubes of L be packed into C orthogonally without any intersection of the interior points between any two cubes of L?

1.2   The 3-Partition Problem

In this paper, we reduce the 3-partition problem to the problem of packing cubes into a cube. The 3-partition problem is known to be NP-complete in the strong sense [6], which is formally defined as follows.

THE 3-PARTITION PROBLEM

INSTANCE: A set of $3m$ ($m > 1$) positive integers, $S = \{a_1, a_2, ..., a_{3m}\}$, and a bound B, such that for every i, $1 \leq i \leq 3m$, $B/4 < a_i < B/2$, and

$$\sum_{i=1}^{3m} a_i = mB.$$

QUESTION: Can S be partitioned into m disjoint subsets $S_1, S_2, ..., S_m$, such that for each j with $1 \leq j \leq m$, the sum of all elements in S_j is exactly B? Note that the constraints of the problem require that each subset S_j should have exactly 3 elements of S.

Note that even if we assume that m is a multiple of 3, the 3-partition problem is still NP-complete in the strong sense [11]. In the rest of this paper, we assume that m is a multiple of 3, and write $m = 3m'$ when it is needed, where m' is a positive integer.

The rest of this paper is organized as follows. In Section 2, we review the related results. In Sections 3, we give new NP-completeness proofs of the cube packing problem in 2-D. This proof is much simplified in comparison with those in [9, 11], and thus it is much easier to be extended to the higher dimensional version. In Section 4, we show that the cube packing problem in $(d > 2)$-D is NP-complete in the strong sense. Section 5 concludes the paper.

2   Review of Related Work

2.1   Li and Cheng's Work on Packing Squares into a Rectangle

In 1989, Li and Cheng [10] proved that the problem of packing squares into a rectangle is NP-complete. Their proof is based on a reduction from the 3-partition problem. Given an instance I of the 3-partition problem, they constructed an instance I_p of packing squares into a rectangle in the following way (see Fig. 1):

$$C = (3A + B, m(A + B/2)) \quad L = \{(A + a_1), (A + a_2), ..., (A + a_{3m})\}$$

where $A > (m-1)B/4$, C is a rectangle of size $(3A + B) \times m(A + B/2)$, and L is a list of $3m$ squares of sizes $A + a_1$, $A + a_2$, ..., $A + a_{3m}$, respectively.

If the instance I has a partition, let $S_i = \{a_{i1}, a_{i2}, a_{i3}\}$, then it is clear that the instance I_p has a packing as shown in Fig. 1.

If the instance I_p has a packing, then at most m squares can be packed into C's vertical dimension of length $m(A + B/2)$. This is because if $m+1$ squares are packed into C's vertical dimension, then the total height of these squares will be bigger than $(m+1)(A + B/4) = m(A + B/2) + (A - (m-1)B/4) > m(A + B/2)$. For similar reason, at most 3 squares can be packed into C's horizontal dimension. Since the total number of squares is $3 \cdot m$ and no overlap among the packed squares is allowed, the only packing that is possible will have m levels of squares with 3 squares per level. Hence, we can obtain a partition that divides the $3m$ squares into m groups $G_i (i = 1, ..., m)$, where each group contains 3 squares.

Let $G_i = \{A + a_{i1}, A + a_{i2}, A + a_{i3}\}$. Then it must be true that

$$A + a_{i1} + A + a_{i2} + A + a_{i3} \leq 3A + B$$

i.e.,

$$a_{i1} + a_{i2} + a_{i3} \leq B$$

Since $\sum_{i=1}^{m}(a_{i1} + a_{i2} + a_{i3}) = mB$, it has to be true that $a_{i1} + a_{i2} + a_{i3} = B$ for every $i = 1, 2, ..., m$, which implies that the 3-partition problem instance I has a partition.

2.2   Leung et al.'s Work on Packing Squares into a Square

In 1990, Leung et al. [9] showed that packing multiple squares into a square is NP-complete. Their proof is also based on a reduction from the 3-partition problem.

Leung et al.'s construction is shown in Fig. 2. In the big square C, multiple small squares are packed. The area α in C is a rectangular area that is similar to Li and Cheng's construction [10] shown in Fig. 1 (the only difference is that the former is rotated by 90 degrees). To the left of the area α, some bigger squares are packed in the area β A very big square is packed into the square area γ that is below the area $\alpha \cup \beta$. Finally, in the area ε that is to the left of the area $\alpha \cup \beta \cup \gamma$ and in the area δ that is on top of the area $\alpha \cup \beta \cup \gamma \cup \varepsilon$, other squares are packed as tightly as possible.

Leung et al. showed that if there is a packing for the constructed instance, then the packing in the area α is just like the one constructed by Li and Cheng, and thus the instance of the 3-partition problem has a partition.

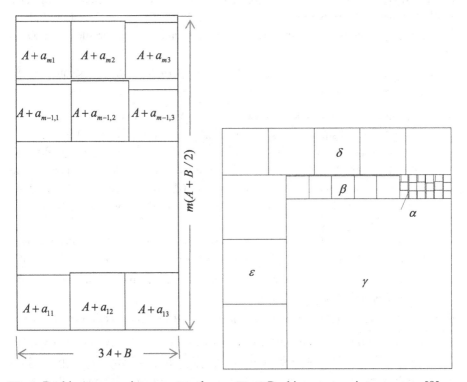

Fig. 1. Packing squares into a rectangle [10]

Fig. 2 Packing squares into a square [9]

2.3   Lu et al.'s Work on Packing Cubes into a Cube in 3-D

Recently, Lu et al. [11] showed that packing multiple cubes into a cube in 3-D is NP-complete. The proof of Lu et al. can be viewed as a 3-D extension of the proof by Leung et al. [9]. In their construction, Lu et al. [11] made some simplification of the proof in [9], and also used one big cubic item in the constructed packing instance, corresponding to the big square γ in Leung et al.'s construction (see Fig. 2).

In the remaining of this paper, we further simplify Leung et al.'s construction [9], in order to make it into the basis of the construction for our proof that can be easily extended to (d≥3)-dimensions. In this simplified construction, unlike the ones in [9, 11], we no longer use the "big" cubic item in the packing instance.

3   Packing Cubes into a Cube in 2-D

We first simplify Leung et al.'s construction for the 2-D case [9]. Given an instance I of the 3-partition problem, $B, S = \{a_1, a_2, ..., a_{3m}\}$, where $m = 3m'$ is a multiple of 3 and $m \geq 8$, we construct an instance I' of the square packing problem in 2-D, as follows.

Let $p = \lceil \sqrt{m/2} \rceil$, $A = 3p(p+1)B+1$, and $l = 3A + B$. Note that because $m \geq 8$, we have $p \geq 2$.

The big square C (the packing space) is of size $p(p+1)l \times p(p+1)l$.

There are 4 types of square items to be packed into C, as follows (see Fig. 3).

◇ $3m$ α-squares whose sizes are $A + a_i, i = 1, 2, ..., 3m$, respectively. Because of their sizes, if I has a partition, then the α-squares can be all packed into the rectangular area α whose size is $l \times m(A + B/2)$. Note that the packing of the α-squares into area α is the same as that constructed by Li and Cheng [10] shown in Fig. 1.

◇ $p(p+1) - m'$ β-squares whose sizes are all of $l_\beta = l - m'(B/2)/(p(p+1) - m')$. Because of their size, the β-squares can clearly be packed into the rectangular area β whose size is $l \times (p(p+1)l - m(A + B/2))$. Note that the size of a β-square is larger than $3A$, because

$$l_\beta = l - m'(B/2)/(p(p+1) - m') = 3A + B - m'(B/2)/(p(p+1) - m')$$
$$> 3A + B - m'(B/2)/(m/2 - m') = 3A + B - m'(B/2)/(3m'/2 - m') = 3A.$$

◇ $p+1$ γ-squares whose sizes are all of pl. All γ-squares can be packed into the rectangular area γ whose size is $pl \times p(p+1)l$, as shown in Fig. 3.

◇ $p(p-1)$ δ-squares whose sizes are all $(p+1)l$. All δ-squares can be packed into the rectangular area δ whose size is $(p-1)(p+1)l \times p(p+1)l$, as shown in Fig. 3.

Lemma 1. The packing instance I' can be constructed from I in polynomial time of m. If I has a partition, then I' has a packing.

Lemma 2. In the instance I', if all α-squares and β-squares can be packed into the rectangular area $\alpha\cup\beta$ (or any rectangular area of size $l\times p(p+1)l$ or $p(p+1)l\times l$), then I has a partition.

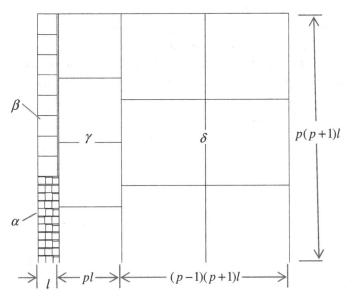

Fig. 3. Packing squares into a square

Proof: Without loss of generality, we assume that the packing area is $\alpha\cup\beta$ as shown in Fig. 3. To simplify the discussion, we imagine a situation in which every β-square is divided into 9 identical squares of size $l_\beta/3$ each. This will create $9p(p+1)-9m'=9p(p+1)-3m$ such squares whose (same) sizes are all larger than A since $l_\beta>3A$. Because each of the $3m$ α-squares is of a size larger than A, we now have totally $9p(p+1)$ squares whose sizes are all larger than A, and they will be packed into the area $\alpha\cup\beta$. We call all these squares the $\underline{A}$-*squares*. Note that if all α-squares and β-squares can be packed into the area $\alpha\cup\beta$, then clearly all the $9p(p+1)$ $\underline{A}$-squares can be packed into the area $\alpha\cup\beta$ as well.

Note that no more than $3p(p+1)$ $\underline{A}$-squares can be packed side by side along a vertical line inside $\alpha\cup\beta$. This is because if $3p(p+1)+1$ $\underline{A}$-squares are packed in this way, then the packing space needed in the vertical direction will be larger than the size of the area $\alpha\cup\beta$ in the vertical dimension, due to

$$(3p(p+1)+1)A=3p(p+1)A+A>3p(p+1)A+3p(p+1)B>p(p+1)l$$

Also, no more than 3 $\underline{A}$-squares can be packed side by side along a horizontal line inside $\alpha \cup \beta$ since $4A > 3A + B = l$. Thus, the number of $\underline{A}$-squares that are packed into the area $\alpha \cup \beta$ cannot be bigger than $3 \cdot 3p(p+1)$ which is exactly the total number of $\underline{A}$-squares. Hence, if a packing of $\alpha \cup \beta$ is achievable, then the $\underline{A}$-squares must be put approximately at the cell positions of a (roughly) uniform grid of size $3 \times 3p(p+1)$, with $3p(p+1)$ layers along the vertical direction and 3 cells along the horizontal direction. Subject to the space constraint in the horizontal direction, this packing means that the $\underline{A}$-squares are partitioned into $3p(p+1)$ sets each of which contains 3 $\underline{A}$-squares whose size sum is no bigger than l.

The $\underline{A}$-squares that are divided from one β-square should actually be packed together, meaning that all $\underline{A}$-squares divided from the β-squares are partitioned among themselves. Thus, the α-squares are partitioned into m sets each of which contains 3 α-squares whose size sum is no bigger than l. By referring to the discussions in Section 2.1, the instance I has a partition.   □

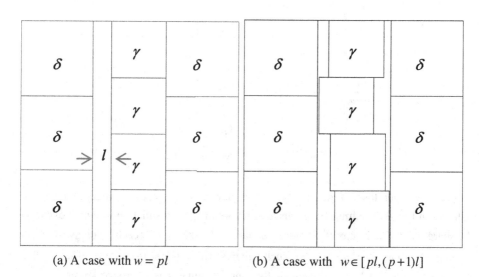

(a) A case with $w = pl$ (b) A case with $w \in [pl, (p+1)l]$

Fig. 4. Examples of packing the γ-squares

Lemma 3. In any achievable packing of the instance I', all γ-squares must be packed into a rectangular area of size $w \times p(p+1)l$ or $p(p+1)l \times w$, where $w \in [pl, (p+1)l]$ (see Fig. 4).

Proof: In this proof, we consider only the γ-squares and the δ-squares. In this situation, if a packing of I' is achievable, then clearly all $p+1$ γ-squares and all $p(p-1)$ δ-squares can be packed into C without any overlapping in their interior.

Note that the space of C can hold at most p^2 δ-squares. We call a packing that packs p^2 δ-squares into C a δ-packing. Suppose we first construct a δ-packing (a δ-packing is certainly achievable). Then at most p δ-squares can be removed from C to make room for packing the $p+1$ γ-squares (otherwise, the number of δ-squares packed into C will be smaller than $p(p-1)$).

Assume C is first packed tightly using p^2 δ-squares (i.e., a δ-packing), and there are still γ-squares to be packed into C in a certain manner. We now analyze how many δ-squares must be removed from C.

As shown in Fig. 5(a), suppose γ-squares $\gamma_1, \gamma_2, ...$ are to be packed into C which is already tightly packed by p^2 δ-squares. We want to determine the lower bound N_δ of how many δ-squares must be removed from C to avoid any interior overlap. We can shift each γ-square that is packed into C to the left (and to the bottom) until it touches the boundary of C or another γ-square without affecting the value of N_δ. Note that if a γ-square is shifted to the left (bottom), it will not cause more δ-squares to be removed and it will leave more space for the δ-square(s) to its right (top) such that they are less likely to be removed. For example, the N_δ value of Fig. 5(a) is no bigger than the N_δ value of Fig. 5(b). To determine N_δ, we can enumerate γ-squares row by row from left to right.

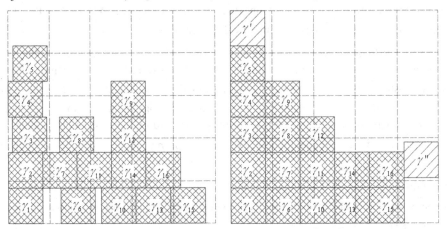

(a) γ-squares are packed in a δ-packing.   (b) γ-squares are shifted to low-left.

Fig. 5. Packing γ-squares into a δ-packing

More specifically, we focus on each row of γ-squares and see how many δ-squares should be removed. Let m_j be the number of γ-squares in the jth row, and n_j be the lower bound of the number of δ-squares that must be removed due to the packing of the jth row of γ-squares. Then, $m_j pl \leq n_j(p+1)l$ must hold (otherwise, there will not be enough space for m_j γ-squares in the jth row). When $j = 1$, we have

$$n_1 = \begin{cases} m_1 & , \text{if } m_1 < p+1 \\ m_1 -1, \text{if } m_1 = p+1 \end{cases}. \tag{1}$$

Note that because of the edge lengths of γ-squares and δ-squares, for any $j < p+1$, the jth row of γ-squares overlaps with the jth row of δ-squares. Thus Equ. (1) also holds for all $j < p+1$. But, for $j = p+1$, the jth row of γ-squares can use the space provided by δ-squares in the $(j-1)^{th}$ row; thus no more δ-square is needed to be removed due to the jth row of γ-squares.

Thus, we have

$$n_j = \begin{cases} m_j & , \text{if } m_j < p+1 \text{ and } j < p+1 \\ m_j -1, \text{if } m_j = p+1 \text{ and } j < p+1 \\ 0 & , \text{if } j = p+1 \end{cases}. \tag{2}$$

We can explain Equ. (2) in the following way: For packing one γ-square into a δ-packing, we generally need to remove one δ-square; but for every fully packed row or column, one δ-square can be saved. For example, in Fig. 5(b), if there is a γ'-square at the top of the first column, then it will not increase N_δ; neither the γ''-square located at the $(p+1)^{th}$ column will increase N_δ.

Now returning to our 2-D square packing problem, there are $p^2 - p$ δ-squares and $p+1$ γ-squares to be packed into C. This is equivalent to that we remove from C at most p δ-squares from a δ-packing for the packing of the $p+1$ γ-squares into C (in other words, we can save one δ-square). Note that this is possible only if we pack all these γ-squares into one row or one column, i.e., all $p+1$ γ-squares must be packed into a rectangular area of size $w \times p(p+1)l$ or $p(p+1)l \times w$, where $w \in [pl, (p+1)l]$.   □

Lemma 4. If there is a packing for the instance I', then all α-squares and β-squares can be packed into the rectangular area $\alpha \cup \beta$ (or any area of size $l \times p(p+1)l$ or $p(p+1)l \times l$).

Theorem 1. The problem of packing multiple squares into a square is NP-complete in the strong sense.

Proof: The problem of packing multiple squares into a square is in NP because a non-deterministic Turing machine can guess a solution and check whether the squares overlap in polynomial time of the square numbers, which is a polynomial of m. The 3-partition problem is NP-complete in the strong sense. Based on Lemmas 1, 2, 3, and 4, the theorem follows.   □

4   Packing Cubes into a Cube in $(d > 2)$-D

4.1   Notation

In $(d > 2)$-D, we use $(r_1, r_2, ..., r_d)$ to represent a boxy region (or boxy item) whose size in the ith dimension is r_i ($i = 1, 2, ..., d$). When the location of the region is significant, we use $(e_1, e_2, ..., e_d)(r_1, r_2, ..., r_d)$ to represent a boxy region defined in the ith dimension by the open interval $(e_i, e_i + r_i)$. We call $(e_1, e_2, ..., e_d)$ the origin of the region. We also use $r_1 \times r_2 \times ... \times r_d$ to represent the boxy region $(r_1, r_2, ..., r_d)$ when we focus only on its size.

For a point $X = \{x_1, x_2, ..., x_d\}$, we say X is inside a region $(e_1, e_2, ..., e_d)$ $(r_1, r_2, ..., r_d)$ if $x_i \in (e_i, e_i + r_i)$ for all $i = 1, 2, ..., d$.

A region $(e_1, e_2, ..., e_d)(r_1, r_2, ..., r_d)$ is defined as a set of (infinitely many) points in ($d > 3$)-D $\{X = \{x_1, x_2, ..., x_d\} \mid \forall x_i \in (e_i, e_i + r_i)\}$. We say that two regions overlap if and only if they share some common inside points. Region R_1 is inside region R_2 if and only if all inside points of R_1 are also inside points of R_2.

We say that the union of two boxy regions is also a boxy region if and only if they have different sizes in no more than one dimension. For example, the union of two boxy regions

$$\varepsilon_1 = (0, 0, ..., 0)(r_1, ..., r_{i-1}, r_i', r_{i+1}, ..., r_d) \qquad \text{and}$$

$$\varepsilon_2 = (0, ..., 0, r_i', 0, ..., 0)(r_1, ..., r_{i-1}, r_i'', r_{i+1}, ..., r_d)$$

is the boxy region $\varepsilon_1 \bigcup \varepsilon_2 = (0, 0, ..., 0)(r_1, ..., r_{i-1}, r_i' + r_i'', r_{i+1}, ..., r_d)$ because they have different sizes only in the ith dimension. In the rest of this paper, we only discuss unions of boxy regions if the result is also a boxy region.

For simplicity, we will continue to use r to represent a cube's size (i.e., its size in every dimension is r), and call the direction of the ith dimension the ith direction.

4.1.1   The Packing Instance in ($d > 2$)-D

Given an instance I of the 3-partition problem, $B, S = \{a_1, a_2, ..., a_{3m}\}$, where $m = 3m'$ is a multiple of 3 and $m \geq 8$, we construct an instance I''' of the cube packing problem in $(d > 2)$-D, as follows.

    I.   **Parameters**

        Let $p = \lceil \sqrt{m/2} \rceil, A = 3p(p+1)B + 1$, $l = 3A + B$, $h = A + B/2$, and $w = 3h$.

    II.   **The packing space**

        The packing space C is a cubic region of size $q = p(p+1)l$. Without loss of generality, let the origin of C be $(0, 0, ..., 0)$. Writing in the boxy region notation, $C = (0,0,...,0)(q,q,...,q)$. We construct C by a series of unions of the following boxy regions:

  ✧ The region $\alpha = (0,0,...,0)(l,mh,w,w,...,w)$, which is the union of the following regions:

$$a_1 = (0,0,0,...,0)(l,mh,h,...,h),$$
$$a_2 = (0,0,h,0,...,0)(l,mh,2h,h,...,h),$$
$$a_3 = (0,0,0,h,0,...,0)(l,mh,w,2h,h,...,h),$$
$$...,$$
$$a_{d-1} = (0,0,0,...,0,h)(l,mh,w,...,w,2h).$$

It can be verified that $((\alpha_1 \cup \alpha_2) \cup ...) \cup \alpha_{d-1} = \alpha$.

  ✧ The Region β which includes the following regions (not by the union operations):

$$\beta_0 = (0,mh,0,...,0)(l,q-mh,w,...,w,w),$$
$$\beta_1 = (0,0,w,0,...,0)(l,q,q-w,w,...,w),$$
$$\beta_2 = (0,0,0,w,0,...,0)(l,q,q,q-w,w,...,w),$$
$$...,$$
$$\beta_{d-2} = (0,0,0,0...,0,w)(l,q,q,q,...,q-w).$$

  ✧ The Region γ: $\gamma = (l,0,0,...,0)(pl,q,q,...,q)$.
  ✧ The Region δ: $\delta = (l+pl,0,0,...,0)((p-1)(p+1)l,q,q,...,q)$.

The union of the above regions is $C = (0,0,...,0)(q,q,...,q)$.

III.   The cubic item list L

The list L of packing cubes includes the following items.

  ✧ The α-cubes: $3^{d-1}m$ cubes, which include:
  ➤ The α_1-cubes: $3m$ cubes of sizes $A+a_i$, $i=1,2,...,3m$, respectively;
  ➤ The α_2-cubes: $3^{d-2}m-m$ cubes, all of size $A+B$;
  ➤ The α_3-cubes: $2 \cdot 3^{d-2}m-2m$ cubes, all of size A.
  ✧ The β-cubes, which consist of:
  ➤ The β_0-cubes: $p(p+1)-m'$ cubes, all of size $(q-mh)/(p(p+1)-m')$;
  ➤ The β_1-cubes: $p(p+1)(p(p+1)-1)$ cubes, all of size $l_\beta = (q-w)/(p(p+1)-1)$;
  ➤ The β_2-cubes: $p^2(p+1)^2(p(p+1)-1)$ cubes, all of size l_β;
  ...   ...
  ➤ The β_{d-2}-cubes: $p^{d-2}(p+1)^{d-2}(p(p+1)-1)$ cubes, all of size l_β.
  ✧ The γ-cubes: $(p+1)^{d-1}$ cubes, all of size pl.
  ✧ The δ-cubes: $p^d - p^{d-1}$ cubes, all of size $(p+1)l$.

Lemma 5. The packing instance I''' can be constructed in polynomial time of m, and if I has a partition, then I''' has a packing.

Lemma 6. In the packing instance I''', if all α-cubes and all β-cubes can be packed into the region $\alpha \cup \beta$ (or any region of size $(l,q,...,q)$, $(q,l,q,...,q)$, ..., or $(q,...,q,l)$), then the instance I has a partition.

Lemma 7. In any realizable packing of the instance I''', all γ-cubes must be packed into a boxy region of size $w \times p(p+1)l \times ... \times p(p+1)l$, $p(p+1)l \times w \times p(p+1)l \times ... \times p(p+1)l$,..., or $p(p+1)l \times ... \times p(p+1)l \times w$, where $w \in [pl,(p+1)l]$.

Lemma 8. If the instance I''' has an achievable packing, then all α-cubes and all β-cubes can be packed into the region $\alpha \cup \beta$ (or any region of size $l \times p(p+1)l \times ... \times p(p+1)l$, $p(p+1)l \times l \times p(p+1)l \times ... \times p(p+1)l$, ..., or $p(p+1)l \times ... \times p(p+1)l \times l$).

Theorem 2. The problem of packing multiple cubes into a cube in $(d>3)$-D is NP-complete in the strong sense.

5   Conclusions

We have proved that the problem of packing cubes into a cube in $(d>3)$-D is NP-complete in the strong sense. Together with the 2-D and 3-D results in [9, 11], this completely settles the open question posted in [5, 7].

Acknowledgement. The research of D. Z. Chen was supported in part by NSF under Grant CCF-1217906.

References

1. Bansal, N., Correa, J.R., Kenyon, C., Sviridenko, M.: Bin Packing in Multiple Dimensions: Inapproximability Results and Approximation Schemes. Mathematics of Operations Research **31**(1), 31–49 (2006)
2. Caprara, A., Lodi, A., Monaci, M.: Fast Approximation Schemes for Two-stage, Two-dimensional Bin Packing. Mathematics of Operations Research **30**, 136–156 (2005)
3. Chung, F.R.K., Garey, M.R., Johnson, D.S.: On Packing Two-dimensional Bins. SIAM Journal on Algebraic and Discrete Methods **3**, 66–76 (1982)
4. Correa, J.R., Kenyon, C.: Approximation schemes for multidimensional packing. In: Proc. 15th ACM–SIAM Symposium on Discrete Algorithms, pp. 179–188 (2004)
5. Epstein, L., van Stee, R.: Online Square and Cube Packing. Acta Informatica **41**(9), 595–606 (2005)
6. Garey, M., Johnson, D.: Computer and Intractability – A Guide to the Theory of NP-Completeness. Freeman, New York (1979)

7. Harren, R.: Approximation Algorithms for Orthogonal Packing Problems for Hypercubes. Theoretical Computer Science **410**(44), 4504–4532 (2009)
8. Kohayakawa, Y., Miyazawa, F.K., Raghavan, P., Wakabayashi, Y.: Multidimensional Cube Packing. Algorithmica **40**, 173–187 (2004)
9. Leung, J.Y.-T., Tam, W.T., Wong, C.S., Chin, F.Y.L.: Packing Squares into a Square. Journal of Parallel and Distributed Computing **10**, 271–275 (1990)
10. Li, K., Cheng, K.H.: Complexity of resource allocation and job scheduling problems in partitionable mesh connected systems. In: Proc. of 1st Annual IEEE Symposium of Parallel and Distributed Processing, pp. 358–365. Silver Spring, MD (1989)
11. Lu, Y., Chen, D.Z., Cha, J.: Packing Cubes into a Cube is NP-complete in the Strong Sense. Journal of Combinatorial Optimization **29**(1), 197–215 (2015)
12. Miyazawa, F.K., Wakabayashi, Y.: Cube Packing. Theoretical Computer Science **297**, 355–366 (2003)

Towards Flexible Demands in Online Leasing Problems

Shouwei Li, Alexander Mäcker, Christine Markarian[(✉)],
Friedhelm Meyer auf der Heide, and Sören Riechers

Heinz Nixdorf Institute and Computer Science Department,
University of Paderborn, Fürstenallee 11, 33102 Paderborn, Germany
{shouwei.li,alexander.maecker,christine.markarian,
fmadh,soeren.riechers}@uni-paderborn.de

Abstract. We consider online leasing problems in which demands arrive over time and need to be served by leasing resources. We introduce a new model for these problems such that a resource can be leased for K different durations each incurring a different cost (longer leases cost less per time unit). Each demand i can be served anytime between its arrival a_i and its deadline $a_i + d_i$ by a leased resource. The objective is to meet all deadlines while minimizing the total leasing costs. This model is a natural generalization of Meyerson's PARKINGPERMITPROBLEM (FOCS 2005) in which $d_i = 0$ for all i. We propose an online algorithm that is $\Theta(K + \frac{d_{max}}{l_{min}})$-competitive where d_{max} and l_{min} denote the largest d_i and the shortest available lease length, respectively. We also extend the SETCOVERLEASING problem by deadlines and give a competitive online algorithm which also improves on existing solutions for the original SETCOVERLEASING problem.

Keywords: Online algorithms · Leasing · Infrastructure problems · Parking permit problem · Deadlines

1 Introduction

Typical infrastructure problems consider scenarios where one has to buy certain resources (e.g., facilities, network nodes, or network connections) in order to generate or improve a given infrastructure (e.g., a supply network). Classical examples of these problems are FACILITYLOCATION, SETCOVER, and STEINERTREE. These problems have a large number of applications in networks, business, logistics and planning. From a theoretical point of view, they are not only widely studied in the offline, but also in the online setting, where the decision of when and which resources to buy must be taken as soon as demands are revealed without knowledge about future demands [2–4]. A common assumption

This work was partially supported by the German Research Foundation (DFG) within the Collaborative Research Center 'On-The-Fly Computing' (SFB 901) and the International Graduate School 'Dynamic Intelligent Systems'.

© Springer International Publishing Switzerland 2015
D. Xu et al. (Eds.): COCOON 2015, LNCS 9198, pp. 277–288, 2015.
DOI: 10.1007/978-3-319-21398-9_22

in related literature is that a resource that is bought once can be used forever without additional costs. This, however, is not true in many real-world scenarios. Instead, the leasing concept is gaining more and more importance in practice. Consider, for example, cloud vendors providing resources to clients. Using leasing, clients are being provided with cheaper, more flexible resources and cloud providers are increasing their profits by reducing their upfront and administrative costs [5]. In 2005, Meyerson [1] introduced the first theoretical LEASING MODEL with the PARKINGPERMITPROBLEM. Here, each day, we want to use the car if it is raining, whereas we walk if it is sunny. If we take the car, we must have a valid parking permit, and there are K different types of parking permits (leases), each with its own duration and cost. The goal is to buy a set of parking permits in order to cover all rainy days and minimize the total cost of purchases (without using weather forecasts). Similar models were later studied for infrastructure problems including FACILITYLOCATION, SETCOVER, and STEINERTREE [1,6–9].

We introduce a new model where, in contrast to related work, demands do not have to be served immediately. As a natural extension, demands can be postponed up to some fixed period of time resulting in a deadline for each demand. For instance, consider a travel agency that offers guided tours to tourists. Each day, new tourists may arrive with deadlines that represent the day their vacation ends. Tourists arriving want to attend the guided tour before leaving. Now, the travel agency pays for each time a guide (tour) is needed. To optimize its profit, the agency must make wise decisions regarding when to hire a guide and for how long such that the longer (more consecutive days) a guide is hired, the lower the costs per day will be. We will represent this by different lease types representing different lease lengths. The decision of buying a lease cannot be modified during the process, i. e. we cannot tell the guide to stay shorter or longer after having hired her.

Another application can be found in the cloud, where cloud clients, who are flexible regarding when to use the resources (e.g., any day within two weeks will do), will be happy to be offered better resource prices for a later day by cloud vendors.

Our Contribution.   In the light of capturing such scenarios, we extend the line of leasing by introducing deadlines. We propose the ONLINELEASINGWITHDEADLINES problem (OLD) for which we give a deterministic (optimal) algorithm. OLD captures the scenario introduced above, where a travel agency has to provide guided tours for tourists arriving and leaving their vacation at different times. We give an $\mathcal{O}(K + \frac{d_{max}}{l_{min}})$-competitive online algorithm for the OLD problem, where K is the number of leases, d_{max} the longest client length (i. e., the longest vacation of a tourist) and l_{min} the shortest lease length (i. e., the minimum length a guide can be hired for).

In the second part of the paper, we also extend the SETCOVERLEASING model [6,9], and introduce the SETCOVERLEASINGWITHDEADLINES problem (SCLD) which we solve by extending techniques we develop for OLD. Here, we give an $\mathcal{O}(\log(m \cdot (K + \frac{d_{max}}{l_{min}})) \log l_{max})$-competitive online algorithm. Our results also imply an improvement for the SETCOVERLEASING problem by removing the time dependency from the competitive factor.

Organization of the Paper. Section 2 presents the state of the art and gives some preliminaries. Section 3 introduces the new model and gives a deterministic algorithm. Section 4 introduces SCLD and presents a randomized algorithm for the latter. Section 5 concludes the paper with suggestions for future work.

2 Related Work and Preliminaries

In this section, we give an overview of leasing problems and present some preliminaries.

Related Work. Meyerson [1] gave a deterministic $\mathcal{O}(K)$-competitive and a randomized $\mathcal{O}(\log K)$-competitive algorithm along with matching lower bounds for the PARKINGPERMITPROBLEM. Inspired by Meyerson's work, Anthony and Gupta [6] generalized his idea by introducing the first leasing variants of FACILITYLOCATION, SETCOVER, and STEINERTREE: FACILITYLEASING, SETCOVERLEASING, and STEINERTREELEASING, respectively. They gave offline algorithms for these problems, resulting from an interesting relationship between infrastructure leasing problems and stochastic optimization. They achieved $\mathcal{O}(K)$, $\mathcal{O}(\log n)$, and $\mathcal{O}(\min(K, \log n))$-approximations for the offline variants of metric FACILITYLEASING, SETCOVERLEASING, and STEINERTREELEASING, respectively. Nagarajan and Williamson [7] later improved the $\mathcal{O}(K)$-approximation to a 3-approximation and gave an $\mathcal{O}(K \log n)$-competitive algorithm for the online variant of metric FACILITYLEASING. Kling et al. [8] improved the results by Nagarajan and Williamson [7] for FACILITYLEASING by removing the dependency on n (and thereby on time), where n is the number of clients. They gave an $\mathcal{O}(l_{max} \log(l_{max}))$-competitive algorithm where l_{max} is the maximum lease length. Abshoff et al. [9] gave the first online algorithm for SETCOVERLEASING and improved previous results for online non-leasing variants of SETCOVER.

Preliminaries. Meyerson [1] showed that it is sufficient to have a simplified LEASING MODEL in which (1) lease lengths l_k are powers of two and (2) leases of the same type do not overlap. Instances/solutions having these properties are said to obey the *interval model*. Similar properties have been used in related settings [1,8] and we detail these insights for completeness' sake. The following lemma states the effect of this model on the approximation guarantee.

Lemma 1. *Any c-competitive algorithm for the interval model can be transformed to a 4c-competitive algorithm for the general leases model.*

Proof. Consider a leasing problem instance I for general leases and construct a new instance I' by rounding each lease length $l_k \in \mathbb{N}$ to the next larger power of two. That is, the lease lengths of I' are $l'_k := 2^{\lceil \log l_k \rceil}$. Let S' denote the solution constructed by the c-approximation algorithm for the interval model when given I'. From S', we construct a solution S for I as follows: for each lease of type k bought at time t in solution S', buy two consecutive leases of type k at times t and $t + l_k$. Since $l_k + l_k \geq l'_k$, any lease pair in S covers at least all the demands covered by the original lease in S'. Moreover, we have $\mathrm{cost}(S) = 2\,\mathrm{cost}(S') \leq 2c \cdot \mathrm{cost}(\textsc{Opt}')$, where $\textsc{Opt}'$ denotes an optimal solution for I' in the interval model. Now, note that an optimal solution $\textsc{Opt}$ for I in the general model yields a solution $\tilde{S}$ for I' in the interval model as follows: for each lease of type k bought at time t in solution $\textsc{Opt}$, buy two leases of type k at times $\lfloor t/l'_k \rfloor \cdot l'_k$ and $\lceil t/l'_k \rceil \cdot l'_k$. These leases cover at least all the demands of the original lease and obey the interval model. Thus, we get $\mathrm{cost}(\textsc{Opt}') \leq \mathrm{cost}(\tilde{S}) = 2\,\mathrm{cost}(\textsc{Opt})$. The lemma's statement follows by combining both inequalities.

3   Online Leasing with Deadlines

In this section, we introduce the ONLINELEASINGWITHDEADLINES problem (OLD) and give a deterministic primal-dual algorithm.

3.1   Problem Definition

On each day t, a number of clients with deadlines $t + d_i$ (we say a client with *interval* $[t, t + d_i]$, where each day corresponds to a distance of 1 in the interval) arrives. There are K different types of leases, each with its own duration and cost (longer leases tend to cost less per day). A client arriving on day t with deadline $t + d$ is *served* if there is a lease which covers at least one day of its interval. This also implies that we can replace all clients arriving on a day t by only the client with the lowest deadline that arrives on that day. Thus, throughout this chapter, we will assume without loss of generality that on every day t, either (i) no client or (ii) only one client with deadline $t + d$ arrives. The goal is to buy a set of leases such that all arriving clients are served while minimizing the total cost of purchases.

A lease of type k has cost c_k and length l_k. l_{min} and l_{max} denote the shortest and the longest lease length, respectively. We denote by d_{max} and d_{min} the longest and the shortest interval length of the clients, respectively. An online algorithm now does not only need to serve clients while minimizing cost, but also needs to decide when to serve a client. Since resources expire after some time, decisions regarding when to serve a client are critical. An online algorithm may decide to serve a client on some day just to realize later on that postponing it would have been a better choice because a later lease could have served more clients. Or, the opposite is true, where an online algorithm may decide to postpone serving a client whereas serving it earlier by enlarging a lease that has been bought would have cost less.

We formulate OLD using integer linear programming (ILP) (see Figure 1). We refer to a type k lease starting at time t as (k, t), a client arriving at time

t with deadline $t + d$ as (t, d), and an interval $[a, a + b]$ as I_a^b. The collection of all leases is L and the collection of all clients is D. We denote by L_t all leases covering day t. We say a lease $(k, t') \in L$ is a *candidate* to client $(t, d) \in D$ if $I_t^d \cap I_{t'}^{l_k} \neq \emptyset$. The sum in the objective function represents the costs of buying the leases. The indicator variable $X_{(k,t)}$ tells us whether lease (k, t) is bought or not. The primal constraints guarantee that each client $(t, d) \in D$ is served. A dual variable $Y(t, d)$ is assigned to each client (t, d).

$$\min \sum_{(k,t) \in L} X_{(k,t)} \cdot c_k$$

$$\text{Subject to: } \forall (t, d) \in D : \sum_{(k,t') \in L, I_t^d \cap I_{t'}^{l_k} \neq \emptyset} X_{(k,t')} \geq 1$$

$$\forall (k, t) \in L : X_{(k,t)} \in \{0, 1\}$$

$$\max \sum_{(t,d) \in D} Y_{(t,d)}$$

$$\text{Subject to: } \forall (k, t) \in L : \sum_{(t',d) \in D, I_{t'}^d \cap I_t^{l_k} \neq \emptyset} Y_{(t',d)} \leq c_k$$

$$\forall (t, d) \in D : Y_{(t,d)} \geq 0$$

Fig 1. ILP Formulation of OLD

3.2 Deterministic Algorithm

In this section, we present a deterministic primal-dual algorithm for OLD, for which we give the analysis in the following section.

We adopt the *interval model* (Lemma 1) in which leases of type k are available only at times t that are a multiple of the corresponding lease length l_k. Thus, any day t can be covered by exactly K different leases. Therefore, when a client $(t, d) \in D$ arrives, our algorithm needs to decide on which day $t' \in [t, t + d]$ to serve it and it needs to specify one of the K leases in $L_{t'}$. For every lease $(k, t) \in L$, we define its *contribution* to be the sum of the values of the dual variables corresponding to clients having (k, t) as a candidate. We say (t', d) *contributes* to (k, t) if (k, t) is a candidate of (t', d) and $Y_{(t',d)} > 0$. Two clients (t', d') and (t, d) with $t' < t$ *intersect* if their corresponding intervals $I_{t'}^{d'}$ and I_t^d intersect at $t' + d'$.

Algorithm. When a client (t, d) arrives, if it does not intersect any client (t', d') with a non-zero dual variable where $t' < t$, we perform the following two steps.

Step 1: We increase the dual variable $Y_{(t,d)}$ of the client until the constraint of some candidate (k, t') becomes tight, i.e.,

$$\sum_{(t,d)\in D:I_t^d\cap I_{t'}^{l,k}\neq\emptyset} Y_{(t,d)} = c_k.$$

We then buy all the leases in L_t with a tight constraint (we set their primal variable to 1). At this point, the following proposition holds.

Proposition 1. *There exists at least one lease with a tight constraint that covers t.*

Proof. Assume, for contradiction, that there is no lease with a tight constraint in L_t. Then according to the algorithm, there must be a lease with a tight constraint in L_j, $j \in [t+1, t+d]$ (we do not stop increasing the client's dual variable until some constraint becomes tight). Moreover, before (t, d) arrives, the contribution to every lease in L_t is at least the contribution to its corresponding lease in L_j, $j \in [t+1, t+d]$. To show that the latter is true, assume, for contradiction, that there is a lease (k, t') in L_j, $j \in [t+1, t+d]$, with a contribution greater than that of its corresponding lease (k, t'') in L_t. Then, there must be a client which has contributed to (k, t') and not to (k, t'') (a client contributes the same to all its candidates). This is only possible if this client has arrived after (t, d), which is a contradiction. Hence, if the constraint of some lease in L_j, $j \in [t+1, t+d]$, becomes tight when (t, d) arrives, then the constraint of its corresponding lease in L_t must become tight as well (at any day, there are exactly K lease types). □

Step 2: By the proposition above, we have that the algorithm buys at least one lease in L_t. Even though the client is now served, we do one more step. We buy the lease(s) from L_{t+d} which correspond(s) to what is bought in Step 1 from L_t (we set the primal variable(s) to 1).

3.3 Analysis

We show that the primal-dual algorithm above is $\mathcal{O}(K)$-competitive for *uniform* OLD and $\mathcal{O}(K+\frac{d_{max}}{l_{min}})$-competitive for *non-uniform* OLD. We also show that the analysis of our algorithm is tight. This also implies an $\mathcal{O}(K)$-competitive factor for the PARKINGPERMITPROBLEM (we just set d_{max} to 0) which coincides with the tight result given by Meyerson [1].

Proposition 2. *Both the primal and the dual solutions constructed by the algorithm are feasible.*

Proof. It is easy to see that the dual constraints are never violated since the algorithm stops increasing the dual variables as soon as some constraint becomes tight. As for the primal solution, we show that each client $(t, d) \in D$ is served. When a client (t, d) arrives, we have two possibilities: either (t, d) intersects a previous client or it does not. If it does not, then our algorithm makes sure it is served in Step 1. Otherwise if it intersects a previous client (t', d) with $Y_{(t',d)}$ being zero, our algorithm makes sure it serves (t, d) in Step 1. If $Y_{(t',d)}$ is greater than zero, then our algorithm already covered days t' and $t' + d$ to serve (t', d). Since (t, d) and (t', d) intersect at $t' + d$, (t, d) is therefore served as well.   □

Theorem 1. *The primal-dual algorithm achieves an optimal $\mathcal{O}(K)$- and an $\mathcal{O}(K + \frac{d_{max}}{l_{min}})$-competitive ratio for uniform and non-uniform OLD respectively.*

Proof. Let $P \subseteq L$ denote the primal solution constructed by the algorithm. Because the dual constraint is tight for each $(k, t) \in P$, we have

$$c_k = \sum_{(t',d)\in D: I_{t'}^d \cap I_t^{lk} \neq \emptyset} Y_{(t',d)}.$$

Hence,

$$\sum_{(k,t)\in P} c_k = \sum_{(k,t)\in P} \sum_{(t',d)\in D: I_{t'}^d \cap I_t^{lk} \neq \emptyset} Y_{(t',d)} = \sum_{(t',d)\in D} Y_{(t',d)} \sum_{(k,t)\in P: I_{t'}^d \cap I_t^{lk} \neq \emptyset} 1.$$

Whenever the algorithm buys leases to serve $(t', d) \in D$, it only buys candidates from $L_{t'}$ (Step 1) and $L_{t'+d}$ (Step 2). Since there are exactly K leases at any day, it therefore buys at most $2K$ candidates. If the algorithm does not buy any further candidates of (t', d), we get an $\mathcal{O}(K)$-competitive ratio by weak duality theorem (both primal and dual solutions are feasible) since

$$\sum_{(k,t)\in P: I_{t'}^d \cap I_t^{lk} \neq \emptyset} 1 \leq 2K.$$

This will be the case for *uniform* OLD since any client sharing common candidates with (t', d) intersects (t', d) at $t' + d$ thus being served at $t' + d$ and the algorithm does not buy any further candidates of (t', d). As for *non-uniform* OLD, the algorithm may buy more of (t', d)'s candidates when new clients sharing common candidates with (t', d) arrive in the coming days. We upper bound the total number of these candidates as follows.

$$\sum_{(k,t)\in P: I_{t'}^d \cap I_t^{lk} \neq \emptyset} 1 \leq \sum_{i=t'}^{t'+d} |L_i| \leq \sum_{j=1}^{K} \left\lceil \frac{d_{max}}{l_j} \right\rceil.$$

By Lemma 1 we have that l_j's are increasing and powers of two. Hence, the right sum above can be bounded by the sum of a geometric series with ratio half.

$$\sum_{j=1}^{K} \left\lceil \frac{d_{max}}{l_j} \right\rceil \leq K + d_{max} \left\lceil \frac{1}{l_1} \left(\frac{1 - (1/2)^K}{1 - 1/2} \right) \right\rceil = K + d_{max} \left\lceil \frac{2}{l_1} (1 - (1/2)^K) \right\rceil.$$

Since $K \geq 1$ we have

$$K + d_{max} \left[\frac{2}{l_1} \left(1 - (1/2)^K \right) \right] \leq K + \frac{d_{max}}{l_1}.$$

Therefore,

$$\sum_{(k,t) \in P: I_{t'}^d \cap I_t^{l_k} \neq \emptyset} 1 \leq K + \frac{d_{max}}{l_{min}}$$

since the algorithm can not buy more than $K + \frac{d_{max}}{l_{min}}$ candidates. □

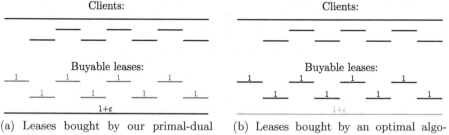

(a) Leases bought by our primal-dual algorithm are marked in red.

(b) Leases bought by an optimal algorithm are marked in green.

Fig. 2. Comparison of our primal-dual and an optimal algorithm for a specific instance of our problem. It is easy to see that the primal-dual algorithm pays almost d_{max}/l_{min} times what the optimal algorithm would pay.

Proposition 3. *The analysis of the aforementioned algorithm is tight.*

Proof. A lower bound of $\Omega(K)$ follows immediately from the lower bound of $\Omega(K)$ for the PARKINGPERMITPROBLEM by setting $d_{max} = 0$. We now give a tight example for $\Omega(d_{max}/l_{min})$ for the non-uniform case. Let d_{max} and l_{min} be arbitrary. For our problem instance, we start with a client $(0, d_{max})$ and add clients $((i-1) \cdot l_{min}, i \cdot l_{min})$ for $i \in \{2, \ldots, \lfloor d_{max}/l_{min} \rfloor\}$. Similarly, we add 2 different lease types, one with length l_{min} and cost 1, and one with length $2^{\lceil \log_2(d_{max}) \rceil}$ and cost $1 + \varepsilon$. See Figure 2 for a visualization. Now, in order to cover client $(0, d_{max})$, the dual variable of this client is increased until

$$\sum_{(t,d) \in D: I_t^d \cap I_{t'}^{l_k} \neq \emptyset} Y_{(t,d)} = c_k, \tag{1}$$

and this happens at the same time for all leases of length l_{min} in the interval $I_0^{d_{max}}$. The algorithm then only buys the leases at the start and at the end point. However, to cover clients $((i-1) \cdot l_{min}, i \cdot l_{min})$ for $i \in \{2, \ldots, \lfloor d_{max}/l_{min} \rfloor\}$, the algorithm buys all the short leases, as constraint (1) is already tight from the prior step. This leads to an overall cost of at least $\lfloor d_{max}/l_{min} \rfloor$, whereas the optimal algorithm only buys the long lease with cost $1 + \varepsilon$. □

4   Application to Set Cover Leasing

In this section, we introduce the SETCOVERLEASINGWITHDEADLINES problem
(SCLD) and give an $\mathcal{O}(\log(m \cdot (K + \frac{d_{max}}{l_{min}})) \log l_{max})$-competitive algorithm.

4.1   Problem Definition

SCLD is a generalization of SETCOVERLEASING in which elements arrive over
time and must be covered by sets from a family of subsets of these elements.
Each set can be leased for K different periods of time. Leasing a set S for a
period k incurs a cost c_S^k and allows S to cover its elements for the next l_k time
steps. The objective is to minimize the total cost of the sets leased, such that
elements arriving at any time t are covered by sets which contain them and are
leased during time t. SCLD extends SETCOVERLEASING by allowing elements to
have deadlines and be covered any time before their deadline. We define SCLD
analogously to *non-uniform* OLD and formulate it using ILP (see Figure 3).

We denote by p the maximum number of sets an element belongs to, by n
the number of elements, and by m the number of sets. We refer to a set S with
lease type k starting on day t as (S, k, t) and an element e arriving on day t with
deadline $t + d$ as (e, t, d). The collection of all set triples is F and the collection
of all element triples is U. We say $(S, k, t') \in F$ is a *candidate* to $(e, t, d) \in U$ if
$e \in S$ and $I_t^d \cap I_{t'}^{l_k} \neq \emptyset$. The sum in the objective function represents the costs
of buying the sets. The indicator variable $X_{(S,k,t)}$ tells us whether (S, k, t) is
bought or not. An element is *covered* if at least one of its candidates is bought.
The primal constraints guarantee that each $(e, t, d) \in U$ is covered.

$$\min \sum_{(S,k,t) \in F} X_{(S,k,t)} \cdot c_S^k$$

$$\text{Subject to: } \forall (e, t, d) \in U : \sum_{(S,k,t') \in F, I_t^d \cap I_{t'}^{l_k} \neq \emptyset, e \in S} X_{(S,k,t')} \geq 1$$

$$\forall (S, k, t) \in F : X_{(S,k,t)} \in \{0, 1\}$$

Fig 3. ILP of SCLD

4.2   Randomized Algorithm

In this section, we present a randomized algorithm for SCLD. We denote by
$F_{(e,t,d)}$ the collection of all candidates of (e, t, d). Our algorithm first solves the
LP of SCLD and then rounds it to solve its ILP. The algorithm maintains for
each set $(S, k, t) \in U$, $2\lceil \log(l_{max})\rceil$ independent random variables $r_{(Skt)(q)}$,
$1 \leq q \leq 2\lceil \log(l_{max})\rceil$, distributed uniformly in the interval $[0, 1]$. We define
$\mu_{Skt} := \min\{r_{(Skt)(q)}\}$.

Algorithm 1. SetCoverLeasingWithDeadlines

When an element (e, t, d) arrives,
(i) (LP solution) *while* $\sum_{(S,k,t)\in F_{(e,t,d)}} X_{(S,k,t)} < 1$;
$$X_{(S,k,t)} = X_{(S,k,t)} \cdot (1 + 1/c_S^k) + \frac{1}{|F_{(e,t,d)}| \cdot c_S^k}$$
(ii) (ILP solution) Round $X_{(S,k,t)}$ to 1 if $X_{(S,k,t)} > \mu_{Skt}$ and if (e,t,d) is not yet covered, buy the cheapest $(S,k,t) \in F_{(e,t,d)}$ (set its primal variable to 1).

4.3 Analysis

We show that the algorithm above is $\mathcal{O}(\log(p \cdot (K + \frac{d_{max}}{l_{min}})) \log l_{max}) = \mathcal{O}(\log(m \cdot (K + \frac{d_{max}}{l_{min}})) \log l_{max})$-competitive for SCLD.

It is easy to see that Algorithm 1 constructs a feasible solution ILP to SETCOVERLEASING. To compute the total expected cost C_{ILP} of ILP, we first bound the cost of the LP solution C_{LP} by $\mathcal{O}(\log(p \cdot (K + \frac{d_{max}}{l_{min}}))) = \mathcal{O}(\log(m \cdot (K + \frac{d_{max}}{l_{min}}))) \cdot$ OPT, where OPT is the optimal solution cost of ILP. Then, we show that C_{ILP} is at most $O(\log l_{max})$ times C_{LP} and hence deduce the expected $\mathcal{O}(\log(m \cdot (K + \frac{d_{max}}{l_{min}})) \log l_{max})$-competitive factor of the algorithm.

To do so, we partition the time horizon into intervals of length l_{max}. Due to the interval model (Lemma 1), all leases of all sets end on days $i : i = 0 \mod l_{max}$. Hence, we bound C_{ILP} over any interval of length l_{max} by $\mathcal{O}(\log(m \cdot (K + \frac{d_{max}}{l_{min}})) \log l_{max}) \cdot \text{OPT}_{l_{max}}$, where $\text{OPT}_{l_{max}}$ is the optimum over the corresponding interval of length l_{max}. Summing up over all such intervals yields our competitive factor for SCLD.

Lemma 2. *The cost $C_{LP(l_{max})}$ of the LP solution over an interval of length l_{max} is at most $\mathcal{O}(\log(p \cdot (K + \frac{d_{max}}{l_{min}}))) \cdot \text{OPT}_{l_{max}} = \mathcal{O}(\log(m \cdot (K + \frac{d_{max}}{l_{min}}))) \cdot \text{OPT}_{l_{max}}$ where $\text{OPT}_{l_{max}}$ is the cost of the optimal solution over this interval.*

Proof. We fix any interval of length l_{max} from our partition. Any set (S_{OPT}, k, t') in the optimum solution over this interval has been a candidate for some element (e, t, d). When (e, t, d) arrives, our algorithm increases the primal variables of (e, t, d)'s candidates until they sum up to one. After $\mathcal{O}(c_{S_{OPT}}^k \cdot \log |F_{(e,t,d)}|)$ increases, $X_{(S_{OPT}, k, t')}$ becomes greater than one and the algorithm makes no further increases. Furthermore, these increases never add a total of more than 2 to the primal variables. This is because

$$\sum_{(S,k,t)\in F_{(e,t,d)}} c_S^k \cdot \left(\frac{X_{(S,k,t)}}{c_S^k} + \frac{1}{c_S^k} \cdot |F_{(e,t,d)}| \right) \leq 2,$$

since $\sum_{(S,k,t)\in F_{(e,t,d)}} X_{(S,k,t)} < 1$ before the increase. The same holds for any other set in the optimum solution over this interval. Using a similar argument as in OLD, we can bound $|F_{(e,t,d)}|$ by $p \cdot (K + \frac{d_{max}}{l_{min}})$ (there are at most $\frac{d_{max}}{l_{min}}$ leases for each of the at most p candidate sets). This completes the proof of the lemma. □

Lemma 3. *The cost $C_{ILP(l_{max})}$ of the ILP solution over an interval of length l_{max} is at most $\mathcal{O}(\log l_{max}) \cdot C_{LP(l_{max})}$, where $C_{LP(l_{max})}$ is the cost of the LP solution over this interval.*

Proof. We fix any interval of length l_{max} from our partition. The probability to buy a set $(S, k, t) \in F$ in this interval is proportional to the value of its primal variable. Hence, $C_{ILP(l_{max})}$ is upper bounded by

$$\sum_{(S,k,t)\in F} 2\log(l_{max}+1) \cdot c_S^k \cdot X_{(S,k,t)}.$$

To guarantee feasibility, every time an element is not covered, the algorithm buys the cheapest candidate, which is a lower bound to $\text{OPT}_{l_{max}}$. The probability that an element is not covered is at most $1/(l_{max})^2$. Since the random variables are drawn independently, we can add the expected costs incurred by the corresponding at most l_{max} elements and deduce a negligible expected cost of $l_{max} \cdot 1/(l_{max})^2 \cdot \text{OPT}_{l_{max}}$ which concludes the proof of the lemma. $\square$

From the two lemmas above, we deduce the following theorem.

Theorem 2. *There is an online randomized algorithm for SCLD with a competitive factor of*

$$\mathcal{O}(\log(p \cdot (K + \frac{d_{max}}{l_{min}}))\log l_{max}) = \mathcal{O}(\log(m \cdot (K + \frac{d_{max}}{l_{min}}))\log l_{max}).$$

SETCOVERLEASING is nothing but a special case of SCLD if we set $d_{max} = 0$. Hence, we deduce the following corollary thereby improving the previous result for SETCOVERLEASING [9] from $\mathcal{O}(\log(m \cdot K)\log n)$ to $\mathcal{O}(\log(m \cdot K)\log l_{max})$ by removing the dependency on n and therefore on time.

Corollary 1. *There is an online randomized algorithm for SETCOVERLEASING that has a time-independent $\mathcal{O}(\log(p \cdot K)\log l_{max}) = \mathcal{O}(\log(m \cdot K)\log l_{max})$-competitive factor.*

5 Conclusion

We have extended the line of leasing by introducing a new model for online leasing problems. As a first infrastructure leasing problem, we have defined SET-COVERLEASING with the new model and proceeding in this direction, we plan to study other infrastructure leasing problems starting with FACILITYLEASING and STEINERTREELEASING.

Our model introduces *flexibility* to demands, thus capturing more general applications. Demands in our model have the flexibility of having a deadline. It will be interesting to extend this work to include models that handle other flexibilities (e.g., can be served on specific days within some period of time). Furthermore, demands in our model require a single day to be served. Allowing demands that require more than one day to be served can be a natural extension of our model. Even though the techniques used in this paper do not carry over directly to this extension, they still give first insights.

Along the same line of leasing lies the important unanswered question of what the price of online leasing is. All algorithms for online leasing problems so far build upon the algorithms for the non-leasing variants of their corresponding problems and the PARKINGPERMITPROBLEM algorithm. Since all these problems generalize the PARKINGPERMITPROBLEM, the only lower bound we have for these problems is $\Omega(K + f(\cdot))$, where K is the lower bound for the PARKINGPERMITPROBLEM and $f(\cdot)$ the lower bound for the underlying non-leasing variant of the problem. It is still not known whether we can prove stronger lower bounds for these problems.

Another interesting direction would be to have a stochastic view of online leasing problems, where demands and/or their deadlines are given according to some probability distribution.

References

1. Meyerson, A.: The parking permit problem. In: Proceedings of the 46th Annual IEEE Symposium on Foundations of Computer Science(FOCS), pp. 274–284 (2005)
2. Meyerson, A.: Online facility location. In: Proceedings of the 42nd Annual IEEE Symposium on Foundations of Computer Science(FOCS), pp. 426–431 (2001)
3. Alon, N., Awerbuch, B., Azar, Y., Buchbinder, N., Naor, J.: The online set cover problem. In: Proceedings of the 35th Annual ACM Symposium on the Theory of Computation(STOC), pp. 100–105 (2003)
4. Berman, P., Coulston, C.: Online algorithms for Steiner tree problems. In: Proceedings of the 29th Annual ACM Symposium on the Theory of Computation(STOC), pp. 344–353 (1997)
5. Ben-Yehuda, O., Ben-Yehuda, M., Schuster, A., Tsafrir, D.: Deconstructing amazon EC2 spot instance pricing. In: Proceedings of the 3rd IEEE International Conference on Cloud Computing Technology and Science(Cloud-Com), pp. 304–311 (2011)
6. Anthony, B.M., Gupta, A.: Infrastructure leasing problems. In: Fischetti, M., Williamson, D.P. (eds.) IPCO 2007. LNCS, vol. 4513, pp. 424–438. Springer, Heidelberg (2007)
7. Nagarajan, C., Williamson, D.P.: Offline and online facility leasing. In: Lodi, A., Panconesi, A., Rinaldi, G. (eds.) IPCO 2008. LNCS, vol. 5035, pp. 303–315. Springer, Heidelberg (2008)
8. Kling, P., Meyer auf der Heide, F., Pietrzyk, P.: An algorithm for online facility leasing. In: Even, G., Halldórsson, M.M. (eds.) SIROCCO 2012. LNCS, vol. 7355, pp. 61–72. Springer, Heidelberg (2012)
9. Abshoff, S., Markarian, C., Meyer auf der Heide, F.: Randomized online algorithms for set cover leasing problems. In: Zhang, Z., Wu, L., Xu, W., Du, D.-Z. (eds.) COCOA 2014. LNCS, vol. 8881, pp. 25–34. Springer, Heidelberg (2014)

Lower Bounds for the Size
of Nondeterministic Circuits

Hiroki Morizumi[(✉)]

Interdisciplinary Graduate School of Science and Engineering,
Shimane University, Matsue, Shimane 690-8504, Japan
morizumi@cis.shimane-u.ac.jp

Abstract. Nondeterministic circuits are a nondeterministic computation model in circuit complexity theory. In this paper, we prove a $3(n-1)$ lower bound for the size of nondeterministic U_2-circuits computing the parity function. It is known that the minimum size of (deterministic) U_2-circuits computing the parity function exactly equals $3(n-1)$. Thus, our result means that nondeterministic computation is useless to compute the parity function by U_2-circuits and cannot reduce the size from $3(n-1)$. To the best of our knowledge, this is the first nontrivial lower bound for the size of nondeterministic circuits (including formulas, constant depth circuits, and so on) with unlimited nondeterminism for an explicit Boolean function. We also discuss an approach to proving lower bounds for the size of deterministic circuits via lower bounds for the size of nondeterministic restricted circuits.

1 Introduction

Proving lower bounds for the size of Boolean circuits is a central topic in circuit complexity theory. The gate elimination method is one of well-known proof techniques to prove lower bounds for the size of Boolean circuits, and has been used to prove many linear lower bounds including the best known lower bounds for the size of Boolean circuits over the basis B_2 [2][3] and the basis U_2 [6][4].

In this paper, we show that the gate elimination method works well also for nondeterministic circuits. For deterministic circuits, it is known that the minimum size of U_2-circuits computing the parity function exactly equals $3(n-1)$ [7]. The proof of the lower bound is based on the gate elimination method and has been known as a typical example that the method is effective. In this paper, we prove a $3(n-1)$ tight lower bound for the size of nondeterministic U_2-circuits computing the parity function, which means that nondeterministic computation is useless to compute the parity function by U_2-circuits and cannot reduce the size from $3(n-1)$.

To the best of our knowledge, our result is the first nontrivial lower bound for the size of nondeterministic circuits (including formulas, constant depth circuits, and so on) with unlimited nondeterminism for an explicit Boolean function.

This work was supported by JSPS KAKENHI Grant Number 15K11986.

© Springer International Publishing Switzerland 2015
D. Xu et al. (Eds.): COCOON 2015, LNCS 9198, pp. 289–296, 2015.
DOI: 10.1007/978-3-319-21398-9_23

In this paper, we show that, for U_2-circuits, a known proof technique (i.e., the gate elimination method) for deterministic circuits is applicable to the nondeterministic case. This implies the possibility that proving lower bounds for the size of nondeterministic circuits may not be so difficult in contrast with the intuition and known proof techniques might be applicable to the nondeterministic case also for other circuits. One of motivations to prove lower bounds for the size of nondeterministic circuits is from some relations between the size of deterministic circuits and nondeterministic circuits. We also discuss an approach to proving lower bounds for the size of deterministic circuits via lower bounds for the size of nondeterministic restricted circuits.

2 Preliminaries

2.1 Definitions

Circuits are formally defined as directed acyclic graphs. The nodes of in-degree 0 are called *inputs*, and each one of them is labeled by a variable or by a constant 0 or 1. The other nodes are called *gates*, and each one of them is labeled by a Boolean function. The *fan-in* of a node is the in-degree of the node, and the *fan-out* of a node is the out-degree of the node. There is a single specific node called *output*.

We denote by B_2 the set of all Boolean functions $f : \{0,1\}^2 \to \{0,1\}$. By U_2 we denote $B_2 - \{\oplus, \equiv\}$, i.e., U_2 contains all Boolean functions over two variables except for the XOR function and its complement. A Boolean function in U_2 can be represented as the following form:

$$f(x,y) = ((x \oplus a) \wedge (y \oplus b)) \oplus c,$$

where $a, b, c \in \{0,1\}$. A U_2-*circuit* is a circuit in which each gate has fan-in 2 and is labeled by a Boolean function in U_2. A B_2-*circuit* is similarly defined.

A *nondeterministic circuit* is a circuit with *actual inputs* $(x_1, \ldots, x_n) \in \{0,1\}^n$ and some further inputs $(y_1, \ldots, y_m) \in \{0,1\}^m$ called *guess inputs*. A nondeterministic circuit computes a Boolean function f as follows: For $x \in \{0,1\}^n$, $f(x) = 1$ iff there exists a setting of the guess inputs $\{y_1, \ldots, y_m\}$ which makes the circuit output 1. In this paper, we call a circuit without guess inputs a *deterministic circuit* to distinguish it from a nondeterministic circuit.

The *size* of a circuit is the number of gates in the circuit. The *depth* of a circuit is the length of the longest path from an input to the output in the circuit. We denote by $size^{dc}(f)$ the size of the smallest deterministic U_2-circuit computing a function f, and denote by $size^{ndc}(f)$ the size of the smallest nondeterministic U_2-circuit computing a function f.

While we mainly consider U_2-circuits in this paper, we also consider other circuits in Section 4. A *formula* is a circuit whose fan-out is restricted to 1. The parity function of n inputs $x_1, \ldots, x_n$, denoted by Parity_n, is 1 iff $\sum x_i \equiv 1$ (mod 2).

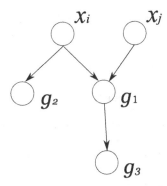

Fig. 1. Proof of Theorem 1

2.2   The Gate Elimination Method

The proof of our main result is based on the gate elimination method, and based on the proof of the deterministic case. In this subsection, we have a quick look at them.

Consider a gate g which is labeled by a Boolean function in U_2. Recall that any Boolean function in U_2 can be represented as the following form:

$$f(x, y) = ((x \oplus a) \wedge (y \oplus b)) \oplus c,$$

where $a, b, c \in \{0, 1\}$. If we fix one of two inputs of g so that $x = a$ or $y = b$, then the output of g becomes a constant c. In such case, we call that g is *blocked*.

Theorem 1 (Schnorr [7]).

$$size^{dc}(\text{Parity}_n) = 3(n-1).$$

Proof. Assume that $n \geq 2$. Let C be an optimal deterministic U_2-circuit computing Parity_n. Let g_1 be a top gate in C, i.e., whose two inputs are connected from two inputs x_i and x_j, $1 \leq i, j \leq n$. Then, x_i must be connected to another gate g_2, since, if x_i is connected to only g_1, then we can block g_1 by an assignment of a constant to x_j and the output of C becomes independent from x_i, which contradicts that C computes Parity_n. By a similar reason, g_1 is not the output of C. Let g_3 be a gate which is connected from g_1. See Figure 1.

We prove that we can eliminate at least 3 gates from C by an assignment to x_i. We assign a constant 0 or 1 to x_i such that g_1 is blocked. Then, we can eliminate g_1, g_2 and g_3. If g_2 and g_3 are the same gate, then the output of g_2 $(= g_3)$ becomes a constant, which means that g_2 $(= g_3)$ is not the output of C and we can eliminate another gate which is connected from g_2 $(= g_3)$. Thus, we can eliminate at least 3 gates and the circuit come to compute Parity_{n-1} or $\neg\text{Parity}_{n-1}$. For deterministic circuits, it is obvious that $size^{dc}(\text{Parity}_{n-1}) = size^{dc}(\neg\text{Parity}_{n-1})$. Therefore,

$$size^{dc}(\text{Parity}_n) \geq size^{dc}(\text{Parity}_{n-1}) + 3$$

$$\vdots$$

$$\geq 3(n - 1).$$

$x \oplus y$ can be computed with 3 gates by the following form:

$$(x \wedge \neg y) \vee (\neg x \wedge y).$$

Therefore, $size^{dc}(\text{Parity}_n) \leq 3(n - 1)$.   □

3   Proof of the Main Result

In this section, we prove the main theorem. For deterministic circuits, there must be a top gate whose two inputs are connected from two (actual) inputs x_i and x_j, $1 \leq i, j \leq n$. However, for nondeterministic circuits, there may be no such gate, since there are not only actual inputs but also guess inputs in nondeterministic circuits. We need to defeat the difficulty. See Section 2.2 for the definition of "block".

Theorem 2.
$$size^{ndc}(\text{Parity}_n) = 3(n - 1).$$

Proof. By theorem 1,

$$size^{ndc}(\text{Parity}_n) \leq size^{dc}(\text{Parity}_n) = 3(n - 1).$$

Assume that $n \geq 2$. Let C be an optimal nondeterministic U_2-circuit computing Parity_n. We prove that we can eliminate at least 3 gates from C by an assignment of a constant 0 or 1 to an actual input.

Case 1. There is an actual input x_i, $1 \leq i \leq n$, which is connected to at least two gates.

Let g_1 and g_2 be gates which are connected from x_i. Since we can block g_1 by an assignment of a constant to x_i, g_1 is not the output of C and there is a gate g_3 which is connected from g_1. See Figure 2.

We prove that we can eliminate at least 3 gates from C by an assignment to x_i. We assign a constant 0 or 1 to x_i such that g_1 is blocked. Then, we can eliminate g_1, g_2 and g_3. If g_2 and g_3 are the same gate, then the output of g_2 $(= g_3)$ becomes a constant, which means that g_2 $(= g_3)$ is not the output of C and we can eliminate another gate which is connected from g_2 $(= g_3)$. Thus, we can eliminate at least 3 gates.

Case 2. Every actual input is connected to at most one gates.

Let g_1 be a gate in C such that one of two inputs is connected from an actual input x_i and the other is connected from a node v whose output is dependent

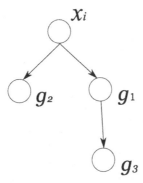

Fig. 2. Case 1

on only guess inputs and independent from actual inputs. (v may be a gate and may be a guess input.) Consider that an assignment to actual inputs and guess inputs is given. Then, if the value of the output of v blocks g_1 by the assignment, then the output of C must be 0, since, if the output of C is 1, then the value of the Boolean function which is computed by C becomes independent from x_i, which contradicts that C computes Parity$_n$. (Note that the output of C can be 0. The difference is from the definition of nondeterministic circuits.) We use the fact above and reconstruct C as follows.

Let c be a constant 0 or 1 such that if the output of v is c, then g_1 is blocked. We fix the input of g_1 from v to $\neg c$ and eliminate g_1. We prepare a new output gate g_2 and connect the two inputs of g_2 from the old output gate and v. g_2 is labeled by a Boolean function in U_2 so that the output of g_2 is 1 iff the input from the old output gate is 1 and the input from v is $\neg c$. Let C' be the reconstructed circuit. See Figure 3.

In the reconstruction, we eliminated one gate (g_1) and added one gate (g_2). Thus, the size of C' equals the size of C. In C', if the output of v is c, then the output of C' becomes 0 by g_2. If the output of v is $\neg c$, then the output of C' equals the output of the old output gate and g_1 has been correctly eliminated since we fixed the input of g_1 from v to $\neg c$ in the reconstruction. Thus, C and C' compute a same Boolean function. We repeat such reconstruction until the reconstructed circuit satisfies the condition of Case 1. The repetition must be ended, since one repetition increases continuous gates whose one input is dependent on only guess inputs (i.e., g_2) at the output. Note that g_1 is not included in the continuous gates, since the output of C must depend on at least two actual inputs.

Thus, we can eliminate at least 3 gates for all cases and the circuit come to compute Parity$_{n-1}$ or $\neg$Parity$_{n-1}$.

Lemma 1.

$$size^{\mathrm{ndc}}(\mathrm{Parity}_{n-1}) = size^{\mathrm{ndc}}(\neg\mathrm{Parity}_{n-1}).$$

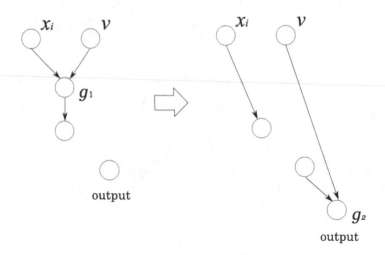

Fig. 3. Case 2

Proof. Let C' be a nondeterministic U_2-circuit computing Parity_{n-1}. For nondeterministic circuits, notice that negating the output gate in C' does not give a circuit computing $\neg\text{Parity}_{n-1}$. We negate an arbitrary actual input x_i, $1 \leq i \leq n$, in C' and obtain a nondeterministic U_2-circuit which computes $\neg\text{Parity}_{n-1}$ and has the same size to C', which can be done by relabeling each Boolean function of all gates which are connected from x_i. $\qquad\square$

Therefore,

$$size^{\text{ndc}}(\text{Parity}_n) \geq size^{\text{ndc}}(\text{Parity}_{n-1}) + 3$$

$$\vdots$$

$$\geq 3(n-1).$$

Thus, the theorem holds. $\qquad\square$

4   Discussions

In this paper, we proved a $3(n-1)$ lower bound for the size of nondeterministic U_2-circuits computing the parity function. To the best of our knowledge, this is the first nontrivial lower bound for the size of nondeterministic circuits with unlimited nondeterminism for an explicit Boolean function. In this section, as one of motivations to prove lower bounds for the size of nondeterministic circuits, we discuss an approach to proving lower bounds for the size of deterministic circuits via lower bounds for the size of nondeterministic restricted circuits.

It is known that the Tseitin transformation [8] converts an arbitrary Boolean circuit to a CNF formula. We restate the Tseitin transformation as the form of the following theorem.

Theorem 3. *Any B_2-circuit of n inputs and size s can be converted to a nondeterministic 3CNF formula of n actual inputs, s guess inputs, and size $O(s)$.*

Proof. We prepare one guess input for the output of each gate in the B_2-circuit, and use the Tseitin transformation.   □

Thus, if we hope to prove a nonlinear lower bound for the size of deterministic general circuits, (which is a major open problem in circuit complexity theory,) then it is enough to prove a nonlinear lower bound for the size of nondeterministic circuits in the theorem. The nondeterministic circuit in Theorem 3 is a constant depth circuit with depth two and some restrictions. In this paper, we saw that, for U_2-circuits, a known proof technique (i.e., the gate elimination method) for deterministic circuits is applicable to the nondeterministic case. It remains future work whether many known ideas or techniques for constant depth circuits are applicable to nondeterministic circuits.

The basic idea of the proof of Theorem 3 and the Tseitin transformation can be widely applied. We show another example in which guess inputs are prepared for a part of gates in the circuit.

Theorem 4. *Any B_2-circuit of n inputs, size $O(n)$ and depth $O(\log n)$ can be converted to a nondeterministic formula of n actual inputs, $O(n/\log\log n)$ guess inputs, and size $O(n^{1+\epsilon})$, where $\epsilon > 0$ is an arbitrary small constant.*

Proof. Let C be such a B_2-circuit. It is known that we can find $O(n/\log\log n)$ edges in C whose removal yields a circuit of depth at most $\epsilon\log n$ ([9], Section 14.4.3 of [1]). We prepare $O(n/\log\log n)$ guess inputs for the edges, and one guess input for the output of C. Consider that an assignment to actual inputs and guess inputs is given. It can be checked whether the value of each guess input corresponds to correct computation by $O(n/\log\log n)$ nondeterministic formulas of size n^ϵ, since the depth is at most $\epsilon\log n$. We construct a nondeterministic formula so that it outputs 1 iff all guess inputs are correct and the guess input which corresponds to the output of C is 1.   □

For the case that the number of guess inputs is limited, there is a known lower bound for the size of nondeterministic formulas [5].

References

1. Arora, S., Barak, B.: Computational Complexity - A Modern Approach. Cambridge University Press (2009)
2. Blum, N.: A boolean function requiring 3n network size. Theor. Comput. Sci. **28**, 337–345 (1984)
3. Demenkov, E., Kulikov, A.S.: An elementary proof of a $3n-o(n)$ lower bound on the circuit complexity of affine dispersers. In: Murlak, F., Sankowski, P. (eds.) MFCS 2011. LNCS, vol. 6907, pp. 256–265. Springer, Heidelberg (2011)
4. Iwama, K., Morizumi, H.: An explicit lower bound of 5n - o(n) for boolean circuits. In: Proc. of MFCS, pp. 353–364 (2002)

5. Klauck, H.: Lower bounds for computation with limited nondeterminism. In: Proc. of CCC, pp. 141–152 (1998)
6. Lachish, O., Raz, R.: Explicit lower bound of 4.5n - o(n) for boolean circuits. In: Proc. of STOC, pp. 399–408 (2001)
7. Schnorr, C.: Zwei lineare untere schranken für die komplexität boolescher funktionen. Computing **13**(2), 155–171 (1974)
8. Tseitin, G.S.: On the complexity of derivation in propositional calculus. In: Slisenko, A.O. (ed.) Studies in Constructive Mathematics and Mathematical Logic, pp. 115–125 (1968)
9. Valiant, L.G.: Graph-theoretic properties in computational complexity. J. Comput. Syst. Sci. **13**(3), 278–285 (1976)

Computing Minimum Dilation Spanning Trees in Geometric Graphs

Aléx F. Brandt, Miguel F.A. de M. Gaiowski,
Pedro J. de Rezende$^{(\boxtimes)}$, and Cid C. de Souza

Institute of Computing, University of Campinas, Campinas, Brazil
rezende@ic.unicamp.br

Abstract. Let P be a set of points in the plane and $G(P)$ be the associated geometric graph. Let T be a spanning tree of $G(P)$. The dilation of a pair of points i and j of P in T is the ratio between the length of the path between i and j in T and their Euclidean distance. The dilation of T is the maximum dilation among all pairs of points in P. The *minimum dilation spanning tree problem* (MDSTP) asks for a tree with minimum dilation. So far, no exact algorithm has been proposed in the literature to compute optimal solutions to the MDSTP. This paper aims at filling this gap. To this end, we developed an algorithm that combines an integer programming model, a geometric preprocessing and an efficient heuristic for the MDSTP. We report on computational tests in which, for the first time, instances of up to 20 points have been solved to proven optimality.

Keywords: Minimum dilation · Stretch factor · Geometric Graphs

1 Introduction

In many practical applications, one is faced with problems requiring the construction of a low cost network connecting a set of locations that satisfy certain quality requirements. Depending on how the cost and the quality are measured, different optimization problems emerge.

In this paper, we consider one such case, known as the *minimum dilation spanning tree problem*. Let P be a set of n points in the plane. For every $u, v \in P$, d_{uv} denotes the Euclidean distance between u and v. The *geometric graph*, $G(P) = (P, E)$, is the weighted complete graph whose vertex-set is P and whose edge weights are given by the Euclidean distances between the corresponding endpoints. Consider a spanning subgraph H of $G(P)$. Let u, v be two vertices in H. The *distance* on H between u, v is defined as the length of a shortest path $\pi_H(u, v)$ connecting u to v in H, denoted $|\pi_H(u, v)|$, or infinite when v is unreachable from u. As usual, the length of a path is defined as the sum of the weights of its edges. The *dilation* (also known as *stretch factor* or *distortion* [1])

Research supported by grants: CNPq (302804/2010-2, 477692/2012-5, 311140/2014-9, 139107/2012-6), FAPESP (2015/08734-9, 2012/17965-6), and FAEPEX.

© Springer International Publishing Switzerland 2015
D. Xu et al. (Eds.): COCOON 2015, LNCS 9198, pp. 297–309, 2015.
DOI: 10.1007/978-3-319-21398-9_24

of H is defined as $\rho(H) = \max\limits_{u,v \in P} |\pi_H(u,v)|/d_{uv}$, see [2]. Since $\rho(G(P)) = 1$, the interesting problems arise when a proper subgraph of $G(P)$ is sought.[1] An often considered situation is when we limit to some constant k the number of edges that can be used to build the network while seeking to obtain the minimum dilation. As this problem only makes sense when the resulting subgraph is connected, the smallest value for k is $n - 1$, restricting the plausible subgraphs to the spanning trees of $G(P)$. This gives rise to the *minimum dilation spanning tree problem* (MDSTP), which is known to be NP-hard [3]. Figure 1 shows an example. Moreover, thus far, no $o(n)$ approximate algorithm for the MDSTP is known [3].

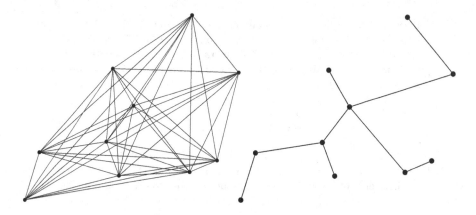

Fig. 1. A geometric graph $G(P, E)$ and its minimum dilation spanning tree

This paper focuses on the design of an exact algorithm for the MDSTP, which is effective in practice for applications that require actual optimal solutions.

Motivation. The investigation of minimum dilation problems is prompted by their applications to robotics, network topology design and many others (see [1,4]). While the literature does not yet include exact algorithms for the MDSTP, we believe that producing optimal solutions may enable us to identify geometric properties that can lead to algorithms that are more effective in practice. Studying the MDSTP may reveal relevant information for solving similar problem for other classes of graphs.

Related Work. Experimental works published on dilation problems attempt to find, for a given t, (sub)optimal t-spanners minimizing the total edge weight, vertex degrees, the number of edges for a fixed dilation, the spanner diameter or the number of crossings ([5,6]). In contrast to what is done here, none of these look for a subgraph that minimizes the dilation *within a given family*.

Some important bounds for the minimum dilation of spanning trees are the worst case lower bound $\Omega(n)$ achieved by the vertex set of a regular n-gon [1]

---

[1] The reader will find ample material on the related concept of t-spanners in [1,2].

and the upper bound $n - 1$ guaranteed by the minimum weight spanning tree [4]. Also, Cheong, Haverkort and Lee [3] (Klein and Kutz [7]) presented a set P of five (seven) points in the plane for which any MDST has self-intersections.

Our Contribution. This work presents the first exact algorithm for the MDSTP. The method employs a mixed integer linear programming (MILP) model that may be computed by state-of-the-art solvers. However, as the MILP formulation, can easily grow to an unmanageable size, we developed a preprocessing phase that exploits the geometry of the problem to significantly reduce the model. Moreover, while as the preprocessing requires the knowledge of good viable solutions, we designed a metaheuristic that yields high quality solutions. As a convenient byproduct, these fine primal bounds also allow the MILP solver to do early pruning of the enumeration tree during the branch-and-bound algorithm. As a result of this strategy, we were able to find provably optimal solutions for instances of the MDSTP with up to 20 points, which is unprecedented and, from our experience, extremely challenging as the largest instances took 30 minutes of CPU time. To enable future comparisons, we also make available our benchmark of thousands of instances, as well as scripts for generating pseudo-random point sets based on either fixed or image-based distributions.

Text Organization. In Section 2 a MILP formulation for the problem is given. Section 3 shows how any known primal bounds can be used in preprocessing routines that, *a priori*, retain or discard edges from participating in an optimal solution or from being in a path joining two vertices in such a solution. These routines strongly influence the performance of our method since, ultimately, they reduce the size of the MILP model. A GRASP based heuristic that we developed for the MDSTP is presented in Section 4. The computational tests are reported in Section 5 and Section 6 discusses future research directions.

2   A MILP Model for the MDSTP

The MILP formulation aims at building paths with limited dilation between all pairs of vertices in such a way that the union of these paths is a spanning tree. This dilation is then minimized in the objective function.

The characterization of the paths is achieved by means of a multicommodity flow network model in which there is a special unitary flow between each distinct pair of vertices. This makes it possible to obtain the length of a particular path and to bound its dilation. The flow model is defined on the directed graph $DG(P) = (P, A)$ obtained from $G(P)$ by replacing each of its edges by two directed arcs. A binary variable x_{ij}^{ab} is introduced corresponding to the flow of the commodity ab (the one associated to the path from a to b) through the directed arc (i, j). Naturally, x_{ij}^{ab} is set to one if and only if the arc (i, j) belongs to the path from a to b in an optimal solution. It is easy to see that a full characterization of a spanning tree can be determined from the subset of variables of the form x_{ij}^{ij} since, once one of these is set to 1, the (unique) path from i to j is comprised solely by the edge $\{i,j\}$. The last element of the model is a continuous variable ρ that represents the optimum. The formulation reads:

$$\min \quad \rho \tag{1}$$

$$\sum_{i \in V \setminus \{a\}} x_{ia}^{ab} - \sum_{i \in V \setminus \{a\}} x_{ai}^{ab} = -1 \qquad \forall a, b \in V \tag{2}$$

$$\sum_{i \in V \setminus \{j\}} x_{ij}^{ab} - \sum_{i \in V \setminus \{j\}} x_{ji}^{ab} = 0 \qquad \forall j \in V \setminus \{a,b\}, \ \forall a, b \in V \tag{3}$$

$$\sum_{i \in V \setminus \{b\}} x_{ib}^{ab} - \sum_{i \in V \setminus \{b\}} x_{bi}^{ab} = 1 \qquad \forall a, b \in V \tag{4}$$

$$x_{ij}^{ab} + x_{ji}^{ab} - x_{ij}^{ij} \leq 0 \qquad \forall \{i,j\} \in E, \ \forall a, b \in V \tag{5}$$

$$\sum_{\{i,j\} \in E} x_{ij}^{ij} = |V| - 1 \tag{6}$$

$$\sum_{(i,j) \in A} \frac{d_{ij}}{d_{ab}} x_{ij}^{ab} - \rho \leq 0 \qquad \forall a, b \in V \tag{7}$$

$$x \in \mathbb{B}^{|A| \cdot |E|} \tag{8}$$

The objective function (1) simply asks for a solution of minimum dilation. For a given pair of points a and b, (2), (3) and (4) impose that the variables x^{ab} describe a path from a to b in the final solution. Once we have the description of the paths between all vertex pairs it remains to ensure that we have a connected structure with only $n - 1$ edges, implying that the resulting subgraph is acyclic. This is done by two sets of constraints: (5) which does the coupling of the arcs in the paths and the edges in the tree and (6) that restricts the number of edges in the solution. Finally, (7) constrains the dilation of every path to be no larger than ρ. As the objective function minimizes ρ, it is clear that, in the solution, this variable is equal to the maximum dilation among all pairs of vertices.

This formulation is based on the *extended multicommodity flow* model described in [8] for the spanning tree problem. An important feature of that model is that it gives a complete description of the convex hull of the integer solutions of the spanning tree problem, called the *spanning tree polytope*. Besides, the number of constraints and variables in this linear system is polynomial in the size of the input graph, i.e., the model is compact. Basically, in this formulation one chooses an arbitrary vertex as the root of the tree, in our case vertex zero, and flows a especial commodity to each other vertex. In reality, the original coupling constraints are slightly more complex than the one presented here as they also ensure that every pair of commodities flows in the same direction in each arc. This is done by enforcing the inequalities $x_{ij}^{0k} + x_{ji}^{0k'} - x_{ij}^{ij} \leq 0$, for all k, k' in $V \setminus \{0\}$ and $\{i,j\}$ in E. Our formulation relinquishes this full description in favor of a more compact and still correct model by using (5).

One may have noticed that the model (1)–(8) already has an enormous number of constraints and variables, both of which are $\Theta(n^4)$. The massive size of the model severely affects the performance of MILP solvers. It is worth noting that other options regarding the coupling constraints exist, some of which leading to tighter linear relaxations. However, these alternative formulations are even larger than the one given here. Besides, in early experiments where they were used, we observed little or no actual gain in the dual bounds – the professed benefit of using such relaxations – despite a noticeable increase in execution times.

In the next section, we discuss how a known primal (upper) bound can be used to accelerate the search for a proven optimal solution. Essentially, this is done by exploiting geometric properties that allow us to fix some of the variables of our MILP model. As we mentioned in the previous paragraph, the size of the formulation can be a bottleneck for the direct usage of our method. Therefore, the effectiveness of the preprocessing is a key step for the success of the approach.

3 Preprocessing

Suppose that a feasible solution for the MDSTP has been computed by some heuristic, whose dilation is ρ'. Now, based on geometric properties, we devise routines that allow the identification of edges of the geometric graph $G(P)$ that must or must not be in any tree with dilation less than or equal to ρ'. Alternatively, we may be able to reach the weaker but relevant conclusion that an edge may be required to or forbidden from being part of the path between a specific pair of vertices in any solution with dilation not exceeding ρ'. As said earlier, these findings allow us to reduce the size of the MILP model as discussed below.

Consider the first case, where we conclude that an edge $\{i, j\}$ is or is not part of a tree with dilation bounded from above by ρ'. The x_{ij}^{ij} variable is either set to one or to zero. Due to (5), the latter case also forces all the $O(n^2)$ path variables associated to arc (i, j) to be set to zero. On the other hand, if $\{i, j\}$ is forced to be in the optimal tree, all the $O(n^2)$ path variables x_{kl}^{ij} for $(k, l) \in A \setminus (i, j)$ are required to take value zero by (2)–(4). Of course, in this case, the $O(n)$ equations describing the path from i to j can also be removed from the model.

Let us focus now on the case where we conclude that a given arc (i, j) is or is not in the path from a to b, with $\{i, j\} \neq \{a, b\}$. When this arc is identified as a required part of the path, not only the variable x_{ij}^{ab} but also the corresponding edge variable x_{ij}^{ij} is set to one. As discussed earlier, the latter assignment propagates the setting of variables to arc variables (a, b). On the other hand, if (i, j) is not in the path from a to b and no other information is at hand, only the flow variable can be set to zero. Thus, in contrast to the previous analysis, there is no direct propagation of variable assignments. But, as we shall see, depending on the distribution of the points of P on the plane and on the primal bound, a large number of variables may be set, considerably reducing the model size.

Below, we describe three routines that trigger the setting of some variable. We assume that a, b, u and v are four distinct points in P, and ρ' is an upper bound for the minimum dilation of the spanning trees of $G(P)$.

The Ellipse-based Elimination Routine. Let T be a spanning tree and suppose u is a vertex on the path $\pi_T(a, b)$, i.e., $\pi_T(a, b) = \pi_T(a, u) \cup \pi_T(u, b)$. The dilation of the pair $\{a, b\}$ in T is given by $\rho_T(a, b) = \frac{|\pi_T(a,b)|}{d_{ab}}$. From the triangle inequality, we have $\pi_T(a, u) \geq d_{au}$ and $\pi_T(u, b) \geq d_{ub}$ and $\rho_T(a, b) \geq \frac{d_{au}+d_{ub}}{d_{ab}}$. Hence, if T has dilation at most ρ', the following constraint must hold:

$$\frac{d_{au} + d_{ub}}{d_{ab}} \leq \rho_T(a, b) \leq \rho' , \qquad (9)$$

which means that u must lie on or inside the ellipse with foci a and b having a major axis of length $\rho' \cdot d_{ab}$. In other words, given an upper bound ρ' for the dilation and two vertices a and b, only the points in the ellipse defined by (9) can be on the path joining a and b in a tree with dilation not exceeding ρ'. A similar argument appears in [7], where the MDSTP is proven to be NP-hard.

Furthermore, all $O(n)$ arcs incoming to or outgoing from any discarded vertex are precluded from being part of the path between a and b and may, therefore, be removed from the model, i.e., the corresponding variables may be set to zero.

The Long Arc Elimination Routine. Once the vertices inside the ellipse with foci a and b have been identified, it may still be possible to discard arcs joining pairs of them from the path from a to b. Consider two such vertices, say, u and v, with $\{u, v\} \neq \{a, b\}$. If the arc (u, v) is in a solution with dilation bounded by ρ', the triangle inequality requires that the following inequality holds:

$$\frac{d_{au} + d_{uv} + d_{vb}}{d_{ab}} \leq \rho_T(a, b) \leq \rho' . \qquad (10)$$

Clearly, when the above inequality is not satisfied, the arc (u, v) cannot belong to the path joining a and b in a solution with dilation limited to ρ'. Since the flow is directed from a to b, using the arc (u, v) is not the same as including (v, u) in the path. So, (10) must be applied to every arc of A, i.e., both directions of the respective edge have to be considered.

This routine can be viewed as an extension of the previous one, in the sense that it evaluates subpaths with two intermediate vertices instead of just one. Longer subpaths could be analyzed in almost the same way. However, the complexity of such tasks would increase quickly rendering them impractical.

The Edge Fixing Routine. A very handy situation occurs when an ellipse with foci a and b contains no other points of P in its interior. We will refer to it as an "empty" ellipse. Of course, if the ellipse's major axis has length $\rho' \cdot d_{ab}$ the only way to connect a and b in a tree with dilation no greater than ρ' is through the edge ab. In other words, this edge is fixed (i.e., must belong to that tree).

As edges are ascertained to be part of the tree, the subgraph they induce is a forest containing connected components of two or more vertices. Two types of reductions of the MILP model can then occur. Firstly, any edge between two vertices on the same component that is not already fixed corresponds to a variable that can be set to zero; otherwise a cycle would be formed. The same holds for the path variables associated to the arcs corresponding to those edges. Similarly, every two vertices in a connected component induced by fixed edges have their paths already defined. Hence, the flow variables on this route can be set to one while, at the same time, all other flow variables associated to arcs on alternative paths can be set to zero.

4   A GRASP for the MDSTP

Greedy Randomized Adaptive Search Procedure (GRASP) is a metaheuristic widely applied to combinatorial optimization problems. Since it is reported to produce high quality solutions within a short computing time [9], we employ it to find good primal bounds for the MDSTP. We assume that the reader is familiar with the workings of a basic GRASP algorithm. So, to explain its use in the context of the MDSTP, just the construction and the local search phases are described.

Firstly, a solution is built by randomizing Prim's algorithm [10] for building a minimum spanning tree (MST) from the (complete) geometric graph $G(P)$. This randomization simply expands the purely greedy choice of a vertex to be added to a partially built MST to a uniformly distributed selection of a vertex from a *Restricted Candidate List* (RCL) of vertices. This list is constructed in the following way. Let d_m (D_m) denote the length of the shortest (longest) edge connecting the current (incomplete) MST to the vertices that are not yet part of this MST. Given an α in the interval $[0, 1]$ we take into the RCL each vertex of $G(P)$ whose distance to the current (incomplete) MST is within the range $[d_m, (1 - \alpha)d_m + \alpha D_m]$. Clearly, the smaller α is, the shorter the RCL will be, and the choice of $\alpha = 0$ leads to a purely greedy algorithm, while $\alpha = 1$ turns it into a randomized one. In our implementation, at each iteration of the GRASP's construction phase, a new value for α is chosen from the set $\{0.0, 0.1, \ldots, 1.0\}$ with equal probability, leading to a likely diverse sequence of viable solutions.

Secondly, in the local search phase, attempts are made to improve known solutions. For the sake of efficiency, the perturbations defining the neighborhood of a solution should be quickly computable. That is why, for the MDSTP the naive approach of merely replacing a tree edge for another is undesirable. After all, that might drastically alter the dilation between unpredictably many pairs of vertices in an arbitrary way and even the tree's. For this reason, we devised a controlled method for navigating within a neighborhood of a given solution, called *Triangle*. Let u and v two non-adjacent vertices in a spanning tree T and z be a vertex in the unique path from u to v in T. Suppose that a and b are the vertices adjacent to z in this path. Denote by T_A (T_B) the subtree containing vertex a (b) and obtained from T by removing the edge (a, z) ((b, z)). Replacing (a, z) by (a, b) changes T in such a way that the dilation of the pair $\{u, v\}$ decreases. On the other hand, by the triangle inequality, we know exactly which pairs of vertices may have their dilation worsened. To see this, define A and B as the set of vertices in T_A and T_B, respectively, and $Z = P - (A \cup B)$. Clearly, the pairs of vertices from $A \times Z$ are the only ones for which the dilation needs to be recalculated. Although there may still be $O(n^2)$ such pairs, in practice, significant reductions in computing times were observed when we applied this method. A homologous analysis holds if we replace (b, z) by (b, a), instead.

To overcome the drawback of the *Triangle* method, which generates only two neighbor solutions, we devised an extended version called *Path*. This new neighborhood iterates the *Triangle* local search through consecutive triples of vertices along a path joining two vertices with maximum dilation. *Path* achieves a compromise between neighborhood size and the time needed to explore it. The attentive reader will realize that, given a path of length k, the size of the *Path* neighborhood is $2k$, instead of only 2 for *Triangle*. Although exploring more solutions is, of course, more expensive, our tests showed that the slight increase in running time was compensated by the gain in quality of the yielded solutions.

Two strategies to halt local search are commonly applied: *first improvement* and *best improvement* (see [11]). Experimentation lead us to chose the former, since best improvement proved too time consuming for the minute benefit it generates after its inherently exhaustive search.

Our implementation also applies *path relinking* [11] to a pool of *elite solutions*. This pool is created throughout the iterations to store the best known solutions, cost wise, as well as those whose costs are within a certain threshold of the best primal bound. Moreover, diversity, measured according to the number of elements in the symmetric difference between two solutions, is also favored. To describe the path relinking process employed here, given two elite solutions S and S', we need to outline how to iteratively perturb S into S', giving us a path in the search space within which a new improved solution is likely to be found [11]. Suppose that S and S' are the starting and target trees, respectively. The path relinking S and S' is traversed as follows. An edge $e \in S' - S$ is added to S creating a cycle. The removal of another edge in $S - S'$ from this cycle leads to a new tree which is a step closer to S'. Clearly, k such iterations create k new trees among which there may be one with lower dilation than both S and S'.

Lastly, notice that the size of the elite set influences the overall performance of GRASP. Too many solutions makes the algorithm run too slow, whereas too few solutions may thwart its ability to generate any improvement. In our tests, we found that a good strategy is to store all the solutions found during the GRASP iterations that resulted in an update of the best known solution up to that point. In our implementation, *path relinking* is executed after 1000 iterations of the loop construction phase/local search phase have been completed.

We conclude this section reiterating that, in this paper, GRASP must be viewed solely as a tool to help in attaining optimality. An in depth investigation on the theme of heuristics, in its own right, for the MDSTP is deserving of future attention.

5   Computational Results

This section discusses the experiments we carried out. We begin by describing the characteristics of our benchmark and of the computational environment. We continue the section with the presentation and analyses of the results obtained.

Instances. We generated a benchmark consisting of instances comprised of uniformly distributed points on a 10×10 square. The coordinates of the points and the distances between two points were both rounded to six decimal places in order to avoid arithmetic pitfalls and to facilitate future comparisons. To promote the fulfillment of the latter goal, the entire benchmark is made available for download at www.ic.unicamp.br/~cid/Problem-instances/Dilation. The set consists of 30 instances for each number of points from the sequence $10, 12, 14, 16, 18, 20$.

Computing Environment. The results reported here were obtained on identical machines featuring: Intel® Xeon® CPU E3-1230 V2 @ 3.30GHz (4 cores and 8 threads) processors; 32GB of RAM; running OS Ubuntu 12.04; using a g++ 4.6.3 compiler and the MILP solver IBM® ILOG® CPLEX® Optimization Studio 12.5.1. The solver was allowed to use all 8 threads simultaneously and to run for at most 30 minutes. Most CPLEX parameters were left at default values, although some changes are noteworthy: *(i)* the use of *Traditional Branch & Cut* search method instead of *Dynamic Search*; *(ii)* higher branching priorities were enforced on the edge variables in detriment of the arc ones; and *(iii)* the relative and absolute gaps were set to 10^{-7}.

Results. We now discuss the relevant tests we ran to assess the efficiency of our algorithm. Since the application of our approach depends on the computation of good primal bounds, our first analysis focus on the performance of the GRASP metaheuristic. To evaluate that, we present in Table 1 statistics relative to the execution of the heuristic on instances for which the optimum was found. The first column of this table displays the number of points per instance. Two groups of four columns follow reporting minimum, average, standard deviation and maximum values, respectively for the gap relative to the optima and for the running times (in milliseconds).

As seen from these results, GRASP solutions are of very high quality, at least for the instance sizes considered in our tests. Despite some fluctuation in the gaps, one can perceive a slight loss in quality as the sizes of the instances increase. Since the main purpose of this work is to compute optimal solutions and, as we will see, we are still unable to prove optimality for instances of more than 20 points, we decided not to invest more time in improving the GRASP heuristic at this time. Clearly, since the MDSTP is NP-hard, pursuing this goal would be an interesting investigation in its own right. But, for now, let us just remark that the CPU times spent by the heuristic, as is, are insignificant on our benchmark.

Next, we analyze the roles played by the primal information and by the preprocessing in the computation of optimal solutions. To accomplish this task, five different variants of the MILP solver were tested: *(i) None*: corresponds to the execution of CPLEX with default options and the complete model given in Section 2; *(ii) Sol*: same as in *(i)*, but with CPLEX also fed with the upper bound as well as an optimal solution; *(iii) PreP*: same as in *(i)* except that, in this case, the model is reduced through the preprocessing discussed in Section 3; *(iv) UB+PreP*: same as in *(iii)* but, besides the preprocessing, the optimum is also given as an upper bound to CPLEX; and, finally, *(v) Sol+PreP*: same as in *(iv)* but adding as the part of the input an optimal solution.

At first glance, the usage of optimal solutions and bounds in these initial tests may sound strange. However, this information was given instead of GRASP outcomes because, as said before, the heuristic results present some oscillation even when equal sized instances are compared. Besides, the quality of the heuristic solution was also seen to deteriorate slightly as the instance size increases. By using optimal information, we intend to minimize the effects of these phenomena which, otherwise, could lead to biased conclusions. The influence of the degradation of the quality of the primal information in the computation of optimal solutions is evaluated later in this section.

The strategy to assess the contributions of the present work should now be clear from the choice of these variants. An obvious way to speed up the running time of an enumeration algorithm is to provide as input a primal bound and a solution with cost equal to this bound. A comparison between the performances of *None* and *Sol* answers the question on how much is gained by applying this standard technique to the MDSTP.

Notice that state-of-the art solvers like CPLEX, are equipped with powerful algebraic preprocessing routines that, when fed with primal information, can dramatically reduce the MILP formulation. Hence, to evaluate the importance of *our* geometric preprocessing, we tested the three remaining variants, all of which include this preprocessing. In *PreP* we constrain ourselves to the model

Table 1. Statistics for the GRASP metaheuristic

Size	Dilation Gap to Opt. Sols (%)				Grasp Runtime (ms)			
	Min	Average	Std Dev.	Max	Min	Average	Std Dev.	Max
10	0	0,07	0,41	2,23	22	44,3	8,8	60
12	0	0,14	0,76	4,15	48	65,8	10,9	90
14	0	0,05	0,22	1,12	65	93,7	17,7	140
16	0	0,01	0,06	0,31	100	125,5	14,4	157
18	0	0,38	1,47	7,38	113	148,2	23,3	198
20	0	0,47	1,62	7,37	147	186,9	16,6	230

Table 2. Statistics for five variations of the MILP solver

Size	Method	# Opt Sols	Avg Exec Time (s)	# Feas Sols	% Avg Gap	Wins	Avg Fxd Edges	Avg free vars (%)
10	Sol+PreP	30	0.2	0	–	23	4.0	32.3
	UB+PreP	30	0.2	0	–	7	4.0	32.3
	PreP	30	0.3	0	–	0	4.0	32.3
	Sol	30	1.5	0	–	0	0.0	100.0
	None	30	5.0	0	–	0	0.0	100.0
12	Sol+PreP	30	1.5	0	–	29	4.5	35.4
	UB+PreP	30	1.8	0	–	1	4.5	35.4
	PreP	30	1.9	0	–	0	4.5	35.4
	Sol	30	22.5	0	–	0	0.0	100.0
	None	30	83.1	0	–	0	0.0	100.0
14	Sol+PreP	30	4.9	0	–	27	4.4	39.6
	UB+PreP	30	7.6	0	–	3	4.4	39.6
	PreP	30	7.7	0	–	0	4.4	39.6
	Sol	30	121.0	0	–	0	0.0	100.0
	None	27	588.0	3	4.3	0	0.0	100.0
16	Sol+PreP	30	42.3	0	–	24	4.1	45.8
	UB+PreP	30	54.6	0	–	4	4.1	45.8
	PreP	30	64.3	0	–	2	4.1	45.8
	Sol	20	387.0	10	25.6	0	0.0	100.0
	None	3	1661.0	27	39.7	0	0.0	100.0
18	Sol+PreP	27	222.1	3	43.5	19	4.7	46.6
	UB+PreP	26	211.1	2	27.4	5	4.7	46.6
	PreP	26	208.4	3	38.8	3	4.7	46.6
	Sol	12	647.4	18	37.8	0	0.0	100.0
	None	0	–	5	72.1	0	0.0	100.0
20	Sol+PreP	21	261.5	9	44.7	12	4.1	49.1
	UB+PreP	23	466.7	0	–	5	4.1	49.1
	PreP	22	400.3	4	46.5	6	4.1	49.1
	Sol	4	1328.6	26	41.3	0	0.0	100.0
	None	0	–	13	95.2	0	0.0	100.0

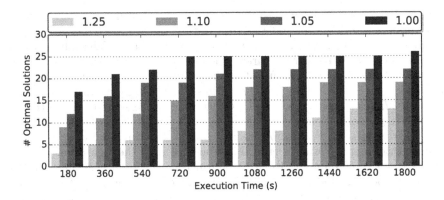

Fig. 2. Variation of the number of optima found as the primal bound deteriorates

reduction, leaving out the primal information. To measure the effect of adding the latter, piece by piece, we first provide only the upper bound in *UB+PreP* and, finally, the complete information in *Sol+PreP*.

The results obtained by the MILP variants are summarized in Table 2. For each instance size displayed in the first column, there are five rows, each corresponding to one variant, as shown in the second column. The third column exhibits the number of optima obtained and the next column shows the average execution time achieved by the corresponding variant, over the instances for which it reached a proven optimum. The fifth column shows the number of instances for which optimality was not proved but that the solver found a feasible solution. In the following column, the value corresponds to the average gaps of the latter solutions. The seventh column gives the total number of instances where the variant outperformed the remaining ones. The eighth and ninth columns allow an assessment of the model reductions resulting from the geometric preprocessing. They include, first, reports of the average number of variables selected by the edge fixing routine, followed by the average percentage of variables that remain in the model after all preprocessing routines have been executed. Percentages are calculated in terms of the total number of variables in the complete MILP model.

Consider the 90 instances of the three largest sizes. Variant *None* could only solve 3 of them to optimality, suggesting that the MILP model, alone, is not very useful. A substantial gain was obtained when the primal information was made available in *Sol* as 36 instances were now solved to optimality. Finally, the contribution of the preprocessing can be fully appreciated when we compare these numbers to those of the three last variants. One can see that any of them solved at least twice as many instances as the other variants where no geometric preprocessing was performed.

Once we have established the relevance of our geometric preprocessing, the 3 variants that use it should still be compared. The numbers of proven optima were 48, 49 and 48 for *Sol+PreP*, *UB+PreP* and *PreP*, respectively. This might suggest a small advantage for *UB+PreP*. However, a closer analysis of the data shows that the number of feasible solutions (including optimal ones) obtained by these variants were 180, 171 and 175, respectively, leaving the *UB+PreP* variant behind the other two and *Sol+PreP* in a better position. The advantage of *Sol+PreP* becomes even more apparent when we turn our attention to the overall performance of the algorithm, including the evaluation of the running times. As can be seen on column *"Wins"*, *Sol+PreP* has a total of 134 instances, compared to 25 of *UB+PreP* and 11 of *PreP*. Concerning the execution times (fourth column), the benefit of employing the routines in Section 3 speaks for itself: just contrast *None* vs. *PreP* and *Sol* vs. *Sol+PreP*. Therefore, one concludes that the geometric preprocessing is crucial for the success of the algorithm.

The last two columns of Table 2 confirm that the preprocessing is indeed a powerful tool for model reduction. In this context, the primal information is fundamental since, without an upper bound, no preprocessing could have been done, in the first place. Moreover, the knowledge of a solution with cost equal to this bound accelerates the convergence of the algorithm, emphasizing the importance of obtaining good primal solutions, as done by our GRASP.

Since GRASP is a heuristic, we cannot guarantee the quality of the primal information yielded by the procedure. Therefore, another relevant issue to consider is how much the overall algorithm's performance deteriorates as the upper bound used for preprocessing and given as input to the solver worsens. This analysis can be carried out based on the bar graph shown in Fig. 2. The data refer to the executions of the MILP variant *UB+PreP* over the 30 instances of size 18 with the upper bounds given by the dilation of the optimal solution multiplied by 1.00, 1.05, 1.10 and 1.25. The bars reflect the cumulative number of optimal solutions found for each of these multiplying factors and measured at every 180 seconds for up to half an hour. For example, after 720 seconds, the number of instances solved was 25, 19, 15 and 6 for each of the aforementioned multiplying factors. From this graph, it is quite clear that the algorithm's performance declines rapidly as the upper bound decreases. From the previous analyses, this effect is likely to be a repercussion of the loss of efficiency of the preprocessing.

6   Future Directions

Some issues are currently being investigated to improve the method presented here, including: *(i)* the customization of the choice of the branch variable according to geometric properties; *(ii)* the use of GRASP and geometric preprocessing in all nodes of the enumeration tree; and, *(iii)* the use of other MILP formulations.

References

1. Narasimhan, G., Smid, M.: Geometric Spanner Networks. Cambridge University Press, New York (2007)
2. Peleg, D., Schäfer, A.A.: Graph spanners. Journal of Graph Theory **13**(1), 99–116 (1989)
3. Cheong, O., Haverkort, H., Lee, M.: Computing a minimum-dilation spanning tree is NP-hard. Comput. Geom. Theory Appl. **41**(3), 188–205 (2008)
4. Aronov, B., de Berg, M., Cheong, O., Gudmundsson, J., Haverkort, H., Smid, M., Vigneron, A.: Sparse geometric graphs with small dilation. Computational Geometry **40**(3), 207–219 (2008)

5. Sigurd, M., Zachariasen, M.: Construction of minimum-weight spanners. In: Albers, S., Radzik, T. (eds.) ESA 2004. LNCS, vol. 3221, pp. 797–808. Springer, Heidelberg (2004)
6. Farshi, M., Gudmundsson, J.: Experimental study of geometric t-spanners. J. Exp. Algorithmics **14**, 3:1.3–3:1.39 (2009)
7. Klein, R., Kutz, M.: Computing geometric minimum-dilation graphs Is NP-hard. In: Kaufmann, M., Wagner, D. (eds.) GD 2006. LNCS, vol. 4372, pp. 196–207. Springer, Heidelberg (2007)
8. Magnanti, T.: Wolsey: Optimal trees. CORE discussion paper. Center for Operations Research and Econometrics (1994)
9. Resende, M.G.C., Ribeiro, C.C.: Greedy randomized adaptive search procedures: advances, hybridizations, and applications. In: Glover, F., Kochenberger, G.A. (eds.) Handbook of Metaheuristics, vol. 57. International Series in Operations Research and Management Science. second edn., pp. 219–249. Springer (2009)
10. Prim, R.C.: Shortest connection networks and some generalizations. The Bell Systems Technical Journal **36**(6), 1389–1401 (1957)
11. Ribeiro, C.C., Resende, M.G.C.: Path-relinking intensification methods for stochastic local search algorithms. Computers and Operations Research **37**, 498–508 (2010)

Speedy Colorful Subtrees

W. Timothy J. White[1]($\boxtimes$), Stephan Beyer[2], Kai Dührkop[1],
Markus Chimani[2], and Sebastian Böcker[1]

[1] Chair for Bioinformatics, Friedrich-Schiller-University, Jena, Germany
{tim.white,kai.duehrkop,sebastian.boecker}@uni-jena.de
[2] Institute of Computer Science, University of Osnabrück, Osnabrück, Germany
{stephan.beyer,markus.chimani}@uni-osnabrueck.de

Abstract. Fragmentation trees are a technique for identifying molecular formulas and deriving some chemical properties of metabolites—small organic molecules—solely from mass spectral data. Computing these trees involves finding exact solutions to the NP-hard MAXIMUM COLORFUL SUBTREE problem. Existing solvers struggle to solve the large instances involved fast enough to keep up with instrument throughput, and their performance remains a hindrance to adoption in practice.

We attack this problem on two fronts: by combining fast and effective reduction algorithms with a strong integer linear program (ILP) formulation of the problem, we achieve overall speedups of 9.4 fold and 8.8 fold on two sets of real-world problems—without sacrificing optimality. Both approaches are, to our knowledge, the first of their kind for this problem. We also evaluate the strategy of solving *global* problem instances, instead of first subdividing them into many *candidate* instances as has been done in the past. Software (C++ source for our reduction program and our CPLEX/Gurobi driver program) available under LGPL at https://github.com/wtwhite/speedy_colorful_subtrees/.

1 Introduction

Metabolites—small molecules involved in cellular reactions—provide a direct functional signature of cellular state. Untargeted metabolomics aims to identify all such compounds present in a biological or environmental sample, and the predominant technology in use is mass spectrometry (MS). This remains a challenging problem, in particular for the many compounds that cannot be found in any spectral library [17,18]. Here we consider tandem mass spectra (MS^2), which measure the masses and abundances of fragments of an isolated compound.

A first step toward full structural elucidation of a compound is the identification of its molecular formula. While it is possible to derive the molecular formula for a given exact mass, measurement inaccuracies have to be considered. Even for high-accuracy instruments, when using an appropriate error range for the mass measurement there may be thousands of possible molecular formulas for a given mass [7]. Approaches for identifying the correct formula include isotope pattern analysis [3], fragmentation pattern analysis [2], or a combination of both [8,9,11,12,15].

© Springer International Publishing Switzerland 2015
D. Xu et al. (Eds.): COCOON 2015, LNCS 9198, pp. 310–322, 2015.
DOI: 10.1007/978-3-319-21398-9_25

Computation of fragmentation trees [12] is a highly powerful method for fragmentation pattern analysis: In the 2013 CASMI (Critical Assessment of Small Molecule Identification) Challenge for identifying molecular formulas, a combination of fragmentation tree and isotope pattern analysis was selected "best automated tool" [6,10]. In addition, fragmentation tree structure can help to derive information about an unknown compound's structure [13,16]. Peaks in the spectrum are annotated with molecular formulas by looking for *consistent explanations*, using knowledge of possible fragmentation events and their probabilities. This translates into finding exact solutions to the NP-hard MAXIMUM COLORFUL SUBTREE (MCS) problem, described later. Unfortunately the problem instances generated can contain over 100,000 edges, and the performance of existing approaches cannot keep up with the throughput of the MS instruments, sometimes limiting the method's appeal in practice. Heuristics often fail to find the optimal solution, and a simple integer linear program (ILP) has been identified as the fastest exact method [14].

We attack this problem on two fronts: by combining fast and effective reduction algorithms with facet-defining inequalities for the ILP formulation of the problem, we achieve overall speedups of 9.4 fold and 8.8 fold on two sets of real-world problems—without sacrificing optimality. Both approaches are, to our knowledge, the first of their kind for this problem. We also evaluate the strategy of solving *global* problem instances, instead of first subdividing them into many *candidate* instances as has been done in the past. Here, we will not evaluate the quality of solutions, as these are identical for any exact method; also, we will assume the edge weights of the MCS problem to be given [5].

1.1 Fragmentation Trees Are Maximum Colorful Subtrees

Consider an MS^2 spectrum containing k peaks $p_1, \ldots, p_k$, having mass-to-charge (m/z) ratios m_i and peak intensities q_i for $1 \leq i \leq k$, listed in decreasing m/z order. Following Böcker and Rasche [2], we use the Round-Robin algorithm [1] to find all possible *explanations* of the parent peak—that is, all candidate molecular formulas having m/z approximately equal to m_1. Each such formula becomes the 1-colored root vertex in a separate MCS instance graph. Within each MCS instance, i-colored vertices are added for each possible explanation of peak p_i, for all $2 \leq i \leq k$. Whenever the molecular formula of v is a subformula of the formula of u, indicating that v could possibly be generated by fragmenting u, we add a directed edge (u, v) and assign an edge weight (which may be positive, negative or zero) according to a probabilistic model of fragmentation. Intuitively, a rooted colorful subtree T in one of these graphs maps each peak to at most one molecular formula in such a way that all formulas in T are consistent with fragmentation of the candidate formula at the root, with the tree of highest total weight corresponding to the best such explanation. By calculating the weights of these optimal trees for all MCS instances and ranking them, the best candidate formula for the spectrum can be determined. Fig. 1 shows an example.

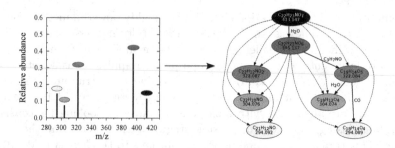

Fig. 1. Example MS^2 spectrum and resulting MCS instance. Nodes (peak explanations) show their molecular formulas and weights. Edges (fragmentation events) in the optimal subtree are solid and labeled with their neutral losses. Edge weights not shown.

The full version of this paper discusses a technique for solving a single *global* MCS instance representing the entire problem, instead of multiple *candidate* instances.

Formally, an instance of the MCS problem is given by (V, E, C, w, c, r) where V is the set of vertices, $E \subsetneq V^2$ is the set of directed edges, C is the set of colors, $w \colon E \to \mathbb{R}$ is the weight function on edges, $c \colon V \to C$ is the function defining colors for each vertex, and $r \in V$ is a distinguished vertex called the *root*. The graph (V, E) is acyclic, and there is a path from r to every $v \in V$.

A subgraph $G' \subseteq G$ is *colorful* iff all vertices in G' have different colors. The weight of an edge $e = (u, v) \in E$ is given by $w(u, v)$, and we define $w(u, v) = -\infty$ when $(u, v) \notin E$. We further extend this function to operate on any subgraph G' in the usual way, by summing over all edges in G'. A subgraph $X \subseteq G$ *dominates* a subgraph $Y \subseteq G$ iff $w(X) \geq w(Y)$.

We would like to assign weights to both edges and vertices: the former to reflect the likelihood of the specific neutral loss in question; the latter to capture peak-specific or explanation-specific information such as peak intensity, mass deviation between measurement and prediction, and estimates of formula plausibility. In order to represent a weight function $w' \colon V \cup E \to \mathbb{R}$ on both vertices and edges using a weight function $w \colon E \to \mathbb{R}$ on edges only, we can simply set $w(u, v) := w'(u, v) + w'(v)$ for each $(u, v) \in E$, since every valid subtree containing v must contain exactly one incoming edge (u, v).

Our goal is to find a maximum colorful r-rooted subtree T of G: that is, among all subtrees rooted at r and in which at most one vertex of any given color appears, a subtree having maximum total weight. This problem is NP-hard. It remains NP-hard even if G is a tree with unit edge weights [2], or if color constraints are dropped [14].[1]

We say that a subgraph G' is *below* a vertex u iff there is a path of zero or more edges from u to every vertex in G'. We denote by G_u the unique maximal

---

[1] When edge weights are constrained to be nonnegative and color constraints are dropped, all leaves will appear in some optimal solution and the problem reduces to the polynomial-time-solvable maximum spanning tree problem.

subgraph of G below u. A color i is below u iff there exists a path from u to a vertex of color i. Furthermore a color i is below a color i' iff there exists an i'-colored vertex u such that i is below u. A subtree T of a graph G is *full in* G iff it is rooted at some vertex u of G, and every edge in G below u is in T. We use the term "cost" to describe a (typically negative) quantity that is to be *added* to a weight to produce another weight. We also declare a vertex to be an ancestor of itself, and use the adjective *strict* to denote non-self ancestors.

Let $n := |V|$, $m := |E|$, and $k := |C|$. Let $\delta^-(U) := \{(v, u) \mid u \in U, v \notin U\}$, $\delta^+(U) := \{(u, v) \mid u \in U, v \notin U\}$, and $\delta(U) := \delta^-(U) \cup \delta^+(U)$. When $U = \{u\}$ we dispense with the braces. We also define $V_i := \{v \in V \mid c(v) = i\}$.

2 Data Reduction

Our data reduction rules seek to shrink an MCS instance X to a smaller instance X' by deleting edges that are provably unnecessary—that is, edges that are simultaneously absent from some optimal solution to X. Here we outline our rules and their computationally efficient implementations.

Vertex Upper Bounds. The following sections describe upper bounds $U(\cdot)$ on the maximum-weight subtree rooted at some vertex u. Particular upper bounds are named by subscripting U; when just U with no subscript appears, it means that any arbitrary upper bound can be substituted. Trivially we have that $U(u) \geq 0$ for all u, since the 0-weight tree containing just u and no edges is a subtree rooted at u. A computationally useful property of all vertex upper bounds is that they remain valid in the face of edge deletions, enabling reductions to safely delete multiple edges in between bound updates.

Child Upper Bound. A simple upper bound $U_\chi(\cdot)$ for a given vertex u can be obtained by considering the upper bounds of u's children and the edges leading from u to them. Specifically we may choose, among all u's outgoing edges to i-colored children $(u, v) \in \delta^+(u) \cap \delta^-(V_i)$, either the edge (u, v) that maximises $w(u, v) + U(v)$ or no edge if this expression is negative. Summing across colors i yields equation (1). This bound tends to become very loose for vertices near the top of the graph, since high-weight edges near the bottom of the graph will usually be visited by large numbers of paths. Nevertheless it is capable of eliminating many edges near the bottom of the graph when applied to the vertex upper bound reduction rule. It can be considerably strengthened by incorporating other vertex upper bounds, such as the Colorful Forest upper bound.

$$U_\chi(u) = \sum_{i \in \{c(v) \mid (u,v) \in E\}} \max\Big\{0, \max_{\substack{(u,v) \in E, \\ c(v) = i}} \big(w(u, v) + U(v)\big)\Big\} \qquad (1)$$

We calculate this bound in $O(m \log k)$ time and $O(n+k)$ space using dynamic programming.

Colorful Forest Upper Bound. We next describe an upper bound $U_\lambda(\cdot)$ obtained by relaxing the subtree constraint. Consider a vertex u, and the subgraph G_u below u. Suppose that for each color i below u we choose either no edge, or some edge $(v, v') \in E(G_u)$ such that $c(v') = i$. All colorful forests in G_u may be generated by choosing incoming edges in this way, and this set of subgraphs contains the set of all colorful subtrees rooted at u, so the problem of finding a maximum-weight colorful forest in G_u is a relaxation of the u-rooted MCS problem. The optimal solution to the relaxed problem is easily found by choosing, for each color i, the maximum-weight incoming edge when this is positive and no edge otherwise, yielding an upper bound on the weight of a colorful subtree rooted at u. This is given in equation (2).

$$U_\lambda(u) = \sum_{i \in C} \max\left\{0, \max_{\substack{(v,v') \in E(G_u), \\ c(v')=i}} w(v, v')\right\} \tag{2}$$

Dynamic programming permits calculation in $O(km)$ time and $O(kn)$ space.

Strengthening the Colorful Forest Bound. The Colorful Forest bound can be strengthened by noticing that whenever the forest that it constructs fails to be a tree, we can determine an upper bound on the cost that must be incurred to transform it into one. This upper bound can be added to the weight of the forest to produce a new, stronger vertex upper bound $U_\Lambda(\cdot)$. Here we merely mention that careful implementation allows this stronger bound to be computed in the same time complexity as the original; for a full description, see the full version of this paper.

Anchor Lower Bound. Given that a vertex u is in the solution T, what is a lower bound $L_a(u, v)$ on the cost of forcing in a given vertex v? Here we assume that T does not already contain a $c(v)$-colored vertex, and only consider attaching v to a vertex in T by a single edge.

If v is a child of u, then clearly $w(u, v)$ is a possibility. Regardless, it may still be possible to attach v to a strict ancestor of u. Specifically, since the "anchor" vertex u is in T by assumption, either $u = r$ or one of the parents of u is also in T. Therefore to form a lower bound, we have the option of attaching v to u if this is possible, or to the worst of u's parents, recursively:

$$L_a(u, v) = \begin{cases} \max\left\{w(u, v), \min_{(p,u) \in E} L_a(p, v)\right\}, & u \neq r \\ w(r, v), & u = r \end{cases}$$

recalling that we define $w(u, v) = -\infty$ whenever $(u, v) \notin E$. ($L_a(u, v)$ will produce $-\infty$ iff there is some path from r to u that contains no vertex with an edge to v.)

$L_a(\cdot, \cdot)$ can be computed via dynamic programming in $O(n^2)$ time and space.

It is also helpful to define $L_{a'}(u, v) := \min_{(p,u) \in E} L_a(p, v)$. This variant of $L_a(\cdot, \cdot)$ excludes any direct edge from u to v from consideration.

Slide Lower Bound. Suppose we have a solution T which contains a vertex u. We want to calculate a lower bound $L_s(u, v)$ on the cost of changing T into a new solution T' by replacing u with another given vertex v of the same color as u. We call this the *Slide lower bound* because in the usual representation of fragmentation graphs, vertices of the same color occupy the same row, so forcing v into and u out of T is akin to horizontally sliding the endpoint of an edge in T from u to v. Such a modification may in general completely change the vertices and edges in the tree below u, subject to the important restriction that it respects color usage: that is, it only ever transforms a subtree T_u into a subtree T'_v such that $c(T'_v) \subseteq c(T_u)$. This reflects the fact that we cannot safely insert vertices of new colors, because these colors may already be in use by other parts of the solution. The full version of this paper describes how to compute $L_s(u, v)$ by dynamic programming in $O(mn_k)$ time and space, where n_k is the maximum number of vertices of any color.

Vertex Upper Bound Rule. If for some edge (u, v) we have that $w(u, v) + U(v) \leq 0$, then clearly any solution containing (u, v) is dominated by a solution in which (u, v) and any subtree below it have been deleted, implying that (u, v) can be safely deleted. Applying this rule before other rules removes certain uninteresting special cases from consideration.

Slide Rule. Whenever two edges exist from a vertex u to distinct vertices v and v' of the same color, there is an opportunity to apply the Slide reduction rule. If

$$w(u, v') - w(u, v) + L_s(v, v') > 0 \tag{3}$$

holds then any solution T containing (u, v) can be improved by sliding (u, v) to (u, v'). This rule can be strengthened by replacing the first term with $L_a(u, v')$, which affords us the chance to connect v' to an ancestor of u. We may then usefully allow $v' = v$ to eliminate edges (u, v) that can always be replaced with a better edge (a, v), where a is a strict ancestor of u.[2]

Dominating Path Rule. The idea behind the Slide rule can be taken further: instead of trying to replace an edge (u, v) with another single edge from an ancestor of u to a vertex of the same color as v, we can replace it with a chain of d edges connecting vertices $v_1, \ldots, v_{d+1}$, with the starting point v_1 an ancestor of u and the endpoint v_{d+1} obeying $c(v_{d+1}) = c(v)$ as before. However we must now pay a price for forcing in each internal vertex v_j for all $2 \leq j \leq d$ in this chain, because the solution may already contain some different vertex of color $c(v_j)$ that needs to be dealt with. This can be done for each such internal vertex by using the Slide lower bound. Suppose the path we wish to force in contains some i-colored vertex x, but the solution already contains a conflicting

---

[2] The full version of this paper discusses a subtlety regarding floating-point arithmetic and comparisons for equality.

vertex—an i-colored vertex $y \neq x$. The solution can be patched up by deleting the incoming edge to y and sliding any subtree below y so that it appears below x for a total cost of $L_s(y, x) - w(p, y)$, where p is y's parent in the solution. Since we do not know, for any color i, which i-colored vertex (if any) is already in the solution, we must take the worst case over all i-colored vertices and all their possible incoming edges:

$$L_{\text{force}}(x) = \min_{y \in V_{c(x)}} \left(L_s(y, x) - \max_{(p,y) \in E} w(p, y) \right) \tag{4}$$

It is now possible to state a recursion to calculate an upper bound on the cost to force in a given vertex x, assuming that a vertex u is already in the solution:

$$f(u, x) = \min \left\{ 0, \alpha, \alpha + L_{\text{force}}(x) \right\} \tag{5}$$

$$\alpha = \max \left\{ L_a(u, x), \max_{p,\, (p,x) \in E} \big(f(u, p) + w(p, x) \big) \right\} \tag{6}$$

We now have that an edge (u, v) can be deleted if there exists an edge (x, z) such that $c(z) = c(v)$ and $w(x, z) + f(u, x) + L_s(v, z) > w(u, v)$. $f(u, x)$ can be calculated in $O(n)$ space because its first argument never varies during recursion.

Two further reduction rules, the Implied Edge rule and the Color Combining rule, are described in the full version of this paper.

3   Integer Linear Programming

Rauf *et al.* [14] surveyed different methods to obtain optimal solutions of the MCS problem, including an integer linear program (ILP). We extend this to obtain a strictly stronger LP relaxation, and solve the resulting ILP using the cutting plane method.

The ILP of Rauf *et al.* [14] is equivalent to

$$\max \quad \sum_{(u,v) \in E} w(u, v)\, x_e \tag{7a}$$

$$\text{s.t.} \quad \sum_{e \in \delta^-(V_i)} x_e \leq 1 \qquad \forall i \in C, \tag{7b}$$

$$\sum_{e \in \delta^-(v)} x_e \geq x_{(v,u)} \quad \forall (v, u) \in E, v \neq r \tag{7c}$$

$$x_e \in \{0, 1\} \qquad \forall e \in E \tag{7d}$$

where x_e is assigned 1 iff the directed edge e is included in the solution. For each $v \in V \setminus \{r\}$ their formulation also includes a constraint $\sum_{e \in \delta^-(v)} x_e \leq 1$, but these constraints are redundant due to the *colorful forest constraints* (7b), which ensure that every color is contained in the solution at most once and that there is at most one incoming directed edge for each vertex. The *connectivity*

(a) the input graph, each vertex has a different color

(b) optimal solution of LP relaxation of (7) with objective value 1.5

(c) optimal solution of LP relaxation of (7)∧(8) with objective value 1

Fig. 2. An example showing that the LP relaxation of (7) including (8) is strictly stronger than without (8)

constraints (7c) say that for each non-root vertex of V, there may only be outgoing directed edges if there is an incoming directed edge. Note that the ILP has a linear number of constraints and variables, so its linear relaxation can be solved as-is without separation.

In this paper, we add the constraints

$$\sum_{e \in \delta^-(V_i)} x_e \le \sum_{f \in \delta^-(S)} x_f \quad \forall i \in C \quad \forall S \subseteq V, V_i \subseteq S \tag{8}$$

that prohibit splits and joins of fractional values. The constraints are valid for the ILP since they only forbid the case where the left-hand side is 1 and the right-hand side is 0, which could only happen if the result is not connected. However, the constraints make the LP relaxation strictly stronger, as can be seen in Fig. 2: in Fig. 2(b) the incoming value of v_1 is 0.5 and the incoming value of v_3 is 1 which is forbidden by (8) for $S = \{v_1, v_2, v_3\}$ and $i = c(v_3)$.

We first solve the LP for a subset of the constraints. Then, we solve the *separation problem*: we search (8) for one or more violated constraints, add them to the LP, and re-solve, iterating the process until there are no further violated constraints. Here, the separation problem can be answered by finding, for each $i \in C$, a minimum r-V_i-cut in the solution network (V, E, x) and testing if it is less than $\sum_{e \in \delta^-(V_i)} x_e$. Although (8) contains an exponential number of constraints, the separation problem can be solved in polynomial time using a Maximum Flow algorithm, and only a small number of iterations are typically needed to find a feasible LP solution.

Theorem 1. (7b) *and* (8) *provide facet-defining inequalities of the problem polytope and are the only necessary ones.*

The proof is given in the full version of this paper. Although just these inequalities suffice for correctness, we also keep (7c) for evaluation in practice because they do not need to be separated.

4   Results and Discussion

We tested the performance of our reductions and ILP improvements on a spectral dataset containing 1232 compounds that appear in the KEGG http://www.kegg.jp metabolite database. From this we selected hard instances where the "classic" ILP from Rauf *et al.* [14]—previously being the fastest exact method for the MCS problem—showed poor running times. We computed fragmentation graphs for each compound and built two datasets for evaluation:

- graphs100: A set containing the 10 hardest candidate instances as well as a random sample of a further 90 hard candidate instances. We use this dataset to measure the performance of our reductions and ILP improvements.
- fmm1: A set of 20 hard global instances, comprising 86358 candidate instances in total. We use this dataset to compare the heretofore typical strategy of solving all candidate MCS instances separately, to solving a single global instance. Results for this dataset are given in the full version of this paper.

Rauf *et al.* [14] found that 95 % of MCS instances could be solved by ILP in under 5 seconds, while some took up to 5.6 minutes. To this end, it is sufficient to consider the hard instances in our comparison. The full version of this paper describes both the datasets and our results in more detail.

We implemented our reductions in ft_reduce, a C++ program that understands a simple language for describing the sequence of reductions to perform, affording flexibility in testing different orders and combinations of reductions. We selected three representative reduction scripts to analyse:

- R1 computes vertex upper bounds using both the Child bound and the Colorful Forest bound, and then applies the Vertex Upper Bound rule.
- R2 does the same, but uses the strengthened Colorful Forest bound.
- R3 applies R2 and then all remaining reduction rules.

Each script iterates until no more edges can be removed. The full version of this paper gives the complete scripts.

We implemented our new ILP formulation using a C++ driver program linked with CPLEX 12.6.0 (http://www.ibm.com/software/integration/optimization/cplex-optimization-studio/). Our new facet-defining cuts can be turned on or off using a command-line argument. In the remainder, we call the solver with these cuts turned on "CPLEX+Cuts", and the solver with them turned off "CPLEX" or "stock CPLEX". For the separation of the split-and-join constraints, we use the Maximum Flow code by Cherkassky and Goldberg [4]. We also performed tests using Gurobi 5.5.0 (http://www.gurobi.com/), although we were not able to implement the cuts efficiently using its callback framework.

All computational experiments were performed on a cluster of four 12-CPU 2.4GHz E5645 Linux machines with 48 GB RAM each. All reductions and all ILP solver runs for the graphs100 dataset ran to completion with a RAM limit of 4 GB and a time limit of 2 hours in place. For the fmm1 dataset, the memory limit was increased from 4 GB to 12 GB, but some instances failed to run to completion in the 2 hour limit. Our reduction program is single-threaded, and ILP solvers were operated in single-threaded mode. All time measurements are in elapsed (wallclock) seconds, and exclude time spent on I/O.

4.1   Results for graphs100 Dataset

Fig. 3 (left) shows the effectiveness of our reductions in shrinking the graphs100 problem instances. Every R1 or R2 reduction removed at least 11.6 % of the edges, and every R3 reduction removed at least 35.4 %, with the average reductions being 62.4 %, 64.3 % and 70.4 % for R1, R2 and R3, respectively. Many instances produced much larger reductions, and it is clear from Fig. 3 (left) that reduced instance size is only very weakly correlated with original instance size.

Fig. 3 (right) compares the performance of various combinations of reduction scripts and ILP solvers. Two effects are immediately apparent: using the strengthened ILP formulation improves average solution times for CPLEX by at least a factor of 4; and applying either the R1 or R2 reduction script produces anywhere from a 30.9 % decrease (from unreduced to R2 on stock CPLEX) to a 57.6 % decrease (from unreduced to R1 on CPLEX+Cuts). We note with particular interest that applying both techniques is substantially *more* effective than would be expected by performing each separately: assuming their effects on running time to be independent, we would expect that both performing an R1 reduction and changing from stock CPLEX to CPLEX+Cuts would result in instances taking on average $0.69109 * 0.25065 = 0.173$ times as long to solve, but

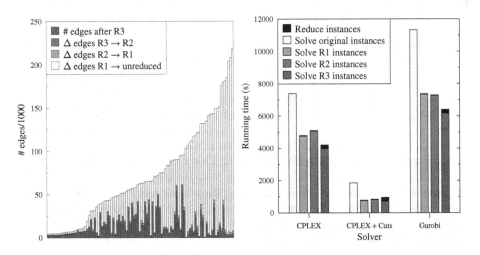

Fig. 3. Left: Comparing unreduced and reduced instance sizes for each graphs100 instance. The bottom bar in each stack gives the number of edges after R3 reduction; higher bars correspond to weaker reductions, with the entire stack indicating the unreduced instance size. Right: Running time evaluation for all graphs100 instances. Each column shows the total elapsed time needed to solve all 100 instances, with reduction time broken out as a black bar at the top.

in fact we find that they take only 0.106 times as long—a relative improvement of 38.7 %, representing a 9.42-fold overall reduction in execution time.

In the other direction, we observe that both stock CPLEX and CPLEX+Cuts take slightly longer to solve the R2 instances than the R1 instances, despite the fact that every R2 instance's edge set is a strict subset of the corresponding R1 instance's edge set, having on average 5.1 % fewer edges. We can only surmise that the additional edges removed by using the strengthened Colorful Forest bound destroyed some structure sought by CPLEX's various heuristics.

The more expensive R3 reductions are a net improvement for the stock CPLEX and Gurobi solvers, but result in an overall slowdown for CPLEX+Cuts.

The highest memory usage on any unreduced instance was 1166MB, 1199MB and 956MB for stock CPLEX, CPLEX+Cuts and Gurobi, respectively. On reduced instances these figures dropped to 853MB, 498MB and 640MB. The highest memory usage by our ft_reduce program was 15MB, 15MB and 31MB for R1, R2 and R3 reductions, respectively.

5   Conclusion

We have presented two highly effective techniques for accelerating the optimal solution of MCS instances, thereby bringing practical *de novo* identification of metabolite molecular formulas a step closer to reality. The two methods complement each other admirably: applying both yields a *larger* speedup than the product of the speedups obtained by applying each separately. Based on our experiments with two real-world datasets, we find that it is essentially always advantageous to use our strengthened ILP formulation and to apply our simple reductions, and frequently advantageous to apply our more complex ones.

The lion's share of the improvement in running times comes from our new, facet-defining cutting planes for ILP solvers. ILP solvers have demonstrated effectiveness across a wide range of hard optimization problems, and we anticipate that they will remain the dominant approach to solving MCS problems. At the same time, the problem reductions we present offer immediately-available speedups (and, often, memory usage reductions) not only for ILP formulations but for any exact or heuristic solution method, such as the "brute force" algorithm of Böcker and Rasche [2] or the Tree Completion heuristic of Rauf *et al.* [14].

We noted above that the use of fragmentation trees goes beyond the determination of molecular formulas [13]: see for instance Shen *et al.* [16] where fragmentation trees are used in conjunction with machine learning to search a molecular structure database using fragmentation spectra. In this analysis pipeline, computing fragmentation trees accounts for more than 90 % of the total running time. To this end, faster methods for this task are highly sought.

References

1. Böcker, S., Lipták, Z.: A fast and simple algorithm for the Money Changing Problem. Algorithmica **48**(4), 413–432 (2007)
2. Böcker, S., Rasche, F.: Towards de novo identification of metabolites by analyzing tandem mass spectra. Bioinformatics **24**, I49–I55 (2008). Proc. of European Conference on Computational Biology (ECCB 2008)
3. Böcker, S., Letzel, M., Lipták, Z., Pervukhin, A.: SIRIUS: Decomposing isotope patterns for metabolite identification. Bioinformatics **25**(2), 218–224 (2009)
4. Cherkassky, B., Goldberg, A.: On implementing push-relabel method for the maximum flow problem. Algorithmica **19**, 390–410 (1997)
5. Dührkop, K., Böcker, S.: Fragmentation trees reloaded. In: Przytycka, T.M. (ed.) RECOMB 2015. LNCS, vol. 9029, pp. 65–79. Springer, Heidelberg (2015)
6. Dührkop, K., Hufsky, F., Böcker, S.: Molecular formula identification using isotope pattern analysis and calculation of fragmentation trees. Mass Spectrom **3**(special issue 2), S0037 (2014)
7. Kind, T., Fiehn, O.: Metabolomic database annotations via query of elemental compositions: Mass accuracy is insufficient even at less than 1 ppm. BMC Bioinformatics **7**(1), 234 (2006)
8. Menikarachchi, L.C., Cawley, S., Hill, D.W., Hall, L.M., Hall, L., Lai, S., Wilder, J., Grant, D.F.: MolFind: A software package enabling HPLC/MS-based identification of unknown chemical structures. Anal Chem **84**(21), 9388–9394 (2012)
9. Meringer, M., Reinker, S., Zhang, J., Muller, A.: MS/MS data improves automated determination of molecular formulas by mass spectrometry. MATCH-Commun Math Co **65**, 259–290 (2011)
10. Nishioka, T., Kasama, T., Kinumi, T., Makabe, H., Matsuda, F., Miura, D., Miyashita, M., Nakamura, T., Tanaka, K., Yamamoto, A.: Winners of CASMI2013: Automated tools and challenge data. Mass Spectrom **3**(special issue 2), S0039 (2014)
11. Pluskal, T., Uehara, T., Yanagida, M.: Highly accurate chemical formula prediction tool utilizing high-resolution mass spectra, MS/MS fragmentation, heuristic rules, and isotope pattern matching. Anal Chem **84**(10), 4396–4403 (2012)
12. Rasche, F., Svatoš, A., Maddula, R.K., Böttcher, C., Böcker, S.: Computing fragmentation trees from tandem mass spectrometry data. Anal Chem **83**(4), 1243–1251 (2011)
13. Rasche, F., Scheubert, K., Hufsky, F., Zichner, T., Kai, M., Svatoš, A., Böcker, S.: Identifying the unknowns by aligning fragmentation trees. Anal Chem **84**(7), 3417–3426 (2012)
14. Rauf, I., Rasche, F., Nicolas, F., Böcker, S.: Finding maximum colorful subtrees in practice. J Comput Biol **20**(4), 1–11 (2013)
15. Rojas-Chertó, M., Kasper, P.T., Willighagen, E.L., Vreeken, R.J., Hankemeier, T., Reijmers, T.H.: Elemental composition determination based on MS^n. Bioinformatics **27**, 2376–2383 (2011)
16. Shen, H., Dührkop, K., Böcker, S., Rousu, J.: Metabolite identification through multiple kernel learning on fragmentation trees. Bioinformatics **30**(12), 157–164 (2014). Proc. of Intelligent Systems for Molecular Biology (ISMB 2014)

17. Tautenhahn, R., Cho, K., Uritboonthai, W., Zhu, Z., Patti, G.J., Siuzdak, G.: An accelerated workflow for untargeted metabolomics using the METLIN database. Nat Biotechnol **30**(9), 826–828 (2012)

18. Wishart, D.S., Knox, C., Guo, A.C., Eisner, R., Young, N., Gautam, B., Hau, D.D., Psychogios, N., Dong, E., Bouatra, S., Mandal, R., Sinelnikov, I., Xia, J., Jia, L., Cruz, J.A., Lim, E., Sobsey, C.A., Shrivastava, S., Huang, P., Liu, P., Fang, L., Peng, J., Fradette, R., Cheng, D., Tzur, D., Clements, M., Lewis, A., Souza, A.D., Zuniga, A., Dawe, M., Xiong, Y., Clive, D., Greiner, R., Nazyrova, A., Shaykhutdinov, R., Li, L., Vogel, H.J., Forsythe, I.: HMDB: A knowledgebase for the human metabolome. Nucleic Acids Res **37**, D603–D610 (2009)

Graph Algorithms II

Algorithmic Aspects
of Disjunctive Domination in Graphs

B.S. Panda[1], Arti Pandey[1(✉)], and S. Paul[2]

[1] Department of Mathematics, Indian Institute of Technology Delhi Hauz Khas,
New Delhi 110016, India
{bspanda,artipandey}@maths.iitd.ac.in
[2] Advanced Computing and Microelectronics Unit, Indian Statistical Institute,
Kolkata 700108, India
paulsubhabrata@gmail.com

Abstract. For a graph $G = (V, E)$, a set $D \subseteq V$ is called a *disjunctive dominating set* of G if for every vertex $v \in V \setminus D$, v is either adjacent to a vertex of D or has at least two vertices in D at distance 2 from it. The cardinality of a minimum disjunctive dominating set of G is called the *disjunctive domination number* of graph G, and is denoted by $\gamma_2^d(G)$. The MINIMUM DISJUNCTIVE DOMINATION PROBLEM (MDDP) is to find a disjunctive dominating set of cardinality $\gamma_2^d(G)$. Given a positive integer k and a graph G, the DISJUNCTIVE DOMINATION DECISION PROBLEM (DDDP) is to decide whether G has a disjunctive dominating set of cardinality at most k. In this article, we first propose a polynomial time algorithm for MDDP in proper interval graphs. Next we tighten the NP-completeness of DDDP by showing that it remains NP-complete even in chordal graphs. We also propose a $(\ln(\Delta^2 + \Delta + 2) + 1)$-approximation algorithm for MDDP, where Δ is the maximum degree of input graph $G = (V, E)$ and prove that MDDP can not be approximated within $(1 - \epsilon)\ln(|V|)$ for any $\epsilon > 0$ unless NP $\subseteq$ DTIME$(|V|^{O(\log \log |V|)})$. Finally, we show that MDDP is APX-complete for bipartite graphs with maximum degree 3.

Keywords: Domination · Chordal graph · Graph algorithm · Approximation algorithm · Np-complete · Apx-complete

1 Introduction

Let $G = (V, E)$ be a graph. For a vertex $v \in V$, let $N_G(v) = \{u \in V | uv \in E\}$ and $N_G[v] = N_G(v) \cup \{v\}$ denote the *open neighborhood* and the *closed neighborhood* of v, respectively. For two distinct vertices $u, v \in V$, the distance $dist_G(u, v)$ between u and v is the length of a shortest path between u and v. A vertex u *dominates* v if either $u = v$ or u is adjacent to v. A set $D \subseteq V$ is called a *dominating set* of $G = (V, E)$ if each $v \in V$ is dominated by a vertex in D, that is, $|N_G[v] \cap D| \geq 1$ for all $v \in V$. The *domination number* of a graph G, denoted by $\gamma(G)$, is the minimum cardinality of a dominating set of G. For a

© Springer International Publishing Switzerland 2015
D. Xu et al. (Eds.): COCOON 2015, LNCS 9198, pp. 325–336, 2015.
DOI: 10.1007/978-3-319-21398-9_26

graph G, the MINIMUM DOMINATION problem is to find a dominating set of cardinality $\gamma(G)$. Domination in graphs is one of the classical problems in graph theory and it has been well studied from theoretical as well as algorithmic point of view [9,10]. Over the years, many variants of domination problem have been studied in the literature due to its application in different fields. The concept of *disjunctive domination* is a recent and an interesting variation of domination [8].

In domination problem, our goal is to place minimum number of sentinels at some vertices of the graph so that all the remaining vertices are adjacent to at least one sentinel. In practice, depending upon the monitoring power, we can have different types of sentinels. To secure the graph with different types of sentinels, we need concept of different variants of domination. Efforts made in this direction have given rise to different types of domination, such as, distance domination, exponential domination, secondary domination. In some cases, it might happen that the monitoring power of a sentinel is inversely proportional to the distance, that is, the domination power of a vertex reduces as the distance increases. Motivated by this idea, Goddard et al. [8] have introduced the concept of *disjunctive domination* which captures the notion of decay in domination with increasing distance. A set $D_d \subseteq V$ is called a *b-disjunctive dominating set* of G if every vertex $v \in V \setminus D_d$ is either adjacent to a vertex in D_d or there are at least b vertices of D_d within a distance of two from v. The minimum cardinality of a b-disjunctive dominating set of G is called the *b-disjunctive domination number* of G and it is denoted by $\gamma_b^d(G)$. A vertex v is said to be *b-disjunctively dominated* by $D_d \subseteq V$ if either $v \in D_d$ or v is adjacent to a vertex of D_d or has at least b vertices in D_d at distance 2 from it. Note that disjunctive domination is more general concept than distance two domination, since the parameter $\gamma_1^d(G)$ is the distance two domination number. For simplicity, 2-disjunctive domination is called disjunctive domination. The disjunctive domination problem and its decision version are defined as follows:

MINIMUM DISJUNCTIVE DOMINATION PROBLEM (MDDP)

Instance: A graph $G = (V, E)$.
Solution: A disjunctive dominating set D_d of G.
Measure: Cardinality of the set D_d.

DISJUNCTIVE DOMINATION DECISION PROBLEM (DDDP)

Instance: A graph $G = (V, E)$ and a positive integer $k \leq |V|$.
Question: Does there exist a disjunctive dominating set D_d of G such that $|D_d| \leq k$?

The concept of disjunctive domination has been introduced recently in 2014 [8] and further studied in [11]. In [8], Goddard et al. have proven bounds on disjunctive domination number for specially regular graphs and claw-free graphs. They have shown that the decision version of b-disjunctive domination is NP-complete for planar and bipartite graphs and also designed a dynamic programming based linear time algorithm to find a minimum b-disjunctive dominating set in a tree. In [11], Henning et al. have studied the relation between domination number and disjunctive domination number of a tree T and proved that $\gamma(T) \leq 2\gamma_2^d(T) - 1$. They have also given a constructive characterization of the trees achieving equality in this bound. On the other hand, a variation of disjunctive domination is also studied in the literature (see [12]).

In this paper, our focus is on algorithmic study of disjunctive domination problem. The rest of the paper is organized as follows. In Section 2, we give some pertinent definitions and notations that would be used in the rest of the paper. In this section, we also observe some graph classes where domination problem is NP-complete but disjunctive domination can be easily solved and vice versa. This motivates us to study the status of the problem in other graph classes. In Section 3, we design a polynomial time algorithm for disjunctive domination problem in proper interval graphs, an important subclass of chordal graphs. In Section 4, we prove that DDDP remains NP-complete for chordal graphs. In Section 5, we design a polynomial time approximation algorithm for MDDP for general graph G with approximation ratio $\ln(\Delta^2 + \Delta + 2) + 1$, where Δ is the maximum degree of G. In this section, we also prove that MDDP can not be approximated within $(1 - \epsilon) \ln(|V|)$ for any $\epsilon > 0$ unless NP $\subseteq$ DTIME($|V|^{O(\log \log |V|)}$). In addition, for bipartite graphs with maximum degree 3, MDDP is shown to be APX-complete in this section. Finally, Section 6 concludes the paper.

2   Preliminaries

2.1   Notations

Let $G = (V, E)$ be a graph. Let $N_G^2(v)$ denote the set of vertices which are at distance 2 from the vertex v in graph G. Let $G[S]$, $S \subseteq V$ denote the induced subgraph of G on the vertex set S. The *degree* of a vertex $v \in V$, denoted by $d_G(v)$, is the number of neighbors of v, that is, $d_G(v) = |N_G(v)|$. The *minimum degree* and *maximum degree* of a graph G is defined by $\delta(G) = \min_{v \in V} d_G(v)$ and $\Delta(G) = \max_{v \in V} d_G(v)$, respectively. A set $S \subseteq V$ is called an *independent set* of G if $uv \notin E$ for all $u, v \in S$. A set $K \subseteq V$ is called a *clique* of G if $uv \in E$ for all $u, v \in K$. A set $C \subseteq V$ is called a *vertex cover* of G if for each edge $ab \in E$, either $a \in C$ or $b \in C$. Let n and m denote the number of vertices and number of edges of G, respectively. In this paper, we only consider connected graphs with at least two vertices.

2.2   Graph Classes

A graph G is said to be a *chordal graph* if every cycle in G of length at least four has a *chord*, that is, an edge joining two non-consecutive vertices of the cycle. Let $\mathscr{F}$ be a family of sets. The *intersection graph* of $\mathscr{F}$ is obtained by taking each set in $\mathscr{F}$ as a vertex and joining two sets in $\mathscr{F}$ if and only if they have a non-empty intersection. A graph G is an *interval graph* if G is the intersection graph of a family $\mathscr{F}$ of intervals on the real line. A graph G is called a *proper interval* graph if it is the intersection graph of a family $\mathscr{F}$ of intervals on the real line such that no interval in $\mathscr{F}$ contains another interval in $\mathscr{F}$ set-theoretically. A vertex $v \in V(G)$ is a *simplicial* vertex of G if $N_G[v]$ is a clique of G. An ordering $\alpha = (v_1, v_2, ..., v_n)$ is a *perfect elimination ordering* (PEO) of G if v_i is a simplicial vertex of $G_i = G[\{v_i, v_{i+1}, ..., v_n\}]$ for all i, $1 \le i \le n$. A graph G has a PEO if and only if G is chordal [7]. A PEO $\alpha = (v_1, v_2, \ldots, v_n)$ of a chordal graph is a *bi-compatible elimination ordering* (BCO) if $\alpha^{-1} = (v_n, v_{n-1}, \ldots, v_1)$, that is, the reverse of α, is also a PEO of G. A graph G has a BCO if and only if G is a proper interval graph [14].

2.3 Domination vs Disjunctive Domination

In this subsection, we make some observations on complexity difference of domination and disjunctive domination problem. It is known that domination problem is NP-complete for split graphs [4] and for graphs with diameter two [2]. But disjunctive domination problem can be easily solved in these graph classes. Because, disjunctive domination number is at most 2 in these classes and $\gamma_2^d(G) = 1$ if and only if G contains a vertex of degree $n - 1$. Next, we define a graph class, called *GC graph*, for which domination problem is easily solvable, but disjunctive domination problem is NP-complete.

Definition 1 (GC graph). *A graph $G' = (V', E')$ is said to be a GC graph if it can be constructed from a general graph $G = (V, E)$ by adding a pendant vertex to every vertex of G. Formally, $V' = V \cup \{w_i \mid 1 \le i \le n\}$ and $E' = E \cup \{v_i w_i \mid 1 \le i \le n\}$.*

Note that, every vertex of a GC graph G' is either a pendant vertex or adjacent to a unique pendant vertex and hence, $\gamma(G') = n$. In Section 4, we show that DDDP is NP-complete for the class of GC graphs.

3 Polynomial Time Algorithm for Proper Interval Graphs

In this section, we present a polynomial time algorithm to find a minimum cardinality disjunctive dominating set in proper interval graphs.

Let $\alpha = (v_1, v_2, \ldots, v_n)$ be a BCO of the proper interval graph G. Let $MaxN_G(v_i)$ denote the maximum index neighbor of v_i with respect to the ordering α. We start with an empty set D. At each iteration i of the algorithm, we update the set D in such a way that the vertex v_i and all the vertices which appear before v_i in the BCO α, are disjunctively dominated by the set D. At the end of nth iteration, D disjunctively dominate all the vertices of graph G. The algorithm DISJUNCTIVE-PIG for finding a minimum cardinality disjunctive dominating set in a proper interval graph is given below.

Next we give the proof of correctness of the algorithm. Let $\alpha = (v_1, v_2, \ldots, v_n)$ be the BCO of a proper interval graph G. Define the set $V_i = \{v_1, v_2, \ldots, v_i\}$, $1 \le i \le n$, and $V_0 = \emptyset$. Also suppose that D_i denotes the set D obtained after processing vertex v_i, $1 \le i \le n$, and $D_0 = \emptyset$. We will prove that D_n is a minimum cardinality disjunctive dominating set of G.

Algorithm 1. DISJUNCTIVE-PIG$(G, \alpha = (v_1, v_2, \ldots, v_n))$

Initialize $D = \emptyset$;
for $i = 1 : n$ **do**
  Compute $N_G[v_i] \cap D$ and $N_G^2(v_i) \cap D$;
  Case 1: Either $N_G[v_i] \cap D \neq \emptyset$, or $|N_G^2(v_i) \cap D| \geq 2$
    No update in D is done;
  Case 2: $N_G[v_i] \cap D == \emptyset$ and $N_G^2(v_i) \cap D == \emptyset$
    Update D as $D = D \cup \{MaxN_G(v_i)\}$;
  Case 3: $N_G[v_i] \cap D == \emptyset$ and $|N_G^2(v_i) \cap D| == 1$
    Find $v_r \in N_G^2(v_i) \cap D$;
    $v_j = MaxN_G[v_i]$; $v_k = MaxN_G[v_j]$;
    $S = \{v_{i+1}, v_{i+2}, \ldots, v_{j-1}\}$;
    Subcase 3.1: For every $v \in S$, either $vv_k \in E$ or $d(v, v_r) = 2$
      Update D as $D = D \cup \{v_k\}$;
    Subcase 3.2: v_s is the least index vertex in S such that
    $d(v_s, v_k) = 2$ and $d(v_s, v_r) > 2$
      Update D as $D = D \cup \{MaxN_G(v_s)\}$;
return D;

Theorem 1. *For each i, $0 \leq i \leq n$, the following statements are true:*

(a) *D_i disjunctively dominates the set V_i.*
(b) *There exists a minimum cardinality disjunctive dominating set D_d^* such that D_i is contained in D_d^*.*

Proof. We prove the theorem by induction on i. The basis step is trivial as $D_0 = \emptyset$. Next assume that the theorem is true for $i - 1$. So, (a) D_{i-1} disjunctively dominates the set V_{i-1}, (b) there exists a minimum cardinality disjunctive dominating set D_d^* such that D_{i-1} is contained in D_d^*.

Next we prove the theorem for i. According to our algorithm, we need to discuss the following three cases.

Case 1: Either $N_G[v_i] \cap D_{i-1} \neq \emptyset$, or $|N_G^2(v_i) \cap D_{i-1}| \geq 2$.
Here $D_i = D_{i-1}$. It is easy to notice that all the conditions of the theorem are satisfied.

Case 2: $N_G[v_i] \cap D_{i-1} = \emptyset$ and $N_G^2(v_i) \cap D_{i-1} = \emptyset$.
Here $D_i = D_{i-1} \cup \{v_j\}$ where $v_j = MaxN_G(v_i)$. Hence, condition (a) of the theorem is trivially satisfied. If $v_j \in D_d^*$, then $D_i \subseteq D_d^*$. Hence both the conditions of the theorem are satisfied, and D_d^* is the required minimum cardinality disjunctive dominating set of G. If $v_j \notin D_d^*$, then there are two possibilities:

(I) There exists a vertex $v_p \in N_G[v_i] \cap D_d^*$.
Define the set $D_d^{**} = (D_d^* \setminus \{v_p\}) \cup \{v_j\}$. Note that $D_i \subseteq D_d^{**}$, and $|D_d^*| = |D_d^{**}|$. Now, to prove condition (b) of the theorem, it is enough to show that D_d^{**} is a disjunctive dominating set of G. Note that $D_{i-1} \cup \{v_j\} \subseteq D_d^{**}$. Now consider an arbitrary vertex v_a of G. If $a < i$, then the vertex v_a is disjunctively dominated

by the set D_{i-1}, and hence by D_d^{**}. If $a \geq i$, and $v_p \in N_G[a]$, then $v_j \in N_G[v_a]$. If $a \geq i$, and $v_p \in N_G^2(v_a)$, then $v_j \in N_G[v_a]$ or $v_j \in N_G^2(v_a)$. This proves that D_d^{**} is a disjunctive dominating set of G.

(II) For $q < s$, vertices $v_q, v_s \in N_G^2(v_i) \cap D_d^*$.
Let $MaxN_G(v_i) = v_j$ and $MaxN_G(v_j) = v_k$. Then $q < s \leq k$. Let $v_t = MaxN_G(v_s)$ and $v_r = MaxN_G(v_t)$. We again consider three possibilities:

(i) $q < s < i$
Here $r \leq j$. Now consider an arbitrary vertex v_a of G. If $a < i$, then the vertex v_a is disjunctively dominated by the set D_{i-1}. If $a \geq i$, and $v_s \in N_G^2(v_a)$ or $v_q, v_s \in N_G^2(v_i)$, then $v_j \in N_G[v_a]$. Hence $(D_d^* \setminus \{v_q, v_s\}) \cup \{v_j\}$ is a disjunctive dominating set of G of cardinality less than $|D_d^*|$, which is a contradiction, as D_d^* is a minimum disjunctive dominating set of G. Therefore, this situation will never arise.

(ii) $q < i < s$
Consider an arbitrary vertex v_a of G. If $a < i$, then the vertex v_a is disjunctively dominated by the set D_{i-1}. If $a \geq i$, and $v_q \in N_G^2(v_a)$, then $v_j \in N_G[v_a]$. If $a \geq i$, and $v_q \notin N_G^2(v_a)$, and either $v_s \in N_G[v_a]$ or $v_s \in N_G^2(v_a)$, then either $v_j \in N_G[v_a]$ or $v_t \in N_G[v_a]$. Hence, if we define $D_d^{**} = (D_d^* \setminus \{v_q, v_s\}) \cup \{v_j, v_t\}$, then D_d^{**} is a minimum cardinality disjunctive dominating set of G and $D_i \subseteq D_d^{**}$. This proves the condition (b) of the theorem.

(iii) $i < q < s$
Here $s \leq k$. Consider an arbitrary vertex v_a of G. If $a < i$, then the vertex v_a is disjunctively dominated by the set D_{i-1}. If $a \geq i$, and $v_q \in N_G[v_a]$ or $v_s \in N_G[v_a]$ or $v_q, v_s \in N_G^2(v_a)$ or $v_s \in N_G^2(v_a)$, then either $v_j \in N_G[v_a]$ or $v_t \in N_G[v_a]$. Hence, if we define $D_d^{**} = (D_d^* \setminus \{v_q, v_s\}) \cup \{v_j, v_t\}$, then D_d^{**} is a minimum cardinality disjunctive dominating set of G and $D_i \subseteq D_d^{**}$. This proves the condition (b) of the theorem.

Case 3: $N_G[v_i] \cap D_{i-1} = \emptyset$ and $|N_G^2(v_i) \cap D_{i-1}| = 1$.
   In this case as well, condition (a) and condition (b) of the theorem are satisfied. Due to space constraints, the proof is omitted.
   Hence our theorem is proved.   □

In view of the above theorem, the set D computed by the algorithm DISJUNCTIVE-PIG is a minimum cardinality disjunctive dominating set of G. Now, we show that the algorithm DISJUNCTIVE-PIG can be implemented in polynomial time. We use the adjacency list representation of the graph. We maintain an array D_{set} for the set D such that $D_{set}[j] = 1$ if $v_j \in D$. We maintain the all pair distance matrix $Dist[1..n, 1..n]$ such that $Dist[i, j]$ is the distance between v_i and v_j. This can be done in $O(n^3)$ time. Now $N_G[v_i] \cap D$ can be computed in $O(n)$ time by looking up $Dist$ matrix and array D_{set}. Similarly, $N_G^2(v_i) \cap D$ can be computed in $O(n)$ time. Also $MaxN_G(v_i)$ can be computed in $O(n)$ time. Hence, in any iteration, all the operations can be done in $O(n^2)$ time. Therefore overall time is $O(n^3)$, as number of iterations are n. Since, BCO of a proper interval graph can be computed in $O(n + m)$ time [15], and all the computations in the algorithm DISJUNCTIVE-PIG can be done in $O(n^3)$ time, we have the following theorem.

Theorem 2. *MDDP can be solved in $O(n^3)$ time in proper interval graphs.*

However, the algorithm DISJUNCTIVE-PIG can be implemented in $O(n + m)$ time using additional data structures. The details are omitted due to space constraints.

4   NP-completeness

In this section, we prove that DDDP is NP-complete for chordal graphs. For that, we first show that DDDP is NP-complete for GC graphs. To prove this NP-completeness result, we use a reduction from another variant of domination problem, namely 2-*domination problem*. For a graph $G = (V, E)$, a set $D_2 \subseteq V$ is called 2-*dominating set* if every vertex $v \in V \setminus D_2$ has at least two neighbors in D_2. Given a positive integer k and a graph $G = (V, E)$, the 2-DOMINATION DECISION PROBLEM (2DDP) is to decide whether G has a 2-dominating set of cardinality at most k. It is known that 2DDP is NP-complete for chordal graphs [13]. The following lemma shows that DDDP is NP-complete for GC graphs.

Lemma 1. *DDDP is NP-complete for GC graphs.*

Proof. Clearly, DDDP is in NP for GC graphs. To prove the NP-hardness, we give a polynomial transformation from 2DDP for general graphs. Let $G = (V, E)$ where $V = \{v_1, v_2, \ldots, v_n\}$, and a positive integer k be an instance of 2DDP. Given G, we construct the graph $G' = (V', E')$ in the following way: $V' = V \cup \{w_i \mid 1 \le i \le n\}$ and $E' = E \cup \{v_i w_i \mid 1 \le i \le n\}$. Clearly G' is a GC graph and it can be constructed from G in polynomial time.

The following claim is enough to complete the proof of the theorem.
Claim 1 G has a 2-dominating set of cardinality at most k if and only if G' has a disjunctive dominating set of cardinality at most k.

Proof. (Proof of the claim) Let D_2 be a 2-dominating set of G of cardinality at most k. Clearly D_2 is a disjunctive dominating set of G'. Because every $v_i \in V'$ either is in D_2 or dominated by at least two vertices of D_2 and every $w_i \in V'$ is either dominated by $v_i \in D_2$ or contains at least two vertices from D_2 at a distance of two. Hence, G' has a disjunctive dominating set of cardinality at most k.

Conversely, suppose that D_d is a disjunctive dominating set of G' of cardinality at most k. Note that, every vertex of G' is either a pendant vertex or a support vertex. Also, the vertex set of graph G is exactly the set of all support vertices of G'. Let P be the set of pendant vertices of graph G', i.e., $P = \{w_i \mid 1 \le i \le n\}$. If a pendant vertex $w_i \in D_d$, then the set $D'_d = (D_d \setminus \{w_i\}) \cup \{v_i\}$ still remains a disjunctive dominating set of G' of cardinality at most k. So, without loss of generality we assume that $D_d \cap P = \emptyset$. Now for every vertex $v_i \in V$, either $v_i \in D_d$ or $|N_G(v_i) \cap D_d| \ge 2$. If not, let there is a vertex $v_i \in V \setminus D_d$ such that

$|N_G(v_i) \cap D_d| \leq 1$. This implies that the vertex $w_i \in V'$ is neither dominated nor has at least two vertices from D_d at a distance of two, contradicting the fact that D_d is a disjunctive dominating set of G'. Hence, D_d is a 2-dominating set of G of cardinality at most k.   □

Hence, it is proved that DDDP is NP-complete for GC graphs.   □

It is easy to observe that, if the graph G is chordal, then the constructed graph G' in Lemma 1 is also chordal. Hence, we have the following theorem.

Theorem 3. *DDDP is NP-complete for chordal graphs.*

5 Approximation Results

5.1 Approximation Algorithm

In this subsection, we propose a $(\ln(\Delta^2 + \Delta + 2) + 1)$-approximation algorithm for MDDP. Our algorithm is based on the reduction from MDDP to the CON-STRAINED MULTISET MULTICOVER (CMSMC) problem. We first recall the definition of the CONSTRAINED MULTISET MULTICOVER problem.

Let X be a set and $\mathcal{F}$ be a collection of subsets of X. The SET COVER problem is to find a smallest sub-collection, say $\mathcal{C}$ of $\mathcal{F}$, such that $\mathcal{C}$ covers all the elements of X, that is, $\cup_{S \in \mathcal{C}} S = X$. The CONSTRAINED MULTISET MULTICOVER problem is a generalization of the SET COVER problem. In this problem, $\mathcal{F}$ is the collection of multisets of X, that is, each element $x \in X$ occurs in a multiset $S \in \mathcal{F}$ with arbitrary multiplicity, and each element $x \in X$ has an integer coverage requirement r_x which specifies how many times x has to be covered. Note that each set $S \in \mathcal{F}$ is chosen at most once. So, for a given set X, a collection $\mathcal{F}$ of multisets of X, and integer requirement r_x for each $x \in X$, the CMSMC problem is to find a smallest collection $\mathcal{C} \subseteq \mathcal{F}$, such that $\mathcal{C}$ covers each element x in X at least r_x times. In the case, when r_x is constant for each $x \in X$, then $\mathcal{C}$ is called a r_x-cover of X, and the CMSMC problem is to find a minimum cardinality r_x-cover of X.

Theorem 4. *The* MINIMUM DISJUNCTIVE DOMINATION PROBLEM *for a graph* $G = (V, E)$ *with maximum degree* Δ *can be approximated with an approximation ratio of* $\ln(\Delta^2 + \Delta + 2) + 1$.

Proof. Let us show the transformation from MDDP to the CMSMC problem.
Construction : Let $G = (V, E)$ be a graph with n vertices and m edges where $V = \{v_1, v_2, \ldots, v_n\}$ (an instance of MDDP). Now we construct an instance of the CMSMC problem, that is, a set X, a family $\mathcal{F}$ of multisets of X, and a vector $R = (r_x)_{x \in X}$ (r_x is a non-negative integer for each $x \in X$) in the following way:

$X = V$, $\mathcal{F} = \{F_1, F_2, \ldots, F_n\}$, where for each i, $1 \leq i \leq n$, F_i is a multiset which contains two copies of each element in $N_G[v_i]$ and one copy of the set of elements which are at distance 2 from the vertex v_i in graph G, $r_x = 2$ for each $x \in X$.

Now we first need to prove the following correspondence.

Claim 2 The set $D = \{v_{i_1}, v_{i_2}, \ldots, v_{i_k}\}$ is a disjunctive dominating set of G if and only if $C = \{F_{i_1}, F_{i_2}, \ldots, F_{i_k}\}$ is a 2-cover of X.

Proof. The proof is omitted due to space constraints.   □

By the above claim, if D_d^* is a minimum cardinality disjunctive dominating set of G and C^* is an optimal 2-cover of X, then $|D_d^*| = |C^*|$. In [16], S. Rajgopalan and V. V. Vazirani gave a greedy approximation algorithm for the CMSMC problem, which achieves an approximation ratio of $\ln(|F_M|) + 1$, where F_M is the maximum cardinality multiset in $\mathcal{F}$. Let C^* be an optimal 2-cover and C' be a 2-cover obtained by greedy approximation algorithm, then $|C'| \leq (\ln(|F_M|) + 1) \cdot |C^*|$. Given a 2-cover of X, we can also obtain a disjunctive dominating set of graph G of same cardinality. Suppose that D_d' is a disjunctive dominating set of G obtained from 2-cover C' of X. Then $|D_d'| \leq (\ln(|F_M|) + 1) \cdot |D_d^*|$. If the maximum degree of the graph G is Δ, then the cardinality of a set in family C will be at most $2(\Delta + 1) + \Delta(\Delta - 1)$, which is equal to $\Delta^2 + \Delta + 2$. Hence $|D_d'| \leq (\ln(\Delta^2 + \Delta + 2) + 1) \cdot |D_d^*|$. This completes the proof of the theorem.   □

5.2   Lower Bound on Approximation Ratio

To obtain the lower bound, we give an approximation preserving reduction from the MINIMUM DOMINATION problem. The following approximation hardness result for the MINIMUM DOMINATION problem is already known.

Theorem 5. *[5] For a graph $G = (V, E)$, the MINIMUM DOMINATION problem can not be approximated within $(1 - \epsilon) \ln |V|$ for any $\epsilon > 0$ unless $NP \subseteq DTIME$ $(|V|^{O(\log \log |V|)})$.*

Theorem 6. *For a graph $G = (V, E)$, MDDP can not be approximated within $(1 - \epsilon) \ln |V|$ for any $\epsilon > 0$ unless $NP \subseteq DTIME(|V|^{O(\log \log |V|)})$.*

Proof. Let us describe the reduction from the MINIMUM DOMINATION problem to MDDP. Let $G = (V, E)$, where $V = \{v_1, v_2, \ldots, v_n\}$ be an instance of the MINIMUM DOMINATION problem. Now, we construct a graph $H = (V_H, E_H)$ an instance of MDDP in the following way: $V_H = V \cup \{w_i, z_i \mid 1 \leq i \leq n\} \cup \{p, q\}$, $E_H = E \cup \{v_i w_i, w_i z_i, z_i p \mid 1 \leq i \leq n\} \cup \{pq\}$.

Fig. 1 illustrates the construction of the graph H from a given graph G. Note that $|V_H| = 3|V| + 2$.

If D^* is a minimum cardinality dominating set of G, then $D^* \cup \{p\}$ is a disjunctive dominating set of H. Hence for a minimum cardinality disjunctive dominating set D_d^* of H, $|D_d^*| \leq |D^*| + 1$.

On the other hand, let D_d be a disjunctive dominating set of H. Consider the vertex w_i. Since w_i is disjunctively dominated by the set D_d, one of the following possibilities may occur:
(i) $v_i \in D_d$, (ii) $w_i \in D_d$ or $z_i \in D_d$, (iii) $|N_H^2(w_i) \cap D_d| \geq 2$, that is, $N_G(v_i) \cap D_d \neq \emptyset$.

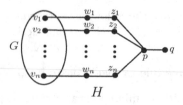

Fig. 1. An illustration to the construction of H from G

If (ii) occurs, then define $D_d = (D_d \setminus \{w_i, z_i\}) \cup \{v_i\}$. Do it for all i, $1 \le i \le n$. Note that the set $D = D_d \cap V$ dominates all the vertices of G, and $|D| \le |D_d|$.

Now suppose that MDDP can be approximated with an approximation ratio of α, where $\alpha = (1 - \epsilon) \ln(|V_H|)$ for some fixed $\epsilon > 0$, by a polynomial time approximation algorithm APPROX-DISJUNCTIVE. Let l be a fixed positive integer. Consider the following algorithm to compute a dominating set of a given graph G.

Algorithm 2. APPROX-DOMINATION(G)

Input: A graph $G = (V, E)$.
Output: A dominating set D of graph G.
begin
    if *there exists a minimum dominating set D' of cardinality $\le l$* **then**
      | $D = D'$;
    else
      Construct the graph H;
      Compute a disjunctive dominating set D_d of H using the
      algorithm APPROX-DISJUNCTIVE;
      for $i = 1 : m$ **do**
        if $w_i \in D_d$ or $z_i \in D_d$ **then**
          $\lfloor$ $D_d = (D_d \setminus \{w_i, z_i\}) \cup \{v_i\}$;
      $D = D_d \cap V$;
    return D;

Clearly, the algorithm APPROX-DOMINATION outputs a dominating set of G in polynomial time. If the cardinality of a minimum dominating set of G is at most l, then it can be computed in polynomial time. So, we consider the case, when the cardinality of a minimum dominating set of G is greater than l. Let D^* denotes a minimum cardinality dominating set of G, and D_d^* denotes a minimum cardinality disjunctive dominating set of H. Note that $|D^*| > l$.

Let D be the dominating set of G computed by the algorithm APPROX-DOMINATION, then $|D| \le |D_d| \le \alpha |D_d^*| \le \alpha(|D^*| + 1) = \alpha(1 + \frac{1}{|D^*|})|D^*| < \alpha(1 + \frac{1}{l})|D^*|$.

Since ϵ is fixed, there exists a positive integer l such that $\frac{1}{l} < \epsilon$. So, $|D| < \alpha(1 + \epsilon)|D^*| = (1 - \epsilon)(1 + \epsilon) \ln(|V_H|)|D^*| = (1 - \epsilon') \ln(|V_H|)|D^*|$.

Since $|V_H| = 3|V| + 1$, and $|V|$ is very large, $\ln(|V_H|) \approx \ln(|V|)$. Hence $|D| < (1 - \epsilon') \ln(|V|)|D^*|$. Hence, the dominating set D computed by the algorithm APPROX-DOMINATION achieves an approximation ratio of $(1 - \epsilon') \ln(|V|)$ for some $\epsilon' > 0$.

By Theorem 5, if the MINIMUM DOMINATION problem can be approximated within a ratio of $(1 - \epsilon') \ln(|V|)$, then NP $\subseteq$ DTIME($|V|^{O(\log \log |V|)}$). This proves that for a graph $H = (V_H, E_H)$, MDDP can not be approximated within a ratio of $(1 - \epsilon) \ln(|V_H|)$ unless NP $\subseteq$ DTIME($|V_H|^{O(\log \log |V_H|)}$). $\qquad \square$

5.3 APX-completeness

In this subsection, we prove that MDDP is APX-complete for bounded degree graphs. To prove this, we need the concept of L-reduction, which is defined as follows.

Definition 2. *Given two NP optimization problems F and G and a polynomial time transformation f from instances of F to instances of G, we say that f is an L-reduction if there are positive constants α and β such that for every instance x of F*

1. *$opt_G(f(x)) \leq \alpha \cdot opt_F(x)$.*
2. *for every feasible solution y of $f(x)$ with objective value $m_G(f(x), y) = c_2$ we can in polynomial time find a solution y' of x with $m_F(x, y') = c_1$ such that $|opt_F(x) - c_1| \leq \beta |opt_G(f(x)) - c_2|$.*

To show the APX-completeness of a problem $\Pi \in$APX, it is enough to show that there is an L-reduction from some APX-complete problem to Π [3].

By Theorem 4, it is clear that MDDP can be approximated within a constant factor for bounded degree graphs. Thus the problem is in APX for bounded degree graphs. To show the APX-hardness of MDDP, we give an L-reduction from the MINIMUM VERTEX COVER PROBLEM (MVCP) for 3-regular graphs which is known to be APX-complete [1].

Theorem 7. *The MINIMUM DISJUNCTIVE DOMINATION PROBLEM is APX-complete for bipartite graphs with maximum degree 3.*

Proof. The proof is omitted due to space constraints.

6 Conclusion

In this article, we have proposed a linear time algorithm for MDDP in proper interval graphs. We have also tightened the NP-completeness of DDDP by showing that it remains NP-complete even in chordal graphs. From approximation point of view, we have proposed an approximation algorithm for MDDP in general graphs and have shown that this problem is APX-complete for bipartite graphs with maximum degree 3. It would be interesting to study the complexity of this problem in other graph classes and also the relation between disjunctive domination number and other domination parameters.

References

1. Alimonti, P., Kann, V.: Some APX-completeness results for cubic graphs. Theoret. Comput. Sci. **237**(1–2), 123–134 (2000)
2. Ambalath, A.M., Balasundaram, R., Rao H., C., Koppula, V., Misra, N., Philip, G., Ramanujan, M.S.: On the kernelization complexity of colorful motifs. In: Raman, V., Saurabh, S. (eds.) IPEC 2010. LNCS, vol. 6478, pp. 14–25. Springer, Heidelberg (2010)
3. Ausiello, G., Crescenzi, P., Gambosi, G., Kann, V., Marchetti-Spaccamela, A., Protasi, M.: Complexity and approximation. Springer, Berlin (1999)
4. Bertossi, A.A.: Dominating sets for split and bipartite graphs. Inf. Process. Lett. **19**(1), 37–40 (1984)
5. Chlebík, M., Chlebíková, J.: Approximation hardness of dominating set problems in bounded degree graphs. Inform. and Comput. **206**, 1264–1275 (2008)
6. Dankelmann, P., Day, D., Erwin, D., Mukwembi, S., Swart, H.: Domination with exponential decay. Discrete Math. **309**, 5877–5883 (2009)
7. Fulkerson, D.R., Gross, O.A.: Incidence matrices and interval graphs. Pacific J. Math. **15**, 835–855 (1965)
8. Goddard, W., Henning, M.A., McPillan, C.A.: The disjunctive domination number of a graph. Quaestiones Math. **37**(4), 547–561 (2014)
9. Haynes, T.W., Hedetniemi, S.T., Slater, P.J.: Fundamentals of domination in graphs. Marcel Dekker Inc., New York (1998)
10. Haynes, T.W., Hedetniemi, S.T., Slater, P.J.: Domination in Graphs, Advanced Topics. Marcel Dekker Inc., New York (1998)
11. Henning, M.A., Marcon, S.A.: Domination versus disjunctive domination in trees. Discrete Appl. Math. (2014)
12. Henning, M.A., Naicker, V.: Disjunctive total domination in graphs. J. Comb. Optim. (2014). doi:10.1007/s10878-014-9811-4
13. Jacobson, M.S., Peters, K.: Complexity questions for n-domination and related parameters. In: Eighteenth Manitoba Conference on Numerical Mathematics and Computing, Winnipeg, MB (1988), Congr. Numer. **68**, 722 (1989)
14. Jamison, R.E., Laskar, R.: Elimination orderings of chordal graphs. In: Combinatorics and applications. ISI, Calcutta, pp. 192–200 (1982, 1984)
15. Panda, B.S., Das, S.K.: A linear time recognition algorithm for proper interval graphs. Inform. Process. Lett. **87**(3), 153–161 (2003)
16. Rajgopalan, S., Vazirani, V.V.: Primal-dual RNC approximation algorithms for set cover and covering integer programs. SIAM J. Comput. **28**, 526–541 (1999)

Algorithmic Aspect of Minus Domination on Small-Degree Graphs

Jin-Yong Lin[1], Ching-Hao Liu[1]([✉]), and Sheung-Hung Poon[2]

[1] Department of Computer Science, National Tsing Hua University, Hsinchu, Taiwan
yongdottw@hotmail.com, chinghao.liu@gmail.com
[2] School of Computing and Informatics, Institut Teknologi Brunei,
Gadong, Brunei Darussalam
sheung.hung.poon@gmail.com

Abstract. Let $G = (V, E)$ be an undirected graph. A *minus dominating function* for G is a function $f : V \to \{-1, 0, +1\}$ such that for each vertex $v \in V$, the sum of the function values over the closed neighborhood of v is positive. The *weight* of a minus dominating function f for G, denoted by $w(f(V))$, is $\sum f(v)$ over all vertices $v \in V$. The *minus domination (MD) number* of G is the minimum weight for any minus dominating function for G. The *minus domination (MD) problem* asks for the minus dominating function which contributes the MD number. In this paper, we first show that the MD problem is $W[2]$-hard for general graphs. Then we show that the MD problem is NP-complete for subcubic bipartite planar graphs. We further show that the MD problem is APX-hard for graphs of maximum degree seven. Lastly, we present the first fixed-parameter algorithm for the MD problem on subcubic graphs, which runs in $O^*(2.3761^{5k})$ time, where k is the MD number of the graph.

1 Introduction

Let $G = (V, E)$ be an undirected graph. A *minus dominating function* for G is a function $f : V \to \{-1, 0, +1\}$ such that for each vertex $v \in V$, the sum of the function values over the closed neighborhood of v is positive, where the *closed neighborhood* of v is the set contains v and all neighbors of v. The *weight* of a minus dominating function f for G, denoted by $w(f(V))$, is $\sum f(v)$ over all vertices $v \in V$. The *minus domination (MD) number* of G, denoted by $\gamma^-(G)$, is the minimum weight for any minus dominating function for G. The *minus domination (MD) problem* asks for the minus dominating function which contributes the MD number.

According to [9], we can see that the function values $+1$ and -1 of the signed domination problem can be formulated as yes-no decisions for social networks. In a similar fashion, the function values $+1$, 0, and -1 of the MD problem can be formulated as yes-uncertain-no decisions for in social networks. In the following, we first mention related work on complexities. Dunbar et al. [4,5] first showed that the MD problem is NP-complete for bipartite graphs and for chordal graphs, and can be solved in linear time for trees. Then, Damaschke [2]

© Springer International Publishing Switzerland 2015
D. Xu et al. (Eds.): COCOON 2015, LNCS 9198, pp. 337–348, 2015.
DOI: 10.1007/978-3-319-21398-9_27

showed that the MD problem is NP-complete for planar graphs of maximum degree 4. They [2] also showed that for every fixed k there is a polynomial-time algorithm, which runs in time $O(3^{8k} \cdot n^{8k})$, for deciding whether a given graph G of maximum degree 4 satisfies $\gamma^-(G) \leq k$. Recently, Faria et al. [6] showed that the MD problem is NP-complete for splitgraphs, and is polynomial for graphs of bounded rankwidth and for strongly chordal graphs.

Next, we survey the related work on parameterized complexities. It is known that Downey et al. [3] showed that the domination problem is $W[2]$-complete. Then, Faria et al. [6] showed that the MD problem has no fixed-parameter algorithm, i.e., not in $W[0]$, unless $P = NP$. They also show that the MD problem is fixed-parameter tractable for planar graphs, when parameterized by the *size* of f, the number of vertices $x \in V$ with $f(x) = 1$, and for d-degenerate graphs, when parameterized by the size of f and by d.

Lastly, we survey the related work on approximation complexities for graphs of bounded degree. First, Alimonti and Kann [1] showed that the domination problem is APX-hard on subcubic graphs. Then, Damaschke [2] showed that the MD number cannot be approximated in polynomial time within a factor $1 + \epsilon$, for some $\epsilon > 0$, for graphs of maximum degree 4, unless $P = NP$.

Outline. We organize the rest of this paper as follows. In Section 2, we show that the MD problem is $W[2]$-hard for general graphs. In Section 3, we show that the MD problem is NP-complete for subcubic bipartite planar graphs. In Section 4, we show the MD problem is APX-hard for graphs of maximum degree 7. In Section 5, with a very involved analysis, we obtain the first FPT-algorithm for the MD problem on subcubic graphs G, which runs in time $O^*(2.3761^{5k})$, where k is the MD number of G.

2 $W[2]$-hardness for General Graphs

In this section, by reducing the domination problem to the MD problem, we show that the MD problem on general connected graphs is $W[2]$-hard. Thus we introduce domination problem as follows. A *dominating set* of a graph $G = (V, E)$ is a vertex set $D \subseteq V$ such that for each vertex $v \in V$, there exists at least one vertex in the closed neighborhood $N_G[v]$ belonging to D. In other words, we can label the vertices in V with $\{0, +1\}$ such that the closed neighborhood of each vertex is positive. The *domination number* of G, denoted by $\gamma(G)$ is the cardinality of the minimum dominating set of G. The *domination problem* asks for a dominating set which contributes the domination number $\gamma(G)$. Downey et al. [3] showed that the domination problem on general connected graphs is $W[2]$-complete.

In our reduction, we need to make use of a graph H as shown in Figure 1(a). If we label the vertices of H as shown in Figure 1(a), where there are four vertices labeled with $+1$ and five vertices labeled with -1, then the weight for such a minus domination function is -1. In the following lemma, we show that -1 is in fact the minimum weight which a legitimate minus dominating function for graph H can provide.

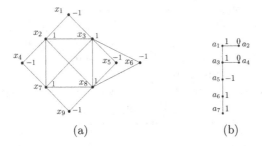

(a)   (b)

Fig. 1. (a) Graph H with weight -1. (b) A key with minimum weight 3.

Lemma 1. *The MD number of graph H shown in Figure 1(a) is at least -1.*

Proof. Suppose to the contrary that $\gamma^-(H) \leq -2$. Let f be a minus dominating function of H whose weight is not more than -2. Let P be the set of vertices labeled with $+1$ in H. If $\gamma^-(H) \leq -2$, then we have $|P| \leq 3$. It is clear that $|P| > 1$. If $|P| = 2$, it is easy to see that the rest vertices should label with 0 in H. Thus we have $|P| \neq 2$. If $|P| = 3$, then there are at most three vertices labeled with -1 in H. Thus we also have $|P| \neq 3$. With such a contradiction, we thus have $\gamma^-(H) \geq -1$. □

With this lemma, now we show the $W[2]$-hardness for the MD problem.

Theorem 1. *The MD problem is $W[2]$-hard for general connected graphs.*

Proof. We reduce the domination problem to our problem as follows. Given a graph $G = (V, E)$, we construct a new graph G' as described below. Initially we set G' to be G. Then we add each vertex v of G' a set of $\deg_G(v) + 1$ keys as shown in Figure 1(b), and connect v to a_7 of each keys in this set, where $\deg_G(v)$ is the degree of vertex v in G. Note that at most one of vertex in each key can be labeled with -1, and thus the weight for each key is greater than or equal to three. We let F be the set of newly added vertices. Next, we add $6m + 2n$ copies of graph H into graph G' and denote them by $H_1, H_2, \ldots, H_{6m+2n}$, respectively. Let vertex set $Y = V(H_1) \cup \ldots \cup V(H_{6m+2n})$. Then we take any previously added key, say ρ. Now, we connect the vertex a_1 of ρ to the corresponding vertex x_2 of each copy $H_i; i = \{1, \ldots, 6m + 2n\}$ of graph H. This completes the construction of the graph G'. Let V' be the set of vertices in G'. As G is a connected graph, graph G' is also a connected graph by our construction. In the following, we show that there is a domination set of size at most k in G if and only if there is a minus dominating function with weight at most k in G'.

Assume that there is a dominating set D of size at most k in G. Then we will show that there is a minus dominating function f with weight at most k of G'. We construct a function f, which labels vertices in F as Figure 1(b), vertices in D with value 0, vertices in $V \setminus D$ with value -1. We also label the corresponding vertices x_2, x_3, x_7, x_8 with value $+1$ and x_1, x_4, x_5, x_6, x_9 with value -1 for each

$H_i; i = \{1, \ldots, 6m + 2n\}$. Now, we claim that f is a minus dominating function of G' with weight at most k. Clearly, the sum weight of the closed neighborhood of any vertex in $F \cup Y$ is positive. Then we consider vertices in $V' \setminus (F \cup Y)$. For any vertex v in $V' \setminus (F \cup Y)$, since D is a dominating set, there is at least one vertex in $N_{G'}[v] \cap D$. Thus more than half of the vertices in $N_{G'}[v]$ are labeled with $+1$ in function f. Thus the weight for induced subgraph $G[N'_G[v]]$ is positive. Hence, f is a legitimate minus dominating function of G'. Since the weight of each H_i is at least -1 by Lemma 1, we obtain that

$$w(f(V')) \leq -(n - k) + (-1)(6m + 2n) + 3(2m + n) = k.$$

Conversely, we can also show that if f is a minus dominating function of G' with weight at most k, then there is a dominating set size at most k in G. The proof is omitted here. Hence, we complete the proof.   □

3   NP-completeness for Subcubic Bipartite Planar Graphs

A graph is called *cubic* if its vertex degrees are degree three, and *subcubic* if its vertex degrees are at most three. It has been shown in [2] that the MD problem on planar graphs of maximum degree four is NP-complete. In this section, we show that the MD problem is NP-complete even for subcubic bipartite planar graphs. We leave open the question that whether the MD problem is NP-complete for cubic bipartite planar graphs. For showing our NP-completeness results, we make use of a lemma presented by Damaschke in [2].

Lemma 2 (Lemma 3 of [2]). *In any graph, we consider a vertex x of degree 1 and the unique neighbor w of x. Then there is an optimal minus dominating function such that $f(x) = 0$ and $f(w) = 1$.*

Then we show the main NP-completeness theorem in the following.

Theorem 2. *The MD problem is NP-complete for subcubic bipartite planar graphs.*

Proof. Clearly, the problem is in NP. We reduce the planar 3SAT problem [7] to this problem. The input instance for the planar 3SAT problem is a set $\{x_1, x_2, \ldots, x_n\}$ of n variables and a Boolean expression with conjunctive normal form $\Phi = c_1 \wedge c_2 \wedge \ldots \wedge c_m$ of m clauses, where each clause consists of exactly three literals, such that the variable clause graph of the input instance is planar. The *planar 3SAT problem* asks for whether there exists a truth assignment to the variables so that the Boolean expression Φ is satisfied. We then describe our polynomial-time reduction as follows.

Variable Gadget. First, we construct the variable gadget V_i for a variable x_i. The variable gadget V_i for x_i is a circular linkage as shown in Figure 2(a). We connect $4m + 2$ keys (of Figure 1(b)) together as shown in Figure 2(a), where $2m + 1$ keys are connected as a *chain of keys* for the upper part of V_i, and the other $2m + 1$ keys are connected as a chain of keys for the lower part of V_i .

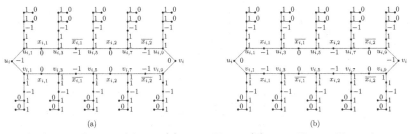

(a)   (b)

Fig. 2. Gadget V_i for variable x_i: (a) $x_i =$ TRUE; (b) $x_i =$ FALSE. Note that vertices $x_{i,j}$ and $\overline{x_{i,j}}$ represents the positive and negative literals for x_i, respectively. They are used to connect to a clause gadget C_j.

The connection of the keys is performed by the introduction of a cycle of length $8m + 4$. See Figure 2(a). We call such a circular linkage a *cycle of keys*. See Figure 2(a). We let u_i and v_i be the leftmost and the rightmost endpoints of V_i. Then we let $u_i, u_{i,1}, u_{i,2}, \ldots, u_{i,4m+1}, v_i$ be the upper chain starting from u_i to v_i. Similarly, we let $u_i, v_{i,1}, v_{i,2}, \ldots, v_{i,4m+1}, v_i$ be the lower chain starting from u_i to v_i. The vertices $u_{i,2}, u_{i,4}, \ldots, u_{i,4m}$ and $v_{i,2}, v_{i,4}, \ldots, v_{i,4m}$ are used for connecting with clause gadgets. That is, $u_{i,2}, u_{i,6}, \ldots, u_{i,4m-2}$ and $v_{i,2}, v_{i,6}, \ldots, v_{i,4m-2}$ are parts for positive literals $x_{i,j}$, and $u_{i,4}, u_{i,8}, \ldots, u_{i,4m}$ and $v_{i,4}, v_{i,8}, \ldots, v_{i,4m}$ are parts for negative literals $\overline{x_{i,j}}$. The interior of the whole variable gadget can be duplicated to make a longer gadget so that there are enough ports on the variable gadget for connecting to the corresponding literal gadgets of the related clauses in the later context.

Next, we first describe the truth assignment of the optimal minus dominating function for a key. The labeling method in Figure 1(b) is optimal, whose weight is 3. There is another way of optimal labeling such that the lowest vertex of the key is labeled with value 0. Since the lowest vertex of a key is connected to the main body of variable gadget, it is always advantageous to use the labeling method shown in Figure 1(b).

Then we describe the truth assignment of the optimal minus dominating function for the whole variable gadget. To attain the assignment of minimum weight, the internal cycle of variable gadget V_i may be labeled as either the way in Figure 2(a) or the way in Figure 2(b). In either way, the sum of the weights $f(x)$ for $x \in V_i$ is $8m$ and such $f(x)$ is the minimum minus dominating function. We use the domination way in Figure 2(a) to represent that $x_i =$ TRUE, and the other domination way in Figure 2(b) to represent that $x_i =$ FALSE.

Clause Gadget. We use clause gadgets to connect to the variable gadgets directly and there is no link gadget. Now we are prepared to construct the clause gadget C_j for clause $c_j = x_i \vee x_k \vee x_\ell$ which contains 28 vertices, that is, $p_1, \ldots, p_8, q_1, \ldots, q_8, r_1, \ldots, r_8, s_1, s_2, s_3, t$, and 33 edges as shown in Figure 3(a). The vertices p_0, q_0 and r_0 lie in variable gadgets V_i, V_k and V_ℓ, respectively. If x_i is TRUE or FALSE, then p_0 connects to a vertex in V_i which represents $x_{i,j}$ or $\overline{x_{i,j}}$, respectively. Similarly, if x_k is TRUE or FALSE, then q_0 connects to a vertex in V_k which represents $x_{k,j}$ or $\overline{x_{k,j}}$, respectively. If x_ℓ is TRUE, then r_0 connects

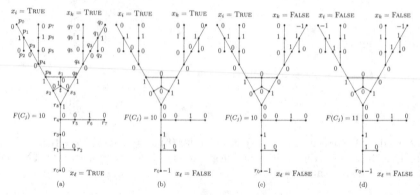

Fig. 3. Gadget C_j for clause c_j: (a) $x_i = x_k = x_\ell = $ TRUE; (b) $x_i = x_k = $ TRUE and $x_\ell = $ FALSE; (c) $x_i = $ TRUE and $x_k = x_\ell = $ FALSE; (d) $x_i = x_k = x_\ell = $ FALSE

to a vertex in V_ℓ which represents $x_{\ell,j}$; otherwise, if x_ℓ is FALSE, then r_0 connects to a vertex in V_ℓ which represents $\overline{x_{\ell,j}}$. This completes the construction of the clause gadget.

We denote the minimum weight for the clause gadget by

$$F(C_j) = \min_f \{ \sum_x f(x) \mid x \in \{p_1, \ldots, p_8, q_1, \ldots, q_8, r_1, \ldots, r_8, s_1, s_2, s_3, t\} \}.$$

By Lemma 2, we assign $f(p_1) = f(q_1) = f(r_1) = 1$, $f(p_2) = f(q_2) = f(r_2) = 0$, $f(p_6) = f(q_6) = f(r_6) = 1$ and $f(p_7) = f(q_7) = f(r_7) = 0$. Then since $\sum_{x \in N[p_4]} f(x) \geq 1$, $\sum_{x \in N[q_4]} f(x) \geq 1$, $\sum_{x \in N[r_4]} f(x) \geq 1$ and $\sum_{x \in N[s_1]} f(x) \geq 1$, thus $F(C_j) \geq 10$. Here we claim that if a clause is TRUE, then $F(C_j)$ meets the lower bound, that is, $F(C_j) = 10$; otherwise, if a clause is FALSE, then $F(C_j) = 11$.

To prove $F(C_j) = 10$ for a TRUE clause gadget, we show that there exists an assignment of $f(x)$ such that $\sum_{x \in N[p_4]} f(x) = 1$, $\sum_{x \in N[q_4]} f(x) = 1$, $\sum_{x \in N[r_4]} f(x) = 1$, and $\sum_{x \in N[t]} f(x) = 1$. To prove $F(C_j) = 11$ for a FALSE clause gadget, we show that there exists an assignment of $f(x)$ such that one of $\sum_{x \in N[p_4]} f(x)$, $\sum_{x \in N[q_4]} f(x)$, $\sum_{x \in N[r_4]} f(x)$, and $\sum_{x \in N[t]} f(x)$ is two, and the rest of them are one.

Analysis of Truth Assignment. We need to analyze totally four cases for the truth assignments for the clause gadget mentioned above. We discuss the case that all three literals are FALSE in the following paragraph since it is the most important case. As for the other three cases, we omit the analysis.

Suppose that all three literals x_i, x_k and x_ℓ are FALSE as shown in Figure 3(d). We prove that $F(C_j) = 10$ is impossible as follows. Suppose that $F(C_j) = 10$. Then $\sum_{x \in N[y]} f(x) = 1$ for $y \in \{p_4, q_4, r_4, t\}$. Since x_i, x_k and x_ℓ are FALSE, we assign $f(x)$ for $x \in N[p_4]$ by setting $f(p_3) = 1$ and $f(p_4) = f(p_5) = 0$, and we perform the similar labeling for $x \in N[q_4]$ and for $x \in N[r_4]$. On the other hand, it is easy to check that the minimum sum of weight of vertices in $\{p_8, q_8, r_8, s_1, s_2, s_3, t\}$ is not less than 2. Hence, $F(C_j) \geq 11$.

In a true assignment of clause c_j, the minimum weight of $F(C_j)$ for the optimal minus dominating function for gadget C_j is 10, which is omitted due to lack of space. Hence, the Boolean expression Φ is satisfied if and only if the constructed graph has a minus dominating function of weight $n(8m+4)+10m$.

$\square$

4   APX-hardness for Graphs of Maximum Degree 7

In this section, we show that the MD problem for graphs of maximum degree 7 is APX-hard. Alimonti and Kann [1] show that the domination problem is APX-hard on subcubic graphs. Here we use a well-known technique called *L-reduction* to show the APX-hardness for the MD problem. See [8] for more details on L-reduction. We show that our problem satisfies the two main properties of L-reduction.

We perform an L-reduction f from an instance of the domination problem on subcubic graphs to the corresponding instance of the MD problem on graphs of maximum degree 7.

Given a subcubic graph $G = (V, E)$, we construct a graph $G' = (V', E')$ as follows. For each vertex v in V, we add $\deg_G(v) + 1$ keys as Figure 1(b) and connect a_7 of each key to v. This completes the construction of G'. For each vertex $v \in V$, we have just added one more edge connecting to the vertex v for each edge adjacent to v in graph G. Since graph G is subcubic, we have that G' is of maximum degree 7. Next, we obtain the following lemma.

Lemma 3. *Let $G = (V, E)$ be a subcubic graph and let the corresponding $G' = (V', E')$ constructed as described above. If D^* is a minimum dominating set of G, and f^* is a minimum minus dominating function of G'. Then $w(f^*(V')) = |D^*| + 6m + 2n$, where $n = |V|$ and $m = |E|$.*

Proof. We construct a function f by assigning the vertices $\{a_1, a_3, a_6, a_7\}$ in each keys with value $+1$, we label D^* and the vertices $\{a_2, a_4\}$ in each keys with value 0, and label other vertices with value -1. Then we verify whether function f is a valid minus dominating function of G', that is, we verify whether $N_G[x] > 0$ for each vertex x in G'.

It is clear that the sum of function values of the neighborhood of each vertex in keys is greater than 0. Now we claim that the same holds for the remaining vertices in V'. For a vertex v labeled with 0, the vertex has at most $\deg_G(v)$ neighbors labeled with -1, since there are $\deg_G(v) + 1$ keys connecting to v, we obtain that $N_G[x] > 0$. Moreover, we consider a u vertex labeled with -1 in V'. Since the corresponding vertex of u in G is not in D^*, there must be at least a neighbor of u in G' labeled with value 0. Hence, the sum of the function values of the closed neighborhood of u is positive. Then we calculate the weight for minus dominating function f, $w(f) = 3(2m+n) - (n - |D^*|) = 6m + 2n + |D^*|$. Thus we have $w(f^*) \le w(f) = 3(2m+n) - (n-k) = 6m + 2n + |D^*|$.

Now let f^* be the minimum minus dominating function of G', let vertex set D in G collect those vertices labeled with 0 and $+1$ in f^*. Then we claim that D is a dominating set of G, in other words, we claim that for each vertex v in

$G \setminus D$, there is at least a neighbor in D. Let v' be a vertex labeled with -1 in G'. Note that there are $2 \deg_G(v) + 2$ vertices in $N_{G'}[v']$. It is clear that the vertices in $N_{G'}[v'] \cap F \setminus \{v\}$ must be labeled with $+1$. Since f^* is a minus dominating function, the sum of $N_{G'}[v']$ must be positive. Hence, there must exist at least one neighbor of $v \notin F$ is labeled with 0 or $+1$. That is, D is a dominating set of G. Then we calculate the size of set D. $|D| \leq w(f^*) - 3(2m + n) + n = w(f^*) - 6m - 2n$. Thus we have $|D^*| \leq |D| \leq w(f^*) - 6m - 2n$. Hence we have $w(f^*(V')) = |D^*| + 6m + 2n$. This completes the proof.   □

Since G is a subcubic graph, we have $m \leq \frac{3n}{2}$. Moreover, according to Lemma 4, $n \leq 5|D^*|$. Hence, $w(f^*) = |D^*| + 6m + 2n \leq |D^*| + 9n + 2n \leq |D^*| + 55|D^*| = 56|D^*|$. Since $w(f^*) \leq 56|D^*|$, then $\alpha = 56$.

Now we consider a minus dominating function f of G', we can construct a dominating set D for G by the above algorithm, then we have $|D| = w(f) - 6m - 2n$. Thus we obtain that $|D| - |D^*| = w(f) - 6m - 2n - (w(f^*) - 6m - 2n) = w(f) - w(f^*)$ Thus $\beta = 1$. Hence, we have proved that f is an L-reduction with $\alpha = 56$ and $\beta = 1$. Finally, we obtain the following theorem.

Theorem 3. *The MD problem is APX-hard for graphs of maximum degree 7.*

5   An FPT-algorithm for Subcubic Graphs

In Section 2, we have shown that the MD problem for subcubic bipartite planar graphs is NP-complete. It thus follows that the MD problem for subcubic graphs is NP-complete. Then it is interesting to study FPT-algorithms for the MD problem on subcubic graphs parameterized by the MD number. Th the best of our knowledge, there is no such FPT-algorithm in the literature.

In this section, we thus present the first FPT-algorithm for the MD problem on subcubic graphs G parameterized by the MD number k. In the following lemma, we begin with showing that our problem has a kernel of size $5k$. Then with an involved analysis, we come up with an FPT-algorithm which runs in time $O^*(2.3761^{5k})$.

Lemma 4. *There is a kernel of size $5k$ for the MD problem on subcubic graphs, and the bound is tight, where k is the weight for the minus dominating function of graph G.*

Proof. Let $G = (V, E)$ be a subcubic graph. For any minus dominating function of G with weight k, we claim that $|V| \leq 5k$. We divide weight k into two parts,

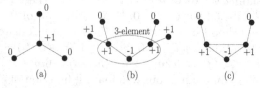

(a)   (b)   (c)

Fig. 4. (a)(b) The two conditions for weight $+1$. (c) A minus dominating function of five vertices with weight 1.

such that one part contains any vertex labeled with $+1$, whose neighbors are all not labeled with -1, and the other part contains any vertex labeled with $+1$, which has a neighbor labeled with -1. For the first part, If every single value one of the weight k is from this part, each value one contains at most four vertices see Figure 4(a), let V_1 be a set contains these vertices, we have $n \leq 4k$. Second, let v be a vertex labeled with value -1, it is clear that v has at least two neighbors labeled with value $+1$, and the distance to any other vertex labeled with -1 is at least three. Thus we obtain a 3-*element*, which contains v labeled with value -1 and its two neighbors see Figure 4(b), which labeled with $+1$. Moreover, for any pair of such subsets, their intersection is empty. Let u be the vertices labeled with $+1$, which is neighboring 3-element, it is clear that u belongs to a 3-element or to V_1. Next we consider the vertices labeled with 0, which connect to 3-element. We can observe that the vertices which labeled with 0 and connect to vertex labeled with -1 in 3-element, the vertices either the vertex connect to $+1$ in 3-element or the vertex labeled 0 in V_1. So the second part contains at most five vertices. If every single value of the weight k is from the second part, we have $n \leq 5k$. Finally, we can observe that all the vertices are concerned in the above two part. Hence, we can obtain a kernel of size $5k$ as an upper bound for the MD problem on subcubic graphs. Furthermore, For a graph with five vertices labeled as in Figure 4(c) is $k = 1$ and $n = 5$, so the bound of kernel is tight.   □

According to this lemma, a naïve FPT-algorithm can be easily obtained via the brute-force method which runs in $O^*(3^{5k}) = O^*(243^k)$ time. Contrary to the brute-force algorithm, we claim that our FPT-algorithm runs in time $O^*(2.3761^{5k}) = O^*(75.7397^k)$, which is a great improvement. Now, our FPT-algorithm is presented in the following theorem. We first give the detailed algorithm and its correctness proof, and then analyze its time complexity.

Theorem 4. *The MD problem for subcubic graphs G can be solved in $O^*(2.3761^{5k})$ time, where k is the MD number of G.*

Proof. Due to Lemma 4, we only need to consider the given subcubic graph G with kernel size of $5k$. Since disconnected components of a graph can be handled separately, we assume that the given graph G is connected in the following context. We also use $N[\cdot]$ to represent $N_G[\cdot]$ for simplicity.

The details of our algorithm are as follows. In our algorithm, we grow a potential optimal minus dominating set D incrementally, where the *minus dominating set D* is a subset of vertices in V labeled with values $+1, 0$ or -1. The *label* of a vertex is the value of vertex assigned by a specific minus dominating function of G. A vertex is called *labeled* if it has been assigned a value; otherwise, it is called *unlabeled*. The weight for the closed neighborhood $N[v]$ of a labeled vertex v is called *valid* (resp. *invalid*) if all the vertices in $N[v]$ are labeled, and the sum of weights of the vertices in $N[v]$ is positive (resp. non-positive). In the process, we maintain a list L of unlabeled vertices which are the neighboring vertices of the currently labeled vertices in D. Initially, L is set to contain one degree-3 vertex of the input graph G, and $D = \emptyset$. During each iteration of our algorithm, we select an arbitrary unlabeled vertex y from list L as the focus vertex, and we assume that x is a labeled vertex adjacent to y in $D \cap N[y]$. We set $\Delta = N[N[y]] \setminus D$. Then our algorithm makes execution branches on all different ways of assigning values to vertices in Δ of new unlabeled vertices in $N[N[y]] \setminus D$, we performing detailed case analysis from *Case 1* to *Case 15*. In the case analysis, our algorithm makes subsequent recursions only on those feasible ways of value assignment in the corresponding cases. For each of such subsequent recursions, the vertices of Δ, which have been labeled, are added into D, resulting in a larger labeled dominating set $D + \Delta$. Then for each vertex v in $D + \Delta$, if all the vertices in $N[v]$ are labeled, we then check whether the weight for $N[v]$ is valid. If we reach any weight for closed neighborhood of a vertex, which is invalid, then the current execution branch is aborted; otherwise, we proceed to update D and L for the next round of execution. We update D by setting $D = D + \Delta$, and then we update list L accordingly by visiting the neighboring vertices of the neighbors of vertices in Δ. More precisely, the vertices in Δ are removed from L, and the unlabeled vertices in the neighborhoods of vertices in Δ are added into L. It is clear that such an update takes only $O(1)$ time. We then proceed to the next execution round with the updated D and L as parameters. We repeat such a selection step until all vertices in G are labeled, that is, L becomes empty. Thus we obtain a candidate minus dominating set D.

In the selection process, we enumerate and store all possible candidates of D according to the above recursive procedure. When all branches of the selection process finished, we obtain a set of candidate minus dominating sets D for the input graph G. We choose the one with minimum weight among all these candidates. This completes our algorithm.

Case analysis for selection step. We divide the analysis into two parts: the initial step and the general selection step.

The initial step. We choose one degree-3 vertex v is placed in D. Then add one neighbor u of v is subsequently into D. In these two beginning steps, there are 8 choices to labeled vertices u and v, since u and v cannot both be of value -1. In any subsequent step, we focus on an unlabeled vertex y, which has a neighbor x in D. Now the degree of x and y can be one, two or three. However, it is easy to see that the worst case happens when both degree of x and y are three. We only need to perform the detailed case analysis for such a case. Thus in the following, we assume that the degree of both x and y are three. Due to the above initial step, we know that x must have another neighbor x_1 in D. Thus we have in total 15 cases to analyze by considering which vertices of $N[N[y]]$ lie in d. See Figure 5.

The selection step. We analyze all possible labeling ways to find the optimal minus dominating function. Due to lack of space, we only provide the analysis of *Case 1* (Figure 5(a)) in the following. The analysis of *Cases 2* to *15* are omitted.

Case 1. Let x_2 be the third neighbor of x, let y_1 and y_2 be the other two neighbors of y, and let z_1, z_2 be the neighbors of y_1, z_3, z_4 be the neighbors of y_2. Since y_1 and y_2 are symmetric, we need to consider 15 cases depending on the content of Δ, the set of unlabeled vertices in the subgraph of G. See Figures 5(a) to (o). $\Delta = \{y, x_2, y_1, y_2, z_1, z_2, z_3, z_4\}$. That is, $\{x, x_1\} \subset D$. See Figure 5(a).

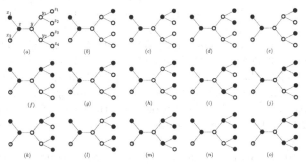

Fig. 5. This figure shows the 15 cases, where the labeled vertices in D are drawn as black disks, the unlabeled vertices in Δ as circles, and the question mark in circle is where we analyze whether the vertex is labeled or unlabeled in this case

In order to reduce the number of cases we need to consider under a specific case, we only need to consider the worst-case scenarios in each of the cases. For such a purpose, we make the following three assumptions. We remark that the three assumptions will be applied to all the subsequent cases in this proof.

(i) If there is an edge not in the subgraph, let u, v be the endpoints. Then we can add two vertices u' and v', where we connect u to u' and connect v to v'. Thus we can label the subgraph with constrain : the value of u and v' is same, so as v and u'. We can observe that the number of feasible ways of labeling the vertices with constrain is less than the vertices without constrains, which we will consider in other cases. Hence, we do not need to consider adding some edges in subgraphs.

(ii) We observe that a vertex labeled with -1 need to have at least two vertices labeled with $+1$ as its neighbors, and a vertex labeled with $+1$ can have vertices labeled with $+1, 0$ or -1 as its neighbors. Thus for a specific case, the worst-case scenario for the number of feasible ways to label the vertices in Δ is when the labeled vertices for the specific case (for instance, vertices x and y in *Case 1*) are all assigned value $+1$.

(iii) We observe that, for a graph G, if there is a connected graph H is delete a vertex v from G, the number choices to labeling vertices in H are less than or equal to number of choices of graph to labeling G with v assigned value $+1$. Hence, we do not need to consider the subgraph which delete some vertices.

For *Case 1*, according to assumption (i), we suppose that there is no edge connecting any pair of vertices in the subgraph, and according to assumption (ii), we suppose that x and x_1 are labeled with $+1$. We have two situations depending on whether x_2 is labeled. In *Case 1(a)*, we first consider x_2 is unlabeled, thus x_2 has at most 3 choices for its labeling, say with value -1, 0 or $+1$. Now we focus on vertex y_1. First we consider to label y_1 with value -1, then $\{z_1, z_2\}$ has at most 3 choices for assigning, and $\{y_2, z_3, z_4$ has at most 14 choices. Thus there are at most 42 choices if y_1 is labeled with -1. Second, we consider to label y_1 with value 0, then $\{z_1, z_2\}$ has at most 6 choices for assigning the values, and $\{y_2, z_3, z_4\}$ has at most 17 choices; hence, there are at most 102 choice if y_1 is labeled 0. Lastly, we consider to label y_1 with value $+1$, then $\{z_1, z_2\}$ has at most 8 choices for assigning the values, and $\{y_2, z_3, z_4\}$ has at most 17 choices. Thus there are at most 136 choice if y_1 is labeled $+1$. After all of $x_2, y_1, y_2, z_1, z_2, z_3, z_4$ are labeled, it is clear that we can label vertex y with an unique minimum value, such that the weights for neighborhoods of x, y, y_1 and y_2 are positive, respectively. By multiplying with the three choices for the value of x_2. Thus there are at most $3 \times (42 + 102 + 136) = 840$ feasible ways in total to label the eight vertices in Δ for *Case 1*(a).

For *Case 1*(b), where x_2 is labeled, we use similar argument as the analysis for *Case 1*(a). But we do not need to multiply three choices for the value of x_2. Thus there are at most 240 feasible ways in total to label the seven vertices in Δ for *Case 1*(b). This finishes the analysis of *Case 1*.

In the above detailed analysis, we obtain the recurrence relation $T(n) \leq 840(n - 8) + O(1)$ for the worst-case running time of *Case 1*(a). By analyzing all 15 cases in Figure 5 and solving the corresponding recurrence relations, we thus have the worst-case running time for the whole algorithm, which occurs at *Case 5*(a) (see Figure 5(e)). Hence we obtain that the total running time of our algorithm is $T(n) = O^*(2.3761^n) = O^*(2.3761^{5k})$.   □

References

1. Alimonti, P., Kann, V.: Hardness of approximating problems on cubic graphs. In: Proc. 3rd Italian Conference on Algorithms and Complexity (CIAC), pp. 288–298 (1997)
2. Damaschke, P.: Minus domination in small-degree graphs. Discrete Applied Mathematics 108(1–2), 53–64 (2001)
3. Downey, R.G., Fellows, M.R.: Parameterized Complexity, 3rd edn. Springer (1999)
4. Dunbar, J., Goddard, W., Hedetniemi, S., McRae, A., Henning, M.A.: The algorithmic complexity of minus domination in graphs. Discrete Applied Mathematics 68(1–2), 73–84 (1996)
5. Dunbar, J., Hedetniemi, S., McRae, A., Henning, M.A.: Minus domination in graphs. Discrete Applied Mathematics 199(1–3), 35–47 (1999)
6. Faria, L., Hon, W.-K., Kloks, T., Liu, H.-H., Wang, T.-M., Wang, Y.-L.: On complexities of minus domination. In: Widmayer, P., Xu, Y., Zhu, B. (eds.) COCOA 2013. LNCS, vol. 8287, pp. 178–189. Springer, Heidelberg (2013)
7. Lichtenstein, D.: Planar formulae and their uses. SIAM Journal on Computing 11(2), 329–343 (1982)
8. Papadimitriou, C.H., Yannakakis, M.: Optimization, approximation, and complexity classes. Journal of Computer and System Sciences 43(3), 425–440 (1991)
9. Zheng, Y., Wang, J., Feng, Q., Chen, J.: FPT results for signed domination. In: Agrawal, M., Cooper, S.B., Li, A. (eds.) TAMC 2012. LNCS, vol. 7287, pp. 572–583. Springer, Heidelberg (2012)

Time-Space Tradeoffs for Dynamic Programming Algorithms in Trees and Bounded Treewidth Graphs

Niranka Banerjee[1], Sankardeep Chakraborty[1], Venkatesh Raman[1(✉)], Sasanka Roy[2], and Saket Saurabh[1]

[1] The Institute of Mathematical Sciences, CIT Campus, Taramani, Chennai 600 113, India
{nirankab,sankardeep,vraman,saket}@imsc.res.in
[2] Chennai Mathematical Institute, H1, SIPCOT IT Park, Siruseri, Chennai 603 103, India
sasanka@cmi.ac.in

Abstract. The well-known Courcelle's theorem states that many graph properties (that are expressible in monadic second order logic) can be solved in linear time on graphs of bounded treewidth. Logspace versions of this using automata theoretic framework are also known. In this paper, we develop an alternate methodology using the standard table-based dynamic programming approach to give a space efficient version of Courcelle's theorem. We assume that the given graph and its tree decomposition are given in a read-only memory. Our algorithms use the recently developed stack-compression machinery and the classical framework of Borie *et al.* to develop time-space tradeoffs for dynamic programming algorithms that use $\mathcal{O}(p \log_p n)$ variables where $2 \leq p \leq n$ is a parameter. En route we also generalize the stack compression framework to a broader class of algorithms, which we believe can be of independent interest.

1 Introduction

The aim of this paper is to demonstrate the power of two class of algorithms – one classical [16] and one recent [10] – and to show that a combination of them provides simple space efficient algorithms for graphs of bounded treewidth even when the tree of the decomposition is given in a read-only memory.

It is well-known [14] that many problems that are NP-hard on general graphs can be solved in linear time, using dynamic programming, on trees and graphs of bounded treewidth. This is captured by the most general theorem due to Courcelle [19] which states that properties that can be expressed in Counting Monadic Second Order Logic (CMSO) can be tested in linear time on graphs of bounded treewidth. Aspvall et al. [9] considered the space required by such dynamic programming algorithms, and showed that the entire computation through a simple post-order traversal requires only $\mathcal{O}(w)$ tables (each of size $f(t)$ for some function

© Springer International Publishing Switzerland 2015
D. Xu et al. (Eds.): COCOON 2015, LNCS 9198, pp. 349–360, 2015.
DOI: 10.1007/978-3-319-21398-9_28

f of the treewidth t) where w is the pathwidth of the tree of the tree decomposition. In particular, as the pathwidth of a tree on n nodes is $\mathcal{O}(\log n)$, this implies that $\mathcal{O}(\log n)$ tables are sufficient. Bodlaender and Telle [15] later show that a set that realizes the optimum solution can be determined in $\mathcal{O}(n \log_s n)$ time using an additional $\mathcal{O}(s)$ words of space where $s \leq \sqrt{cn/2}$ is a parameter where c is an upper bound on the degree of a vertex in the tree decomposition (see Definition 1 in Section 4).

Our main objective in this paper is to implement these standard dynamic programming algorithms in the read-only memory model where the input tree is given in a read-only input. Apart from adjacency of the graph, we can only access the leftmost child, right sibling and the bags of each node in constant time (in particular, we do not assume parent pointers which poses a challenge when doing a bottom-up traversal). We show that the above bounds can be achieved if we use an additional stack that could grow to linear size. Our next step is then to use the recent method of Barba et al.[10] to reduce the space to $\mathcal{O}(\log n)$ words, without too much degradation in time. Towards this end, we first generalize the stack-compression method to a broader class of stack algorithms.

Related Work. Elberfeld et al.[24] showed that Bodlaender's [12] linear time algorithm (to determine whether a given graph has treewidth at most k for a fixed k) and Courcelle's [19] linear time algorithm (to determine the satisfiability of a CMSO expressible property on a bounded treewidth graph) can be implemented in $\mathcal{O}(1)$ words, i.e. using logarithmic bits of space. However, the algorithm of Elberfeld et al. [24] is reasonably complicated, and is based on the automata theoretic framework of Courcelle's theorem, while we use the natural table based approach of the dynamic programming algorithms.

Recently there has been a spate of work in space efficient graph and geometric algorithms [17,23,7,5,2,27,22,11,6,3] due to their theoretical interest and practical motivation for implementation in small hand held devices [4]. For example, for the MAXIMUM INDEPENDENT SET problem covered by our main result in this paper, Bhattacharya et al. [11] gave an $\mathcal{O}(n)$ algorithm using $\mathcal{O}(h)$ word space; here h is the height of the tree that could be as high as n. In contrast, our algorithm gives the optimum value in $\mathcal{O}(n)$ time and $\mathcal{O}(n^\epsilon)$ words of space even for weighted trees and bounded treewidth graphs. It can also output a set realizing the optimum value in the same time and space for unweighted trees. But for weighted trees and bounded treewidth graphs, reporting a solution set takes $\mathcal{O}(n^2)$ time using the same space. We also have algorithms that take $\mathcal{O}(\log n)$ words of space and $\mathcal{O}(n^{1+\epsilon})$ time to output the optimum value. Here ϵ is any fixed positive constant less than 1.

Our results, apart from adding to the growing body of literature on space efficient graph algorithms, bring to light the recent technique to reduce the space requirement of algorithms that use only a stack (that could grow to linear size) and a constant number of auxiliary words.

Organization of the Paper. In Section 2, we first explain the general working scheme of stack based algorithms and also state the main lemma from [10].

Then we provide our generalization of the stack based algorithms, and prove our stack compression lemma for the generalization. In Section 3 we provide algorithms for specific problems on trees using the stack compression machinery, as a warm up for our general result. In Section 4 we develop our main space efficient algorithm for problems expressible in CMSO. We also consider the weighted versions and the modifications necessary to output the solution set. Section 5 contains some concluding remarks.

Model. Our algorithms work in read-only memory model where the input elements cannot be modified or moved. We believe that read-only memory model is a natural model to study algorithms on graphs. Our algorithms work with the standard left-most child, right-sibling based representation for rooted trees [18] (this is slightly weaker in contrast to the doubly connected edge list representation used in [8,11]). I.e. we assume that the tree (of the input or of the tree-decomposition is given in a representation where the children of a node are organized in a linked list. I.e. given a node label, we can find the bag associated with it, its left-most child and its right sibling in constant time. In particular, we do not have parent pointers associated with the nodes, unless stated otherwise.

For the weighted versions of the algorithms and for graphs of bounded treewidth, we assume that weights or the bag sets of the vertices are given in a separate array indexed by the labels of the vertices of the tree (which we assume are in $\{1, 2, \ldots n\}$). For bounded treewidth graphs, apart from the tree decomposition, we also need the graph in read-only memory, represented in a way that adjacency can be checked in constant time. As for output, the algorithms will output values during the execution in a write once array, which cannot be accessed by the algorithm after reporting. We also denote the set of natural numbers by $\mathbb{N}$. When measuring extra space, we count the number of variables or words used by the algorithm, in addition to the input in read-only memory. We assume that standard RAM computations (including additions, multiplications and logical operations) can be done in constant time.

2 Generalized Stack Framework

In what follows, we explain the general scheme of stack based algorithms defined in [10]. Let $\mathcal{A}$ be a class of deterministic algorithms which uses a stack and optionally other auxiliary data structures of constant size. The operations that can be supported are push, pop and accessing the k topmost elements of the stack for a constant k.

Let $X \in \mathcal{A}$ be any algorithm and $a_1, a_2, \cdots, a_n \in I$ be the values of the input, in the order in which they are treated by $\mathcal{A}$. We only assume that given a_i, we can access the next input element in constant time. We will call X a stack based algorithm if, given a set of ordered input I, X processes $a_i \in I$ one by one in order and based on a_i, the top k elements of the stack and the auxiliary data structures' current configuration, it decides to pop some elements off the stack first then push i (or some function of i) along with some constant words of information into the stack. Then the computation moves forward to the next element, until all the elements are exhausted. The final result which is stored in the stack, is then output by popping them one by one until the stack is empty. Any algorithm following this structure is called a *stack based algorithm*. We provide the pseudocode below for completeness [10].

Algorithm 1. Basic scheme of a stack based algorithm

1: Initialize stack and auxiliary data structure DS with $\mathcal{O}(1)$ elements from I
2: **for all** subsequent input $a \in I$ **do**
3:    **while** some-condition(a, DS, STACK.TOP(1),$\cdots$, STACK.TOP(k)) **do**
4:       STACK.POP
5:    **end while**
6:    **if** another-condition(a, DS, STACK.TOP(1),$\cdots$, STACK.TOP(k)) **then**
7:       STACK.PUSH(a)
8:    **end if**
9: **end for**
10: Report(STACK)

For such stack based algorithms, in [10] they prove the following result.

Lemma 1 (Theorem 2, [10]). *Any stack algorithm which takes $\mathcal{O}(n)$ time and $\Theta(n)$ space can be adapted so that, for any parameter $2 \leq p \leq n$, it solves the problem in $\mathcal{O}(n^{1+(1/\log p)})$ time using $\mathcal{O}(p \log_p n)$ variables.*

The main idea of their proof is to divide the input into blocks in the order in which the input is processed. The size of the block depends on the allowed work space. During the execution of the stack algorithm, we store only the first and the last elements of the block which was pushed into the stack (except up to two top blocks that are stored in full). While simulating the algorithm, if we ever need any value not stored in stack, we invoke a reconstruction operation to generate the block containing that element and proceed. For efficiently supporting the reconstruction operation, we also store the content of the auxiliary data structure just after the first element of each block is pushed into the stack.

Our first observation is that we can generalize the above stack based algorithms to a broader class of algorithms and still prove a version of Lemma 1. More specifically,

- while we continue to assume that there can be at most one push while processing an element, we can let it precede and succeed with a sequence of pops (the original framework only allowed it to be preceded with pops.)
- We allow the number of auxiliary variables to be a parameter t, which need not be a constant, and any of the auxiliary variables can be accessed in a constant time. This is possible if we simply assume that the auxiliary variables form an array of t elements. We then capture the time and space also as a function of the number of auxiliary variables.

- We assume that the next element to be processed could depend on the pushes and pops done and hence could take more than a fixed constant time to determine. In fact, as the input is in read-only memory, the algorithm may make several scans over the input (along with the stack and auxiliary storage) and take, say some $g(n, t)$ time (as k is a constant, we ignore its dependence in g) to find the next element to process.
- We will also assume that each element is processed at most once.
- Finally, we allow the condition that decides what to push into the stack or whether to pop the stack element to take time an arbitrary function $h(n, t)$ of n and t and space $s(n)$ words.

Hence each push and pop can take $h(n, t)$ time, and hence the entire algorithm takes $\mathcal{O}(nh(n, t) + ng(n, t))$ time using $s(n) + t$ words apart from a stack. We also allow the algorithm to output elements to a write-once output array along the way. We call such a stack algorithm, a generalized stack based algorithm. We give the pseudocode below.

Algorithm 2. Basic scheme of a generalized stack based algorithm

---

1: Initialize the stack and auxiliary data structure DS of size t
2: Initialize the first input element a to be processed.
3: **repeat**
4:     **while** some-condition(a, input, DS, STACK.TOP(1),$\cdots$, STACK.TOP(k)) **do**
5:         STACK.POP
6:     **end while**
7:     **if** some-condition(a, input, DS, STACK.TOP(1),$\cdots$, STACK.TOP(k)) **then**
8:         STACK.PUSH(a)
9:     **end if**
10:    **while** some-condition(a, input, DS, STACK.TOP(1),$\cdots$, STACK.TOP(k)) **do**
11:        STACK.POP
12:    **end while**
13:    $a \leftarrow next(a)$ ($next(a)$ is computed based on a, all the pops and push done in this step as well as the input set)
14: **until** a is NULL

For such generalized stack based algorithms, we can prove the following theorem:

Theorem 1. *Any generalized stack algorithm that takes $\mathcal{O}(n(h(n, t) + g(n, t)))$ time and $\mathcal{O}(n + t + s(n))$ space can be adapted so that, for any parameter $2 \le p \le n$, it solves the problem in $\mathcal{O}((h(n, t) + g(n, t))n^{1+(1/\log p)})$ time using $\mathcal{O}(pt \log_p n + s(n))$ variables, where $\mathcal{O}(n)$ space is for the explicit stack and $\mathcal{O}(t)$ space is for auxiliary data structures, $s(n)$ space and $h(n, t)$ time to check the conditions for pushes and pops, and $g(n, t)$ time to find the next element to process.*

Proof. (*Sketch*) The space for auxiliary data structures plays a role in Lemma 1 when we compress a block in the stack, and store the 'context' of the auxiliary data structures with the first element of the block. In our case, we can store the values of the t variables into the stack, thereby increasing the storage space by a factor of t.

Also the reconstruction steps in Lemma 1 simulate the original stack algorithm repeatedly (sometimes with recursive calls), and hence the ability to compute the next element of an element after processing that element (which could take more than a constant time) does not affect the analysis in any way. At the base case of the recursive step of reconstruction, a block of size p is reconstructed with full stack space of size p which takes $\mathcal{O}(p(g(n,t)+h(n,t)))$ time (as only the elements of that block are pushed and popped), so $T(p) = \mathcal{O}(p(g(n,t)+h(n,t)))$. The general step remains as before resulting in the earlier recurrence for the total running time as $T(n) = 2pT(\frac{n}{p}) + \mathcal{O}(p)$ which solves to the claimed runtime bound. □

3   Algorithms on Rooted Trees

In the next section we give a generic algorithm for problems expressible in CMSO over graphs of bounded treewidth. In this section we explain our algorithms for specific problems on rooted trees. Apart from serving as a warm up to the general result, this serves to illustrate the main idea that to obtain a space efficient algorithm, all we need to do is to turn a natural dynamic programming based algorithm into a stack based algorithm and then use Theorem 1 to obtain the desired trade-off between space and time. Thus, these algorithms are useful in practice. We exemplify our approach by giving algorithms for Minimum Vertex Cover, Maximum Independent Set and Minimum Weighted Dominating Set (MWDS).

Theorem 2. Minimum Vertex Cover *and* Maximum Independent Set *can be solved on a rooted tree on n vertices in $\mathcal{O}(n^{1+(1/\log p)})$ time using $\mathcal{O}(p \log_p n)$ variables of extra space where $2 \leq p \leq n$. We can also output the corresponding set for each of the above problems in the same time and space. The output is generated through an output array (which cannot be seen/used by the algorithm once the output has been generated) which will list out the vertices.*

Proof. We explore the tree in a depth first fashion by pushing elements into the stack as we traverse until we find the leaf. After we compute the values at the leaf, we pop the stack, and we transfer its value to its parent (which is at the top of the stack). Thus we get around the lack of parent pointer by pushing elements into the stack as we traverse (and so the parent is at the top of the stack when an element is popped), but then this lets the stack grow to size n. Then we apply the stack compression of Theorem 1 to reduce the space. First notice that this algorithm as we described fits into our generalized stack framework (with t, $g(n,t)$, $h(n,t)$ and $s(n)$ as constants) and hence Theorem 1 can be applied. Note that we require the generalized stack framework as the next vertex to be processed in the depth first search is a function of the pushes and pops and hence may take more than a constant time. Also in the depth first search, sometimes a push is followed by the pops (which wasn't allowed in the original stack framework). Below we give specific details of the algorithm for the problems claimed in the theorem.

MINIMUM VERTEX COVER: A standard algorithm to find a minimum vertex cover in a tree is based on the observation that if v is a vertex with degree 1, then there exists a minimum vertex cover that contains v's unique neighbor u.

Hence the algorithm repeatedly includes the neighbor of leaf nodes into the solution and deletes them (and their leaf neighbors) from the tree, until the tree becomes empty. This algorithm can be implemented using a stack as follows: we do a depth first traversal of the tree starting from the root, including every vertex into the stack along with a bit 0 associated with the vertex (to indicate our initial guess to exclude that vertex from the solution) as we visit. The first popping happens when we visit a leaf. While popping out a vertex (after visiting its entire subtree), if its value remains as 0, we make its parent (which would be at the top of the stack) 1. Once a node becomes 1, it stays as 1 during its lifetime in the stack. While popping a node if its associated bit is 1, we output it in the output array.

Clearly this implements the above algorithm as we pop only elements that are leaves or have become leaves after some deletion of vertices and before throwing when a leaf has value 0, we make its parent 1.

Depth first order can be implemented by accessing a node's leftmost child to push at every step. Once we know that a node has no children, then we find its right sibling (the next child of its parent to visit in the depth first order) to add to stack before popping the node. This completes the algorithm, and it is easy to see that this algorithm essentially implements the algorithm outlined above using a stack and a constant number of variables. When a vertex assigned 0 is popped it implies that it is a leaf and hence its parent is assigned 1 which is correct by the observation above. MAXIMUM INDEPENDENT SET can be solved similarly (by switching the role of 1 and 0 in the above algorithm).   □

A Weighted Case. The algorithms we described for MINIMUM VERTEX COVER use the observation that for any leaf, its unique neighbor can be picked up in any minimum vertex cover. This is no longer true in the weighted variants of these problems (see Appendix for definitions). Thus, in this case we resort to a modification of standard dynamic programming algorithm. We exemplify the approach via MINIMUM WEIGHTED DOMINATING SET (MWDS). However, this comes at a cost: now we cannot output the desired set in the same running time, though it is possible to output the set with a linear factor increase in the runtime.

Theorem 3. MWDS *can be solved, on a rooted tree with n vertices, in time* $\mathcal{O}(n^{1+(1/\log p)})$ *using* $\mathcal{O}(p \log_p n)$ *variables of extra space where* $2 \leq p \leq n$.

Proof. For MWDS the standard dynamic programming algorithm (see [26] for an algorithm for MWDS in bounded treewidth graphs) needs the values at each child of a node before computing the value of the node, which may result in some non-trivial storage at each node of the tree. A dynamic programming algorithm works as follows. It computes four quantities $I(v), D(v), DE(v), E(v)$ for each vertex v, which are defined as below.

1. $E(v) =$ The size of a minimum dominating set of the subtree rooted at v, that does not contain the vertex v.
2. $DE(v) =$ The size of a minimum dominating set of the subtree rooted at v that dominates all vertices of the subtree except possibly v.
3. $I(v) =$ The size of a minimum dominating set of the subtree rooted at v that contains vertex v.
4. $D(v) =$ The size of a minimum dominating set of the subtree rooted at v.

Clearly $D(v) = \min\{I(v), E(v)\}$. It is easy to find these values for leaf nodes, and once we have computed these quantities for all the children of a node v, they can be computed for v as follows: let $u_1, u_2, \ldots u_d$ be the children of v. Let $H(v) = \sum_{i=1}^{d} D(u_i)$. Then
$I(v) = w(v) + \sum_{i=1}^{d} DE(u_i)$, $DE(v) = \min\{I(v), H(v)\}$,
$E(v) = \min_{1 \le i \le d}\{I(u_i) + \sum_{k=1, k \ne i}^{d} D(u_k)\}$

Now, all these quantities can be computed in a depth first order as follows. We initialize $DE(u), H(u)$ and $E(u)$ to 0, and $I(u)$ to $w(u)$, as we push a vertex u into the stack. Once we pop a vertex u, we would have visited its entire subtree, and hence computed these values for that vertex. We can then easily compute $D(u)$ for that vertex. Then we update its parent (which is at the top of the stack) with its contribution to all of these quantities of its parent. More precisely, for the parent vertex v, we make $I(v) \leftarrow I(v) + DE(u); H(v) \leftarrow H(v) + D(u)$.

Updating $E(v)$ is a bit tricky, but it becomes easier, if we rewrite the quantity $E(v)$ in the above equation as,

$$E(v) = \left\{ \sum_{i=1}^{d} D(u_i) + \min_{1 \le i \le d} (I(u_i) - D(u_i)) \right\} = H(v) + \min_{1 \le i \le d} (I(u_i) - D(u_i)).$$

To compute $H(v)$, we simply keep track of $\min\{I(u) - D(u)\}$ over its children u of v and also the vertex u that realizes the minimum. When we pop a vertex u, we update the DE value of its parent v as $H(v) = H(v) + D(u)$ and update the $\min\{I(u) - D(u)\}$ at its parent by $(I(u) - D(u))$ if the $(I(u) - D(u))$ quantity is smaller. When all the children are popped, $E(v)$ is set to $E(v) = H(v) + (I(u) - D(u))$. It is easy to see that the above expression correctly computes the four quantities in a postorder traversal using just a stack and the final answer is given by $D(root)$. As the algorithm visits the tree in postorder traversal and uses just a stack, by similar argument to that of Theorem 2, the result follows using Theorem 1. □

4   Algorithms for Graphs of Bounded Treewidth

In this section we design space efficient version of optimization variant of Courcelle's Theorem. We follow the proof of Borie *et al.* [16] and use the machinery of stack compression to obtain the desired theorem. We start with the notations and the language in which we will be working with.

Treewidth. For a rooted tree T and a non-root node $t \in V(T)$, by parent(t) we denote the parent of t in the tree T. For two nodes $u, t \in T$, we say that u is a *descendant* of t, denoted $u \preceq t$, if t lies on the unique path connecting u to the root. Note that every node is thus its own descendant.

Definition 1 (tree decomposition). *A tree decomposition of a graph G is a pair (T, β), where T is a rooted tree and $\beta : V(T) \to 2^{V(G)}$ is a mapping such that:*

- *for each node $v \in V(G)$ the set $\{t \in V(G) | v \in \beta(t)\}$ induces a nonempty and connected subtree of T,*
- *for each edge $e \in E(G)$ there exists $t \in V(T)$ such that $e \subseteq \beta(t)$.*

The set $\beta(t)$ is called the *bag at* t, while sets $\beta(u) \cap \beta(v)$ for $uv \in E(T)$ are called *adhesions*. Following the notation from [25], for a tree decomposition (T, β) of a graph G we define auxiliary mappings $\sigma, \gamma : V(T) \to 2^{V(G)}$ as

$$\sigma(t) = \begin{cases} \emptyset & \text{if t is the root of } T \\ \beta(t) \cap \beta(\text{parent}(t)) & \text{otherwise} \end{cases}$$

$$\gamma(t) = \bigcup_{u \preceq t} \beta(u)$$

We now define a class of graph optimization problems, called MIN/MAX-CMSO$[\psi]$, with one problem for each CMSO sentence ψ on graphs, where ψ has a free vertex (edge) set variable S. We refer to [1,19,20] and the book of Courcelle and Engelfriet [21] for a detailed introduction to CMSO. In [21], CMSO is referred to as CMS_2. The MIN-CMSO problem defined by ψ is denoted by MIN-CMSO$[\psi]$ and defined as follows.

MIN-CMSO$[\psi]$
Input: A graph $G = (V, E)$.
Question: Find the cardinality of a minimum sized subset $S \subseteq V$ ($S \subseteq E$) (if exists) such that $(G, S) \models \psi$.

The definition of MAX-CMSO$[\psi]$ problem is analogous to the MIN-CMSO$[\psi]$ problem. The only difference is that now we try to find a maximum sized subset $S \subseteq V$. Here, we only give an algorithm for MIN/MAX-CMSO$[\psi]$ problems when S is a vertex subset. All of our results can be extended to edge setting in a straightforward way. In particular, an edge set problem on graph $G = (V, E)$ can be transformed to a vertex subset problem on the edge-vertex incidence graph $I(G)$ of G, which is a bipartite graph with vertex bipartition's V and E with edges between vertices $v \in V$ and $e \in E$ if and only if v is incident with e in G. It is well-known that the treewidth of G and $I(G)$ only differ by a factor of 2. To make the translation work in the proof, it is sufficient to use the fact that the property of being an incidence graph of a graph G is expressible in CMSO. From now on *we only concentrate on* MIN/MAX-CMSO[ψ] *problems defined over vertex subsets.*

Now we give a couple of examples of problems, that can be encoded using MIN/MAX-CMSO[ψ]. To express MAXIMUM INDEPENDENT SET we do as follows. Given a graph G and a vertex subset X, a simple constant length formula **indp**(X) that verifies that X is an independent set of G is: $\forall_{x \in X} \forall_{y \in X} \neg \mathbf{adj}(x, y)$. Thus we can express MAXIMUM INDEPENDENT SET using the above logical sentence. Let us consider another example, namely, MINIMUM DOMINATING SET. Given a graph G and a vertex subset X, a simple constant length formula **dom**(X) that verifies that X is a dominating set of G is: $\forall_{x \in V(G)}[x \in X \vee \exists_{y \in X} \mathbf{adj}(x, y)]$.

Dynamic Programming Algorithms over Tree Decompositions. The standard dynamic programming algorithms on bounded treewidth graphs for standard optimization problems, see [13,1,14] proceed by doing a bottom-up traversal of the tree computing tables at every node starting from the leaves. Aspvall et al., in [9] identify the space requirement of the algorithms and argue that $\mathcal{O}(\log n)$ 'open tables' are sufficient in the bottom-up implementation. Bodlaender and Telle [15], building on the work of Aspvall et al., claim that any property expressible in monadic second order logic can be implemented using $O(\log n)$ tables in the bottom-up traversal.

However, implementing this in the read-only memory model without parent pointers, provide challenges in terms of space, and that is our task in this section.

In particular, we provide an alternate $\mathcal{O}(\log n)$ word space algorithm that precludes the need for parent pointers in the input representation, and uses the stack compression machinery recently developed by Barba et al. [10] to prove the following theorem.

Theorem 4. ($\spadesuit$)[1] *Let G be a graph given with a tree decomposition ($T = (V_T, E_T), \beta$) of width k. Then* MIN/MAX-CMSO[ψ] *can be solved in time* $\mathcal{O}(\tau(k) \cdot n^{1+(1/\log p)})$ *time and* $\mathcal{O}(\tau(k) \cdot p \log_p n)$ *space algorithm, for any parameter $2 \leq p \leq n$. Here, $|V| = n$ and τ is a function of k alone.*

In fact we prove a weighted variant of Theorem 4. We also obtain an algorithm that not only outputs the weight of a value of a maximum weighted subset (or a minimum weighted subset) S such that $(G, S) \models \psi$, but also the set S. We call this version of the problem CONSTRUCTIVE-WEIGHTED-MIN/MAX-CMSO[ψ].

Theorem 5. ($\spadesuit$) *Let G be a graph given with a tree decomposition ($T = (V_T, E_T), \beta$) of width k. Then* CONSTRUCTIVE-WEIGHTED-MIN/MAX-CMSO[ψ] *can be solved in time* $\mathcal{O}(\tau(k) \cdot n^{2+(2/\log p)})$ *time and* $\mathcal{O}(\tau(k) \cdot p \log_p n)$ *space algorithm, for any parameter $2 \leq p \leq n$. Here, $|V| = n$ and τ is a function of k alone.*

---

[1] Proofs of results marked with ($\spadesuit$) will appear in full version.

5   Conclusion

We have shown that several optimization problems can be solved on trees and bounded treewidth graphs using logarithmic number of extra variables, in polynomial time even when the input tree is given in a read-only memory. We achieve this by modifying the standard dynamic programming algorithms to use only a stack and using the recent stack compression routine to reduce space. Barba et al. [10] also provide a stack compression scheme that can be used to reduce work space to $\mathcal{O}(1)$ words provided the (full stack) algorithm satisfies what they called a "green" property. The standard dynamic programming algorithms we use are not "green". It would be interesting to see whether our approach can be extended to obtain algorithms using only $\mathcal{O}(1)$ or even $o(\log n)$ words. This would give an alternate $o(\log n)$ (words of) space version of Courcelle's theorem. Another open problem is whether this approach helps to give an alternate logarithmic space version of Bodlaender's theorem [12].

Bodlaender and Telle [15] give a divide and conquer strategy to find the optimum set in $\mathcal{O}(n \log n)$ time (against a naive $O(n^2)$ time) using $\mathcal{O}(\log n)$ words of extra space, when the tree of the tree-decomposition has a constant number of children for each node. Extending this (in the read-only memory model) to the case when each node has an arbitrary number of children, or to obtain a 'nice-tree decomposition' from a general tree-decomposition and implementing their approach in logarithmic (words of) space in read-only memory model, are challenging problems. It would also be interesting to find other applications of our generalized stack compression framework.

References

1. Arnborg, S., Lagergren, J., Seese, D.: Easy problems for tree-decomposable graphs. Journal of Algorithms **12**, 308–340 (1991)
2. Asano, T.: Constant-Working-Space Algorithms for Image Processing. In: Nielsen, F. (ed.) ETVC 2008. LNCS, vol. 5416, pp. 268–283. Springer, Heidelberg (2009)
3. Asano, T.: Constant-Working-Space Algorithms: How Fast Can We Solve Problems without Using Any Extra Array? In: Hong, S.-H., Nagamochi, H., Fukunaga, T. (eds.) ISAAC 2008. LNCS, vol. 5369, pp. 1–1. Springer, Heidelberg (2008)
4. Asano, T.: Designing Algorithms with Limited Work Space. In: Ogihara, M., Tarui, J. (eds.) TAMC 2011. LNCS, vol. 6648, pp. 1–1. Springer, Heidelberg (2011)
5. Asano, T., Doerr, B.: Memory-constrained algorithms for shortest path problem. In: Proceedings of the 23rd Annual Canadian Conference on Computational Geometry, CCCG (2011)
6. Asano, T., Kirkpatrick, D., Nakagawa, K., Watanabe, O.: $\tilde{O}(\sqrt{n})$-Space and polynomial-time algorithm for planar directed graph reachability. In: Csuhaj-Varjú, E., Dietzfelbinger, M., Ésik, Z. (eds.) MFCS 2014, Part II. LNCS, vol. 8635, pp. 45–56. Springer, Heidelberg (2014)
7. Asano, T., Mulzer, W., Rote, G., Wang, Y.: Constant-work-space algorithms for geometric problems. JoCG **2**(1), 46–68 (2011)
8. Asano, T., Mulzer, W., Wang, Y.: Constant-work-space algorithms for shortest paths in trees and simple polygons. J. Graph Algorithms Appl. **15**(5), 569–586 (2011)

9. Aspvall, B., Telle, J.A., Proskurowski, A.: Memory requirements for table computations in partial k-tree algorithms. Algorithmica **27**(3), 382–394 (2000)
10. Barba, L., Korman, M., Langerman, S., Sadakane, K., Silveira, R.: Space-time trade-offs for stack-based algorithms. Algorithmica (2014) (in press)
11. Bhattacharya, B.K., De, M., Nandy, S.C., Roy, S.: Maximum independent set for interval graphs and trees in space efficient models. In: Proceedings of the 26th Canadian Conference on Computational Geometry, CCCG (2014)
12. Bodlaender, H.L.: A linear-time algorithm for finding tree-decompositions of small treewidth. SIAM J. Comput. **25**(6), 1305–1317 (1996)
13. Bodlaender, H.L.: A partial k-arboretum of graphs with bounded treewidth. Theor. Comput. Sci. **209**(1–2), 1–45 (1998)
14. Bodlaender, H.L., Koster, A.M.C.A.: Combinatorial optimization on graphs of bounded treewidth. Comput. J. **51**(3), 255–269 (2008)
15. Bodlaende, H.L., Telle, J.A.: Space-efficient construction variants of dynamic programming. Nord. J. Comput. **11**(4), 374–385 (2004)
16. Borie, R.B.: Gary Parker, R., Tovey, C.A.: Automatic generation of linear-time algorithms from predicate calculus descriptions of problems on recursively constructed graph families. Algorithmica **7**(5&6), 555–581 (1992)
17. Bose, P., Morin, P.: An improved algorithm for subdivision traversal without extra storage. Int. J. Comput. Geometry Appl. **12**(4), 297–308 (2002)
18. Cormen, T.H., Leiserson, C.E., Rivest, R.L., Stein, C.: Introduction to Algorithms, 3rd edn. MIT Press (2009)
19. Courcelle, B.: The monadic second-order logic of graphs I: Recognizable sets of finite graphs. Inform. and Comput. **85**, 12–75 (1990)
20. Courcelle, B.: The expression of graph properties and graph transformations in monadic second-order logic. In: Handbook of Graph Grammars and Computing by Graph Transformation, vol. 1, pp. 313–400. World Sci. Publ., River Edge (1997)
21. Courcelle, B., Engelfriet, J.: Graph Structure and Monadic Second-Order Logic. Cambridge University Press (2012)
22. Datta, S., Limaye, N., Nimbhorkar, P., Thierauf, T., Wagner, F.: Planar graph isomorphism is in log-space. In: Proceedings of the 24th Annual IEEE Conference on Computational Complexity, CCC 2009, pp. 203–214 (2009)
23. De, M., Nandy, S.C., Roy, S.: Convex hull and linear programming in read-only setup with limited work-space. CoRR, abs/1212.5353 (2012)
24. Elberfeld, M., Jakoby, A., Tantau, T.: Logspace versions of the theorems of bodlaender and courcelle. In: 51th Annual IEEE Symposium on Foundations of Computer Science, FOCS 2010, pp. 143–152 (2010)
25. Grohe, M., Marx, D.: Structure theorem and isomorphism test for graphs with excluded topological subgraphs. SIAM J. Comput. **44**(1), 114–159 (2015)
26. Niedermeier, R.: Invitation to Fixed-Parameter Algorithms. Oxford University Press (2006)
27. Reingold, O.: Undirected connectivity in log-space. J. ACM 55(4), 17:1–17:24 (2008)

Reducing Rank of the Adjacency Matrix
by Graph Modification

S.M. Meesum[1]([✉]), Pranabendu Misra[1], and Saket Saurabh[1,2]

[1] Institute of Mathematical Sciences, Chennai, India
{meesum,pranabendu,saket}@imsc.res.in
[2] University of Bergen, Bergen, Norway

Abstract. The main topic of this article is to study a class of graph
modification problems. A typical graph modification problem takes as
input a graph G, a positive integer k and the objective is to add/delete
k vertices (edges) so that the resulting graph belongs to a particular
family, $\mathcal{F}$, of graphs. In general the family $\mathcal{F}$ is defined by forbidden sub-
graph/minor characterization. In this paper rather than taking a struc-
tural route to define $\mathcal{F}$, we take algebraic route. More formally, given a
fixed positive integer r, we define $\mathcal{F}_r$ as the family of graphs where for
each $G \in \mathcal{F}_r$, the rank of the adjacency matrix of G is at most r. Using
the family $\mathcal{F}_r$ we initiate algorithmic study, both in classical and parameterized complexity, of following graph modification problems: r-RANK
VERTEX DELETION, r-RANK EDGE DELETION and r-RANK EDITING.
These problems generalize the classical VERTEX COVER problem and a
variant of the d-CLUSTER EDITING problem. We first show that all the
three problems are NP-Complete. Then we show that these problems are
fixed parameter tractable (FPT) by designing an algorithm with run-
ning time $2^{\mathcal{O}(k \log r)} n^{\mathcal{O}(1)}$ for r-RANK VERTEX DELETION, and an algorithm for r-RANK EDGE DELETION and r-RANK EDITING running in
time $2^{\mathcal{O}(f(r)\sqrt{k} \log k)} n^{\mathcal{O}(1)}$. We complement our FPT result by designing
polynomial kernels for these problems.

1 Introduction

A typical graph modification problem takes as an input a graph G, a positive
integer k and the objective is to add/delete k vertices (edges) so that the resulting
graph belongs to a particular family, $\mathcal{F}$, of graphs. One of the classical way to
define $\mathcal{F}$ is by defining what is called *graph property*. A *graph property* Π is a set
of graphs, which is closed under isomorphism. The property Π is non-trivial if
it includes infinitely many graphs and also excludes infinitely many graphs. The
property Π is called *hereditary* if for any graph $G \in \Pi$, all induced subgraphs
of G are also present in Π. A hereditary property Π has an *induced forbidden
set* characterization if there is a family Forb(Π) of graphs such that, a graph G
is in Π if and only if no induced subgraph of G is in Forb(Π). Another way of
defining $\mathcal{F}$ is by excluding some forbidden *minors*. A graph H is said to be a
minor of a graph G, if H can be obtained from G be deleting some edges and

© Springer International Publishing Switzerland 2015
D. Xu et al. (Eds.): COCOON 2015, LNCS 9198, pp. 361–373, 2015.
DOI: 10.1007/978-3-319-21398-9_29

vertices, and contracting some edges. A hereditary property Π has a *forbidden minor* characterization if there is a family $\mathsf{Forb}(\Pi)$ of graphs such that no graph in Π has a graph in $\mathsf{Forb}(\Pi)$ as a minor. For a graph property Π the Π-GRAPH MODIFICATION problem is defined as follows. Given a graph G and a positive integer k, can we delete (edit) at most k vertices (edges) so that the resulting graph G' is in Π?

Lewis and Yannakakis have shown that for any Π which is non-trivial and hereditary, the corresponding vertex deletion problems are NP-Complete [13,18]. In addition for several graph properties, the corresponding edge editing problem are known to be NP-Complete as well [2]. This motivates the study of these problems in algorithmic paradigms that are meant for coping with NP-hardness, such as approximation algorithms and parameterized complexity. In this paper, we study yet another kinds of graph modification problems in the framework of *parameterized complexity*. In this framework, each instance of the problem is *parameterized*, i.e. assigned a number k, which is called the *parameter*, which represents some property of the instance. For example, a typical parameter is the size of the optimum solution to the instance. The problem is called *fixed parameter tractable* (FPT), if a parameterized instance (x, k) of the problem is solvable in time $f(k)n^{\mathcal{O}(1)}$, where n is the size of the instance. A parameterized problem is said to admit a *polynomial kernel* if there is a polynomial time algorithm that given an instance of the problem, outputs an equivalent instance whose size is bounded by a polynomial in the parameter. It is known that whenever Π has *finite* induced forbidden set characterization (that is, $|\mathsf{Forb}(\Pi)|$ is finite), the corresponding deletion problems are known to be FPT [3]. Similarly, whenever Π is characterized by a *finite* forbidden set of minors, the corresponding problems are FPT by a celebrated result of Robertson and Seymour [16]. These problems include classical problems such as VERTEX COVER, FEEDBACK VERTEX SET, SPLIT VERTEX DELETION. There has also been extensive study of graph modification problems when the corresponding $\mathsf{Forb}(\Pi)$ is not finite, such as CHORDAL VERTEX DELETION, INTERVAL VERTEX DELETION, CHORDAL COMPLETION and ODD CYCLE TRANSVERSAL [4,5,8,15].

In this paper we step aside and study a class of graph modification problems that are defined algebraically rather than structurally. Given a graph G, two most important matrices that can be associated with it are its adjacency matrix A_G or the corresponding laplacian L_G. One could study graph modification problems where after editing edges/vertices, the resulting graph has a certain kind of eigenvalue spectrum or has a certain upper bound on the second eigenvalue or the corresponding adjacency matrix satisfies certain properties. The topic of this paper is one of these kind of problems. In particular we want the adjacency matrix of the resulting graph to be some fixed constant r. To define these problems formally, for a positive integer r, define Π_r to be the set of graphs G such that the rank of the adjacency matrix A_G is at most r. The rank of the adjacency matrix A_G is an important quantity in graph theory. It also has connections to many fundamental graph parameters such as the clique number, diameter, domination number etc. [1]. All these motivate the study of the following problems.

---

r-RANK VERTEX DELETION   **Parameter:** k

Input: A graph G and a positive integer k

Question: Can we delete at most k vertices from G so that $rank(A_G) \leq r$?

We use $\triangle$ to denote the standard symmetric difference of sets.

---

r-RANK EDITING   **Parameter:** k

Input: A graph G and a positive integer k

Question: Can we find a set $F \subseteq V(G) \times V(G)$ of size at most k such that $rank(A_{G'}) \leq r$, where $G' = (V(G), E(G) \triangle F)$?

The set F denotes a set of edits to be performed on the graph G. The $\triangle$ operation acts on the sets as follows: if $(u, v) \in F$ and $(u, v) \in E(G)$ then we delete the corresponding edge from G and if $(u, v) \in F$ and $(u, v) \notin E(G)$ then we add the corresponding edge to G. We also consider a variant of r-RANK EDITING, called r-RANK EDGE DELETION, where we are allowed to only delete edges.

These problems are also related to some well known problems in graph algorithms. Observe that if $rank(A_G) = 0$, then G is an empty graph and if $rank(A_G) = 2$ then G is a complete bipartite graph with some isolated vertices. There are no graphs such that $rank(A_G) = 1$. So for $r = 0$, r-RANK VERTEX DELETION is the well known VERTEX COVER problem. Similarly for $r = 2$, a solution to r-RANK EDGE DELETION is a complement of a solution to MAXIMUM EDGE BICLIQUE, where the goal is to find a bi-clique subgraph of the given graph with maximum number of edges [14]. A graph G is called a *Cluster Graph* if every component is a clique. We may also view the above problems as variants of the d-CLUSTER VERTEX DELETION and d-CLUSTER EDITING. In d-CLUSTER VERTEX DELETION, we wish to delete at most k vertices of the graph, so that the resulting graph is a cluster graph with at most d components. Similarly in d-CLUSTER EDITING we wish to add/delete at most k edges to the graph, so that the resulting graph is a cluster graph with at most d components. These problems are known to be NP-hard and they admit FPT algorithms parameterized by the solution size [7,9,10]. In our problems, instead of reducing the graph to a disjoint union of d cliques, we ask that the graph be reduced to a graph of low rank adjacency matrix. In this paper, we obtain following results about these problems.

- We first show that all the three problems are NP-Complete.
- Then we show that these problems are FPT by designing an algorithm with running time $2^{\mathcal{O}(k \log r)} n^{\mathcal{O}(1)}$ for r-RANK VERTEX DELETION, and an algorithm for r-RANK EDGE DELETION and r-RANK EDITING running in time $2^{\mathcal{O}(f(r)\sqrt{k} \log k)} n^{\mathcal{O}(1)}$. Observe that the edge-editing problem admits a subexponential FPT algorithm.
- Finally, we design polynomial kernels for these problems.

Our results are based on structural observations on graphs with low rank adjacency matrix and applications of some elementary methods in parameterized complexity.

2   Preliminaries

We use $\mathbb{R}$ and $\mathbb{N}$ to denote the set of reals and integers, respectively. For two sets X and Y, we define $X \triangle Y = (X \setminus Y) \cup (Y \setminus X)$, i.e. the set of all elements which are exactly in X or Y but not both. A vector v of length n is an ordered sequence of n values from $\mathbb{R}$. A collection of vectors $\{v_1, v_2, \ldots, v_k\}$ are said to be linearly dependent if there exist values $a_1, a_2, \ldots, a_k$, not all zeros, such that $\sum_{i=1}^{k} a_i v_i = 0$. Otherwise these vectors are called linearly independent. A matrix A of dimension $n \times m$, is a sequence of values (a_{ij}). The i-th row of A is defined as the vector $(a_{i1}, a_{i2}, \ldots, a_{im})$ and the j-th column of A is defined as the vector $(a_{1j}, a_{2j}, \ldots, a_{nj})$. The row set and the column set of A are denoted by $\mathbf{R}(A)$ and $\mathbf{C}(A)$ respectively. For $I \subseteq \mathbf{R}(A)$ and $J \subseteq \mathbf{C}(A)$, we define $A_{I,J} = (a_{ij} \mid i \in I, j \in J)$, i.e. it is the submatrix (or minor) of A with the row set I and the column set J. The *rank* of a matrix is the cardinality of the maximum sized collection of columns which are linearly independent. Equivalently, the rank of a matrix is the cardinality of the maximum sized collection of rows which are linearly independent. It is denoted by $\mathrm{rank}(A)$.

For a graph G, we use $V(G)$ and $E(G)$ to denote the vertex set and the edge set of G. For a set of vertices V and a set of edges E on V, we use (V, E) to denote the graph formed by them. Let the vertex set be ordered as $V(G) = \{v_1, v_2, \ldots, v_n\}$. We only consider simple graphs in this paper, i.e. every edge is distinct and the two endpoints of an edge are also distinct. The complement of a graph G is defined as the graph $\overline{G} = (V(G), \overline{E(G)})$ where $\overline{E(G)} = \{(u, v) \mid (u, v) \in (V(G) \times V(G)) \setminus E(G), u \neq v\}$. The *adjacency matrix* of G, denoted by A_G, is defined as the $n \times n$ matrix whose rows and columns are indexed by $V(G)$, such that $A_G(i, j) = 1$ if and only if $(v_i, v_j) \in E(G)$. For a vertex $v \in V(G)$, we use $R(v)$ to denote the row of A_G corresponding to v. Similarly we define $C(v)$ to denote the column of A_G corresponding to v. For a set of vertices $S \subseteq V(G)$, we use $R(X)$ and $C(Y)$ to denote the set of rows and columns corresponding to the vertices in S. Similarly for $X, Y \subseteq V(G)$, we use $A_{X,Y}$ to denote the submatrix of A_G corresponding to rows in $R(X)$ and columns in $C(Y)$. For a vertex $v \in G$, $N(v) = \{u \in G \mid (u, v) \in E(G)\}$ denotes the neighbourhood of v, and $N[v] = N(v) \cup \{v\}$ denotes the *closed neighbourhood* of v. An *independent set* in a graph G is a set of vertices X such that for every pair of vertices $u, v \in X$, $(u, v) \notin E$. A *clique* on n vertices, denoted by K_n, is a graph on n vertices such that every pair of vertices has an edge between them. A *bi-clique* is a graph G, where $V(G) = V_1 \uplus V_2$ and for every pair of vertices $v_1 \in V_1$ and $v_2 \in V_2$, there is an edge $(v_1, v_2) \in E(G)$. When $|V_1| = |V_2| = n$ we denote the biclique by $K_{n,n}$. We can similarly define complete multipartite graphs. For a set of vertices U and a graph G, $G[U]$ denotes the induced subgraph of G on $U \cap V(G)$, and $G \setminus U$ denotes the graph $G[V(G) \setminus U]$.

Given a graph G, we define a relation $\sim$ on $V(G)$. Two vertices u and v are $u \sim v$ if and only if $N(u) = N(v)$. This definition gives us the following lemma.

Lemma 1. (i) $u \sim v$, if and only if $R(u) = R(v)$ and $C(u) = C(v)$.
(ii) $\sim$ partitions V as $V_1, V_2, \ldots, V_l$ where each V_i is an independent set in G.

(iii) For each pair of V_i and V_j, either there are no edges between V_i and V_j, or $G[V_i \uplus V_j]$ is a bi-clique.

The subsets $V_1, \ldots, V_l$ are called *independent modules* of the graph G. We also refer to V_i as an equivalence class in G. The reduced graph of G is the graph $G^\sim$ as follows. The vertex set is $V(G^\sim) = \{u_1, \ldots, u_l\}$ where l is the number of independent modules of G. The edge set is $E(G^\sim) = \{(u_i, u_j) | G[V_i \uplus V_j]$ is a bi-clique $\}$.

We denote the adjacency matrix of $G^\sim$ by $A_G^\sim$. Observe that every vertex of $G^\sim$ has a distinct neighbourhood, therefore all the columns of $A_G^\sim$ are distinct. We now have the following lemma.

Lemma 2 (Proposition 1. [1]). $\mathsf{rank}(A_G) = \mathsf{rank}(A_G^\sim)$.

Following result by Lovåsz [12] gives us a bound on the number of vertices in $G^\sim$, when $\mathsf{rank}(A_G) = r$.

Theorem 1. *If the rank of the adjacency matrix of a graph G is r then the number of vertices in $G^\sim$ is at most $c \cdot 2^{\frac{r}{2}}$ for an absolute constant c.*

The following corollary of the above theorem is used extensively in our algorithms.

Theorem 2 ($\star^1$). *For every fixed r, the number of distinct reduced graphs $G^\sim$ such that $\mathsf{rank}(G^\sim) = r$, is upper bounded by $c \cdot 2^{2^r}$, for some absolute constant c.*

We also require the following lemma from [1](Proposition 7).

Lemma 3. *The only reduced graph of rank r with no isolated vertices and K_r as a subgraph is K_r itself.*

We also require the following results in the subsequent proofs.

Lemma 4 ($\star$). *$G^\sim$ is isomorphic to an induced subgraph of G.*

Observation 3 ($\star$) *If G' is an induced subgraph of G, then $\mathsf{rank}(A_{G'}) \leq \mathsf{rank}(A_G)$.*

3   Reducing Rank by Deleting Vertices

In this section, we consider r-RANK VERTEX DELETION.

The set of vertices whose deletion reduces the rank to r is called solution set and k the solution size. We call r the *target rank*. We begin with a structural observation on the solution set of a given instance of r-RANK VERTEX DELETION.

---

[1] Due to space constraints, proofs of the results marked $\star$ have been omitted. These will appear in the full version of the paper.

Lemma 5. *Let (G, k) be an instance to r-RANK VERTEX DELETION and $\sim$ be the equivalence relation on $V(G)$. If $S \subseteq V(G)$ is a minimal solution for the instance (G, k) then it either contains all the vertices of an equivalence class defined by $\sim$ on $V(G)$ or none of it.*

Proof. Let $S \subseteq V(G)$ be a minimal solution, such that S contains at least one vertex of an equivalence class V_1 but it does not contain all of the vertices in V_1. Observe that all the vertices of $V_1 \setminus S$ have the same neighbourhood in $G \setminus S$, and so they go to the same equivalence class in $G \setminus S$. Let $S' = S \setminus V_1$. We claim that S' is a valid solution.

Let $V_1', \ldots, V_l'$ be a partition of $V(G \setminus S)$ into independent modules such that $(V_1 \setminus S) \subseteq V_1'$. Now, consider the graph $G \setminus S'$ and the partition of $V(G \setminus S')$ as $V_1'', V_2'', \ldots, V_l''$, where $V_1'' = V_1' \cup V_1$ and $V_i'' = V_i'$ $\forall i \geq 2$. Now observe that for any $u, v \in V_i''$, $N(u) = N(v)$. And for any u and v from different classes $N(u) \neq N(v)$, because $G \setminus S$ is an induced subgraph of $G \setminus S'$, and $(N(u) \cap V(G \setminus S)) \neq (N(v) \cap V(G \setminus S))$. Thus $V_1'', \ldots, V_l''$ is a partition of $V(G \setminus S')$ into independent modules. Further observe that $(V_i'' \cup V_j'')$ induces a complete bi-clique in $G \setminus S'$ if and only if $(V_i' \cup V_j')$ induces a complete bi-clique in $G \setminus S'$. Therefore $G \setminus S$ and $G \setminus S'$ have the same reduced graph. So by Lemma 2, $\mathrm{rank}(A_{G \setminus S}) = \mathrm{rank}(A_{G \setminus S'})$.

So $S' \subsetneq S$ is also a solution. But this contradicts the minimality of S, which implies that no such S exists.   □

We obtain the following useful corollary of the above lemma.

Corollary 1. *A minimal solution to an instance (G, k) of r-RANK VERTEX DELETION is disjoint from any equivalence class of G of cardinality greater than k.*

We also have the following useful observation.

Observation 4 ($\star$) *Let S be a solution to an instance (G, k) of r-RANK VERTEX DELETION. Let $U \subseteq V(G)$. Then $S \cap U$ is a solution to the instance $(G[U], k)$.*

3.1   NP Completeness

Recall that r-RANK VERTEX DELETION may be defined as a node-deletion problem with respect to the graph class Π_r. Since Π_r is non-trivial and hereditary, therefore this problem is NP-Complete [13,18].

Theorem 5 ($\star$). *r-RANK VERTEX DELETION is NP-Complete.*

The proof of the theorem above appears in the full version of the paper, it gives a reduction from the VERTEX COVER problem.

3.2 A Parameterized Algorithm for r-RANK VERTEX DELETION

Let (G, k) be an instance of r-RANK VERTEX DELETION. Let $G^{\sim}$ be the reduced graph of G. We have the following corollary of Lemma 5.

Corollary 2 ($\star$). *Let G' be obtained from G by removing all but $k + 1$ vertices from each equivalence class of vertices in G. Then the instances (G, k) and (G', k) are equivalent instances of r-RANK VERTEX DELETION.*

The rank of a matrix can be defined alternatively in terms of determinant of its square submatrices as follows.

Definition 1. *A matrix A over real numbers has rank equal to r if all the $(r + 1) \times (r + 1)$ submatrices of A have determinant zero and there exists a submatrix of size $r \times r$ such that its determinant is non-zero.*

Let $\mathcal{H}$ be a collection of subsets of a set U. Then $S \subseteq U$ is called a *hitting set* of $\mathcal{H}$, if S intersects every set in the collection $\mathcal{H}$. We shall use the notion of hitting set to show that r-RANK VERTEX DELETION admits a polynomial kernel. Let G be a graph with adjacency matrix $A_{n \times n}$. Let $\mathcal{H}(G) = \{X \cup Y : X, Y \subseteq V(G), |X| = |Y| = r + 1, \mathrm{rank}(A_{X,Y}) = r + 1\}$ be a family of sets over $V(G)$.

Lemma 6. *For any $S \subseteq V(G)$, the rank of the adjacency matrix of $G \setminus S$ is at most r if and only if S is a hitting set of $\mathcal{H}(G)$.*

Proof. Let A be the adjacency matrix of G and let $\mathcal{H} = \mathcal{H}(G)$. For $S \subseteq V(G)$, let $A \setminus S$ denote the adjacency matrix of the graph $G \setminus S$. Observe that $A \setminus S$ is obtained from A by deleting the rows and columns corresponding to the vertices in S.

Let $S \subseteq V(G)$ be such that $A \setminus S$ has rank at most r, but S is not a hitting set of $\mathcal{H}$. Then there is a set in $\mathcal{H}$ which is not hit by S and corresponds to a set of rows and columns whose submatrix has rank equal to $r + 1$, which is present in $A \setminus S$. This is a contradiction.

Conversely, let S be any hitting set of $\mathcal{H}$, but $A \setminus S$ has rank greater than $r + 1$. Then there is a submatrix of $A \setminus S$ of size $(r + 1) \times (r + 1)$ which has rank equal to $r + 1$, which is also a submatrix of A. Then by the definition of $\mathcal{H}$, the set of vertices corresponding to this submatrix is present in $\mathcal{H}$. But then S hits this set, which contradicts the fact that the rows and columns corresponding to these vertices are present in $A \setminus S$. $\square$

As a corollary of the above lemma, we immediately obtain a FPT algorithm for r-RANK VERTEX DELETION by branching on any set $X \cup Y$ in $\mathcal{H}(G)$.

Theorem 6 ($\star$). r-RANK VERTEX DELETION *admits an FPT algorithm running in time $2^{\mathcal{O}(k \log r)} n^{\mathcal{O}(1)}$.*

Now, we show that r-RANK VERTEX DELETION admits a polynomial kernel, by an application of the well known Sunflower lemma. We begin with the definition of a sunflower. A sunflower with k petals and a core Y is a collection of sets $S_1, \ldots, S_k$ such that $S_i \cap S_j = Y$ for all $i \neq j$; the sets $S_i \setminus Y$ are petals, and we require that none of the sets S_i is empty. Note that a family of pairwise disjoint sets is also a sunflower.

Lemma 7 (Sunflower Lemma [11]). *Let $\mathcal{F}$ be a family of sets with each set having cardinality equal to s. If $|\mathcal{F}| > s!(k+1)^s$ then $\mathcal{F}$ contains a sunflower with $k+2$ petals.*

Theorem 7 ($\star$). r-RANK VERTEX DELETION *admits a kernel having at most* $(2(r+1)) \cdot (2(r+1))!(k+1)^{2r+2}$ *vertices.*

4   Reducing Rank by Editing Edges

In this section, we consider the problem of reducing the rank of a given graph by editing it's edge set. As before, we parameterize the problem by the solution size k. For the ease of presentation, we define the following notation. Let G be a graph and F be a set of edits. Let $G' = G \bigtriangleup F$ and $H = G'^{\sim}$. Then each vertex h of H corresponds to an equivalence class V'_h of G'. Let ϕ be a map from $V(G)$ to $V(H)$, which maps a vertex $v \in V(G)$ to the vertex $h \in V(H)$ if $v \in V_h$. Observe that, for a given graph G, F uniquely determines (ϕ, H). And let (ϕ, H) be such that for every vertex $h \in V(H)$ there is a vertex in $v \in V(G)$ such that $\phi(v) = h$. Then we can find a set F of edits to $E(G)$ such that the reduced graph of H is the same as the reduced graph $G \bigtriangleup F$. For each pair of vertices u, v in G such that $\phi(u) \neq \phi(v)$, if $(u, v) \in E(G)$ and $(\phi(u), \phi(v)) \notin E(H)$, then we need to delete the edge (u, v). Therefore we add (u, v) to F. Similarly, if $(u, v) \notin E(G)$ and $(\phi(u), \phi(v)) \in E(H)$ then we need to add the edge (u, v). So we add (u, v) to F. This completes the description of F and observe that it is uniquely determined by ϕ and H.

We now have the following lemma about the structure of minimal solutions of r-RANK EDITING.

Lemma 8. *Let F be a minimal solution to an instance (G, k) of r-RANK EDIT-ING. Then for any two independent equivalence classes V_i and V_j of G, either F contains all the edges $V_i \times V_j$ or it contains none of them.*

Proof. Let G' denote the graph obtained after performing the edits in F on G. Let $H = G'^{\sim}$ and ϕ be the map as defined above. Suppose there is an equivalence class V_1 in G whose vertices go into different equivalence classes of G'. Let u be any vertex in V_1 which received the minimum number of edits among all the vertices in V_1, and let $h_u = \phi(u)$. Pick any vertex $v \in V_1$ which goes into a different equivalence class than that of u in the edited graph G'. Define the map ψ from $V(G)$ to $V(H)$ as $\psi(v) = h_u$ and $\psi(w) = \phi(w)$ for all $w \in V \setminus \{v\}$. Let F'' be the set of edits corresponding to (ψ, H) and $G'' = G \bigtriangleup F''$. Let H' be the induced subgraph of H such that every vertex of H' is the image of some vertex of G under the map ϕ. If $V(H') = V(H)$ then $G'''^{\sim}$ is the graph H. Otherwise, by Lemma 4, $H'^{\sim}$ is an induced subgraph of H'. Since H' is an induced subgraph of H, therefore $H'^{\sim}$ is isomorphic to an induced subgraph of H. Since $G'''^{\sim} = H'^{\sim}$, therefore by Observation 3, $\text{rank}(A_{G''}) \leq \text{rank}(A_H) = \text{rank}(A_{G'})$. Let F_u' be the set of edits in F' with u as one endpoint, and let $F_v'' = \{(v,w)|(u,v) \in F_u', w \neq v\}$. Observe that $F'' = (F' \cup F_v'') \setminus (F_v' \cup \{(u,v)\})$, where F_v' denotes the edits in F' with v as one end point. Then clearly $|F''| \leq |F'|$.

We iteratively perform the above operation for all those vertices in V_1 which are mapped to a different equivalence class in G'. Thus we can find a solution such that all the vertices of V_1 go to the same equivalence class after editing. Moreover, applying this operation ensures that all the vertices in an equivalence class in G receive the same set of edits.

To complete the proof, observe that the graph induced on two equivalence classes is a complete bipartite graph. After editing, suppose V_i and V_j are contained in W_i and W_j respectively, where W_i and W_j are the equivalence classes of the edited graph. If $W_i = W_j$ then all the vertices between V_i and V_j have been deleted. If $W_i \neq W_j$ then all the edges of $V_i \times V_j$ are present if and only if W_i and W_j have an edge.   □

We can show a similar lemma for r-RANK EDGE DELETION.

4.1   NP Completeness

We give a reduction from the d-CLUSTERING problem, which is defined as follows. Given a graph G and two positive integers d and k, find a set F of k edges such that the graph $(V, E \triangle F)$ is the partition of at most d disjoint cliques. This problem is known to be NP-Complete for any $d \geq 2$ [17]. However it is FPT when parameterized by k [7].

Theorem 8. r-RANK EDITING *is* NP-Complete *for any* $r \geq 3$

Proof. It is clear that the problem is in NP, since we can verify any claimed solution in polynomial time.

To show that the problem is NP hard for a fixed $r \geq 3$, we give a reduction from the 2-CLUSTERING problem. Given an instance (G, k); let $V = V(G)$, $E = E(G)$ and $n = |V|$. We define an instance of r-RANK EDITING as follows. Let Z be the complete $(r-2)$-partite graph $K_{k+2,\ldots,k+2}$, where each partition has exactly $k+2$ vertices, and all the edges between any two partitions are present. Observe that it has rank $r-2$ and has K_{r-2} as it's reduced graph. We construct the graph H by taking the complement $\overline{G}$ of G, a copy of the graph Z and adding all edges between the vertices of Z and vertices in $\overline{G}$. Then the instance of r-RANK EDITING is (H, k). Let $\overline{E} = E(\overline{G})$.

In the forward direction, suppose that the instance (G, k) has a minimal solution F of size at most k such that $G' = (V, E \triangle F)$ is a partition of two disjoint cliques X and Y. Then observe that $H' = (V \cup V(Z)$,

$(\overline{E} \bigtriangleup F) \cup E(Z) \cup E(Z, V(G)))$ is a graph with K_r or K_{r-1} as it's reduced graph. Thus H' has rank r or $r - 1$.

In the reverse direction, suppose F is a minimal solution of size k for (H, k). Let $H' = (V \cup V(Z), (\overline{E} \bigtriangleup F) \cup E(Z) \cup E(Z, V(G)))$. Observe that there are no isolated vertices in H', since every vertex has at least $k + 1$ neighbours in H. Since each equivalence class of $H[Z]$ has $k + 2$ vertices and $|F| \leq k$, therefore by Lemma 8, $F \cap \{(E(Z) \cup E(Z, V(G))\} = \phi$. Thus $F \subseteq V(G) \times V(G)$, and any independent equivalence class of H' is contained in either $V(G)$ or $V(Z)$. Now suppose that X and Y are two equivalence classes of H' which are contained in $V(G)$ such that $H'[X \cup Y]$ is a bi-clique. Then observe that the induced subgraph $H'[X \cup Y \cup V(Z)]$ has no isolated vertices and it has K_r as it's reduced subgraph. Therefore by Lemma 3, the reduced graph of H' is also K_r, which implies that X and Y form a partition of $V(G)$. And if we cannot find such an X and Y, then $H'[V(G)]$ contains no edges and so the reduced graph of H' is K_{r-1}. Therefore $H'[V(G)]$ is either an independent set or a bi-clique, and so $\overline{H'[V(G)]}$ can be partitioned into at most 2 cliques. Since $(V, E \bigtriangleup F) = \overline{H'[V(G)]}$, we have that F is a solution to the instance (G, k).   □

For r-RANK EDGE DELETION, we can show the following theorem, by reducing from the MAXIMUM EDGE BI-CLIQUE Problem [14].

Theorem 9 (⋆). *For any fixed $r \geq 2$, r-RANK EDGE DELETION is* NP-Complete.

4.2   A Parameterized Algorithm for r-Rank Editing

We now show that r-RANK EDITING is FPT parameterized by the solution size and admits a polynomial kernel. The results for r-RANK EDGE DELETION follow in a similar manner. We call a vertex $v \in V$, an *affected vertex* with respect to the solution F, if there is some edge in F which is incident on v. Observe that by Lemma 8, if a vertex in an equivalence class is affected with respect to a minimal solution F, then every vertex in V_1 is also affected. In that case, we say that the equivalence class V_1 is affected by F.

Lemma 9 (⋆). *For any instance (G, k) of the r-RANK EDITING problem. If $G^\sim$ has more than $c \cdot 2^{r/2} + 2k$ vertices then the instance has no solution.*

As a corollary of the above result and Lemma 8, we have the following kernel for r-RANK EDITING.

Theorem 10 (⋆). *r-RANK EDITING admits a kernel with $\mathcal{O}((2^{r/2} + 2k) \cdot (k+1))$ vertices.*

Next, let A be the adjacency matrix of G, and let $A_{X,Y}$ be a submatrix of A having rank $r + 1$ where X and Y correspond to a subset of rows and columns respectively. Then observe that any solution to the instance (G, k), must edit an edge contained in $A_{X,Y}$. This observation immediately gives us an FPT algorithm for r-RANK EDITING, in a similar way to Theorem 6.

Theorem 11. r-RANK EDITING *admits an FPT algorithm running in time* $2^{O(k \log r)} n^{O(1)}$.

However we can obtain a sub-exponential algorithm for r-RANK EDITING, by using an algorithm of Damaschke et. al. [6]. We begin with the definition of the *closed neighbourhood relation* on G. We define a relation $\approx$ on vertices of G, where $u \approx v$ if and only if $N[u] = N[v]$ in G. Observe that $\approx$ is an equivalence relationship on vertices of G, where each equivalence class is a clique. Let $V_1, \ldots, V_l$ be a partition of vertices $V(G)$. We define the $\approx$-reduced graph $G^\approx$ as follows: the vertex set is $V(G^\approx) = \{u_1, \ldots, u_l\}$ and the edge set is $E(G^\approx) = \{(u_i, u_j)$ if and only if all the edges between V_i and V_j are present in $E(G)\}$. Note that $G^\approx$ may contain several isolated vertices.

The H-BAG EDITING problem is defined as follows.

H-BAG EDITING   *Parameter: k*
Input: Graphs G, H and an integer k.
Question: Can we find a set F of k edges such that $\approx$-reduced graph of $G' = (V(G), E(G) \triangle F)$ is an induced subgraph of H?

Theorem 12 ([6]). *Any bag modification problem with a fixed graph H can be solved in* $\mathcal{O}^*(2^{O(\sqrt{k} \log k)})$ *time*[2].

We show how to use the above algorithm to obtain a sub-exponential FPT algorithm for r-RANK EDITING.

Lemma 10. *Given a graph G, let $G^\sim$ denote the reduced graph with respect to the excluded neighborhood relation $\sim$ and let $G^\approx$ denote the reduced graph with respect to the closed neighborhood relation $\approx$. Then the graph $(\overline{G})^\approx$ is the complement of $G^\sim$.*

Proof. Observe that it suffices to prove that the equivalence classes of $\approx$ in $\overline{G}$ are exactly the same as the equivalence classes of $\sim$ in G. Consider the forward direction. Suppose $u \sim v$ for $u, v \in V[G]$, we want to prove that $u \approx v$ in $\overline{G}$. As $u \sim v$, $N_G(u) = N_G(v)$ therefore $N_{\overline{G}}[u] = V[G] \backslash N_G(u) = V[G] \backslash N_G(v) = N_{\overline{G}}[v]$, moreover $(u, v) \in E[\overline{G}]$, which implies $u \approx v$ in $\overline{G}$. We can show the reverse direction in a similar way. Since the $\sim$-equivalence classes of G are same as the $\approx$-equivalence classes of $\overline{G}$, their corresponding reduced graphs are complements of each other. □

Corollary 3. *If H is a $\sim$-reduced graph then its complement is a $\approx$-reduced graph.*

Theorem 13 ($\star$). *An instance (G, k) r-RANK EDITING can be solved in* $\mathcal{O}^*(2^{O(\sqrt{k} \log k)})$ *time.*

---

[2] The $\mathcal{O}^*$ notation hides the terms depending on r which is assumed to be a constant, and polynomial multiplicative factors.

5   Conclusion

In this paper we studied the vertex deletion and edge edition problems of reducing the "rank of the graph". We saw that the problems are NP-Complete and they admit FPT algorithms and kernels.

We conclude with a few open problems. Is it possible to obtain improved kernels and algorithms for these problems? In particular, is it possible to improve the exponent of the subexponential algorithm for r-RANK EDITING? Further, what is complexity of the problem of reducing the number of distinct eigenvalues of a graph by deleting a few vertices or editing a few edges?

References

1. Akbari, S., Cameron, P.J., Khosrovshahi, G.B.: Ranks and signatures of adjacency matrices (2004)
2. Burzyn, P., Bonomo, F., Durán, G.: Np-completeness results for edge modification problems. Discrete Applied Mathematics 154(13), 1824–1844 (2006)
3. Cai, L.: Fixed-parameter tractability of graph modification problems for hereditary properties. Information Processing Letters 58(4), 171–176 (1996)
4. Cao, Y., Marx, D.: Chordal editing is fixed-parameter tractable. In: STACS, pp. 214–225 (2014)
5. Cao, Y., Marx, D.: Interval deletion is fixed-parameter tractable. ACM Transactions on Algorithms (TALG) 11(3), 21 (2015)
6. Damaschke, P., Mogren, O.: Editing the simplest graphs. In: Pal, S.P., Sadakane, K. (eds.) WALCOM 2014. LNCS, vol. 8344, pp. 249–260. Springer, Heidelberg (2014)
7. Fomin, F.V., Kratsch, S., Pilipczuk, M., Pilipczuk, M., Villanger, Y.: Tight bounds for parameterized complexity of cluster editing with a small number of clusters. Journal of Computer and System Sciences 80(7), 1430–1447 (2014)
8. Fomin, F.V., Villanger, Y.: Subexponential parameterized algorithm for minimum fill-in. SIAM Journal on Computing 42(6), 2197–2216 (2013)
9. Guo, J.: A more effective linear kernelization for cluster editing. In: Chen, B., Paterson, M., Zhang, G. (eds.) ESCAPE 2007. LNCS, vol. 4614, pp. 36–47. Springer, Heidelberg (2007)
10. Hüffner, F., Komusiewicz, C., Moser, H., Niedermeier, R.: Fixed-parameter algorithms for cluster vertex deletion. Theory of Computing Systems 47(1) (2010)
11. Jukna, S.: Extremal combinatorics: with applications in computer science. Springer Science & Business Media (2011)
12. Kotlov, A., Lovász, L.: The rank and size of graphs. Journal of Graph Theory 23(2), 185–189 (1996)
13. Lewis, J.M., Yannakakis, M.: The node-deletion problem for hereditary properties is np-complete. Journal of Computer and System Sciences 20(2), 219–230 (1980)
14. Peeters, R.: The maximum edge biclique problem is np-complete. Discrete Applied Mathematics 131(3), 651–654 (2003)

15. Reed, B., Smith, K., Vetta, A.: Finding odd cycle transversals. Operations Research Letters **32**(4), 299–301 (2004)
16. Robertson, N., Seymour, P.D.: Graph minors. xiii. the disjoint paths problem. Journal of Combinatorial Theory, Series B **63**(1), 65–110 (1995)
17. Shamir, R., Sharan, R., Tsur, D.: Cluster graph modification problems. Discrete Appl. Math. **144**(1-2) (2004)
18. Yannakakis, M.: Node-and edge-deletion np-complete problems. In: STOC, pp. 253–264. ACM (1978)

Knapsack and Allocation

On the Number of Anchored Rectangle Packings for a Planar Point Set

Kevin Balas[1,2] and Csaba D. Tóth[1,3]([✉])

[1] California State University Northridge, Los Angeles, CA, USA
balask@lamission.edu, cdtoth@acm.org
[2] Los Angeles Mission College, Sylmar, CA, USA
[3] Tufts University, Medford, MA, USA

Abstract. We consider packing axis-aligned rectangles $r_1, \ldots, r_n$ in the unit square $[0, 1]^2$ such that a vertex of each rectangle r_i is a given point p_i (i.e., r_i is *anchored* at p_i); and explore the combinatorial structure of all locally maximal configurations. When the given points are lower-left corners of the rectangles, then the number of maximal packings is shown to be at most $2^n C_n$, where C_n is the nth Catalan number. The number of maximal packings remains exponential in n when the points may be arbitrary corners of the rectangles. Our upper bounds are complemented with exponential lower bounds.

1 Introduction

Let P be a finite set of points $p_1, \ldots, p_n$ in the unit square $[0, 1]^2$. An *anchored rectangle packing* for P is a set of axis-aligned empty rectangles $r_1, \ldots, r_n$, that lie in $[0, 1]^2$, are interior-disjoint, and point p_i is one of the four corners of r_i for $i = 1, \ldots, n$. We say that rectangle r_i is *anchored* at p_i. In a *lower-left anchored rectangle packing* (*L-anchored packing*, for short), p_i is the lower-left corner of r_i for all i.

Anchored rectangle packings have applications in map labeling in geographic information systems [15–17] and VLSI design [18]. A fundamental problem is to find the maximum total area $A(P)$ (resp., $A_L(P)$) of the rectangles in an anchored (resp., L-anchored) rectangle packing of P. Allen Freedman conjectured (c.f. [23,24]) that if $(0, 0) \in P$, then P admits an L-anchored rectangle packing of area at least $1/2$, that is $A_L(P) \geq 1/2$. The currently known best lower bound in this case is $A_L(P) \geq 0.091$ due to Dumitrescu and Tóth [11].

A rectangle r_i with lower-left anchor $p_i = (a_i, b_i)$, can be parameterized by two variables x_i and y_i such that $r_i = [a_i, x_i] \times [b_i, y_i]$. Consequently, the area of an L-anchored rectangle packing is a continuous multivariable function in $2n$ variables $\sum_{i=1}^{n} \text{area}(r_i) = \sum_{i=1}^{n}(x_i - a_i)(y_i - b_i)$, over a domain determined by the geometric constraints of the packing. We call an L-anchored rectangle packing *maximum* (resp., *maximal*) if it attains the global (resp., a local) maximum of this function. We define *maximum* and *maximal* anchored rectangle packing analogously.

© Springer International Publishing Switzerland 2015
D. Xu et al. (Eds.): COCOON 2015, LNCS 9198, pp. 377–389, 2015.
DOI: 10.1007/978-3-319-21398-9_30

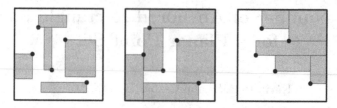

Fig. 1. Left: a set P of 5 points in the unit square $[0,1]^2$ and an anchored rectangle packing for P. Middle: a maximal anchored rectangle packing for P. Right: A maximal L-anchored rectangle packing for P.

For computing the maximum area, $A_L(P)$ or $A(P)$, for a given point set P, it is instrumental to estimate the *number* of maximum packings. It is easily seen that the number of maximum packings is at least exponential in $n = |P|$ if, for example, P contains n points on a diagonal of $[0,1]^2$. The enumeration of locally maximal configurations, which can be computed greedily, combined with reverse search [7] yields a simple strategy for finding the global maximum. In this paper, we control the number of (locally) maximal anchored and L-anchored rectangle packings. For an integer $n \in \mathbb{N}$, let $M(n)$ (resp., $M_L(n)$) denote the largest number of maximal rectangle packings over all sets $P \subset [0,1]^2$ of n noncorectilinear points (two points are *corectilinear* if they have the same x- or y-coordinate).

Results. In this paper, we prove exponential upper and lower bounds for $M_L(n)$ and $M(n)$. Our upper bound for $M_L(n)$ is expressed in terms of the nth Catalan number $C_n = \frac{1}{n+1}\binom{2n}{n} \sim 4^n/(n^{3/2}\sqrt{\pi})$.

Theorem 1. *We have $\Omega(4^n/\sqrt{n}) \le M_L(n) \le C_n 2^n = \Theta(8^n/n^{3/2})$.*

Note that both the lower and upper bounds are larger than C_n. The lower bound follows from an explicit construction. The upper bound is the combination of two tight upper bounds. Each L-anchored rectangle packing induces a subdivision of $[0,1]^2$ into "staircases" (*L-subdivisions*, defined in Sec. 3). We show that the number of L-subdivisions for n points is at most C_n, and this bound is attained when the points form an antichain under the product order. We also show that each L-subdivision is induced by at most 2^{n-1} L-anchored rectangle packings, and this bound is attained when the points form a chain under the product order.

The machinery developed for the proof of Theorem 1 does not extend to general anchored rectangle packings. Nevertheless, we can prove that the number of maximal (any corner) anchored rectangle packings is exponential

Theorem 2. *There exist constants $1 < c_1 < c_2$ such that $\Omega(c_1^n) \le M(n) \le O(c_2^n)$.*

We derive an exponential upper bound using the contact graph of the rectangles in a packing. Specifically, we show that the contact graph can be represented by a planar embedding of the contact graph in which the vertices are points in the rectangles, and the edges are represented by polylines with at most one bend per edge. The number of graphs with such an embedding is known to be exponential [12]. We can encode all maximal anchored rectangle packings for P using one such graph and $O(n)$ bits of additional information. This leads to an exponential upper bound.

Remark. In a maximal anchored or L-anchored rectangle packing, we may assume that all vertices of all rectangles lie on one of the $(n+2)^2$ "grid points" induced by the vertical and horizontal lines passing through the n points in P and the corners of $[0,1]^2$ (cf. Sec. 2). This crucial property discretizes the problem, but is insufficient for establishing an exponential upper bound. By choosing the points (x_i, y_i) among the grid points, we obtain only a weak upper bound of $(n-1)^n$ (resp., $(n!)^2$ for L-anchored packings).

Related Work. Packing axis-aligned rectangles in a rectangular container, albeit without anchors, is the unifying theme of several classic optimization problems. The 2D knapsack problem, strip packing, and 2D bin packing involve arranging a set of given rectangles in the most economic fashion [3,8,14]. The maximum weight independent set for rectangles involves selecting a maximal area packing from a set of given rectangles [4]. These optimization problems are NP-hard, and there is a rich literature on the best approximation algorithms. Our problem setup is fundamentally different: the rectangles have variable sizes, but their location is constrained by the anchors. In this sense, it is reminiscent to classic Voronoi diagrams for n points in the plane. However, the Voronoi cells tile the space without gaps. Area maximization problems arise in the context of Voronoi games [5,9], where two players alternately choose points in a bounding box and wish to maximize the total area of the Voronoi cells of their points.

Combinatorial bounds for the number of some other geometric configurations on n points in the plane have been studied extensively. Determining the maximum number of (geometric) triangulations on n points in the plane captivated researchers for decades. The current best upper and lower bounds are $\Omega(8.65^n)$ and $O(30^n)$ [10,20]. Ackerman et. al. [1,2] established an upper bound of $O(18^n/n^4)$ for the number of *rectangulations* of n points in $[0,1]^2$, where a *rectangulation* is a subdivision of $[0,1]^2$ into $n+1$ rectangles by n axis-parallel segments, each containing a given point. This structure is reminiscent of L-subdivisions, defined in Sec. 3, for which we prove a tight upper bound of $C_n \leq O(4^n/n^{3/2})$. The number of anchored rectangle packings has not been studied before. It is not known if finding the maximum area of an anchored rectangle packing of n given points is NP-hard.

2   Discretization of Maximal Anchored Rectangle Packings

Let $P \subset [0,1]^2$ be a set of noncorectilinear points $p_1, \ldots, p_n$. The vertical and horizontal lines that pass through the points in P and the edges of the bounding box are called *grid lines*. The *grid points* are the intersections of the grid lines.

It is easy to see that all vertices of a maximal L-anchored rectangle packing must be grid points.

Proposition 1. *If an L-anchored rectangle packing for P has maximal area, then all corners of all rectangles are grid points.*

Proof. Consider an anchored rectangle packing of maximal area. The left and bottom edges are on grid lines. This implies that each rectangle may only expand up and to the right. Because the packing is of maximal area, no rectangle can expand (while other rectangles are fixed). The upper and right edges of each rectangle are necessarily in contact with the bottom and left edges of other rectangles or with the bounding box. This places the upper-right vertex at the intersection of two grid lines and thus on a grid point. We have shown that the lower-left and the upper-right corner of every rectangle is a grid point. From the definition of grid lines, this implies that all corners of all rectangles are grid points.   □

The situation is more subtle when the rectangles can be anchored at arbitrary corners. Specifically, a local maximum may be attained at a "plateau" where the configuration can vary continuously while maintaining the same maximal area. A transformation that maintains the total area of the rectangles is called *equiareal*.

Proposition 2. *If an anchored rectangle packing for P has maximal area, then*

- *the local maximum is isolated, and all vertices of all rectangles are grid points, or*
- *there is an equiareal continuous deformation to an anchored rectangle packing in which all vertices of all rectangles are grid points; furthermore, the deformation either creates a contact between two previously disjoint rectangles, or decreases the area of some rectangle to 0.*

Proof. Consider a maximal anchored rectangle packing for P. Suppose that at least one rectangle has a vertical or horizontal edge not on a grid line. Assume first that a vertical edge of a rectangle is not on a grid line. Let ℓ be the vertical line through the leftmost such edge. Denote by L the set of rectangles whose right edges intersect ℓ, and R the set of rectangles whose left edges intersect ℓ. We can deform the rectangles in L and R simultaneously by translating ℓ. The sum of heights of rectangles in L equals the sum of heights of rectangles in R, otherwise translating ℓ in one of the two possible directions increases the total area before ℓ becomes a grid line. When ℓ shifts to the left, the rectangles in L shrink and may potentially reach 0 area; while the rectangles in R expand and may potentially reach another rectangle. However, because of the choice of ℓ, all edges of such a rectangle lie on grid lines. Translate ℓ until the area of a rectangle in L drops to 0, or the left edge of a rectangle in R reaches the boundary of a new rectangle or the bounding box. Repeat this operation for the next leftmost line ℓ until all vertical edges are on grid lines.

Note that horizontal edges were not affected by the above transformations. We can now deform the horizontal edges of the rectangles (independent of the vertical edges) by repeating the argument starting with the topmost horizontal line. Necessarily, all vertices of all rectangles become grid points.   □

In the remainder of the paper, we consider maximal anchored rectangle packings in which the vertices of all rectangles are grid points.

3   Lower-Left Anchored Rectangle Packings

The key tool for the proof of Theorem 1 is a subdivision of the unit square $[0,1]^2$ into staircase polygons, defined below. Let $P = \{p_1, \ldots, p_n\}$ be a set of noncorectilinear points in $[0,1]^2$. We may assume that $(0,0) \notin P$ (by scaling P, if necessary, since maximality is an affine invariant). Let $q = (0,0)$ denote the lower-left corner of $[0,1]^2$.

An *L-shape* for a point p_i $(i = 1, \ldots, n)$ is the union of a horizontal and a vertical segment whose left and bottom endpoint, respectively, is p_i. Refer to Fig. 2(a). An *L-subdivision* for P is formed by n L-shapes for p_i $(i = 1, \ldots, n)$ such that the top and right endpoint of each L-shape lies in another L-shape or the boundary of $[0,1]^2$. The L-shapes subdivide $[0,1]^2$ into $n+1$ simple polygons, called *staircases*. By construction, the lower-left corner of each staircase is either $q = (0,0)$ or a point in P. The upper-right vertices of a staircase are called *steps* of the staircase.

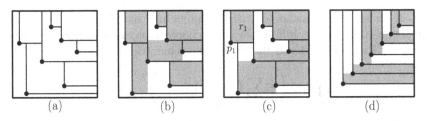

Fig. 2. (a) An L-subdivision for P. (b) An *L-subdivion* induced by a maximal L-anchored rectangle packing. (c) Maximal anchored rectangles in the staircases that do not form a maximal L-anchored rectangle packing: rectangle r_1 could expand. (d) For n points on the line $y = x$, $M_L(P) = 2^{n-1}$.

Proposition 3. *In every L-subdivision for P, the n staircases anchored at the points in P jointly have at most $2n - 1$ steps.*

Proof. Each step of a staircase is either the upper-right corner of $[0,1]^2$, or a top or right endpoint of an L-shape. Every such point is the step of a unique staircase. The n L-shapes yield $2n$ steps, and the upper-right corner of $[0,1]^2$ yields one step. The staircase anchored at $q = (0,0)$ has at least two steps, hence the remaining staircases jointly have at most $2n + 1 - 2 = 2n - 1$ steps.   □

Maximal L-anchored packings versus L-subdivisions

Proposition 4. *For every maximal L-anchored rectangle packing of P, there is an L-subdivision such that rectangle r_i lies in the staircase anchored at p_i for $i = 1, \ldots, n$.*

Proof. Let $r_1, \ldots, r_n$ be an L-anchored rectangle packing for $p_1, \ldots, p_n \in [0,1]^2$. For each $i = 1, \ldots, n - 1$, successively draw an L-shape as follows (refer to Fig. 2(b)). First extend the bottom edge of r_i to the right until it hits the bounding box, the left edge of another rectangle, or a previously drawn L-shape. Similarly, extend the left edge of r_i up until it hits the bounding box, the bottom edge of another rectangle, or a previously drawn L-shape. The n L-shapes form an L-subdivision. By construction, the L-shapes are disjoint from the interior of the rectangles $r_1, \ldots, r_n$, hence each rectangle lies in a staircase. Since the lower-left corner of each staircase is $q = (0,0)$ or a point in P, each staircase with lower-left corner $p_i \in P$ contains the rectangle anchored at p_i. □

In the L-subdivision described in Proposition 4, each rectangle r_i ($i = 1, \ldots, n$) is a maximal rectangle within a staircase polygon. However, the converse is not necessarily true. Choose maximal rectangles, in all staircases, with lower-left corners in P. This need not produce a maximal L-anchored rectangle packing for P. See an example in Fig. 2(c). Nevertheless, we can derive an upper bound for $M_L(P)$.

Proposition 5. *In every L-subdivision for P, $|P| = n$, there are at most 2^{n-1} possible ways to choose a maximal rectangle in each staircase whose lower-left corner is in P. This bound is the best possible.*

Proof. If the staircases anchored at the n points in P have $t_1, \ldots, t_n$ steps, then there are precisely $\prod_{i=1}^{n} t_i$ different ways to choose a maximal anchored rectangle in each. By Proposition 3 and the arithmetic-geometric mean inequality yields

$$\prod_{i=1}^{n} t_i \leq \left(\frac{1}{n} \sum_{i=1}^{n} t_i \right)^n = \left(2 - \frac{1}{n} \right)^n < 2^n. \tag{1}$$

The maximum of $\prod_{i=1}^{n} t_i$ subject to $\sum_{i=1}^{n} t_i = 2n - 1$ and $t_1, \ldots, t_n \in \mathbb{N}$ is attained when the $t_i's$ are distributed as evenly as possible, say, $t_1 = \ldots = t_{n-1} = 2$ and $t_n = 1$. Consequently, $\prod_{i=1}^{n} t_i \leq 2^{n-1}$. This upper bound is attained when the points in P form a chain in the product order (e.g., points on the line $y = x$), then $n - 1$ staircases have 2 steps, and the staircase incident to $(1,1)$ has only 1 step (Fig. 2(d)). □

Let $S(P)$ be the number of all L-subdivisions for a noncorectilinear point set P; and let $S(n) = \max_{|P|=n} S(P)$. By Proposition 5, we have $M_L(P) \leq S(P)2^n$ and $M_L(n) \leq S(n)2^n$.

The Number of L-subdivisions. We prove a tight upper bound for $S(n)$, the maximum number of L-subdivisions for a set of n points in the unit square. Our upper bound is expressed in terms of the nth Catalan number $C_n = \frac{1}{n+1}\binom{2n}{n} \sim 4^n/\sqrt{\pi n^3}$.

Lemma 1. *For every $n \in \mathbb{N}$, we have $S(n) = C_n$.*

Proof. Lower bound. Let P be a set of n points that form an antichain under the product order (e.g., points on the line $y = 1 - x$). In this case, each staircase anchored at a point in P is a rectangle, and it is well known [21,22] that the number of rectangular subdivisions is the Catalan number C_n. Hence $S(P) = C_n$ in this case.

Upper Bound. Let P be an arbitrary noncorectilinear set of n points in $[0,1]^2$. We may assume that the points $p_1, \ldots p_n$ are sorted by their x-coordinates, that is, $x_1 < \ldots < x_n$. If the points form an antichain under the product order, then their y-coordinates are monotone decreasing, and $S(P) = C_n$. Otherwise, we incrementally modify the y-coordinates of the points to become monotone decreasing such that the number of L-subdivisions increases. In each incremental step, we modify the y-coordinate of one point.

Suppose that the points in P do not form an antichain under the product order; and i is the smallest index such that the points with larger indices, $\{p_j \in P : j > i\}$, form an antichain and are incomparable to all other points (refer to Fig. 3). Let Z_0 be the minimum axis-aligned rectangle incident to $(0,1)$ that contains the points $p_1, \ldots, p_{i-1}$; and let Z_1 be the minimum axis-aligned rectangle incident to $(1,0)$ that contains the points $p_{i+1}, \ldots, p_n$. By the choice of i, the boxes Z_0 and Z_1 are on opposite sides of the vertical line $x = x_i$, as well as a horizontal line below $y = \min_{1 \leq k \leq i} y_k$. Let p_i' be the intersection of these two lines.

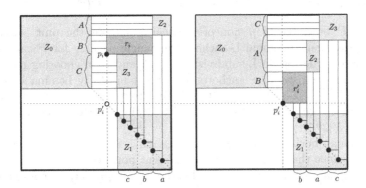

Fig. 3. Left: A schematic image of an L-subdivision D for P. Right: The corresponding L-subdivision D' for the modified point set P'.

We move point p_i to p_i'. Denote by P' the modified point set. In order to show $S(P) \leq S(P')$, we construct an injective map $f : S(P) \to S(P')$. For every L-subdivision D of the point set P, we construct a unique L-subdivision $D' = f(D)$ of the modified point set P'. Let D be an L-subdivision of P (Fig. 3, left). Since no other point dominates p_i, the staircase anchored at p_i is a rectangle, that we denote by r_i. We introduce some notation. Some horizontal segments of L-shapes of points in Z_0 cross the right edge of Z_0: Let A, B, and C, respectively, denote the number of L-shapes whose horizontal segments pass above r_i, hit the left edge of r_i, and pass below r_i. Similarly, some vertical segments of L-shapes of points in Z_1 cross the top edge of Z_1: Let a, b, and c, respectively, denote the number of L-shapes whose vertical segments pass right of r_i, hit the bottom edge of r_i, and end strictly below the bottom edge of r_i. Let Z_2 be the axis-aligned rectangle that contains all intersections between the A horizontal segments passing above r_i and the a vertical segments right of r_i. Similarly, let Z_3 contain the intersections between the C horizontal segments passing below r_i and the c vertical segments that end strictly below r_i.

We can now define the L-subdivision D' for the modified point set P'. The arrangement of L-shapes restricted to the boxes Z_0 and Z_1 remains the same. Consequently, $A + B + C$ horizontal segments exit the right edge of Z_0, and $a + b + c$ vertical segments exit the top edge of Z_1. Draw an L-shape for the point p'_i such that it blocks the B lowest horizontal segments that exit Z_0 and the b leftmost vertical segments that exit Z_1. Group the remaining horizontal (resp., vertical) segments into bundles of size A and C (resp., a and c). Let the groups of size A and a intersect in the same pattern as in Z_2, and the groups of size C and c as in Z_3. This completes the description of the L-subdivision D'. By construction $D' = f(D)$ is a unique L-subdivision, and the function f is injective.   □

We are now ready to prove Theorem 1.

Theorem 1. *We have* $\Omega(4^n/\sqrt{n}) \leq M_L(n) \leq C_n 2^n = \Theta(8^n/n^{3/2})$.

Proof. Let P be a set of n noncorectilinear points in the unit square. By Proposition 4, every maximal L-anchored rectangle packing for P can be constructed by considering an L-subdivision for P, and then choosing a maximal rectangle from each staircase anchored at a point in P. By Lemma 1, we have $S(P) \leq S(n) = C_n$ L-subdivisions for P. By Proposition 5, there are at most 2^{n-1} ways to choose maximal rectangles in the staircases. Consequently, we have $M_L(P) \leq S(P)2^{n-1} \leq C_n 2^{n-1}$.

Even though both Proposition 5 and Lemma 1 are tight, their combination is not tight, since they are attained on different point configurations: n points that form a chain or an antichain under the product order. Our lower bound

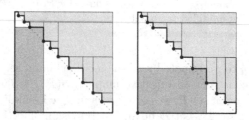

Fig. 4. One point at the origin and $n - 1$ points on the line $y = 1 - x$

is based on the following construction (refer to Fig. 4). Place one point at the origin and $n - 1$ points on the line $y = 1 - x$. The L-shape of the first point is contained in the boundary of $[0, 1]^2$, and the last $n - 1$ points admit $C_{n-1} = \frac{1}{n}\binom{2n-2}{n-1} \sim 4^{n-1}/\sqrt{(n-1)^3\pi}$ L-subdivisions. The first point has a staircase with n steps ($t_1 = n$), all other staircases are rectangles ($t_i = 1$, for $i = 2, \ldots, n$). Consequently, $M_L(P) = S(P) \prod_{i=1}^{n} t_i = C_{n-1} \cdot n = \Theta(4^n/\sqrt{n})$, as required. $\square$

4   General Anchored Rectangle Packings

In this section, we prove Theorem 2. We show that a maximal anchored rectangle packing for a point set P can be reconstructed from the contact graph of the rectangles, and from $O(n)$ bits of additional information. Since a maximal rectangle packing may contain rectangles of 0 area (cf. Proposition 2), we need to be careful defining contact graphs.

The *contact graph* of a rectangle packing is a graph $G = (V, E)$, where V corresponds to the set of vertices, E to the set of edges, and two vertices are connected by an edge iff the corresponding rectangles have positive area and intersect in a nontrivial line segment; or one rectangle has 0 area and lies on the boundary of the other rectangle. It is easy to see that the contact graph of a rectangle packing is planar. However, the number of n-vertex planar graphs is super-exponential [13]. The number of graphs reduces to exponential with suitable geometric conditions. For a set P of n points in the plane, for example, the number of straight-line graphs with vertex set P is only exponential. An $\exp(O(n))$ bound was first shown by Ajtai et al. [6] using the crossing number method. The current best upper bound is $O(187.53^n)$, due to Sharir and Sheffer [20]. The contact graphs of any anchored rectangle packings for P can be embedded in the plane such that the vertex set is P, but these graphs cannot always be realized by straight-line edges. It turns out that a weaker condition will suffice: a *1-bend* embedding of a planar graph $G = (V, E)$ is an embedding in which the vertices are distinct points in the plane, and the edges are polylines with one bend per edge (that is, each edge is the union of two incident line segments). Frankeke and Tóth [12] proved recently that for every n-element point set, the number of such graphs is at most $\exp(O(n))$.

Lemma 2. *Let $P = \{p_1, \ldots, p_n\}$ be a noncorectilinar set in $[0, 1]^2$. The contact graph of every maximal anchored rectangle packing for P has a 1-bend embedding in which the vertex representing rectangle r_i is point p_i for $i = 1, \ldots, n$.*

The proof would be straightforward if the anchors were in the interior of the rectangles. In that case, we could simply choose a *bend point* on the common boundary between two rectangles in contact, and then draw a 1-bend edge between their anchors via the bend point. When the anchors are at corners of the rectangles, we need to be more careful to prevent any overlap between adjacent edges.

Proof. Let $r_1, \ldots, r_n$ be a maximal anchored rectangle packing for P. For every $i = 1, \ldots, n$, point p_i is a corner of the rectangle r_i. For every two rectangles

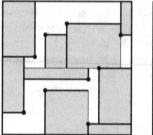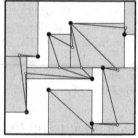

Fig. 5. Left: A maximal anchored rectangle packing for P. Thick lines indicate the L-shapes incident to the points in P. Right: A 1-bend embedding of the contact graph of the rectangles.

in contact, r_i and r_j, choose an arbitrary *preliminary* bend point $q_0(i,j)$ on the common boundary of r_i and r_j.

Define the *L-shape* anchored at p_i as the union of the two edges of r_i incident to p_i. Note that the relative interiors of the n L-shapes are pairwise disjoint, since the points $p_1, \ldots, p_n$ are noncorectilinear. Let the bend point $q(i,j) = q_0(i,j)$ if the preliminary bend point is not on an L-shape or if one of the rectangles has 0 area. Otherwise, assume $q_0(i,j)$ is on the L-shape of p_i. Then choose a bend point $q(i,j)$ in the interior of r_i in a sufficiently small neighborhood of the preliminary point $q_0(i,j)$. Now, for any two rectangles in contact, r_i and r_j, draw a 1-bend edge between p_i and p_j via $q(i,j)$. No two edges cross or overlap, and hence we obtain a 1-bend embedding of the contact graph of the rectangles. $\square$

For a fixed point set P, by Lemma 2, the contact graph of every maximal anchored rectangle packing admits a 1-bend embedding on the vertex set P. However, several maximal anchored rectangle packings may yield the same contact graph (as an abstract graph). We show that all maximal anchored rectangle packings for P can be encoded by their contact graphs and $O(n)$ bits of additional information. By Proposition 2, we may assume that all vertices of a maximal rectangle packing are grid points. Furthermore, we may also assume that there is no equiareal continuous deformation that creates a new contact or reduces the area of a rectangle to 0.

Fix a noncorectilinear point set $P = \{p_1, \ldots, p_n\}$. Every maximal anchored rectangle packing $r_1, \ldots, r_n$ is encoded by the following information:

(1) The contact graph G of the rectangles $r_1, \ldots, r_n$;
(2) for $i = 1, \ldots n$, an indicator variable σ_i such that $\sigma_i = 0$ iff area$(r_i) = 0$;
(3) for $i = 1, \ldots n$, the position of the anchor p_i in r_i (lower-left, lower-right, etc.);
(4) for each edge (i,j) of G, the orientation of the line segment $r_i \cap r_j$.

We now show that we can uniquely reconstruct a maximal anchored rectangle packing from this information.

Lemma 3. *For every noncorectilinear point set P, every code described above determines at most one maximal anchored rectangle packing for P, which can be (re)constructed in polynomial time.*

Proof. We are given the points $p_1, \ldots, p_n$, and for every $i = 1, \ldots, n$, we know which corner of the rectangle r_i is p_i. To reconstruct the rectangles r_i ($i = 1, \ldots, n$), it is enough to find the corner of r_i opposite to p_i, which we denote by (x_i, y_i). That is $r_i = [\min(a_i, x_i), \max(a_i, x_i)] \times [\min(b_i, y_i), \max(b_i, y_i)]$. We determine the parameters x_i (resp., y_i) with the following strategy.

> Consider a rectangle r_i, and assume without loss of generality that p_i is the lower-left corner of r_i. If r_i is not in contact with any rectangle r_j such that $r_i \cap r_j$ is vertical and $a_i < a_j$, then $x_i = 1$ (that is, r_i extends to the right edge of the bounding box $[0,1]^2$). If r_i is in contact with a rectangle r_j such that the segment $r_i \cap r_j$ is vertical, $a_i < a_j$, and the anchor p_j is the lower-left or upper-left corner of r_j, then we have $x_i = a_j$. Analogous conditions determine y_i in some cases.

We now show that our assumptions from Proposition 2 ensure that the above strategy determines x_i and y_i for all $i = 1, \ldots, n$. If the above strategy fails to find x_i, then r_i is in contact with a rectangle r_j such that the segment $r_i \cap r_j$ is vertical, $a_i < a_j$, but p_j is the lower-right or upper-right corner of r_j. In this case, we call (r_i, r_j) a *horizontal pair*. Analogously, if the strategy does not find y_i, then r_i is part of some *vertical pair* (r_i, r_j). The horizontal (resp., vertical) pairs define a subgraph of the contact graph, that we denote by G_H (resp., G_V). Each connected component C of the graph G_H (resp., G_V) corresponds to rectangles whose left or right edge lies on some common vertical (resp., horizontal) line ℓ.

Consider a component C of G_H (the argument is analogous for G_V). The line ℓ must be right of all lower-left and upper-left anchors of rectangles in C, and left of all lower-right and upper-right anchors. Suppose that there exists a maximal anchored rectangle packing that satisfies these constraints. Denote by $L \subset C$ (resp., $R \subset C$) the set of rectangles whose right (resp., left) edges lie on ℓ. Similarly to the proof of Proposition 2, we deform the rectangles in L and R simultaneously by translating ℓ. If the sum of heights of rectangles in L and R differ, then translating ℓ in one of the two possible directions increases the total area, contradicting maximality. If the sum of heights of rectangles in L and R are equal, then translating ℓ in any direction is an equiareal deformation. We can now translate ℓ left until the area of a rectangle in L drops to 0 or a rectangle in R is in contact with a new rectangle on the left of ℓ. This contradicts our assumption that equiareal deformations create neither new contacts nor new rectangles of 0 area. Consequently, G_H (resp., G_V) is the empty graph, there are neither horizontal nor vertical pairs, and the above strategy uniquely determines x_i and y_i for all $i = 1, \ldots, n$. $\square$

Theorem 2. *There exist constants $1 < c_1 < c_2$ such that $\Omega(c_1^n) \leq M(n) \leq O(c_2^n)$.*

Proof. The combination of Lemmas 2 and 3 yields the upper bound. Theorem 1 gives the lower bound.

5 Conclusions

We have considered two variants of anchored rectangle packings: the anchors p_i were required to be either the lower-left or arbitrary corners of the rectangles r_i. We could consider a variant that we call *relaxed anchored rectangle packing*, where the anchors p_i are contained in the rectangles r_i. In this case, the maximum area of a rectangle packing is always 1, since the bounding box can be subdivided into n parallel strips, each containing a point in P. Note that a rectangle $r_i = [x_i, x_i'] \times [y_i, y_i']$ is now described by 4 variables. In a relaxed anchored rectangle packing, however, a local maximum need not attain the global maximum. Nevertheless, the technique of Section 4 extends to this variant: each maximal rectangle packing can be reconstructed from the contact graphs of the rectangles (which has an embedding using polylines with at most one bend per edge), and $O(1)$ bits of additional information per rectangle. Consequently, the number of locally maximal packings for an n-element point set is bounded by $\exp(O(n))$.

Analogous problems arise for anchored packings with other simple geometric shapes, such as circular disks or positive homothets of some convex body. For packings with object of bounded description complexity, the configuration space can be parameterized with $O(n)$ variables, and some of the techniques developed here do generalize. However, several crucial steps in our work have relied on properties of axis-aligned rectangles. Determining the maximum area covered by a packing remains open for both anchored and L-anchored rectangle packings. For other geometric shapes (e.g., circular disks), finding the maximum area covered by relaxed anchored variants is already a challenging problem.

References

1. Ackerman, E.: Counting problems for geometric structures: rectangulations, floorplans, and quasi-planar graphs, PhD thesis, Technion (2016)
2. Ackerman, E., Barequet, G., Pinter, R.: On the number of rectangulations of a planar point set. J. Combin. Theory, Ser. A **113**(6), 1072–1091 (2006)
3. Adamaszek, A., Wiese, A.: Approximation schemes for maximum weight independent set of rectangles. In: Proc. 54th FOCS. IEEE (2013)
4. Adamaszek, A., Wiese, A.: A quasi-PTAS for the two-dimensional geometric knapsack problem. In: Proc. 26th SODA. SIAM (2015)
5. Ahn, H.-K., Cheng, S.-W., Cheong, O., Golin, M., van Oostrum, R.: Competitive facility location: the Voronoi game. Theoret. Comput. Sci. **310**, 457–467 (2004)
6. Ajtai, M., Chvátal, V., Newborn, M., Szemerédi, E.: Crossing-free subgraphs. Annals Discrete Math. **12**, 9–12 (1982)
7. Avis, D., Fukuda, K.: Reverse search for enumeration Discrete Appl. Math. **65**, 21–46 (1996)
8. Bansal, N., Khan, A.: Improved approximation algorithm for two-dimensional bin packing. In: Proc. 25th SODA, pp. 13–25. SIAM (2014)
9. Cheong, O., Har-Peled, S., Linial, N., Matoušek, J.: The one-round Voronoi game. Discrete Comput. Geom. **31**, 125–138 (2004)

10. Dumitrescu, A., Schulz, A., Sheffer, A., Tóth, C.D.: Bounds on the maximum multiplicity of some common geometric graphs. SIAM J. Discrete Math. **27**(2), 802–826 (2013)

11. Dumitrescu, A., Tóth, C.D.: Packing anchored rectangles. In: Proc. 23rd SODA, pp. 294–305. SIAM (2012); and Combinatorica **35**(1), 39–61 (2015)

12. Francke, A., Tóth, C.D.: A census of plane graphs with polyline edges. In: Proc. 30th SoCG, pp. 242–250. ACM Press (2014)

13. Giménez, O., Noy, M.: Asymptotic enumeration and limit laws of planar graphs. J. AMS **22**, 309–329 (2009)

14. Harren, R., Jansen, K., Prädel, L., van Stee, R.: A $(5/3 + \varepsilon)$-approximation for strip packing. Comput. Geom. **47**(2), 248–267 (2014)

15. Kakoulis, K.G., Tollis, I.G.: Labeling algorithms, chap. 28. In: Tamassia, R. (ed.) Handbook of Graph Drawing and Visualization. CRC Press (2013)

16. Knuth, D., Raghunathan, A.: The problem of compatible representatives. SIAM J. Discete Math. **5**, 36–47 (1992)

17. van Kreveld, M., Strijk, T., Wolff, A.: Point labeling with sliding labels. Comput. Geom. **13**, 21–47 (1999)

18. Murata, H., Fujiyoshi, K., Nakatake, S., Kajitani, Y.: VLSI module placement based on rectangle-packing by the sequence-pair. IEEE Trans. CAD Integrated Circuits and Systems **15**(12) (1996)

19. Santos, F., Seidel, R.: A better upper bound on the number of triangulations of a planar point set. J. Combin. Theory, Ser. A **102**, 186–193 (2003)

20. Sharir, M., Sheffer, A.: Counting plane graphs: cross-graph charging schemes. Combinat. Probab. Comput. **22**, 935–954 (2013)

21. Stanley, R.: Problem k^8 in Catalan addendum to Enumerative Combinatorics, vol. 2, May 25, 2013. http://www-math.mit.edu/~rstan/ec/catadd.pdf

22. Thomas, H.: New combinatorial descriptions of the triangulations of cyclic polytopes and the second higher Stasheff-Tamari posets. Order **19**(4), 327–342 (2002)

23. Tutte, W.: Recent Progress in Combinatorics: Proceedings of the 3rd Waterloo Conference on Combinatorics, May 1968. Academic Press, New York (1969)

24. Winkler, P.: Packing rectangles. In: Mathematical Mind-Benders, pp. 133–134, A.K. Peters Ltd., Wellesley (2007)

Approximate Truthful Mechanism Design for Two-Dimensional Orthogonal Knapsack Problem

Deshi Ye[✉] and Guochuan Zhang

College of Computer Science, Zhejiang University, Hangzhou 310027, China
{yedeshi,zgc}@zju.edu.cn

Abstract. This paper provides a technique for designing truthful mechanisms for a combinatorial optimization problem, which requires composition algorithms. We show that the composition algorithm $A \circ B$ is monotone if the algorithm A and the algorithm B are both monotone. Then, we apply this technique to the two-dimensional orthogonal knapsack problem with provable approximation bounds, improving the previous results in [5].

Keywords: Mechanism design · Knapsack auction · Approximation algorithms

1 Introduction

Traditional optimization problems assume that the input data are available to the algorithm designer. However, in many Internet applications, such as combinatorial auction, the algorithms whose inputs are provided by selfish agents prefer to lie if there are benefits for themselves. Mechanism design is to deal with such selfish settings. The mechanism designer proposes allocation and payment algorithms for all agents beforehand, the agents then decide to report their own input data. Without loss of generality, assume that agents are rational and attempt to maximize their own utilities. Most of previous research work concerns on incentive compatible or truthful mechanisms, in which a dominate strategy for an agent is to report the true input data. Designing efficient truthful mechanisms that approximate the optimal social welfare was first considered by Nisan and Ronen [27]. Two different social objective functions of mechanisms were studied in this approach. One is about the utilitarian optimization problems, such as combinatorial auction and the knapsack problem. The other is to minimize the makespan of parallel machines with private speeds.

Technique designing for combinatorial auctions was well studied. Mu'alem and Nisan [26] provided several ways to combine two allocations algorithms, such as the MAX operator and the If-Then-Else operator. They studied the

D. Ye—Research was supported in part by China Scholarship Council.

G. Zhang—Research was supported by NSFC(11271325).

© Springer International Publishing Switzerland 2015
D. Xu et al. (Eds.): COCOON 2015, LNCS 9198, pp. 390–401, 2015.
DOI: 10.1007/978-3-319-21398-9_31

mechanisms for restricted combinatorial auctions where the subset of items of each bidder is known and only the valuation of these items is unknown by the mechanism (the single parameter case), a 2-approximation mechanism based on the greedy method for the knapsack problem was provided. Briest et al. [10] designed a new approach in rounding scheme that leads to monotone FPTASs, and therefore a monotone FPTAS for knapsack problem. Moreover, the problems considered in [10] are of multi-parameter. Chekuri and Gamzu [12] studied the greedy iterative packing truthful mechanism for the multiple knapsack problem. By presenting a property of loser-independent, they gave a truthful $(2+\varepsilon)$-approximate mechanism among single-minded agents, and $(e/(e-1)+\varepsilon)$-approximate mechanism for knapsacks with identical capacity. When the number of knapsacks is a fixed constant, Briest et al. [10] presented a monotone PTAS for the multiple knapsack problem. Grandoni et al. [19] designed monotone truthful multi-criteria FPTASs for multi-objective problems, which implies a monotone truthful FPTAS for the multi-dimensional knapsack problem. However, their FPTASs may violate each budget constraint by a factor $(1+\varepsilon)$. In general, various techniques appeared for the multi-dimensional packing problems, such as the convex-decomposition technique [24] as well as maximal-in-range [17].

In this work we aim at a technique for designing truthful mechanisms for the *two-dimensional orthogonal knapsack* with composition algorithms. In two-dimensional orthogonal knapsack(2DOK) (or Rectangle packing (RP)), it is asked to pack a set of rectangles into a bin (or a larger rectangle), where each rectangle is associated with a value. The goal is to maximize the total value of the selected rectangles that can be packed into the bin. This problem is motivated by the scenario in the advertising auction. In web applications there is a rectangle space available for advertisements. A set of advertisers would like to bid a room for displaying their graphical advertisements. An advertisement is usually a rectangle. The auctioneer (the owner of the space) has to choose a set of rectangles that can be packed into the space while the social welfare is maximized. If all the advertisements have the same width, the problem is reduced to the classical knapsack auction. Note that our problem differs from the multi-dimensional knapsack auction [19] in the fact our problem involves geometric constraints. Babaioff and Blumrosen [5] dealt with selling advertisement space on a newspaper page that can be modelled by packing convex figures in a plane, in which they show an $O(R)$-approximation truthful mechanism if convex figures are rectangles, where R is the ratio of the maximum diameter of a rectangle and the minimum width of a rectangle.

1.1   Related Work

The maximization problem of rectangle packing was considered in the literature. Jansen and Zhang [21] designed a $(2+\varepsilon)$-approximation algorithm. For the special case of maximizing the number of packed rectangles, Jansen and Zhang provided an asymptotic FPTAS (AFPTAS) as well as a PTAS. Resource augmentation was studied by Fishkin et al. [18]. For the multiple knapsack problem, FPTAS was ruled out even for two knapsacks unless P=NP [11,13]. Kellerer [22] devised a PTAS if the knapsacks are identical. Chekuri and Khanna [13] provided a PTAS for the general multiple knapsack problem. Jansen [20] showed that there exists an EPTAS. However, there is no EPTAS for the two-dimensional knapsack problem [23].

1.2 Our Contribution

Our main result is to design truthful mechanisms for the two-dimensional orthogonal knapsack problem via a composable algorithm. We show that the composition algorithm $A \circ B$ is monotone if the algorithm A (selection part) is monotone and the algorithm B (allocation part) is monotone. Suppose that $\varepsilon > 0$ is a given number. We first provide a monotone truthful and deterministic algorithm $A \circ B$ and show that the approximation ratio is at most $7 + \varepsilon$. If rotation of 90 degrees is allowed, we obtain an approximation ratio of at most $3 + \varepsilon$. Moreover, for multiple knapsacks, we derive a $(9 + \varepsilon)$-approximation algorithm for fixed number of knapsacks and a 14.2378-approximation algorithm for any arbitrary number of knapsacks. Again, with rotation, the bound can be improved to $7.5 + \varepsilon$. For square packing we can achieve a better bound of $3 + \varepsilon$. Therefore we give small constant truthful approximation bounds and improve the previous bounds in [5].

2 The Two-Dimensional Orthogonal Knapsack Problem

The optimization version of the two-dimensional orthogonal knapsack problem is to select a set of rectangles with maximum value such that they can be packed into a given knapsack (a rectangle). Suppose that the knapsack has capacity $C = (a, b)$, where C is a rectangle with width a and height b. In the view of mechanism design, items are controlled by selfish agents, and each rectangle (or item) j has its true type (a_j, b_j, v_j), where $0 < a_j \le a$ is the width of rectangle j and $0 < b_j \le b$ is the height of rectangle j, and v_j is the value of this item. Each agent j sends her bid (a'_j, b'_j, v'_j) to a mechanism, and then the mechanism computes the output O and the payments for every agent based on the bids of all agents. Thus, the mechanism for two-dimensional orthogonal knapsack problem is an allocation algorithm A and a payment function p^A. The mechanism's goal is to maximize the social welfare, i.e., the total value of selected items assigned in the knapsack.

Let $d = (d_1, d_2, \ldots, d_n)$ be the bidders of all agents, where d_j is the declaration of agent j and n is the number of agents. If agent j reports her true type, then $d_j = (a_j, b_j, v_j)$. We consider *single-minded* version introduced by Lehmann, Oćallaghan, and Shoham [25]. Let $O(d)$ be the allocations of the mechanism based on the reporting d and each $O_j(d) \in O(d)$ is the allocation for each agent j. Each $O_j(d)$ is a rectangle, and $O_j(d)$ is empty if agent j is not selected. We define $(a_j, b_j) \le O_j(d)$ if a_j is no more than the width of $O_j(d)$ and b_j is no more than the height of $O_j(d)$, i.e., the rectangle (a_j, b_j) can be packed inside the rectangle $O_j(d)$. The value function $v_j(O_j(d))$ for each output $O_j(d)$ is given as below.

$$v_j(O_j(d)) = \begin{cases} v_j, \ if (a_j, b_j) \le O_j(d) \\ 0, \ otherwise. \end{cases} \tag{1}$$

In addition, we only consider *unknown* size of this problem, meaning that the size of rectangle is only known by the corresponding agent. Each agent j may declare any value of a_j and b_j. An agent j's utility in a mechanism (A, p^A) is

$$u_j(d) = v_j(O_j(d)) - p_j^A$$

by the bidding of d, where p_j^A is the payment for agent j. Each agent attempts to maximize her utility, and thus might manipulate the mechanism by declaring a false type. A mechanism is *truthful* or *incentive compatible*, if no agent j would increase her utility by any false declaration, i.e.

$$u_j((a_j, b_j, v_j), (a'_{-j}, b'_{-j}, v'_{-j})) \ge u_j((a'_j, b'_j, v'_j), (a'_{-j}, b'_{-j}, v'_{-j}))$$

for any declaration (a'_j, b'_j, v'_j).

For any instance I, let $SC(I)$ be the total value obtained by a mechanism, and $OPT(I)$ be the total value obtained by an optimal solution, then the mechanism is ρ-approximation if $\frac{OPT(I)}{SC(I)} \le \rho$.

2.1   The Mechanism

A bid (a'_j, b'_j, v'_j) of agent j is a *winning declaration* if $(a_j, b_j) \le O_j$ (this item is selected in the knapsack), otherwise, it is a *losing declaration*. For a bid (a'_j, b'_j, v'_j), a declaration (a''_j, b''_j, v''_j) is said to be a *higher declaration* if $a''_j \le a'_j$ and $b''_j \le b'_j$, and $v''_j \ge v'_j$. For the simplification we let $d = (a, b, v) = ((a_1, b_1, v_1), (a_2, b_2, v_2), \ldots, (a_n, b_n, v_n))$. Let (a_{-j}, b_{-j}, v_{-j}) denote the declaration without agent j, which can be represented as

$$((a_1, b_1, v_1), \ldots, (a_{j-1}, b_{j-1}, v_{j-1}), (a_{j+1}, b_{j+1}, v_{j+1}), \ldots, (a_n, b_n, v_n)).$$

Definition 2.1. *(Monotone) We say that an algorithm A for two-dimensional orthogonal knapsack problem is monotone if, for any agent bidder j, (a'_j, b'_j, v'_j) is a winning declaration then any higher declaration also wins.*

From the property of monotone, we observe that algorithm A defines a *critical value* θ_j^A, which is the minimum value v'_j such that (a'_j, b'_j, v'_j) is a winning declaration if we fix the declaration of $(a'_{-j}, b'_{-j}, v'_{-j})$ and declaration (a'_j, b'_j). We say that algorithm A is *exact* if $O_j(d') = (a'_j, b'_j)$ or $O_j(d') = \emptyset$ for each declaration (a'_j, b'_j, v'_j) of d'.

Definition 2.2. *The payment p^A associated with the monotone allocation algorithm A that is based on the critical value is defined by $p_j^A = \theta_j^A$ if agent j wins, and $p_j^A = 0$ otherwise.*

A mechanism $M_A = (A, p^A)$ is *normalized* , if its payment p^A is defined as in Definition 2.2, i.e. agents that are not selected pay 0.

Theorem 2.3. [10] *Let A be a monotone and exact algorithm for some utilitarian problem and single-minded agents. Then the normalized mechanism $M_A = (A, p^A)$ is truthful.*

Proof. Briest et al. [10] showed that the theorem is valid for a single dimensional knapsack problem, generalized from the combinatorial auction problem [25]. This result can be easily extended to our 2-dimensional orthogonal knapsack problem as it is utilitarian. For the sake of completeness, we give the sketch of the proof.

For any agent j, let us fix the declarations of any other agents. The true type of agent j is (a_j, b_j, v_j). The first step is to prove that the utility function of the declaration (a'_j, b'_j, v_j) is at least that of the arbitrary declaration (a'_j, b'_j, v'_j) for any v'_j. It is worth to mention that the critical value is independent of $j's$ declaration of v'_j. Let θ_j be the critical value of declaration (a'_j, b'_j, v'_j). If agent j is selected or not selected in both declarations, the utilities are the same. If the agent j is not selected in declaration (a'_j, b'_j, v_j) and selected in declaration (a'_j, b'_j, v'_j), we have $v'_j \geq \theta_j > v_j$. Thus, the utility of agent j in declaration of (a'_j, b'_j, v'_j) is negative, while the utility of agent j in declaration of (a'_j, b'_j, v_j) is zero. Conversely, if the agent j is selected in declaration (a'_j, b'_j, v_j) and not selected in declaration (a'_j, b'_j, v'_j), we have the utility of agent j is non-negative and zero for these two declarations, respectively.

The second step is to show the utility of declaration of (a_j, b_j, v_j) is no less than (a'_j, b'_j, v_j) for any (a'_j, b'_j). If $a'_j < a_j$ or $b'_j < b_j$, from the exactness, the value of agent j is zero and hence the utility is non positive. Observe that the utility of a truth declaration is non-negative. Therefore, the utility is not increasing by such a kind of lying.

Let us focus on the lying that $a'_j \geq a_j$ and $b'_j \geq b_j$. Let θ_j and θ'_j be the critical values according to the true declaration and the lying declaration, respectively. We have $\theta_j \leq \theta'_j$ from the monotonicity of the allocation algorithm A. If the agent j is not selected by lying declaration, its utility is zero, while the true declaration is non-negative. If the agent j is selected in both declaration, the utility of true declaration is $v_j - \theta_j$ that is no less than the utility of lying declaration $v_j - \theta'_j$. Now, we only consider that the agent j is not selected in true declaration. In this case we have $v_j < \theta_j \leq \theta'_j$, which indicates the utility of agent j is negative if it is selected by lying to (a'_j, b'_j, v_j). □

Hence it is sufficient to design a monotone allocation algorithm to obtain a truthful mechanism for our problem by Theorem 2.3. To this end, in the following, we provide a composition algorithm.

2.2 Composition Algorithm

Given two algorithms A and B, we define the composition of algorithm $A \circ B$ in the following way: For any given input I, run the algorithm A on I and let O_1

be the set of winners. Then run the algorithm B on O_1 and let O_2 be the set of winners. The output of algorithm $A \circ B$ is the allocation of items in O_2.

For any instance I of the two-dimensional knapsack problem (a, b, v), we define a new instance I', which is a one-dimensional knapsack problem (s, v), where $s_j = a_j \cdot b_j$ is the area of the rectangle item j, and the capacity of the knapsack is $C = a \cdot b$. A bit overuse of notations, in the remainder of this paper, C is the area of the rectangle (a, b) when we refer to the capacity of the one-dimensional knapsack problem, and C is the rectangle (a, b) when we refer to the two-dimensional knapsack.

Composition Algorithm $A \circ B$

1. Run 1-dimensional knapsack algorithm A for the new instance I', return the set of selected items O_1.
2. Run algorithm B on the instance O_1 for 2-dimensional orthogonal knapsack problem, return the set of selected items O_2.

Theorem 2.4. *If the algorithm A and the algorithm B are both monotone, then its composition algorithm $A \circ B$ is monotone.*

Proof. Suppose that the agent j with true type (a_j, b_j, v_j) is a winning declaration. Then we need to prove that a higher declaration $j' = (a'_j, b'_j, v'_j)$ is also a winner, i.e. $a_j \geq a'_j$, $b_j \geq b'_j$ and $v_j \leq v'_j$. Since j is a winner, then $j \in O_2$ and we have $j \in O_1$ too. It holds that $j' \in O_1$ since algorithm A is monotone. We know $j' \in O_2$ because of the algorithm B is monotone, which therefore j' is also a winner. $\square$

2.3 Monotone Algorithm

An item R_j is called *big* if $a_j > a/2$ and $b_j > b/2$; it is *wide* if $a_j > a/2$ and $b_j \leq b/2$; it is *tall* if $a_j \leq a/2$ and $b_j > b/2$; and it is small if $a_j \leq a/2$ and $b_j \leq b/2$.

Lemma 2.5. [21] *If the total area of a set T of items is at most $C/2$ and there are no tall items (or there are no wide items), then the items in T can be packed into a bin with capacity C.*

Now we are ready to show a monotone algorithm for the two-dimensional orthogonal knapsack problem.

Lemma 2.6. *Algorithm 1 is monotone.*

Proof. Note that Algorithm 1 is composable, which consists of algorithms A and B. For any agent j that is a winner, if its declaration is higher, we have $s'_j \leq s_j$ and $v'_j \geq v_j$. The algorithm A in line 2 has already been proved to be monotone in the accordingly references.

For algorithm B, the monotone is shown as below. If the output of original declaration is due to G_1, i.e., it returns the m items with largest value, then, clearly, it is monotone since item j will also be a winner from $v'_j \geq v_j$.

Algorithm 1.. Allocation algorithm for 2-dimensional knapsack problem

---

1: For any input I (a_i, b_i, v_i), the new instance I' is (s_i, v_i), where $s_i = a_i \cdot b_i$, and the capacity of the knapsack is $C = ab$, where (a, b) is the capacity of the knapsack in input I.

2: **Algorithm A**: Run a monotone algorithm A for 1-dimensional knapsack problem on this new instance I'. Let O_1 be the output, and $V(O_1)$ be the value of total selected items. Specifically, in case of single knapsack, we adopt the monotone FPTAS [10] for the 1-dimensional knapsack problem as algorithm A. For multiple knapsack problem with fixed number of knapsacks, we adopt the monotone PTAS [10] for the general assignment problem as algorithm A. For multiple knapsack problem with arbitrarily number of knapsacks, we adopt the monotone algorithm in [12] as algorithm A.

3: **Algorithm B**: Let m be the number of knapsacks. Select m items with maximal value from O_1. Denote these m items as G_1. Let α be a constant that will be given later.

4: **if** $V(G_1) \geq \alpha V(O_1)$, **then**

5:     **return** Assign each item in G_1 to a different knapsack respectively.

6: **else**

7:     Remove all the big items from O_1.

8:     Then consider two sets of remaining items, T_1 consists of wide items and small items, T_2 consists of tall items and small items. We choose the set T_h with larger value, i.e. $V(T_h) = \max(V(T_1), V(T_2))$.

9:     We adopt the monotone algorithm A for 1-dimensional knapsack problem on the instance T_h. However, we set the capacity of the knapsacks to be $C/2$. Denote the selected items in this step as G_2.

10:     For the output G_2, we adopt the 2-dimensional packing algorithm as indicated in Lemma 2.5 by Jansen and Zhang [21] to pack these items in the original knapsack with capacity $C = (a, b)$.

11: **end if**

If the output of original declaration is G_2, fixing other declarations, agent j reports a higher declaration j'. We denote it as d'. The output of d' is either G_1' or G_2'. If it is G_1', i.e. the total value of the largest m items is larger than $\alpha V(O_1)$, noting that the total value of the largest m items in original declarations is smaller than $\alpha V(O_1)$ due to the output is G_2, then item j must be selected.

If the output of d' is G_2', we know that all big items are not included in T_h. The item j is a wide item (or tall) or a small item. If j is a small item, its higher bidder j' is also a small item. If j is a wide (or tall) item, its higher bidder either is a wide (or tall) item or a small item, i.e. the item with a high declaration is also in T_h. A higher declaration of item j ensures that item j must be selected due to the monotone of the algorithm A. Thus, the algorithm B is monotone. Since, algorithm A and B are both monotone, by Theorem 2.4, Algorithm 1 is monotone.   ☐

Lemma 2.7. *Let the approximation ratio of algorithm A be $\rho_A \geq 1$. For any given $\varepsilon > 0$, the approximation ratio of Algorithm 1 for $m = 1$ is $7\rho_A^2$ by letting*

$\alpha = 1/7$. *The approximation ratio of Algorithm 1 for $m \geq 2$ is $9\rho_A^2$ by letting* $\alpha = 9$.

Proof. Let V^{opt} be the value of selected items achieved by any optimal algorithm. We know $V^{opt} \leq \rho_A V(O_1)$. Let $V(O_2)$ be the value of selected items achieved by Algorithm 1. If the algorithm B outputs the m items with the maximum value, we have $V(O_2) \geq \alpha V(O_1)$, and then $V^{opt} \leq \frac{\rho_A}{\alpha} V(O_2)$ follows.

If the final accepted items are due to G_2, we show its value is at least $\alpha V(O_1)$. In this case, the maximum value among these items is no more than $\alpha V(O_1)$.

Note that in T_h, there are no both wide items and tall items. W.l.o.g, we assume there is no tall item, i.e. each item has height at most of $b/2$. According to the algorithm B, we apply the algorithm A to select items among T_h with the capacity of a knapsack $C/2$.

Now we are going to find a lower bound of the optimal value of selecting items in T_h with capacity $C/2$, which is denoted by $V^*(T_h)$. For the O_1 items in each knapsack, it is a feasible solution for a single knapsack with capacity C. Let us consider each knapsack j. Denote the value of items in the knapsack j to be $V(T_h, j)$, and we have $V(T_h) = \sum_{j=1}^{m} V(T_h, j)$. We split the knapsack j into two identical sub-knapsacks, each with capacity $C/2$ with width a and height $b/2$. See Figure 1 for an illustration. We get three sets of items, the items below the divided line, the item crossing the divided line, and the items above the divided line, respectively. It is worth to note that the area of all these three sets is at most of $C/2$. Then select one set with maximum value in each knapsack implies that the optimal value is at least $V(T_h, j)/3$ for any m knapsacks.

On the other hand, if we remove $\gamma \geq 2$ big items in the knapsack j, then the total area in this knapsack is at most of $C/2$, and all items from T_h in knapsack j can be selected by an optimal algorithm with capacity $C/2$.

In all, the value of the selected items in an optimal solution in knapsack j is at least $V(T_h, j)/3$ if $\gamma \leq 1$, or at least $V(T_h, j)$ if $\gamma \geq 2$, where γ is the number of big items in knapsack j from O_1.

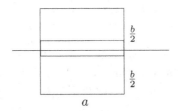

Fig. 1. Illustration of the split knapsack

Regarding the approximation ratios, let us first consider $m = 1$, i.e., the single knapsack problem. Suppose that γ big items are removed, the value lost is at most $\gamma \alpha V(O_1)$. In case of $m = 1$, we have $\gamma \leq 3$. Hence the total value of items in O_1 without counting removed items is at least $(1 - \gamma\alpha)V(O_1)$, then the value of the selected items T_h is $V(T_h) = \max(V(T_1), V(T_2)) \geq \frac{1-\gamma\alpha}{2}V(O_1)$.

We have $V^*(T_h) \geq V(T_h)/3 \geq \frac{1-\alpha}{6}V(O_1)$ if $\gamma \leq 1$, and $V^*(T_h) \geq V(T_h) \geq \frac{(1-3\alpha)}{2}V(O_1)$ if $\gamma \geq 2$. It holds that $V^*(T_h) \geq \min\{\frac{1-\alpha}{6}V(O_1), \frac{(1-3\alpha)}{2}V(O_1)\}$. Let $\alpha = 1/7$. Note that $\min\{\frac{1-\alpha}{6}, \frac{1-3\alpha}{2}\} = 1/7$.

$$V(O_2) \geq V^*(T_h)/\rho_A$$
$$\geq \min\{\alpha, \frac{1-\alpha}{6}, \frac{1-3\alpha}{2}\}V(O_1)/\rho_A$$
$$= \frac{1}{7}V(O_1)/\rho_A \geq \frac{1}{7\rho_A^2}V^{opt}.$$

Now we consider the approximation ratios for multiple knapsacks, in which $m \geq 2$. W.l.o.g, we assume that each of the first k knapsacks removes at most one big item. Then in the left $m-k$ knapsacks, each will remove at least two big items. Hence, we know $V^*(T_h) \geq \frac{1}{3}\sum_{j=1}^{k} V(T_h, j) + \sum_{j=k+1}^{m} V(T_h, j) \geq V(T_h)/3$. On the other hand, $2V(T_h) \geq V(O_1) - \sum_{j=1}^{m} V(B, j)$, where $V(B, j)$ is the total value of big jobs in knapsack j in O_1. From the fact that the final output is G_2, the value of the largest m items is at most $\alpha V(O_1)$, and there are at most $3m$ big items, we have $\sum_{j=1}^{m} V(B, j) \leq 3\alpha V(O_1)$. Thus, $V^*(T_h) \geq V(T_h)/3 \geq \frac{1-3\alpha}{6}V(O_1)$.

The approximation ratio $9\rho_A^2$ follows from the following inequality by letting $\alpha = 9$.

$$V(O_2) \geq \min\{\alpha V(O_1), V^*(T_h)/\rho_A\}$$
$$\geq \min\{\alpha, \frac{1-3\alpha}{6}\}V(O_1)$$
$$\geq \min\{\alpha, \frac{1-3\alpha}{6}\}V^{opt}/\rho_A^2.$$

$\square$

Theorem 2.8. *Let the mechanism employ Algorithm 1 as the allocation algorithm and adopt the associated critical value payment scheme. Then it is an approximation truthful mechanism for a number of knapsack problems including the 2-dimensional orthogonal knapsack problem, the multiple orthogonal knapsack problem with a constant number of knapsacks, and the multiple orthogonal knapsack problem with arbitrary number of knapsacks. The respective approximation ratios are at most $7 + \varepsilon$, $9 + \varepsilon$, and $9e/(e-1) + \varepsilon$, for any fixed $\varepsilon > 0$.*

Proof. The monotonicity has been proved from Lemma 2.6. Let us consider the approximation ratios by Lemma 2.7. For $m = 1$, let $\delta > 0$ be an arbitrarily small positive number, we adopt the monotone FPTAS [10] as the algorithm A, hence $\rho_A = 1 + \delta$. Thus, let $\varepsilon > 0$, and select δ such that $14\delta + 7\delta^2 \leq \varepsilon$, the approximation ratio for single 2-dimensional knapsack problem is at most $7 + \varepsilon$.

For $m \geq 2$, Briest, Krysta, and Vöcking [10] provided a monotone PTAS for the generalized assignment problem with constant number of knapsacks, where knapsack problem is a special case of generalized assignment problem. Hence, for any $\varepsilon > 0$, we obtain a $(9 + \varepsilon)$-approximation algorithm for any fixed m knapsacks. While, for arbitrary number of knapsacks, Chekuri and Gamzu [12] provided $e/(e-1) + \varepsilon$ monotone algorithm for multiple knapsack problem with identical capacity, which indicates a $9e/(e-1) + \varepsilon \approx 14.2378 + \varepsilon$ for our problem. □

2.4 Packing Square Items or Allowing Rotation

In this section, we consider a special case of packing rotatable items, i.e., the allocation algorithm allows to rotate an item by 90 degrees. In particular, we consider the items are squares, i.e. for each item i we have $a_i = b_i$. These two special cases have a common property that an algorithm can always ensure that there are no both wide items and tall items. In square items version, all items are either a big item or a small item. In the rotatable version, one can rotate a tall item or a wide item such that only one kind of such item remains.

The key idea to improve the approximation ratio is that we do not need to choose between wide items and tall items. We revise allocation algorithm B in Algorithm 1 as below: Apply algorithm A to solve the instance O_1 after rotating all tall items into wide items if necessary, while the capacity of the knapsack is set to $C/2$.

Corollary 2.9. *If all the items are squares or rotation of items are allowed in the allocation algorithm, then for any given $\varepsilon > 0$, assuming the approximation ratio of algorithm A in Algorithm 1 to be ρ_A, the approximation ratio for the multiple 2-dimensional knapsack problem is $3\rho_A^2$.*

Proof. The monotonicity of the revised algorithm is analogous to Lemma 2.6, because any higher declaration of an item from T_h still belongs to T_h because there are no both width items and tall items.

Lemma 2.5 guarantees the feasibility of packing of the selected items. Let us consider the approximation ratio. We have the optimal value $V^*(T_h)$ for the knapsack problem with capacity $C/2$ is at least $V(O_1)/3$ from the proof of Lemma 2.7. Let $V(O_2)$ be the value of final selected items by our algorithm, and V^{opt} be the optimal value. Clearly, $V^{opt} \leq \rho_A \cdot V(O_1)$ and $V(O_1)/3 \leq V^*(T_h) \leq \rho_A \cdot V(O_2)$. Therefore, we have $V^{opt} \leq 3\rho_A^2 \cdot V(O_2)$, which gives the approximation ratio at most of $3\rho_A^2$. □

Remark: For $m = 1$ and multiple knapsacks with constant number of knapsacks, the approximation ratio for the case of square items or with rotation is $3+\varepsilon$ due to algorithm A is $1+\varepsilon$ approximated [10]. For multiple knapsacks with an arbitrary number of knapsacks, the approximation ratio is $3(e/(e-1))^2 + \varepsilon \approx 7.5 + \varepsilon$ due to algorithm A [12] is $e/(e-1) + \varepsilon$ approximated.

3 Concluding Remarks

We have presented techniques for designing truthful mechanisms for the two-dimensional orthogonal knapsack problem. We hope that our technique can be flexible and extended to other problems with composition algorithms. Note that for the two-dimensional orthogonal knapsack problem, if item sizes are public, then the $(3 + \varepsilon)$-approximation algorithm provided by Jansen and Zhang [21] is a truthful mechanism. Recall that the non-strategic algorithm achieves an upper bound of $2+\varepsilon$ [21]. It is worth to see a better truthful mechanism. Another interesting problem is to see if it is possible to design a truthful PTAS for the multiple one-dimensional knapsack problem with an arbitrary number of knapsacks, which therefore reduces the approximation ratio for the multiple two-dimensional orthogonal knapsack problems automatically.

References

1. Andelman, N., Azar, Y., Sorani, M.: Truthful approximation mechanisms for scheduling selfish related machines. In: Diekert, V., Durand, B. (eds.) STACS 2005. LNCS, vol. 3404, pp. 69–82. Springer, Heidelberg (2005)
2. Archer, A., Tardos, É.: Truthful mechanisms for one-parameter agents. In: Proceedings of the 42nd IEEE Symposium on Foundations of Computer Science (FOCS), pp. 482–491 (2001)
3. Archer, A.F.: Mechanisms for discrete optimization with rational agents. Ph.D. thesis, Cornell University (2004)
4. Auletta, V., De Prisco, R., Penna, P., Persiano, G.: Deterministic truthful approximation mechanisms for scheduling related machines. In: Diekert, V., Habib, M. (eds.) STACS 2004. LNCS, vol. 2996, pp. 608–619. Springer, Heidelberg (2004)
5. Babaioff, M., Blumrosen, L.: Computationally-feasible truthful auctions for convex bundles. Games and Economic Behavior **63**(2), 588–620 (2008)
6. Bougeret, M., Dutot, P.-F., Jansen, K., Otte, C., Trystram, D.: A fast 5/2-approximation algorithm for hierarchical scheduling. In: D'Ambra, P., Guarracino, M., Talia, D. (eds.) Euro-Par 2010, Part I. LNCS, vol. 6271, pp. 157–167. Springer, Heidelberg (2010)
7. Bougeret, M., Dutot, P.F., Jansen, K., Otte, C., Trystram, D.: Approximating the non-contiguous multiple organization packing problem. In: Calude, C.S., Sassone, V. (eds.) TCS 2010. IFIP AICT, vol. 323, pp. 316–327. Springer, Heidelberg (2010)
8. Bougeret, M., Dutot, P.F., Jansen, K., Otte, C., Trystram, D.: Approximation algorithms for multiple strip packing. In: Bampis, E., Jansen, K. (eds.) WAOA 2009. LNCS, vol. 5893, pp. 37–48. Springer, Heidelberg (2010)
9. Bougeret, M., Dutot, P.F., Trystram, D.: An extention of the 5/2-approximation algorithm using oracle. Research Report (2010)
10. Briest, P., Krysta, P., Vöcking, B.: Approximation techniques for utilitarian mechanism design. SIAM Journal on Computing **40**(6), 1587–1622 (2011)
11. Caprara, A., Kellerer, H., Pferschy, U.: The multiple subset sum problem. SIAM Journal on Optimization **11**, 308–319 (2000)
12. Chekuri, C., Gamzu, I.: Truthful mechanisms via greedy iterative packing. In: Dinur, I., Jansen, K., Naor, J., Rolim, J. (eds.) Approximation, Randomization, and Combinatorial Optimization. LNCS, vol. 5687, pp. 56–69. Springer, Heidelberg (2009)
13. Chekuri, C., Khanna, S.: On multi-dimensional packing problems. In: Proceedings of the 10th annual ACM-SIAM symposium on Discrete Algorithms (SODA), pp. 185–194 (1999)

14. Christodoulou, G., Kovács, A.: A deterministic truthful ptas for scheduling related machines. In: Proceedings of the 21th Annual ACM-SIAM Symposium on Discrete Algorithms (SODA), pp. 1005–1016 (2010)
15. Coffman, E.G., Garey, M.R., Johnson, D.S., Tarjan, R.E.: Performance bounds for level oriented two-dimensional packing algorithms. SIAM Journal on Computing 9, 808–826 (1980)
16. Dhangwatnotai, P., Dobzinski, S., Dughmi, S., Roughgarden, T.: Truthful approximation schemes for single-parameter agents. In: Proceedings of 49th Annual IEEE Symposium on Foundations of Computer Science (FOCS), pp. 15–24 (2008)
17. Dughmi, S., Roughgarden, T.: Black-box randomized reductions in algorithmic mechanism design. In: Proceedings of the 51st Annual IEEE Symposium on Foundations of Computer Science (FOCS), pp. 775–784 (2010)
18. Fishkin, A.V., Gerber, O., Jansen, K., Solis-Oba, R.: On packing rectangles with resource augmentation: Maximizing the profit. Algorithmic Operations Research 3(1), 1–12 (2008)
19. Grandoni, F., Krysta, P., Leonardi, S., Ventre, C.: Utilitarian mechanism design for multi-objective optimization. In: Proceedings of the Twenty-First Annual ACM-SIAM Symposium on Discrete Algorithms (SODA), pp. 573–584 (2010)
20. Jansen, K.: Parameterized approximation scheme for the multiple knapsack problem. SIAM Journal on Computing 39, 1392–1412 (2009)
21. Jansen, K., Zhang, G.: Maximizing the total profit of rectangles packed into a rectangle. Algorithmica 47(3), 323–342 (2007)
22. Kellerer, H.: A polynomial time approximation scheme for the multiple knapsack problem. In: Hochbaum, D.S., Jansen, K., Rolim, J.D.P., Sinclair, A. (eds.) RANDOM 1999 and APPROX 1999. LNCS, vol. 1671, pp. 51–62. Springer, Heidelberg (1999)
23. Kulik, A., Shachnai, H.: There is no eptas for two-dimensional knapsack. Information Processing Letters 110(16), 707–710 (2010)
24. Lavi, R., Swamy, C.: Truthful and near-optimal mechanism design via linear programming. Journal of the ACM 58(6), 25 (2011)
25. Lehmann, D., Oćallaghan, L.I., Shoham, Y.: Truth revelation in approximately efficient combinatorial auctions. Journal of the ACM (JACM) 49(5), 577–602 (2002)
26. Mu'Alem, A., Nisan, N.: Truthful approximation mechanisms for restricted combinatorial auctions. Games and Economic Behavior 64(2), 612–631 (2008)
27. Nisan, N., Ronen, A.: Algorithmic mechanism design. In: Proceedings of the thirty-first annual ACM Symposium on Theory of Computing (STOC), pp. 129–140 (1999)
28. Steinberg, A.: A strip-packing algorithm with absolute performance bound 2. SIAM Journal on Computing 26, 401–409 (1997)
29. Ye, D., Han, X., Zhang, G.: Online multiple-strip packing. Theoretical Computer Science 412(3), 233–239 (2011)
30. Ye, D., Zhang, G.: Coordination mechanisms for selfish parallel jobs scheduling - (extended abstract). In: Agrawal, M., Cooper, S.B., Li, A. (eds.) TAMC 2012. LNCS, vol. 7287, pp. 225–236. Springer, Heidelberg (2012)
31. Zhuk, S.: Approximate algorithms to pack rectangles into several strips. Discrete Mathematics and Applications 16(1), 73–85 (2006)

Online Integrated Allocation of Berths and Quay Cranes in Container Terminals with 1-Lookahead

Jiayin Pan[1,2]($\boxtimes$) and Yinfeng Xu[1,2]

[1] School of Management, Xi'an Jiaotong University, Xi'an 710049, China
panjy1991@stu.xjtu.edu.cn
[2] The State Key Lab for Manufacturing Systems Engineering, Xi'an 710049, China
yfxu@mail.xjtu.edu.cn

Abstract. This paper studies an online over-list model of the integrated allocation of berths and quay cranes in container terminals with 1-lookahead ability. The objective is to minimize the maximum completion time of container vessels, i.e., the makespan. We focus on two different types of vessels, three berths and a small number of QCs in the hybrid berth layout, with 1-lookahead information. We propose a $(1 + \sqrt{2})/2$-competitive algorithm for the case with 4 cranes and a 5/4-competitive algorithm for the case with 5 cranes, respectively. Both of the algorithms are proved to be optimal.

Keywords: Scheduling · Online algorithm · Lookahead information · Container terminal

1 Introduction

Recently, the optimization problems involved in seaside operations planning attract increasing attention in the operation research and transportation research literatures. The seaside operations planning in container terminals basically comprises the berth allocation problem (BAP), the quay crane assignment problem (QCAP), and the quay crane scheduling problem (QCSP). Since BAP and QCAP are much interrelated in practice, a trend towards an integrated solution of berth and QC(quay crane) resources is observed in the recent literatures (see Bierwirth and Meisel, 2010; Carlo *et al.*, 2013)

Most literatures investigated the offline version of the BAP-QCAP problem such that the scheduler knows the complete information of all vessels at the beginning. Park and Kim (2003) provided pioneering integration approaches, they decided on the berthing position, the berthing time, and the number of cranes to assign to each vessel together. Lokuge and Alahakoon (2007) studied a dynamic BAP-QCAP problem with hybrid berths, aiming at minimizing total waiting time and minimizing total tardiness. They developed a multi-agent system which constitutes a feedback loop integration of the BAP and the QCAP. Giallombardo *et al.*(2010), Blazewicz *et al.*(2011) and Zhen *et al.*(2011) presented a mixed integer programming and gave a tatu search algorithm for a

© Springer International Publishing Switzerland 2015
D. Xu et al. (Eds.): COCOON 2015, LNCS 9198, pp. 402–416, 2015.
DOI: 10.1007/978-3-319-21398-9_32

dynamic BAP-QCAP problem in the discrete berth layout. Liang *et al.*(2011) considered a dynamic discrete BAP-QCAP problem as a non-linear bi-objective programming model. Chen *et al.* (2012) presented a mixed integer programming to solve the problem which the objective was minimizing tardiness.

Zhang *et al.*(2008) studied the online version of container vessel scheduling at the earliest. They considered the QC scheduling problem with non-crossing constraints, and considered both online over-list and over-time model, with the objective of minimizing the makespan. Zhang *et al.* developed an algorithm that proved a $m/\lceil \log 2(m+1) \rceil$-competitive for the over-list model and an algorithm that proved a 3-competitive for the general model of over-time model where m is the number of QCs.

There are some related studies about online model with lookahead information in the field of scheduling. Mandelbaum *et al.* (2010) studied online parallel machines scheduling with k-lookahead, where k is the number of lookahead jobs. For the objective of minimizing the makespan, they found a 1-competitive online algorithm when there were only two job types. Zheng *et al.*(2013) studied online single machine scheduling, also with k-lookahead jobs case, showed that the lookahead ability can effective improve the competitive performance of online strategy.

In this paper, we investigate the online vesion of the BAP-QCAP problem with 1-lookahead in the hybrid berth layout (Imai *et al.*, 2005), focus on the over-list model that vessels are released one by one. As in practice, the vessels dynamically arrive at the container terminal. The scheduler can foresee the information of the vessels that arrive in the next a couple of days, and produce a schedule to be executed in the next day for the vessels based on FCFS rule. Namely, the scheduler makes the decision on the assignment of a current vessel with the information of the current vessel and next k vessels. Hence, we model the above FCFS-based assignment pattern as the online over-list BAP-QCAP problem with k-lookahead information, and evaluate the performance of such kind of strategies. More precisely, we consider the cases with three berths, four or five QCs, and have 1-lookahead information. We prove that there exist optimal online algorithms with competitive ratios of $(1 + \sqrt{2})/2$ and $5/4$, respectively.

The rest of this work is organized as follows. Section 2 describes the problem and gives some notations. In Section 3 and 4, we deal with the case with 4 and 5 QCs respectively, and present a matching upper and lower bound for each case. Finally Section 5 concludes this work.

2 Problem Statement and Basic Notations

There are some vessels that request service (loading or unloading containers) by berths and QCs in the quay. In this paper, we only consider two types of vessels, each type corresponds to a service request with uniform vessel size and processing load. More precisely, the small request corresponds to small vessel size which needs one berth to assign, and its processing load is equal to one. However, the large request corresponds to large vessel size which needs two consecutive berths to assign, and its processing load is equal to Δ (≥ 2). For simplicity, let $r_i = 1$ and $r_i = \Delta$ denote a small request and a large request respectively. Considering the physical restrictions imposed on the number of QCs to be assigned to a vessel, we further assume that a small (or large) request can be serviced by two (or four) QCs at most simultaneously. Namely, the processing time of a small (or large) request r_i is equal to $1/m_i$ (or Δ/m_i) units of time where m_i is the number of QCs for processing the request. A list of requests $\mathcal{I} = \{r_1, r_2, \ldots, r_n\}$ ($n \geq 1$) are to be released one by one. The scheduler must immediately allocate the service combination of berth and QC as well as starting time. In addition to the information about the current request in list, the scheduler has all the information about the next 1 request in the list. The objective is to minimize the makespan, i.e., the end time of the last completed request.

As both berth and QC are expensive resources in the container terminal, generally, many ports in China have very limited number of berths and QCs. Hence, we focus on the scenario with three consecutive berths. Which serve either three small requests or one small request and one large request simultaneously, implying there at most need 6 QCs. According to the actual situation, in this paper, we consider 4 and 5 QCs. The QCs move in the same rail, satisfying the non-cross constraint. We denote the three berths by b_1, b_2 and b_3 from left to right, and by $q_1, q_2, \ldots, q_m$ the m ($4 \leq m \leq 6$) QCs from left to right in the quay.

For a request r_i, we denote its start and end time by s_i and e_i. $t_{i,m}$ ($m = \{1, 2, 3\}$) denotes the earliest time by which at least m consecutive berths have completed all of their currently allocated requests. $C_{i,j}$ denotes the earliest time by which berth b_j has completed all of its currently allocated requests.

Adopting the quadruple notation scheme in Bierwirth and Meisel (2010), we denotes this problem as $hybr \,|online - over - list, \, LD = 1 \,|BAP - QCAP|$ $C_{\max}$, where LD means the number of lookahead requests.

To evaluate the performance of an online strategy $\mathcal{A}$, we often use the competitive ratio (Borodin and El-Yaniv, 1998). For any request input instance $\mathcal{I}$, let $C_{\mathcal{A}}(\mathcal{I})$, $C^*(\mathcal{I})$ be the makespan of schedule produced by an online algorithm $\mathcal{A}$ and that of an optimal schedule respectively. Then algorithm $\mathcal{A}$ is ρ-competitive where

$$\rho = \inf_r \left\{ r \left| \frac{C_{\mathcal{A}}(\mathcal{I})}{C^*(\mathcal{I})} \leq r \right. \right\}$$

3 The Case with Four QCs

In this section we focus on the case with 3 berths and 4 QCs in the problem $hybr \,|online - over - list, \, LD = 1 \,|BAP - QCAP| \, C_{\max}$. Firstly, we analyze the lower bound, i.e., no online algorithm can perform better than it from a worst-case point of view. Secondly, we present an online algorithm named MLIST. Before analyzing the lower bound and algorithm, we first introduce two definitions.

Definition 1. *Waste time segment* and *available time segment:* For an idle time segment $[s_i, e_i)$, let q be the number of idle QCs. If $0 < q(e_i - s_i) < 1$, then $[s_i, e_i)$ is a waste time segment, denoted by T_w. Otherwise, $[s_i, e_i)$ is an available time segment, denoted by T_a.

The segment of each request r_i may induce a forced idle time segment, i.e., an idle time segment $[s_i, e_i)$, only for during $[s_i, e_i)$, there exist idle QCs. If $0 < q(e_i - s_i) < 1$, where q is the number of the idle QCs, even a small request later on cannot be satisfied within the time segment, and thus $[s_i, e_i)$ is a waste time segment. Otherwise, $[s_i, e_i)$ is an available time segment. Set $T_a=[t_1,t_2)$ where $t_1=s_i$, $t_2=e_i$, and $T_a=[0,0)$ initially. And let $|T_a|$ denote the total available time, $|T_w|$ denote the total waste time.

3.1 Lower Bound

Theorem 1. *For problem hybr* $|online - over - list, LD = 1$ $|BAP - QCAP|$ $C_{\max}$, *any online scheduling algorithm has a competitive ratio of at least* $(1 + \sqrt{2})/2$.

Proof. To prove the theorem, we construct a request input sequence $\mathcal{I}$. $\mathcal{I}$ contains at least two requests. Let $C_{max}(\mathcal{I})$ and $C^*(\mathcal{I})$ be the makespan of schedule produced by $\mathcal{A}$ and an optimal offline algorithm OPT, respectively.

According to the requires of the request for the berth and quay cranes, if request $r_i = \Delta$, then $\mathcal{A}$ has four choices: processes r_i on two consecutive berths with one, two, three or four QCs. For processing r_i with one QC, there always exists at least one idle QC during $[s_i, e_i)$, $\mathcal{A}$ won't select it. If request $r_i = 1$, then $\mathcal{A}$ has two choices: processes r_i on a single berth with one or two QCs.

Assume $\mathcal{I}$ contains the first request $r_1 = \Delta$, with lookahead $r_2 = 1$. If $\mathcal{A}$ processes r_1 with four QCs on berths b_1, b_2, no more requests arrive, which implying $C_{\max}(\mathcal{I}) \geq 1/2 + \Delta/4$. While OPT processes the requests r_1 and r_2 with three and one QC respectively, at time 0. $C^*(\mathcal{I}) = \max\{\Delta/3, 1\}$. Except $\Delta \geq 6$, OPT processes the request r_1 and r_2 with four and two QCs respectively, at time 0. $C^*(\mathcal{I}) = 1/2 + \Delta/4$. If $\mathcal{A}$ processes r_1 with three QCs on berth b_1, b_2, offline adversary releases request $r_3 = 1$ with no more requests, implying $C_{\max}(\mathcal{I}) \geq \min\{\max\{\Delta/3, 2\}, 1/2 + \Delta/3\}$. While OPT processes the request r_1 with four QCs on berths b_1, b_2, r_2 with two QCs on berth b_1, and r_3 with two QCs on berth b_2, at time 0. $C^*(\mathcal{I}) = 1/2 + \Delta/4$. If $\mathcal{A}$ processes r_1 with two QCs on berth b_1, b_2, then no more requests arrive. $C_{\max}(\mathcal{I}) \geq \Delta/2$, while OPT processes the request r_1 and r_2 with three and one QC respectively, at time 0. $C^*(\mathcal{I}) = \max\{\Delta/3, 1\}$. Except $\Delta \geq 6$, OPT processes the request r_1 and r_2 with four and two QCs respectively, at time 0, $C^*(\mathcal{I}) = 1/2 + \Delta/4$.

After processed. We have $\rho \geq (1 + \sqrt{2})/2$, setting $\Delta = (6 + 12\sqrt{2})/7$. For the sake of completeness, this middle paragraph is presented in Appendix A. The theorem follows. $\square$

3.2   Algorithm MLIST

On the release of any request r_i $(i \geq 1)$, if $i = n$, then the scheduler can lookahead r_n is the last request of consequence $\mathcal{I}$, so we can optimal assign this request to minimize $C_{\max}$. Otherwise, $1 \leq i \leq n - 1$, MLIST behaves as follows.

Case 1. $r_i = \Delta, r_{i+1} = 1$. There are two subcases in the following:
- Case 1.1. When $(2 + 2\sqrt{10})/3 \leq \Delta \leq (6 + 12\sqrt{2})/7$, for the case with $|T_a| \geq 3$, assign r_i to the two idle consecutive berths with three QCs, and reset T_a. Otherwise, set $s_i = \max\{C_{i,1}, C_{i,2}\}$, and assign r_i to the leftmost two berths with three QCs.
- Case 1.2. When $2 \leq \Delta \leq (2 + 2\sqrt{10})/3$ or $\Delta \geq (6 + 12\sqrt{2})/7$, assign r_i to the leftmost two berths with four QCs. Set $s_i = \max\{C_{i,1}, C_{i,2}\}$.

Case 2. $r_i = \Delta, r_{i+1} = \Delta$. For the case with $|T_a| \geq 3$, assign r_i to the two idle consecutive berths with three QCs, and reset T_a. Otherwise, assign r_i to the leftmost two berths with four QCs. Set $s_i = \max\{C_{i,1}, C_{i,2}\}$.

Case 3. $r_i = 1, r_{i+1} = 1$. When $T_a = [0, 0)$, assign r_i to the berth b_1 with two QCs. Set $s_i = C_{i,1}$. Otherwise, assign r_i to the idle berth, reset T_a.

Case 4. $r_i = 1, r_{i+1} = \Delta$. There are two subcases below:
- Case 4.1. When $(2 + 2\sqrt{10})/3 \leq \Delta \leq (6 + 12\sqrt{2})/7$, for the case with $T_a = [0, 0)$, assign r_i to the berth b_1 with one QC. Set $s_i = \max\{C_{i,1}, C_{i,2}, C_{i,3}\}$. For the other case, assign r_i to the idle berth, reset T_a.
- Case 4.2. when $2 \leq \Delta \leq (2 + 2\sqrt{10})/3$ or $\Delta \geq (6 + 12\sqrt{2})/7$, for the case with $T_a = [0, 0)$, assign r_i to the berth b_1 with two QCs, set $s_i = C_{i,1}$. Otherwise, assign r_i to the idle berth, reset T_a.

Theorem 2. *For problem hybr* $|online - over - list,\ LD = 1\ |BAP - QCAP|$ $C_{\max}$, *with 3 berths and 4 QCs, MLIST is* $(1 + \sqrt{2})/2$ *-competitive.*

Proof. Given any request input sequence $\mathcal{I} = \{r_1, r_2, ..., r_n\}$. Let $C_\sigma(\mathcal{I})$ and $C^*(\mathcal{I})$ be the makespan of a schedule produced by MLIST and by an optimal offline algorithm OPT. Based on the interval of Δ, we consider two cases.

Case 1. $(2 + 2\sqrt{10})/3 \leq \Delta \leq (6 + 12\sqrt{2})/7$, there are four subcases in the following.

Case 1.1. The last two requests are $r_{n-1} = \Delta,\ r_n = 1$.

Case 1.1.1. $T_a = [0, 0)$, we have $C^*(\mathcal{I}) = C_\sigma(\mathcal{I}) = C_{n-2}(I) + max\{\frac{\Delta}{3}, 1\}$.

Case 1.1.2. $|T_w| = k(\Delta/3 - 1)$ or $3k(1 - \Delta/3)$ $(k \geq 1)$, it indicates that the sequence contains pairs of $(r_i = \Delta, r_{i+1} = 1)$ or $(r_i = 1, r_{i+1} = \Delta)$ before r_{n-1}, and those requests consist of k large and k small requests. Assume that excluding those $\in\|$ requests, $C_{\max} = t$ before r_{n-1}, then $C_\sigma(\mathcal{I}) = t + (k+1)max\{\Delta/3, 1\}$, $C^*(\mathcal{I}) \geq t + (k+1)(\Delta + 1)/4$, $C_\sigma(\mathcal{I})/C^*(\mathcal{I}) \leq (8\sqrt{10} - 20)/5$.

Case 1.1.3. $|T_w| = k(\Delta/3 - 1)$ or $3k(1 - \Delta/3)$ $(k \geq 0)$ and $|T_a| = 3$, it indicates that the sequence contains pairs of $(r_i = \Delta, r_{i+1} = 1)$ or $(r_i = 1, r_{i+1} = \Delta)$ before r_{n-1}, those requests consist of k large and k small requests. And $r_{n-2} = 1$. Assume that excluding those 2k requests, $C_{\max} = t$ before r_{n-2}, then $C_\sigma(\mathcal{I}) = t + k \max\{\Delta/3, 1\} + \Delta/3 + 1/2$, $C^*(\mathcal{I}) \geq t + (k+1)(\Delta+1) + 1/4$, $C_\sigma(\mathcal{I})/C^*(\mathcal{I}) \leq (1 + \sqrt{2})/2$.

Case 1.2. The last two requests are $r_{n-1} = \Delta, r_n = \Delta$.

Case 1.2.1. $T_a = [0,0)$, we have $C^*(\mathcal{I}) = C_\sigma(\mathcal{I}) = C_{n-2}(\mathcal{I}) + \Delta/2$.

Case 1.2.2. $|T_w| = k(\Delta/3 - 1)$ or $3k(1 - \Delta/3)$ $(k \geq 1)$, it indicates that the sequence contains pairs of $(r_i = \Delta, r_{i+1} = 1)$ or $(r_i = 1, r_{i+1} = \Delta)$ before r_{n-1}, and those requests consist of k large and k small requests. Assume that excluding those 2k requests, $C_{max} = t$ before r_{n-1}, then $C_\sigma(\mathcal{I}) = t + k \max\{\Delta/3, 1\} + \Delta/2$, $C^*(\mathcal{I}) \geq t + [k(\Delta+1) + 2\Delta]/4$, $C_\sigma(\mathcal{I})/C^*(\mathcal{I}) \leq (8\sqrt{10} - 20)/5$.

Case 1.2.3. $|T_w| = k(\Delta/3 - 1)$ or $3k(1 - \Delta/3)$ $(k \geq 0)$ and $|T_a| = 3$, it indicates that the sequence contains pairs of $(r_i = \Delta, r_{i+1} = 1)$ or $(r_i = 1, r_{i+1} = \Delta)$ before r_{n-1}, those requests consist of k large and k small requests. And $r_{n-2} = 1$. Assume that excluding those 2k requests, $C_{max} = t$ before r_{n-2}, then $C_\sigma(\mathcal{I}) = t + (k+1) \max\{\Delta/3, 1\} + \Delta/4$, $C^*(\mathcal{I}) \geq t + [(k+1)(\Delta+1) + \Delta]/4$, $C_\sigma(\mathcal{I})/C^*(\mathcal{I}) \leq (8\sqrt{10} - 20)/5$.

Case 1.3. The last two requests are $r_{n-1} = 1, r_n = 1$.

Case 1.3.1. $T_a = [0,0)$, we have $C^*(\mathcal{I}) = C_\sigma(\mathcal{I}) = C_{n-2}(\mathcal{I}) + 1/2$.

Case 1.3.2. $|T_a| = 1$, it indicates that the list doesn't contain $(r_i = \Delta, r_{i+1} = 1)$ or $(r_i = 1, r_{i+1} = \Delta)$ sequence before r_{n-1}, then $C^*(\mathcal{I}) = C_\sigma(\mathcal{I}) = C_{n-2}(\mathcal{I}) + 1/2$, $C_\sigma(\mathcal{I})/C^*(\mathcal{I}) = 1$.

Case 1.3.3. $|T_w| = k(\Delta/3 - 1)$ or $3k(1 - \Delta/3)$ $(k \geq 1)$, it indicates that the sequence contains pairs of $(r_i = \Delta, r_{i+1} = 1)$ or $(r_i = 1, r_{i+1} = \Delta)$ before r_{n-1}, and those requests consist of k large and k small requests. Assume that excluding those 2k requests, $C_{max} = t$ before r_{n-1}, then $C_\sigma(\mathcal{I}) = t + 1/2 + k \max\{\Delta/3, 1\}$, $C^*(\mathcal{I}) \geq t + [k(\Delta+1) + 2]/4$, $C_\sigma(\mathcal{I})/C^*(\mathcal{I}) \leq (8\sqrt{10} - 20)/5$.

Case 1.3.4. $|T_a| = 1$ and $|T_w| = k(\Delta/3 - 1)$ or $3k(1 - \Delta/3)$ $(k \geq 1)$, it indicates that the sequence contains pairs of $(r_i = \Delta, r_{i+1} = 1)$ or $(r_i = 1, r_{i+1} = \Delta)$ before r_{n-1}, those requests consist of k large and k small requests. And $r_{n-2} = 1$. Assume that excluding those 2k requests, $C_{max} = t$ before r_{n-2}, then $C_\sigma(\mathcal{I}) \leq t + 1 + k \max\{\Delta/3, 1\}$, $C^*(\mathcal{I}) \geq t + [k(\Delta+1) + 3]/4$, $C_\sigma(\mathcal{I})/C^*(\mathcal{I}) \leq (28 - 4\sqrt{10})/13$.

Case 1.3.5. $|T_a| = \Delta/3$ and $|T_w| = k(\Delta/3 - 1)$ or $3k(1 - \Delta/3)$ $(k \geq 0)$, it indicates that the sequence contains pairs of $(r_i = \Delta, r_{i+1} = 1)$ or $(r_i = 1, r_{i+1} = \Delta)$ before r_{n-1}, those requests consist of k large and k small requests. And $r_{n-2} = \Delta$. Assume that excluding those 2k requests, $C_{max} = t$ before r_{n-2}, then $C_\sigma(\mathcal{I}) = t + 1/2 + k \max\{\Delta/3, 1\}$, $C^*(\mathcal{I}) \geq t + [(k+1)(\Delta+1) + 1]/4$, $C_\sigma(\mathcal{I})/C^*(\mathcal{I}) \leq (8\sqrt{10} - 20)/5$.

Case 1.4. The last two requests are $r_{n-1} = 1, r_n = \Delta$.

Case 1.4.1. $T_a = [0,0)$, we have $C^*(\mathcal{I}) = C_\sigma(\mathcal{I}) = C_{n-2}(\mathcal{I}) + max\{\frac{\Delta}{3}, 1\}$.

Case 1.4.2. $|T_a| = \Delta/3$, it indicates $r_{n-2} = \Delta$. Assume $C_{max} = t$ before r_{n-2}, then $C^*(\mathcal{I}) = C_\sigma(I) = t + \Delta/4 + \max\{\Delta/3, 1\}$, $C_\sigma(\mathcal{I})/C^*(\mathcal{I}) = 1$.

Case 1.4.3. $|T_w| = k(\Delta/3 - 1)$ or $3k(1 - \Delta/3)$ $(k \geq 1)$, it indicates that the sequence contains pairs of $(r_i = \Delta, r_{i+1} = 1)$ or $(r_i = 1, r_{i+1} = \Delta)$ before r_{n-1}, and those requests consist of k large and k small requests. Assume that excluding those $2k$ requests, $C_{\max} = t$ before r_{n-1}, then $C_\sigma(\mathcal{I}) = t + (k+1)\max\{\Delta/3, 1\}$, $C^*(\mathcal{I}) \geq t + (k+1)(\Delta+1)/4$, $C_\sigma(\mathcal{I})/C^*(\mathcal{I}) \leq (8\sqrt{10}-20)/5$.

Case 1.4.4. $|T_a| = \Delta/3$ and $|T_w| = k(\Delta/3-1)$ or $3k(1-\Delta/3)$ $(k \geq 1)$, it indicates that the sequence contains pairs of $(r_i = \Delta, r_{i+1} = 1)$ or $(r_i = 1, r_{i+1} = \Delta)$ before r_{n-1}, and those requests consist of k large and k small requests, and $r_{n-2} = \Delta$. Assume that excluding those $2k$ requests, $C_{\max} = t$ before r_{n-2}, then $C_\sigma(\mathcal{I}) = t + (k+1)\max\{\Delta/3, 1\} + \Delta/4$, $C^*(\mathcal{I}) \geq t + [(k+1)(\Delta+1) + \Delta]/4$, $C_\sigma(\mathcal{I})/C^*(\mathcal{I}) \leq (8\sqrt{10}-20)/5$.

Case 1.4.5. $|T_a| = 1$ and $|T_w| = k(\Delta/3-1)$ or $3k(1-\Delta/3)$ $(k \geq 0)$, it indicates that the sequence contains pairs of $(r_i = \Delta,\ r_{i+1} = 1)$ or $(r_i = 1,\ r_{i+1} = \Delta)$ before r_{n-1}, and those requests consist of k large and k small requests, and $r_{n-2} = 1$. Assume that excluding those $2k$ requests, the $C_{\max} = t$ before r_{n-2}, then $C_\sigma(\mathcal{I}) = t + 1/2 + k\max\{\Delta/3, 1\} + \Delta/4$, $C^*(\mathcal{I}) \geq t + [(k+1)(\Delta+1) + 1]/4$, $C_\sigma(\mathcal{I})/C^*(\mathcal{I}) \leq (8\sqrt{10}-20)/5$.

Case 2. Otherwise, $2 \leq \Delta \leq (2+2\sqrt{10})/3$ or $\Delta \geq (6+12\sqrt{2})/7$. In this condition, the algorithm is the same as that in Zheng *et al.*, we can refer to their proof(see Appendix B). With the limit of interval of Δ, we can get $C_\sigma(\mathcal{I})/C^*(\mathcal{I}) \leq (1+\sqrt{2})/2$.

The theorem follows.   □

4   The Case with Five QCs

In this section we focus on the case with 3 berths and 5 QCs in the problem $hybr\,|online - over - list,\ LD = 1\,|BAP - QCAP|\,C_{\max}$, analyze the lower bound and present an online algorithm named GTR. Similarly, before analyzing the lower bound and algorithm, we first introduce two definitions (Zheng *et al.*).

Definition 2. *Strict waste time segment* and *strict available time segment:* For an idle time segment $[e_k, s_i)$, with r_i is a large request, and request r_k is scheduled before the request r_i. If $0 < s_i - e_k < 1/2$, the idle time segment $[e_k, s_i)$ is a strict waste time segment, denoted by T_w^*. Otherwise, $[e_k, s_i)$ is a strict available time segment, denoted by T_a^*.

The segment of each large request r_i may induce a forced idle time segment, i.e., an idle time segment $[e_k, s_i)$ (for some $1 \leq k < i$, request r_k is scheduled before the request r_i, no matter which berths and QCs assigned)only for during $[e_k, s_i)$, there exists idle QCs,. If $0 < s_i - e_k < 1/2$, a small request later on cannot be satisfied within the time segment, thus it is a waste time segment. Otherwise, it is an available time segment. Set $T_a^* = [t_1, t_2)$ where $t_1 = e_k$, $t_2 = s_i$, and $T_a^* = [0, 0)$ initially.

4.1   Lower Bound

Theorem 3. *For problem* $hybr\,|online - over - list,\ LD = 1\,|BAP - QCAP|\ C_{\max}$, *any online scheduling algorithm has a competitive ratio of at least* $5/4$.

Proof. To prove the theorem, we construct a request input sequence $\mathcal{I}$. $\mathcal{I}$ contains at least two requests. Let $C_{max}(\mathcal{I})$ and $C^*(\mathcal{I})$ be the makespan of schedule produced by $\mathcal{A}$ and an optimal offline algorithm OPT, respectively.

The first request $r_1 = 1$, and lookahead $r_2 = 1$. If A processes r_1 with two QCs on berth b_1, offline adversary releases request $r_3 = \Delta$ and $r_4 = \Delta$, $C_{\max}(\mathcal{I}) \geq 1/2 + \Delta/2$. While OPT processes the request r_1 with one QC on berth b_1 at time 0, r_2 with one QC on berth b_1 after completing r_1, and r_3 with four QCs on berth b_2, b_3 at time 0, r_4 with four QCs on berth b_2, b_3 after completing r_3. $C^*(\mathcal{I}) = \max\{2, \Delta/2\}$. Setting $\Delta = 4$, we have $C_{\max}(\mathcal{I})/C^*(\mathcal{I}) \geq 5/4$. If A processes r_1 with one QC on berth b_1, no more requests arrive. $C_{\max}(\mathcal{I}) \geq 1$. While OPT processes r_1, r_2 with two QCs on different berths at time 0. $C^*(\mathcal{I}) = 1/2$. $C_{\max}(\mathcal{I})/C^*(\mathcal{I}) \geq 2$. The theorem follows. □

4.2   Algorithm GTR

On the release of any request r_i ($i \geq 1$), if $i = n$, then the scheduler can lookahead r_n is the last request of consequence $\mathcal{I}$, so we can optimal assign this request to minimize $C_{\max}$. Otherwise, $1 \leq i \leq n - 1$, $\mathcal{A}$ behaves as follows.

Case 1. $r_i = \Delta, r_{i+1} = 1$. There are four subcases in the following:
- Case 1.1. $t_{i,1} = t_{i,2} = t_{i,3}$. If $\Delta \geq 3$, assign r_i to the two leftmost berths with four QCs. If $3 \geq \Delta \geq 2$, assign r_i to the rightmost two berths with three QCs. Set $s_i = t_{i,1}$.
- Case 1.2. $t_{i,1} = t_{i,2} < t_{i,3}$. If $C_{i,1} = t_{i,1}$, assign r_i to the leftmost two berths with four QCs. If $C_{i,1} = t_{i,3}$, assign r_i to the rightmost two berths with three QCs. Set $s_i = t_{i,3}$.
- Case 1.3. $t_{i,1} < t_{i,2} = t_{i,3}$. If $\Delta \geq 3$, assign r_i to the leftmost two berths with four QCs. If $3 \geq \Delta \geq 2$, when $C_{i,1} = t_{i,1}$, assign r_i to the rightmost two berths with three QCs. When $C_{i,1} = t_{i,3}$, assign r_i to the leftmost two berths with four QCs. Set $s_i = t_{i,3}$.
- Case 1.4. $t_{i,1} < t_{i,2} < t_{i,3}$. Assign r_i to the leftmost two berths with four QCs, set $s_i = t_{i,3}$.

case 2. $r_i = \Delta, r_{i+1} = \Delta$. If $t_{i,1} = t_{i,2} < t_{i,3}$, and $C_{i,1} = t_{i,3}$, assign r_i to the rightmost two berths with three QCs. Otherwise, assign r_i to the leftmost two berths with four QCs. Set $s_i = t_{i,3}$.

case 3. $r_i = 1, r_{i+1} = 1$. If $T_a^* = [t_1, t_2) \neq [0, 0)$, set $s_i = t_2$ and assign r_i to idle berth. Otherwise if $T_a^* = [0, 0)$, there are the following four subcases.
- Case 3.1. $t_{i,1} = t_{i,2} = t_{i,3}$. Assign r_i to the leftmost berth with two QCs, set $s_i = t_{i,1}$.
- Case 3.2. $t_{i,1} = t_{i,2} < t_{i,3}$. Assign r_i to the middle berth with two QCs, set $s_i = t_{i,1}$.
- Case 3.3. $t_{i,1} < t_{i,2} = t_{i,3}$. When $C_{i,1} = t_{i,3}$, if $t_{i,2} - t_{i,1} < 1/2$, assign r_i to the leftmost berth with two QCs, set $s_i = t_{i,3}$. If $t_{i,2} - t_{i,1} \geq 1/2$, assign r_i to the rightmost berth with one QC, set $s_i = t_{i,1}$. When $C_{i,1} = t_{i,1}$, assign r_i to the leftmost berth with two QCs, set $s_i = t_{i,1}$.

- Case 3.4. $t_{i,1} < t_{i,2} < t_{i,3}$. Assign r_i to the middle berth with two QCs, set $s_i = t_{i,1}$.

case 4. $r_i = 1, r_{i+1} = \Delta$. If $T_a^* = [t_1, t_2) \neq [0,0)$, set $s_i = t_2$ and assign r_i to the idle berth. Otherwise if $T_a^* = [0,0)$, there are the following four subcases.

- Case 4.1. $t_{i,1} = t_{i,2} = t_{i,3}$. If $\Delta \geq 3$, assign r_i to the rightmost berths with one QC. If $3 \geq \Delta \geq 2$, assign r_i to the leftmost berth with two QCs. Set $s_i = t_{i,1}$.
- Case 4.2. $t_{i,1} = t_{i,2} < t_{i,3}$. When $\Delta \geq 3$, if $C_{i,1} = t_{i,3}$, $t_{i,2} - t_{i,1} = 1/2$ and $\Delta \geq 6$, assign r_i to the middle berth with two QCs. If $C_{i,1} = t_{i,1}$, assign r_i to the rightmost berth with one QC. Otherwise, assign r_i to the leftmost berth with two QCs. When $3 \geq \Delta \geq 2$, assign r_i to the middle berth with two QCs.
- Case 4.3. $t_{i,1} < t_{i,2} = t_{i,3}$. If $\Delta \geq 3$, assign r_i to the rightmost berths with one QC. If $3 \geq \Delta \geq 2$, when $C_{i,1} = t_{i,1}$, assign r_i to the leftmost berth with two QCs. When $C_{i,1} = t_{i,3}$, assign r_i to the rightmost berth with one QC. Set $s_i = t_{i,1}$.
- Case 4.4. $t_{i,1} < t_{i,2} < t_{i,3}$. Assign r_i to the middle berth with two QCs, set $s_i = t_{i,1}$.

Lemma 1. *When $\Delta \geq 3$, if there exists a T_w^* segment in schedule σ (the schedule produced by GTR), then the schedule σ should have the sequence $r_{i-2} = 1$, $r_{i-1} = 1$, $r_i = \Delta$, $r_j = \Delta$ $(j > i)$, with the condition $t_{i-2,1} = t_{i-2,2} = t_{i-2,3}$ and $\Delta \leq 6$. When $3 \geq \Delta \geq 2$, if there exists a T_w^* segment in schedule σ, then the schedule σ should have the sequence $r_{i-1} = 1$, $r_i = \Delta$ or $r_{i-1} = \Delta$, $r_i = 1$, $r_j = \Delta$ $(j > i)$, with the condition $t_{i-1,1} = t_{i-1,2} = t_{i-1,3}$.*

Proof. By the definition of T_w^* and the description of the algorithm, T_w^* exists because the assignment of at least two large requests, denoted by r_i, r_j $(j > i)$, and r_i is assigned to the rightmost berths with three QCs, r_j is assigned to the leftmost berths with four QCs. T_w^* is on b_1.

When $\Delta \geq 3$, by the description of the algorithm, only in case 1.2 $(C_{i,1} = t_{i,3})$ and 2.1, the large request will be assigned to the rightmost berths with three QCs. For the case 1.2 $(C_{i,1} = t_{i,3})$ and 2.1, $t_{i,1} = t_{i,2} < t_{i,3}$ and $C_{i,1} = t_{i,3}$, it occurs by $r_{i-2} = 1$ and $r_{i-1} = 1$, r_{i-2} assigned to the leftmost berth. For the case r_{i-2} assigned to berth b_1, should satisfy the condition that $t_{i-2,1} = t_{i-2,2} = t_{i-2,3}$ or $t_{i-2,1} < t_{i-2,2} = t_{i-2,3} = C_{i-2,1}$ with $t_{i-2,2} - t_{i-2,1} < 1/2$. For the case r_{i-1} assigned to berth b_1, should satisfy the condition that $C_{i-2,1} = t_{i-2,1} = t_{i-2,2} < t_{i-2,3}$, $t_{i-2,2} - t_{i-2,1} < 1/2$ and $\Delta \leq 6$. So we can find that if there exists a T_w^* segment, the schedule σ should have the sequence $r_{i-2} = 1$, $r_{i-1} = 1$, $r_i = \Delta$, $r_j = \Delta$ $(j > i)$, $t_{i-2,1} = t_{i-2,2} = t_{i-2,3}$ and $\Delta \leq 6$.

When $3 \geq \Delta \geq 2$, the same proof as $\Delta \geq 3$, we can find that if it exists a T_w^* segment, the schedule σ should have the sequence $r_{i-1} = 1$, $r_i = \Delta$ or $r_{i-1} = \Delta$, $r_i = 1$, $r_j = \Delta$ $(j > i)$ and $t_{i-1,1} = t_{i-1,2} = t_{i-1,3}$.

The lemma follows.   □

Lemma 2. *There is at most one T_a^* segment in schedule σ, and it is by the same condition as T_w^*.*

Proof. By the definition of T_a^*, it is easy to find T_a^* segment formed by the same condition as T_w^* segment. But different with T_w^* segment, $0 < s_i - e_k < 1/2$. In a T_a^* segment, $s_i - e_k \geq 1/2$. That means, once the schedule σ has a T_a^* segment, the after small requests will be assigned in the idle berths, until $0 < s_i - e_k < 1/2$. So there is at most one T_a^* segment in schedule σ. The lemma follows.   □

Theorem 4. *For problem* $hybr \,|online - over - list, \; LD = 1\,|BAP - QCAP|$ $C_{\max}$ *with 3 berths and 5 QCs, GTR is 5/4 -competitive.*

Proof. Given any request input sequence $\mathcal{I} = \{r_1, r_2, ..., r_n\}$. Let $C_\sigma(\mathcal{I})$ and $C^*(\mathcal{I})$ be the makespan of a schedule produced by GTR and by an optimal offline algorithm OPT. Based on the interval of Δ, we consider two cases .

Case 1. $\Delta \geq 3$, we consider four subcases by the existence of T_a^*, T_w^*.

Case 1.1. $T_w^* = \phi$, $T_a^* = [0, 0)$. We consider five subcases by the assignment of r_{n-1}, r_n.

Case 1.1.1. $t_{n-1,1} = t_{n-1,2} = t_{n-1,3}$. It is easy to prove $C_\sigma(\mathcal{I}) = C^*(\mathcal{I})$.

Case 1.1.2. $r_{n-1} = \Delta$, $r_n = 1$. When r_{n-1} is assigned by algorithm cases 1.2($C_{i,1} = t_{i,3}$), 1.3, 1.4, for $T_a^* = [0, 0)$, $T_w^* = \phi$, thus the left four QCs on leftmost two berths are kept busy during $[0, C_\sigma(\mathcal{I}))$, we have $C_\sigma(\mathcal{I}) \leq t_{n-1,3} + \max\{1, \Delta/4\}$, $C^*(\mathcal{I}) \geq t_{n-1,3} + (1 + \Delta)/5 - (t_{n-1,3} - t_{n-1,1})/5$, implying $C_\sigma(\mathcal{I})/C^*(\mathcal{I}) \leq 5/4$. When r_{n-1} is assigned by algorithm case 1.2 ($C_{i,1} = t_{i,1}$), implying $\Delta \leq 6$, $t_{n-1,3} - t_{n-1,1} = 1$ and $t_{n-1,3} \geq 1$, thus before assign r_{n-1}, $T_w^* = \phi$, and maybe exist one T_a^* segment with $t_2 - t_1 = 1/2$ or not. If before assign r_{n-1}, $T_a^* = [0, 0)$, then $C_\sigma(\mathcal{I}) = \max\{t_{n-1,1} + \Delta/3, t_{n-1,3} + 1/2\}$, $C^*(\mathcal{I}) \geq t_{n-1,1} + (1 + \Delta)/5 + (t_{n-1,3} - t_{n-1,1})/5$, implying $C_\sigma(\mathcal{I})/C^*(\mathcal{I}) \leq 5/4$. If before assign r_{n-1}, $T_a^* = [t_1, t_2)$ with $t_2 - t_1 = 1/2$, thus $C_\sigma(\mathcal{I}) = \max\{t_{n-1,1} + \Delta/3, t_{n-1,3}\}$, $C^*(\mathcal{I}) \geq t_{n-1,1} + \Delta/5 + (t_{n-1,3} - t_{n-1,1})/5$. Since there exist a T_a^* segment, then $t_{n-1,1} \geq \Delta/3$, we get $C_\sigma(\mathcal{I})/C^*(\mathcal{I}) \leq 20/17$.

Case 1.1.3. $r_{n-1} = \Delta$, $r_n = \Delta$. For $T_w^* = \phi$, $T_a^* = [0, 0)$ and thus the left QCs on leftmost two berths are kept busy during $[0, C_\sigma(\mathcal{I}))$, we have $C^*(\mathcal{I}) \geq 4C_\sigma(\mathcal{I})/5$.

Case 1.1.4. $r_{n-1} = 1$, $r_n = 1$. Before assign r_{n-1}, $T_w^* = \phi$, but there maybe exist a T_a^* segment with $t_2 - t_1 = 1$ or $t_2 - t_1 = 1/2$. If before assign r_{n-1}, $T_a^* = [0, 0)$, then we have $C_\sigma(I) \leq t_{n-1,3} + 1/2$ and $C^*(\mathcal{I}) \geq t_{n-1,3} + 2/5 - (t_{n-1,3} - t_{n-1,1})/5 \geq 4t_{n-1,3}/5 + 2/5$, so $C_\sigma(\mathcal{I})/C^*(\mathcal{I}) \leq 5/4$. If before assign r_{n-1}, $T_a^* = [t_1, t_2)$ and $t_2 - t_1 = 1/2$, then the last completed request before r_{n-1} is a large request. Notice that there is a waste of QC utility equal to 1 in T_a^*, thus $C^*(\mathcal{I}) \geq t_{n-1,3} + 2/5 - (t_{n-1,3} - t_{n-1,1})/5 - 1/5$. By the case condition, meaning $\Delta = 3(k + 1)/2 \geq 9/2$, $C_\sigma(\mathcal{I}) = t_{n-1,3}$, then $C_\sigma(\mathcal{I})/C^*(\mathcal{I}) < 5/4$. If before assign r_{n-1}, $T_a^* = [t_1, t_2)$ and $t_2 - t_1 = 1$. Notice that there is a waste of QC utility equal to 2 in T_a^*, so $C_\sigma(\mathcal{I}) = t_{n-1,3}$ and $C^*(\mathcal{I}) \geq t_{n-1,3} + 2/5 - (t_{n-1,3} - t_{n-1,1})/5 - 2/5$, then $C_\sigma(\mathcal{I})/C^*(\mathcal{I}) \leq 5/4$.

Case 1.1.5. $r_{n-1} = 1$, $r_n = \Delta$. For $T_w^* = \phi$, $T_a^* = [0, 0)$ and thus the left four QCs on leftmost two berths are kept busy during $[0, C_\sigma(\mathcal{I}))$, we have $C^*(\mathcal{I}) \geq 4C_\sigma(\mathcal{I})/5$.

Case 1.2. $T_a^* = [0,0)$ and $T_w^* = \phi$. Assume there are p T_w^* segments, for each $T_w^* = [t_1^*, t_2^*)$, $t_2^* - t_1^* = 1/2$, then $t_{n,1} \geq pt_2$. we have $C^*(\mathcal{I}) \geq C_\sigma(\mathcal{I}) - (C_\sigma(\mathcal{I}) - t_{n,1})/5 - p(t_2 - t_1)/5 \geq 4C_\sigma(\mathcal{I})/5$.

Case 1.3. $T_a^* = [t_1, t_2) \neq [0,0)$ and $T_w^* = \phi$. By the case condition, $t_{n,1} \geq t_2$, $C^*(\mathcal{I}) \geq C_\sigma(\mathcal{I}) - (C_\sigma(\mathcal{I}) - t_{n,1})/5 - (t_2 - t_1)/5 \geq 4C_\sigma(\mathcal{I})/5$.

Case 1.4. $T_a^* = [t_1, t_2) \neq [0,0)$ and $T_w^* = [t_1^*, t_2^*) \neq \phi$. By the lemma condition, implying $t_2 - t_1 = \Delta/3 - k/2 > 1/2$, $t_2^* - t_1^* = \Delta/3 - k^*/2 < 1/2$ and $t_{n,1} \geq t_2$. Assume there are p T_w^* segment, then $t_1 \geq p$. $C^*(\mathcal{I}) \geq C_\sigma(\mathcal{I}) - (C_\sigma(\mathcal{I}) - t_{n,1})/5 - (t_2 - t_1)/5 - p(t_2^* - t_1^*)/5 \geq 4C_\sigma(\mathcal{I})/5$.

Case 2. $3 > \Delta \geq 2$, we consider four cases by the existence of T_a^*, T_w^* .

Case 2.1. $T_w^* = \phi$, $T_a^* = [0,0)$, we consider four subcases by the assignment of r_{n-1}, r_n.

Case 2.1.1. $r_{n-1} = \Delta$, $r_n = 1$. There are two situations in this case, before assign r_{n-1}, $T_a^* = [0,0)$ or $T_a^* = [t_1, t_2)$, $t_2 - t_1 = 1/2$. If before assign r_{n-1}, $T_a^* = [0,0)$, then $C_\sigma(\mathcal{I}) \leq t_{n-1,3} + \Delta/3$, $C^*(\mathcal{I}) \geq t_{n-1,3} + (1+\Delta)/5 - (t_{n-1,3} - t_{n-1,1})/5 \geq 4t_{n-1,3}/5 + (1+\Delta)/5$, thus $C_\sigma(\mathcal{I})/C^*(\mathcal{I}) < 5/4$. If before assign r_{n-1}, $T_a^* = [t_1, t_2)$, $t_2 - t_1 = 1/2$. By the case condition, the last completed request in σ before r_{n-1}. So $C_\sigma(\mathcal{I}) \leq t_{n-1,3} + \Delta/4$, $C^*(\mathcal{I}) \geq t_{n-1,3} + (1+\Delta)/5 - (t_{n-1,3} - t_{n-1,1})/5 - 1/5 \geq 4t_{n-1,3}/5 + \Delta/5$. Thus $C_\sigma(\mathcal{I})/C^*(\mathcal{I}) < 5/4$.

Case 2.1.2. $r_{n-1} = \Delta$, $r_n = \Delta$. For $T_a^* = [0,0)$, $T_w^* = \phi$ and thus the left QCs on leftmost two berths are kept busy during $[0, C_\sigma(\mathcal{I}))$, we have $C^*(\mathcal{I}) \geq 4C_\sigma(\mathcal{I})/5$.

Case 2.1.3. $r_{n-1} = 1, r_n = 1$. Before assign r_{n-1}, $T_w^* = \phi$, and there may exist a T_a^* segment with $t_2 - t_1 = 1$ or $t_2 - t_1 = 1/2$. If before assign r_{n-1}, $T_a^* = [0,0)$, then we have $C_\sigma(\mathcal{I}) \leq t_{n-1,3} + 1/2$, $C^*(\mathcal{I}) \geq t_{n-1,3} + 2/5 - (t_{n-1,3} - t_{n-1,1})/5 \geq 4t_{n-1,3}/5 + 2/5 = 4C_\sigma(\mathcal{I})/5$. If before assign r_{n-1}, $T_a^* = [t_1, t_2)$ and $t_2 - t_1 = 1/2$, then the last completed request before r_{n-1} is a large request. Notice that there is a waste of QC utility equal to 1 in T_a^*, thus $C_\sigma(\mathcal{I}) = \max\{t_{n-1,1} + 1, t_{n-1,3}\}$, $C^*(\mathcal{I}) \geq t_{n-1,3} + 2/5 - (t_{n-1,3} - t_{n-1,1})/5 - 1/5$. When $t_{n-1,3} \leq t_{n-1,1} + 1$, by the lemma condition, implying $t_{n-1,3} - t_{n-1,1} = \Delta/4$, $t_{n-1,1} \geq k\Delta/3$, $k(\Delta/3 - 1/2) = p/2$ and $p \geq 1$. We can get $C_\sigma(\mathcal{I})/C^*(\mathcal{I}) \leq 8/7$. When $t_{n-1,3} > t_{n-1,1} + 1$, $C^*(\mathcal{I}) \geq t_{n-1,3} + 2/5 - (t_{n-1,3} - t_{n-1,1})/5 - 1/5 \geq 4C_\sigma(\mathcal{I})/5$. If before assign r_{n-1}, $T_a^* = [t_1, t_2)$ and $t_2 - t_1 = 1$. Notice that there is a waste of QC utility equal to 2 in T_a, so $C_\sigma(\mathcal{I}) = t_{n-1,3}$ and $C^*(\mathcal{I}) \geq t_{n-1,3} + 2/5 - (t_{n-1,3} - t_{n-1,1})/5 - 1/5 \geq 4C_\sigma(\mathcal{I})/5$.

Case 2.1.4. $r_{n-1} = 1$, $r_n = \Delta$. For $T_a^* = [0,0)$, $T_w^* = \phi$ and thus the left QCs on leftmost two berths are kept busy during $[0, C_\sigma(\mathcal{I}))$, we have $C^*(\mathcal{I}) \geq 4C_\sigma(\mathcal{I})/5$.

Cases 2.2, 2.3, 2.4 are the same as cases 1.2, 1.3, 1.4 respectively.

The theorem follows.   □

5   Conclusion

This paper considers an online integrated allocation of berths and quay cranes in container terminal with 1-lookahead. We focus on the hybrid layout with three berths and four, five cranes, and present an online deterministic algorithm respectively, which are proved to be optimal in competitiveness. In the future, we will focus on the case of six cranes, and consider k-lookahead case.

Acknowledgments. The authors would like to acknowledge the financial support of Grants (No.61221063.) from NSF of China and (No.IRT1173) from PCSIRT of China.

Appendix

A: Lower bound of the case with four QCs

we show the complete calculation process as follow.

If A processes r_1 with two QCs on berth b_1, we get:

$$C_{max}(\mathcal{I}) \geq 1/2 + \Delta/4 \tag{1}$$

$$\Delta \geq 6, \ C^*(\mathcal{I}) = 1/2 + \Delta/4 \tag{2}$$

$$6 \geq \Delta \geq 3, \ C^*(\mathcal{I}) = \Delta/3 \tag{3}$$

$$3 \geq \Delta \geq 2, \ C^*(\mathcal{I}) = 1 \tag{4}$$

If $\mathcal{A}$ processes r_1 with three QCs on berth b_1, b_2, we get:

$$\Delta \geq 6, \ C_{max}(\mathcal{I}) \geq \Delta/3 \tag{5}$$

$$6 \geq \Delta \geq 9/2, \ C_{max}(\mathcal{I}) \geq 2 \tag{6}$$

$$9/2 \geq \Delta \geq 2, \ C_{max}(\mathcal{I}) \geq 1/2 + \Delta/3 \tag{7}$$

$$C^*(\mathcal{I}) = 1/2 + \Delta/4 \tag{8}$$

If $\mathcal{A}$ processes r_1 with two QCs on berth b_1, b_2, we get:

$$C_{max}(\mathcal{I}) \geq \Delta/2 \tag{9}$$

$$\Delta \geq 6, \ C^*(\mathcal{I}) = 1/2 + \Delta/4 \tag{10}$$

$$6 \geq \Delta \geq 3, \ C^*(\mathcal{I}) = \Delta/3 \tag{11}$$

$$3 \geq \Delta \geq 2, \ C^*(\mathcal{I}) = 1 \tag{12}$$

After processed Eqs.(1-4),we have:

$$\Delta \geq 6, \ \rho \geq 1 \tag{13}$$

$$6 \geq \Delta \geq 3, \ \rho \geq \max\left\{\frac{6 + 3\Delta}{4\Delta}\right\} \tag{14}$$

$$3 \geq \Delta \geq 2, \ \rho \geq \max\left\{\frac{2 + \Delta}{4}\right\} \tag{15}$$

After processed Eqs.(5-8),we have:

$$\Delta \geq 6, \ \rho \geq \max\left\{\frac{4\Delta}{6+3\Delta}\right\} \geq 1 \tag{16}$$

$$6 \geq \Delta \geq 9/2, \ \rho \geq \max\left\{\frac{8}{2+\Delta}\right\} \tag{17}$$

$$9/2 \geq \Delta \geq 2, \ \rho \geq \max\left\{\frac{6+4\Delta}{6+3\Delta}\right\} \tag{18}$$

After processed Eqs.(9-12),we have:

$$\Delta \geq 6, \ \rho \geq \max\left\{\frac{2\Delta}{2+\Delta}\right\} \geq 1 \tag{19}$$

$$6 \geq \Delta \geq 3, \ \rho \geq \frac{3}{2} \tag{20}$$

$$3 \geq \Delta \geq 2, \ \rho \geq \frac{\Delta}{2} \tag{21}$$

From Eqs.(13-21),we get:

$$\Delta \geq 6, \ \rho \geq 1 \tag{22}$$

$$6 > \Delta \geq \frac{9}{2}, \ \rho \geq \max\left\{\frac{3\Delta+6}{4\Delta}\right\}, \ \rho \geq \frac{13}{12} \tag{23}$$

$$\frac{9}{2} > \Delta \geq \frac{6+12\sqrt{2}}{7}, \ \rho \geq \max\left\{\frac{3\Delta+6}{4\Delta}\right\}, \ \rho \geq \frac{1+\sqrt{2}}{2} \tag{24}$$

$$\frac{6+12\sqrt{2}}{7} > \Delta \geq 3, \ \rho \geq \max\left\{\frac{4\Delta+6}{3\Delta+6}\right\}, \ \rho \geq \frac{1+\sqrt{2}}{2} \tag{25}$$

$$3 > \Delta \geq \frac{2+2\sqrt{10}}{3}, \ \rho \geq \max\left\{\frac{4\Delta+6}{3\Delta+6}\right\}, \ \rho \geq \frac{6}{5} \tag{26}$$

$$\frac{2+2\sqrt{10}}{3} > \Delta \geq 2, \ \rho \geq \frac{4+\sqrt{10}}{6} \tag{27}$$

Setting $\Delta = (6+12\sqrt{2})/7$, then $\rho \geq (1+\sqrt{2})/2$.

B: Theorem 2

As in case 2, namely, $2 \leq \Delta \leq (2+2\sqrt{10})/3$ or $\Delta \geq (6+12\sqrt{2})/7$, the algorithm is the same as that in Zheng *et al.*, we can refer to their proof. For the sake of completeness, we present their proof here:

Proof. Given any request input sequence $\mathcal{I} = \{r_1, r_2, ..., r_n\}$. let σ be the schedule produced by LIST. Let $C_\sigma(\mathcal{I})$ be the makespan of σ, and $C^*(\mathcal{I})$ that of a schedule produced by an optimal onine algorithm OPT. We consider two cases below.

Case 1. The last request is a large one, i.e., $r_n = \Delta$. For the case with $T_a^* = [0,0)$, we have $C^*(\mathcal{I}) = C_\sigma(\mathcal{I}) = C_{n,1} + \Delta/4$ since neither LIST nor OPT makes any waste of QC utility within $[0, C_\sigma(\mathcal{I}))$.

For the other case with $T_a^* = [t_1, t_2) \neq [0,0)$, let $k \geq 1$ be the number of large requests after time t_2. $C_\sigma(\mathcal{I}) = t_2 + k\Delta/4$. If $t_2 = 1/2$ and $k = 1$, implying $n = 2$, $r_1 = 1$ and $r_2 = \Delta$, then $C^*(\mathcal{I}) \geq \min\{1/2 + \Delta/4, \max\{1, \Delta/3\}\}$ and thus $C_\sigma(\mathcal{I})/C^*(\mathcal{I}) \leq 5/4$; otherwise if either $t_2 \geq 1$ or $k \geq 2$, $C^*(\mathcal{I}) \geq t_2 + k\Delta/4 - 1/4$ and $C_\sigma(\mathcal{I})/C^*(\mathcal{I}) \leq 6/5$.

Case 2. The last request is a small request, i.e., $r_n = 1$. We claim that $T_a^* = [0,0)$ in σ since otherwise r_n would have been processed in the T_a^*. Divide this case into two subcases by whether $C_{n,1} = C_{n,2}$.

Case 2.1. $C_{n,1} = C_{n,2}$, In this case $C_\sigma(\mathcal{I}) = C_{n,1} + 1/2$. If either $n = 1$ or $n = 3$ with $r_1 = r_2 = r_3 = 1$, we have $C_{n,1} \leq 1/2$ and $C_\sigma(\mathcal{I}) = C^*(\mathcal{I})$; if $n = 2$ and $r_1 = \Delta$, then $C_{n,1} = \Delta/4$ and $C^*(\mathcal{I}) \geq \min\{1/2 + \Delta/4, \max\{1, \Delta/3\}\}$, implying $C_\sigma(\mathcal{I})/C^*(\mathcal{I}) \leq 5/4$; otherwise if $C_{n,1} \geq 1$, then $C^*(\mathcal{I}) \geq C_{n,1} + 1/4$, implying a ratio of at most $6/5$.

Case 2.2. $C_{n,1} < C_{n,2}$ (or $C_{n,2} < C_{n,1}$). Then we have by previous analysis that $C_{n,2} = C_{n,1} + 1/2$ (or $C_{n,1} = C_{n,2} + 1/2$), and thus $C_\sigma(\mathcal{I}) = C^*(\mathcal{I})$.

The theorem follows. $\qquad\qquad\square$

References

1. Bierwirth, C., Meisel, F.: A survey of berth allocation and quay crane scheduling problems in container terminals. European Journal of Operational Research **202**(3), 615–627 (2010)
2. Blazewicz, J., Cheng, T.C.E., Machowiak, M., Oguz, C.: Berth and quay crane allocation: a moldable task scheduling model. Journal of The Operational Research Society **62**, 1189–1197 (2011)
3. Borodin, A., El-Yaniv, R.: Online Computation and Competitive Analysis. Cambridge University Press, New York (1998)
4. Carlo, H., Vis, I., Roodbergen, K.: Seaside operations in container terminals: literature overview, trends, and research directions. Flexible Services and Manufacturing Journal, 1–39 (2013)
5. Chen, J.H., Lee, D.H., Cao, J.X.: A combinatorial benders cuts algorithm for the quayside operation problem at container terminals. Transportation Research Part E: Logistics and Transportation Review **48**(1), 266–275 (2012), select Papers from the 19th International Symposium on Transportation and Traffic Theory
6. Giallombardo, G., Moccia, L., Salani, M., Vacca, I.: Modeling and solving the tactical berth allocation problem. Transportation Research Part B: Methodological **44**(2), 232–245 (2010)
7. Imai, A., Sun, X., Nishimura, E., Papadimitriou, S.: Berth allocation in a container port: using a continuous location space approach. Transportation Research Part B: Methodological **39**(3), 199–221 (2005)
8. Liang, C., Guo, J., Yang, Y.: Multi-objective hybrid genetic algorithm for quay crane dynamic assignment in berth allocation planning. Journal of Intelligent Manufacturing **22**(3), 471–479 (2011)

9. Lokuge, P., Alahakoon, D.: Improving the adaptability in automated vessel scheduling in container ports using intelligent software agents. European Journal of Operational Research **177**(3), 1985–2015 (2007)
10. Mandelbaum, M., Shabtay, D.: Scheduling unit length jobs on parallel machines with lookahead information. Journal of Scheduling **14**(4), 335–350 (2011)
11. Park, Y.M., Kim, K.H.: A scheduling method for berth and quay cranes. OR Spectrum **25**(1), 1–23 (2003)
12. Zhang, L., Khammuang, K., Wirth, A.: On-line scheduling with non-crossing constraints. Operations Research Letters **36**(5), 579–583 (2008)
13. Zhen, L., Chew, E.P., Lee, L.H.: An integrated model for berth template and yard template planning in transshipment hubs. Transportation Science **45**(4), 483–504 (2011)
14. Zheng, F., Cheng, Y., Liu, M., Xu, Y.: Online interval scheduling on a single machine with finite lookahead. Computers & Operations Research **40**(1), 180–191 (2013)
15. Zheng, F., Qiao, L., Liu, M., Chu, C.: Online integrated allocation for small numbers of berths and quay cranes in container terminals (working paper)

Disjoint Path Allocation with Sublinear Advice

Heidi Gebauer[1], Dennis Komm[2], Rastislav Královič[3], Richard Královič[4], and Jasmin Smula[2]([✉])

[1] Institute of Applied Mathematics and Physics, Zurich University of Applied Sciences, Winterthur, Switzerland
geba@zhaw.ch
[2] Department of Computer Science, ETH Zürich, Zürich, Switzerland
{dennis.komm,jasmin.smula}@inf.ethz.c
[3] Department of Computer Science, Comenius University, Bratislava, Slovakia
kralovic@dcs.fmph.uniba.sk
[4] Google Inc., Zürich, Switzerland

Abstract. We study the disjoint path allocation problem. In this setting, a path P of length L is given, and a sequence of subpaths of P arrives online, one in every time step. Each such path requests a permanent connection between its two end-vertices. An online algorithm can admit or reject such a request; in the former case, none of the involved edges can be part of any other connection. We investigate how much additional binary information (called "advice") can help to obtain a good solution. It is known that, with roughly $\log_2 \log_2 L$ advice bits, it can be guaranteed that a $\log_2 L$-competitive solution is computed. In this paper, we prove the surprising result that, with $L^{1-\varepsilon}$ advice bits, it is not possible to obtain a solution with a competitive ratio better than $(\delta \log_2 L)/2$, where $0 < \delta < \varepsilon < 1$. This shows an interesting threshold behavior of the problem. A fairly good competitive ratio, namely $\log_2 L$, can be obtained with very few advice bits. However, any increase of the advice does not help any further until an almost linear number of advice bits is supplied. Then again, it is also known that linear advice allows for optimality.

1 Introduction

The input of an *online problem* arrives piecewise as a sequence of n requests $x_1, \ldots, x_n$ in consecutive time steps. An *online algorithm* ALG computes a sequence of answers $y_1, \ldots, y_n$, where each answer y_i, $1 \le i \le n$, must be given in the ith time step while only depending on the requests $x_1, \ldots, x_i$ that are known up to this point. If the given online problem is a maximization problem, we assess the solution quality of ALG by comparing the *gain* of its solution to the one hypothetically reachable if the whole input sequence were known in advance; this is modeled by an offline algorithm OPT that has this knowledge. More formally, ALG is called *c-competitive* if there is a non-negative constant (with respect to the input length) α such that, on any instance $I = (x_1, \ldots, x_n)$ of the given

Partially funded by SNF grant 200021-141089 and VEGA grant 1/0979/12.

© Springer International Publishing Switzerland 2015
D. Xu et al. (Eds.): COCOON 2015, LNCS 9198, pp. 417–429, 2015.
DOI: 10.1007/978-3-319-21398-9_33

online maximization problem, we have $\mathrm{gain}(\mathrm{Opt}(I)) \leq c \cdot \mathrm{gain}(\mathrm{Alg}(I)) + \alpha$; the smallest c for which this holds is called the *competitive ratio* of Alg. If the above inequality even holds with $\alpha = 0$, we call Alg *strictly c-competitive*. This framework, called *competitive analysis*, has been around for three decades now [17] and has become the standard tool to investigate the performance of online algorithms. For a detailed introduction to online algorithms and competitive analysis, we refer the reader to the literature [6]. Similar to the approximation ratio that measures what is lost when computing a solution to a hard offline problem in polynomial time, the competitive ratio tells us what is lost due to incomplete knowledge of the input at hand. However, since a complete absence of knowledge about the input is unrealistic in many real-world environments, it is reasonable to ask to what extent some certain additional information about the yet unrevealed requests can be exploited. While there are many models such as *semi-online* problems [10] where some specific parameters of the input are known, the *advice complexity* of an online problem tries a more general approach. Here, we augment an online algorithm with an *advice tape* that may contain any binary information about the input. We may think of the advice as being prepared by an oracle that sees the whole input in advance. An online algorithm Alg with such an additional resource is called an *online algorithm with advice*. Alg is called *c-competitive with advice complexity b* if, for every input I of the given problem, there is some advice string ϕ such that Alg has a competitive ratio of at most c while never accessing more than the first b bits of ϕ. Note that we assume that the advice string is infinitely long [4,13]. This way, Alg cannot determine itself when the advice "ends," which may carry some additional information. Consequently, online algorithms with advice generalize many other approaches that assume additional information. Here, lower bounds are of particular interest, i. e., statements of the sort that some specific competitive ratio can never be reached with some given amount of additional information, no matter what this information actually is.

In this paper, we continue the study of the *disjoint path allocation problem on paths*, DPA for short. Here, we are given a path P of length L, i. e., with $L + 1$ vertices. A request is equal to a non-empty subpath of L. Alg must answer any such request by either admitting or rejecting it; this decision is final. If the request is admitted, a permanent connection between the two end-vertices of the subpath is established. After that, all involved edges are busy and cannot be part of any other connection. Therefore, a feasible solution corresponds to a set of *edge-disjoint paths*, namely the admitted subpaths of L. We call a request *blocked* if it cannot be admitted as the consequence of an earlier admission; an example is shown in Fig. 1. Note that, if L is known in advance (as a parameter of the problem), we can easily set the constant α of the definition of the competitive ratio to $L - 1$, which implies that every online algorithm that admits at least one request (e. g., a simple greedy algorithm) is indeed 1-competitive; this is, of course, undesirable for a serious analysis. In this paper, we therefore assume that L is given with the first request, and is thus a part of the input.

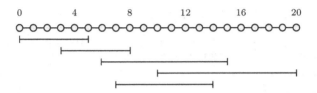

Fig. 1. An example of a DPA instance for $L = 20$. The requests arrive from top to bottom. If, e. g., an online algorithm admits the first one in time step 1, it cannot admit the second one. It is easy to see that no algorithm can admit more than 2 requests.

The advice complexity is usually a function b of the input length n. However, since the competitive ratio of algorithms for DPA is commonly a function of the graph size and we consider paths of length L only, we define b as a function of L.

We now discuss known results, describe our contribution, and put it into context. Due to space constraints, some of the technical details are omitted.

1.1   Related Work

A first model of online computation with advice was given by Dobrev et al. [9]. The model used in this paper was introduced by Hromkovič et al. [13] and first applied by Böckenhauer et al. [4]. There is an alternative model of computing with advice introduced by Emek et al. [11]. In this setting, the advice is not read from a tape, but it is supplied in every time step, and the number of advice bits is the same in every time step. Both models have so far been used to study a large number of problems, including the paging problem [4,14], the k-server problem [3,11,12,16], or metrical task systems [11]. One of the first online problems studied in the model of computing with advice as we use it in this paper was the DPA problem [4]. However, the authors of [4] mainly studied both the advice complexity and the reachable competitive ratio with respect to the input length n. The authors gave a lower bound of roughly $n/(2c)$ advice bits to obtain a strictly c-competitive solution. In this paper, we use the length L of the underlying path as a measurement, which is more consistent with the classical work [6]. With respect to L, the bound presented by Böckenhauer et al. [4] translates to roughly $(\log_2 L)/c$, which is improved exponentially by our main result. Böckenhauer et al. [5] noted that there is a $\log_2 L$-competitive online algorithm with advice that uses $\lceil \log_2 \lceil \log_2 L \rceil \rceil$ advice bits; this is a direct consequence from the "classify-and-randomly-select" algorithm from Awerbuch et al. [1]. Barhum et al. [2] generalized this technique and combined it with advice yielding online algorithms with advice that use a small amount of advice bits and obtain a solution of high quality. Moreover, they showed that $L - 1$ advice bits are both sufficient and necessary to be optimal. So far, no lower bound on the competitive ratio (except for optimality) that is achievable when reading $\omega(\log_2 L)$ advice bits is known.

1.2 Outline, Techniques, and Results

The remainder of this paper is devoted to giving non-trivial lower bounds on the number of advice bits necessary to obtain a certain competitive ratio. The result implies two interesting bounds.

1. For any c smaller than $1/2 \cdot (\log_2 L)/(\log_2 \log_2 L)^{1/4}$, any c-competitive online algorithm with advice needs to read at least $\Omega\big(L/(4^c \, c^4)\big)$ advice bits. Note that this bound is more general than the one presented by Barhum et al. [2], which only gives a statement for $b \le \log_2 \log_2(L/2)$.
2. For any δ, $0 < \delta < 1$, any $(\delta/2 \cdot \log_2 L)$-competitive online algorithm with advice needs to read at least $\omega(L^{1-\varepsilon})$ advice bits, for any constant ε with $\delta < \varepsilon < 1$. This complements the upper bound of $\log_2 L$ using $\lceil \log_2 \lceil \log_2 L \rceil \rceil$ advice bits [4]. The result is particularly surprising as it shows that we need almost double exponentially more advice to be $(\delta/2 \cdot \log_2 L)$-competitive instead of $\log_2 L$-competitive. Indeed, the number of advice bits necessary is almost linear. Then again, with a linear number of advice bits, it is possible to compute an optimal solution [2].

We prove the result by constructing a set $\mathcal{I}$ of instances such that any deterministic online algorithm can achieve the competitive ratio c only on a small fraction of the instances from $\mathcal{I}$. The number of these instances is bounded by some probabilistic arguments. All these instances can be organized as the leaves of a tree, such that paths from the root to some inner vertex v correspond to the instances that are leaves of the subtree rooted at v. Then we show that any online algorithm with advice needs to read many advice bits to achieve a competitive ratio of at most c on all instances from $\mathcal{I}$.

2 The Main Result

We start with some technical preliminaries that we need for the analysis of the given online algorithm with advice. For our calculations, we need Bernoulli's inequality, which states the following [7].

Fact 1. For every $x \in \mathbb{R}^{\ge -1}$ and every $n \in \mathbb{N}^{\ge 0}$, we have $(1+x)^n \ge 1+nx$. $\square$

The following argumentation involves a random variable with hypergeometric distribution. Therefore, we now establish a result that follows from a well-known bound for the tail of the hypergeometric distribution. First, let us recall that a random variable with hypergeometric distribution with parameters M, N, and n counts the number of black balls drawn from an urn containing N balls, out of which exactly M are black, when drawing n balls uniformly at random without replacement (see, e. g., [15]). The following bound was established by Chvátal [8].

Fact 2. Consider a discrete random variable X with hypergeometric distribution with parameters M, N, and n, i. e.,

$$\Pr(X = i) = \binom{M}{i}\binom{N-M}{n-i}\Big/\binom{N}{n}.$$

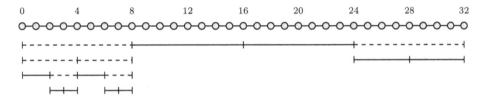

Fig. 2. An example of an instance from the constructed instance set $\mathcal{I}$ for a path of length $L = 32$ and $h = 3$. There are four phases with $L/2^h = 4$ requests each. Open requests are depicted by dashed lines. All other requests are closed (and therefore belong to the optimal solution).

Then, with $\mathrm{e} = 2.71828\ldots$ being Euler's number, we have

$$\mathbb{E}(X) = n \cdot M/N \quad \text{and} \quad P(X \le \mathbb{E}(X) - tn) \le \mathrm{e}^{-2t^2 n}, \quad \text{for any } t \ge 0. \qquad \Box$$

We have to adapt this result slightly for our purposes.

Corollary 1. *Let X be a discrete random variable with hypergeometric distribution with parameters M, N, and n, and let $t \ge 0$. Then, for every $M' \le M$, we have*

$$\Pr\left(X \le n \cdot \frac{M'}{N} - tn\right) \le \mathrm{e}^{-2t^2 n}.$$

Now let us describe how to construct the set $\mathcal{I}$ of instances for DPA. For the sake of simplicity, let L be a power of 2. Furthermore, let $h := h(L)$ be a parameter depending on L with

$$h \in \mathbb{N}^{\ge 1} \quad \text{and} \quad h \le \log_2 L - 1. \tag{1}$$

Then the requests are presented to the algorithm in $h + 1$ phases. In each phase i, with $1 \le i \le h+1$, the algorithm is given $L/2^h$ edge-disjoint requests of length 2^{h-i+1}. Hence, in the first phase, $L/2^h$ edge-disjoint subpaths of length 2^h are presented, whose concatenation forms the complete path P. Half of the requests from phase i, with $1 \le i \le h$, are so-called *closed* requests, for which no intersecting requests will be presented anymore, and which hence belong to the optimal solution computed by an optimal algorithm OPT. The other half of these requests are *open*, i.e., they are split into two edge-disjoint requests of length 2^{h-i} each, which are then presented in phase $i+1$. Finally, in phase $h+1$, the algorithm is given $L/2^h$ subpaths of length 1 each, which all belong to the optimal solution; for an example, see Figs. 2 and 3.

Observation 1. The optimal solution on any instance I from $\mathcal{I}$ has a gain of

$$\text{gain}(\text{OPT}(I)) = \frac{(h+2)L}{2^{h+1}}.$$

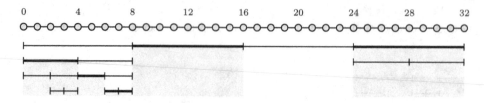

Fig. 3. When given the instance from Fig. 2 as an input, a deterministic algorithm ALG might admit the requests as depicted in this picture. Admitted requests are marked by thick lines. By admitting a request, all requests that intersect with this request become blocked, which is depicted by areas shaded in gray.

Proof. For every phase i with $1 \le i \le h$, there are $L/2^{h+1}$ requests that belong to the optimal solution, and additionally, there are $L/2^h$ ones from phase $h+1$, which yields

$$h \cdot \frac{L}{2^{h+1}} + \frac{L}{2^h} = \frac{hL}{2^{h+1}} + \frac{2L}{2^{h+1}} = \frac{(h+2)L}{2^{h+1}} . \qquad \square$$

Let us introduce another parameter $f := f(c)$, such that $f > 0$. Both parameters f and h must be chosen according to the competitive ratio that an algorithm is supposed to achieve. In the remainder of this chapter, we prove the following general theorem.

Theorem 1. *Any online algorithm for DPA with a competitive ratio of*

$$c := \frac{h+2}{2 \cdot \left(1 + \frac{h}{f}\right)}$$

needs to read at least $b := L/(2^h f^2) \cdot \log_2 e - \log_2 h$ *advice bits.*

After that, we choose concrete values for f and h to obtain more tangible lower bounds, which are formulated as corollaries at the end of this section.

We start our argumentation by making the following observation. The set $\mathcal{I}$ of instances can naturally be represented by a $\binom{L/2^h}{L/2^{h+1}}$-ary tree of depth h (i. e., with $h+1$ levels), as depicted in Fig. 4. The root is on level 0 and corresponds to all instances from $\mathcal{I}$. There are $\binom{L/2^h}{L/2^{h+1}}$ instances on level 1, each of them representing all instances with the same particular set of open requests from phase 1. Any vertex on level i represents the set of all instances with the same particular sets of open requests from phases $1, \dots, i$. Hence, for all instances that are represented by the same vertex on level i, the requests presented in the first $i+1$ phases are exactly the same. Every leaf is located on level h and corresponds to a single instance from $\mathcal{I}$.

Now consider some vertex v on level i with $0 \le i \le h$ and some arbitrary instance I_v represented by v. Then, any deterministic algorithm ALG, given I_v as its input, is always in the same state at the beginning of phase $i+1$, independently

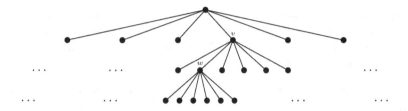

Fig. 4. An example of an instance tree. Figs. 2 and 3 are both placed in a scenario with a path of length 32 and $h + 1 = 4$ phases (hence, with $L/2^h = 4$ requests presented per phase). In this scenario, there are 6 possibilities to choose $L/2^{h+1} = 2$ out of the 4 requests to be open in every phase. Hence, every inner vertex of the corresponding instance tree has exactly 6 children (most of which are only indicated by dots due to space restrictions). The root represents all instances, the vertex v represents all instances in which the same set of requests from phase 1 are open, and each leaf on level h represents all instances in which the same set of requests from phases $1, \ldots, h$ are open, hence, each leaf represents a single instance.

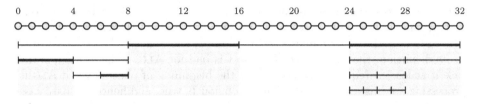

Fig. 5. In the instance tree from Fig. 4, on every level there must be one vertex that contains the instance from Fig. 3. Without loss of generality, on level 1, let this vertex be v. Then, v also contains the instance depicted in this figure. In both instances, the same requests from phase 1 are open, and thus, the requests presented in the first two phases are the same for *all* instances represented by v. Hence, at the beginning of and also throughout phase 2, each fixed deterministic algorithm ALG must be in the same state given any instance corresponding to v as its input, having seen and admitted the exact same requests so far.

of the instance that it gets as its input, i. e., it has seen and admitted the same requests so far on every instance represented by v; see Fig. 5.

From now on, let ALG be an arbitrary, but fixed deterministic algorithm for DPA. For a given vertex v on level i, let $\gamma^{(i)}$ be the gain of ALG on any instance represented by v during phase i, hence, the number of admitted requests during this phase. Moreover, let $\hat{\gamma}^{(i)}$ be the gain of ALG during all phases up to and including phase i, hence, $\hat{\gamma}^{(i)} := \sum_{j=1}^{i} \gamma^{(j)}$.

Let us introduce the following notion of bad phases and vertices. We call a phase i a *bad* phase for ALG if, at the beginning of this phase, at least

$$d_{i-1} := \hat{\gamma}^{(i-1)} - (i-1) \cdot \frac{L}{2^h \cdot f} \tag{2}$$

requests from phase i are already blocked. Furthermore, let us call a vertex v on level $i - 1$ *bad* for ALG if, when ALG is given any instance corresponding to v as its input, phase i is bad for ALG. Phases and vertices that are not bad are called *good*. Moreover, let us define the set of requests from phase i that are blocked at the beginning of phase $i + 1$ (including those that were admitted in phase i, which are blocking themselves) to be R_i.

Lemma 1. *If at least $d_i/2$ requests from R_i are open, then phase $i + 1$ is bad for ALG.*

Proof. Underneath each open request from phase i, two requests appear in phase $i + 1$. If $d_i/2$ requests from R_i are open, then at least d_i requests are presented in phase $i+1$ that are already blocked at the beginning of phase $i+1$. This matches the definition of a bad phase.   □

Lemma 2. *The fraction of bad vertices on level i is at least*

$$\left(1 - e^{-L/\left(2^h \cdot f^2\right)}\right)^i .$$

Proof. We prove the claim by induction on i. On level $i = 0$, there is only one vertex, namely the root representing all instances from $\mathcal{I}$. Obviously, the algorithm did not admit any requests before phase 1, and hence, $\hat{\gamma}^{(0)} = 0$. According to its definition, the root is bad if phase 1 is bad for ALG. This, in turn, is the case if at least 0 requests are blocked at the beginning of phase 1 when ALG is processing any instance; see (2). This is obviously true, and hence the base case is covered.

Let us now assume that the claim holds for some level $i - 1$. We will show that the claim then also holds for level i. From now on, let v be some bad vertex on level $i-1$. First we prove that the fraction of bad vertices among the children of v is at least $1 - e^{-L/(2^h \cdot f^2)}$.

Since v is bad, phase i must be bad for ALG, given any instance I_v corresponding to v as its input. Therefore, for each such instance I_v, at least d_{i-1} requests from phase i are already blocked at the beginning of phase i. As ALG admits $\gamma^{(i)}$ further requests in this phase, at least $d_{i-1} + \gamma^{(i)}$ requests from phase i are blocked at the beginning of phase $i + 1$, including the admitted requests from phase i. This set of requests corresponds to the set R_i defined earlier. From Lemma 1, we know that, if at least $d_i/2$ requests from R_i are open, then phase $i+1$ is bad, which implies that at the beginning of phase $i+1$, at least d_i requests from phase $i + 1$ are already blocked. Since the set of instances that correspond to a child w of v is a subset of the set of instances that correspond to v, and since we just showed that phase $i+1$ is bad for an arbitrary instance I_v, phase $i + 1$ is also bad for each instance I_w; Fig. 6 gives an example.

Hence, a sufficient condition for w to be bad is that at least $d_i/2$ requests from R_i are open when giving ALG an instance I_w as its input. Thus, we have the following scenario. There are $N := L/2^h$ requests in phase i, as in every phase. Out of these, $M \geq M' := d_{i-1} + \gamma^{(i)}$ are blocked at the beginning of phase $i + 1$. The set of these requests is R_i. Each child w of v corresponds to the set of instances in which the same set of $n := L/2^{h+1}$ requests from phase i

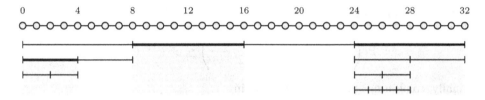

Fig. 6. This instance, like those from Figs. 3 and 5, corresponds to the vertex v from Fig. 4. The deterministic algorithm ALG admits $\hat{\gamma}^{(1)} = 2$ requests in phase 1. Hence, $d_1 = \hat{\gamma}^{(1)} - 1 \cdot L/(2^h \cdot f) = 2 - 32/(2^3 \cdot f) = 2 - 4/f < 2$. Since 2 requests from phase 2 are already blocked at the beginning of this phase, v is a bad vertex. Wlog, let w from Fig. 4 be the vertex on level 2 containing the instance from this picture. The vertex w is bad if, out of all requests from phase 3, at least $d_2 = \hat{\gamma}^{(2)} - 2 \cdot 32/(2^3 \cdot f) = 3 - 8/f < 3$ are blocked at the beginning of phase 3. Hence, in this instance, out of the 3 requests from phase 2 that are blocked after phase 2, at least $d_2/2 < 1.5$ must be open. This is clearly the case, since 2 such requests are open; thus, w is bad.

are open requests. We are interested in the fraction p of children w of v that correspond to instances in which at least $d_i/2$ requests from R_i are open. This is equivalent to the following. We have an urn containing N balls (i.e., requests), out of which $M \geq M'$ are black (i.e., in R_i), we draw n balls (i.e., open n requests) without replacement, and we are interested in the probability that the number of black balls drawn (i.e., open requests from R_i) is at least $d_i/2$.

Let X be a random variable that counts the number of open requests from R_i in this scenario. Note that X has a hypergeometric distribution with parameters $M \geq d_{i-1} + \gamma^{(i)}$, $N = L/2^h$, and $n = L/2^{h+1}$, and we are interested in $\Pr(X \geq d_i/2)$. With

$$\frac{d_i}{2} = \frac{1}{2}\left(\hat{\gamma}^{(i)} - i \cdot \frac{L}{2^h \cdot f}\right) = \frac{d_{i-1}}{2} + \frac{\gamma^{(i)}}{2} - \frac{L}{2^{h+1} \cdot f}$$

we obtain

$$\Pr\left(X \geq \frac{d_i}{2}\right) \geq \Pr\left(X > \frac{d_i}{2}\right) = 1 - \Pr\left(X \leq \frac{d_{i-1}}{2} + \frac{\gamma^{(i)}}{2} - \frac{L}{2^{h+1} \cdot f}\right). \quad (3)$$

Corollary 1 gives us a means to bound

$$\Pr\left(X \leq n \cdot \frac{M'}{N} - t \cdot n\right) = \Pr\left(X \leq \frac{L}{2^{h+1}} \cdot \frac{d_{i-1} + \gamma^{(i)}}{\frac{L}{2^h}} - t \cdot \frac{L}{2^{h+1}}\right)$$

$$= \Pr\left(X \leq \frac{d_{i-1} + \gamma^{(i)}}{2} - t \cdot \frac{L}{2^{h+1}}\right)$$

from above for any $t \geq 0$. Hence, choosing $t := 1/f$ yields

$$\Pr\left(X \leq \frac{d_{i-1} + \gamma^{(i)}}{2} - t \cdot \frac{L}{2^{h+1}}\right) = \Pr\left(X \leq \frac{d_{i-1}}{2} + \frac{\gamma^{(i)}}{2} - \frac{L}{2^{h+1} \cdot f}\right).$$

Then, according to Corollary 1, we get

$$\Pr\left(X \le \frac{d_{i-1} + \gamma^{(i)}}{2} - \frac{L}{2^{h+1}f}\right) \le e^{-L/(2^h f^2)} . \tag{4}$$

Finally, combining (3) and (4), we obtain

$$\Pr\left(X \ge \frac{d_i}{2}\right) \ge 1 - \Pr\left(X \le \frac{d_{i-1}}{2} + \frac{\gamma^{(i)}}{2} - \frac{L}{2^{h+1} \cdot f}\right) \ge 1 - e^{-L/(2^h f^2)} .$$

Hence, we have now shown that, for each bad vertex v on level $i - 1$, the fraction of bad vertices among its children is at least $1 - e^{-L/(2^h f^2)}$.

At this point, we are almost done. The only thing that remains to do is to exhibit a connection to the number of bad vertices on level i. All vertices on level $i-1$ have the same number of children and due to the induction hypothesis, for every bad vertex on level $i - 1$, a fraction of at least $1 - e^{-L/(2^h f^2)}$ of its children is bad. Hence, the fraction of bad vertices on level i is at least

$$\left(1 - e^{-L/(2^h f^2)}\right)^{i-1} \cdot \left(1 - e^{-L/(2^h f^2)}\right) = \left(1 - e^{-L/(2^h f^2)}\right)^i . \qquad \Box$$

A direct consequence from this result that many vertices are bad is that many instances are bad for ALG.

Corollary 2. *For any deterministic online algorithm* ALG, *the fraction of instances in* $\mathcal{I}$ *which are bad for* ALG *is at least*

$$\left(1 - e^{-L/(2^h f^2)}\right)^h .$$

Proof. Every single instance corresponds to a leaf in the instance tree, and is thus located at level h. Plugging in the result of Lemma 2 proves the statement. $\Box$

We have now shown that there are many bad instances for a given deterministic algorithm ALG for DPA. What we will show next is that the choice of the term "bad" was indeed justified for these instances, i.e., that ALG can actually only admit few requests on any bad instance.

Lemma 3. *Let* ALG *be an arbitrary but fixed deterministic algorithm for DPA, and let* $I \in \mathcal{I}$ *be a bad instance for* ALG. *Then, the gain of* ALG *on* I *is at most*

$$\mathrm{gain}(\mathrm{ALG}(I)) \le \frac{L}{2^h}\left(1 + \frac{h}{f}\right) .$$

Proof. According to the definition of bad vertices (2), an instance (corresponding to a vertex on level h of the instance tree) is bad if there are at least $d_h = \hat{\gamma}^{(h)} - h \cdot L/(2^h f)$ requests from phase $h + 1$ that are already blocked at the beginning of phase $h + 1$. In this last phase, ALG is presented $L/2^h$ requests, and thus, the number of requests ALG can admit in this phase is

$$\gamma^{(h+1)} \leq \frac{L}{2^h} - \left(\hat{\gamma}^{(h)} - h \cdot \frac{L}{2^h f}\right) = \frac{L}{2^h} \cdot \left(1 + \frac{h}{f}\right) - \hat{\gamma}^{(h)} .$$

For the number of admitted intervals at the end of the computation and thus, for the total gain of ALG on any bad instance I, we obtain

$$\hat{\gamma}^{(h)} + \gamma^{(h+1)} \leq \hat{\gamma}^{(h)} + \frac{L}{2^h} \cdot \left(1 + \frac{h}{f}\right) - \hat{\gamma}^{(h)} = \frac{L}{2^h} \cdot \left(1 + \frac{h}{f}\right) . \qquad \square$$

All in all, we have shown that, for a fixed deterministic algorithm ALG, there are many instances on which ALG has only small gain. We now combine these results to prove Theorem 1.

Proof of Theorem 1. Consider an arbitrary but fixed deterministic algorithm ALG for DPA. The competitive ratio of ALG on an arbitrary bad instance I is, according to Lemma 3 and Observation 1,

$$c = \frac{\text{gain}(\text{OPT}(I))}{\text{gain}(\text{ALG}(I))} \geq \frac{\frac{(h+2)L}{2^{h+1}}}{\frac{L}{2^h} \cdot \left(1 + \frac{h}{f}\right)} = \frac{2^h}{2^{h+1}} \cdot \frac{h+2}{\left(1 + \frac{h}{f}\right)} = \frac{h+2}{2\left(1 + \frac{h}{f}\right)} .$$

Now consider an arbitrary online algorithm $\mathcal{A}$ with advice for DPA that reads b advice bits. We can interpret $\mathcal{A}$ in the usual way as a set of 2^b deterministic algorithms, $\mathcal{A} = \{\text{ALG}_1, \ldots, \text{ALG}_{2^b}\}$ [3,4]. From Corollary 2, we know that, for every such deterministic algorithm ALG_i, the fraction of good instances from $\mathcal{I}$, and hence the fraction of instances on which ALG_i has a competitive ratio of at most $(h + 2)/(2(1 + h/f))$, is at most

$$1 - \left(1 - \frac{1}{e^{L/(2^h f^2)}}\right)^h \leq \frac{h}{e^{L/(2^h f^2)}} ,$$

where we used Bernoulli's inequality (Fact 1), plugging in the values $n := h$ and $x := -1/e^{L/(2^h f^2)}$. Note that this is legitimate as long as $2^h > 0$ and $f > 0$, since then $L/(2^h f^2) \geq 0$ and hence $x \geq -1$.

Obviously, the best case for $\mathcal{A}$ is met if the good instances of all ALG_i's, $1 \leq i \leq 2^b$ are pairwise disjoint. Therefore, the number of deterministic algorithms that are necessary to guarantee a competitive ratio of at most $(h+2)/(2(1+h/f))$ for every instance from $\mathcal{I}$ is at least $(e^{L/(2^h f^2)})/h$.

To be able to distinguish this many different deterministic strategies, the number of advice bits the online algorithm ALG has to read is at least

$$\log_2 \left(\frac{e^{L/(2^h f^2)}}{h}\right) = \frac{L}{2^h f^2} \cdot \log_2 e - \log_2 h . \qquad \square$$

Now that we have established a general lower bound that gives a minimum number of advice bits necessary to achieve a specific competitive ratio, we use Theorem 1 to get two concrete lower bounds by choosing concrete values for h and f that are in accordance with (1).

Corollary 3. *For any* $c = c(L)$ *with* $1 < c \leq 1/2 \cdot (\log_2 L)/(\log_2 \log_2 L)^{1/4}$, *any online algorithm for DPA that achieves a competitive ratio of* c *needs to read at least* $\Omega\big(L/(4^c\, c^4)\big)$ *advice bits.*

Finally, from Theorem 1 we can also derive a more concrete result on the number of advice bits necessary to achieve competitive ratios in the order of $\log_2 L$.

Corollary 4. *Let* δ *be an arbitrary constant with* $0 < \delta < 1$. *Any online algorithm for DPA that achieves a competitive ratio of* $\delta/2 \cdot \log_2 L$ *needs to read at least* $\omega(L^{1-\varepsilon})$ *advice bits, for any constant* ε *with* $\delta < \varepsilon < 1$.

Acknowledgments. The authors would like to thank Hans-Joachim Böckenhauer and Juraj Hromkovič for very valuable discussions.

References

1. Awerbuch, B., Bartal, Y., Fiat, A., Rosén, A.: Competitive non-preemptive call control. In: Proc. of SODA 1994, pp. 312–320. SIAM (1994)
2. Barhum, K., Böckenhauer, H.-J., Forišek, M., Gebauer, H., Hromkovič, J., Krug, S., Smula, J., Steffen, B.: On the Power of Advice and Randomization for the Disjoint Path Allocation Problem. In: Geffert, V., Preneel, B., Rovan, B., Štuller, J., Tjoa, A.M. (eds.) SOFSEM 2014. LNCS, vol. 8327, pp. 89–101. Springer, Heidelberg (2014)
3. Böckenhauer, H.-J., Komm, D., Královič, R., Královič, R.: On the advice complexity of the k-server problem. In: Aceto, L., Henzinger, M., Sgall, J. (eds.) ICALP 2011, Part I. LNCS, vol. 6755, pp. 207–218. Springer, Heidelberg (2011)
4. Böckenhauer, H.-J., Komm, D., Královič, R., Královič, R., Mömke, T.: On the Advice Complexity of Online Problems. In: Dong, Y., Du, D.-Z., Ibarra, O. (eds.) ISAAC 2009. LNCS, vol. 5878, pp. 331–340. Springer, Heidelberg (2009)
5. Böckenhauer, H.-J., Komm, D., Královič, R., Královič, R., Mömke, T.: Online algorithms with advice. Technical Report 614, Department of Computer Science. ETH Zurich (2009)
6. Borodin, A., El-Yaniv, R.: Online Computation and Competitive Analysis. Cambridge University Press (1998)
7. Carothers, N.L.: Real analysis. Cambridge University Press (2000)
8. Chvátal, V.: The tail of the hypergeometric distribution. Discrete Mathematics **25**(3), 285–287 (1979)
9. Dobrev, S., Královič, R., Pardubská, D.: Measuring the problem-relevant information in input. Theoretical Informatics and Applications (RAIRO) **43**(3), 585–613 (2009)
10. Epstein, L., Favrholdt, L.M.: Optimal preemptive semi-online scheduling to minimize makespan on two related machines. Operations Research Letters **30**(4), 269–275 (2002)

11. Emek, Y., Fraigniaud, P., Korman, A., Rosén, A.: Online computation with advice. Theoretical Computer Science **412**(24), 2642–2656 (2011)
12. Gupta, S., Kamali, S., López-Ortiz, A.: On Advice Complexity of the k-server Problem under Sparse Metrics. In: Moscibroda, T., Rescigno, A.A. (eds.) SIROCCO 2013. LNCS, vol. 8179, pp. 55–67. Springer, Heidelberg (2013)
13. Hromkovič, J., Královič, R., Královič, R.: Information Complexity of Online Problems. In: Hliněný, P., Kučera, A. (eds.) MFCS 2010. LNCS, vol. 6281, pp. 24–36. Springer, Heidelberg (2010)
14. Komm, D., Královič, R.: Advice complexity and barely random algorithms. Theoretical Informatics and Applications (RAIRO) **45**(2), 249–267 (2011)
15. Rice, J. A.: Mathematical Statistics and Data Analysis. Duxbury Press, 3rd edn. (2007)
16. Renault, M.P., Rosén, A.: On Online Algorithms with Advice for the k-Server Problem. In: Solis-Oba, R., Persiano, G. (eds.) WAOA 2011. LNCS, vol. 7164, pp. 198–210. Springer, Heidelberg (2012)
17. Sleator, D.D., Tarjan, R.E.: Amortized efficiency of list update and paging rules. Communications of the ACM **28**(2), 202–208 (1985)

Graph Algorithms III